Multiresolution Time Domain Scheme for Electromagnetic Engineering

Multiresolution Time Domain Scheme for Electromagnetic Engineering

YINCHAO CHEN
Department of Electrical Engineering
University of South Carolina
Columbia, South Carolina

QUNSHENG CAO
AHPCRC, Institute of Technology
University of Minnesota
Minneapolis, Minnesota

RAJ MITTRA
Electrical Engineering Department
Pennsylvania State University
State College, Pennsylvania

WILEY-INTERSCIENCE

A JOHN WILEY & SONS, INC., PUBLICATION

Published by John Wiley & Sons, Inc., Hoboken, New Jersey.
Published simultaneously in Canada.

For general information on our other products and services please contact our Customer Care Department within the U.S. at 877-762-2974, outside the U.S. at 317-572-3993 or fax 317-572-4002.

Wiley also publishes its books in a variety of electronic formats. Some content that appears in print, however, may not be available in electronic format.

Library of Congress Cataloging-in-Publication Data

Chen, Yinchao.
Multiresolution time domain scheme for electromagnetic engineering / Yinchao Chen, Qunsheng Cao, Raj Mittra.
p. cm.
Includes bibliographical references and index.
ISBN 0-471-27230-2 (cloth)
1. Electromagnetic interference—Mathematical models. 2. Electromagnetism. 3. Time-domain analysis. 4. Wavelets (Mathematics) 5. Multivariate analysis. 6. Very high speed integrated circuits. I. Cao, Qunsheng. II. Mittra, Raj. III. Title.
TK7867.2.C48 2005
621.3—dc22

2004054806

Printed in the United States of America.

10 9 8 7 6 5 4 3 2 1

Contents

Preface

Although the multiresolution time domain (MRTD) scheme has seen dramatic growth in the last decade, few texts and research books have focused on this subject. This may be due to the fact that the comprehensive mathematical content of the multiresolution analysis (MRA) is quite challenging for most electrical and electronic engineers; meanwhile, mathematicians may not be familiar with engineering applications, in particular, those in the areas of applied electromagnetic wave engineering. The major objective of this book is to bridge the two aspects of electromagnetic wave engineering and to provide an efficient means for electrical and electronic engineers to quickly master the MRTD while acquiring a comprehensive understanding of the content.

After investigation and research on the MRTD scheme for several years, we feel obligated to write a MRTD book by targeting the introduction of fundamental concepts of the MRA and MRTD applications for electromagnetic wave engineering. First, by spending much time researching references and testing numerical implementations, we have struggled to clarify various MRA concepts and to model the MRTD accurately. This research and experimentation are certainly valuable to those who are interested in learning the technique quickly, since the materials we provide will save them a tremendous amount of time. Second, the subject of the MRA and MRTD certainly can be presented as a stand-alone course for senior undergraduates and graduate students in electrical engineering, physics, and applied mathematics, since there is currently no textbook available on the market. We hope that this book can serve this objective. Finally, we have tried to organize the book as a self-contained reference book that engineers and scientists can use to learn advanced mathematical topics of the MRA and its application through an efficient MRTD scheme—without needing to search for complicated mathematical references.

This book presents a combination of theoretically advanced mathematical topics and their applications in time domain Maxwell solution techniques; in particular, it is characterized by the following unique features:

- *Concepts of Signal Space, the MRA, and Scaling and Wavelet Functions.* From an engineer's point of view, we systematically describe the concepts of signal space, MRA, scaling functions, and wavelets. These complex concepts are illustrated with many examples, derivations, and plots.
- *Construction of MRA Families.* We show how to generate various MRA families that are frequently used in current MRTD applications, such as the B-spline MRA family, Battle–Lemarié spline family, Coiflet family, and biorthogonal MRA and biorthogonal CDF families. Also, we demonstrate Daubechies' procedure of MRA construction and the Daubechies family.
- *Interconnection Among the MRTD, FDTD, and MoM.* We present the interconnection among the MRTD, finite difference time domain (FDTD), and method of moments (MoM), which is explained and demonstrated by analyzing a pair of traditional transmission equations.
- *MRTD Boundary Truncations.* Due to coupling between cells in the MRTD scheme, a boundary cannot be truncated directly by assigning it a tangential value for both open and shielded structure. The book describes the boundary truncation techniques in conjunction with the anisotropic perfectly matched layer (APML) and the multiple image technique (MIT) for open and shielded boundaries, respectively.
- *MRTD Scattering Analysis.* The book presents a MRTD near-to-far-zone field transformation technique for the efficient calculation of far fields involving two-dimensional and three-dimensional scattering targets, demonstrating its great potential for modeling large and complex geometries.
- *MRTD Applications.* We include extensive MRTD applications in the book, such as one-dimensional TEM wave propagation, two-dimensional printed transmission lines and guided-wave propagation structures, and three-dimensional microwave devices and integrated circuits.
- *Comprehensive Collection and Derivation of Integral Relations.* For the reader's convenience, we present many integral relations used in the MRTD implementation and provide detailed derivations of the MRTD update equations.
- *Generalized Differential Matrix Operators (GDMOs).* This book introduces a matrix technique to efficiently formulate vector differential equations that is very useful in dealing with anisotropic materials.

Given the unique features described above, this book offers an efficient instructional tool for learning MRA theory and its MRTD applications for professional engineers, physicists, and applied mathematicians. It also may serve as a textbook for teaching MRA and MRTD to undergraduates and graduates in those fields. Also, it serves as a reference resource for electrical engineers, numerical researchers, and physicists, who are interested in knowing more about the MRA concepts and MRTD applications in efficient computational electromagnetics.

YINCHAO CHEN
Columbia, South Carolina

QUNSHENG CAO
Minneapolis, Minnesota

RAJ MITTRA
State College, Pennsylvania

Acknowledgments

We would like to express our sincere thanks to our colleagues and associates who have provided encouragement and help with preparation of the manuscript. In particular, we would like to thank Prof. Ted Simpson (EE, USC), Prof. Benjamin Beker (TI), Prof. Paul Huray (EE, USC), and Prof. Richard Mellitz (Intel), for their discussions about fundamental transmissions and antennas, high-speed circuit signal integrity, and highly involved applied electromagnetic wave boundary-value problems. We are grateful to Prof. Xian Wu (Math, USC) for his comments and revisions on Chapters 2 and 3 of the manuscript. We thank Ms. Yanjie Zhu for her contribution in generating the distribution plots of the scaling and wavelet functions and the tables of coefficients of various scaling and wavelet functions.

We sincerely appreciate Ms. Loralee Donath's (Linguistics, USC) help in editing the entire manuscript.

Finally, we would likc to express our gratitude to our families for their long-term support and encouragement. It would not have been possible to complete this manuscript without their understanding and patience during our absences on weekends and holidays.

YINCHAO CHEN
QUNSHENG CAO
RAJ MITTRA

CHAPTER ONE

Introduction

1.1 PROLOGUE

The rapid development of computer techniques and information technologies in recent decades has fueled the need for efficient tools for electromagnetic modeling of microwave and millimeter-wave integrated circuits, high-speed and high-density very large scale integration (VLSI) circuits including computer chips, and wireless communication applications. A number of computational techniques are currently used for electromagnetic modeling by practicing microwave and antenna engineers, including the method of moments (MoM), the finite difference time domain (FDTD) method, and the finite element method (FEM). Among these, the FDTD, introduced by K. S. Yee in 1966 [1], appears to be one of the most widely used methods for many engineering applications [1–8].

The FDTD method can be implemented in a relatively straightforward manner. It can easily deal with complex geometrical features as well as different material properties; in addition, it can efficiently handle communication waveforms with wide bandwidths. Various modifications and extensions of the original FDTD algorithm have been implemented to enhance its performance, including hybrid methods [9, 10], higher-order FDTD schemes [11, 12], and Berenger's perfectly matched layer (PML) [13]. The latter enables us to truncate the FDTD mesh with boundaries that are very close to the structures being modeled.

Although the FDTD technique is simple and versatile, it places a heavy burden on computer resources when it involves modeling a complex problem that occupies a large computational volume. This is because it uses a fine spatial resolution in the frequency band of interest, usually 20 cells or more per wavelength, in order to achieve a reasonable degree of accuracy. Although progress in computer technology has made it possible to use the conventional FDTD technique to

Multiresolution Time Domain Scheme for Electromagnetic Engineering
By Yinchao Chen, Qunsheng Cao, and Raj Mittra
ISBN 0-471-27230-2

solve relatively large problems, it is important to enhance its capabilities further in order to tackle even larger problems.

The multiresolution time domain (MRTD) scheme, recently introduced by Krumpholz and Katehi [14, 15], shows excellent potential for achieving this goal by reducing the grid density to a level close to the Nyquist sampling rate. The use of scaling and wavelet functions and the application of the multiresolution analysis—in conjunction with the MoM-based discretization of Maxwell equations—form the cornerstones of the MRTD scheme. In this scheme, the electromagnetic fields are represented in terms of expansions of scaling and wavelet functions for spatial variation, and rectangular pulse functions for time variation.

The technique has a number of advantages, including a unified and higher-order field expansion structure for both the FDTD and MoM techniques. To put it slightly differently, the FDTD method is the simplest version of the MRTD scheme. We will show how the users of the MRTD scheme, described herein, can significantly reduce the requirements of computation time and memory for a class of electromagnetic simulation problems. We will also present numerous examples to demonstrate that the MRTD scheme is developing rapidly and that it has been applied successfully to the analysis of a broad variety of microwave and antenna structures [14–26].

1.2 OBJECTIVES

Our objectives here are: (1) to detail the essential characteristics and applications of the MRTD scheme, which are based mainly on the field expansions of scaling and high-level wavelet functions of the cubic spline Battle–Lemarié, Haar, Daubechies, and biorthogonal Cohen–Daubechies–Feauveau (CDF) as well as the biorthogonal interpolating wavelet families; (2) to develop a unified theoretical framework suitable for the electromagnetic modeling of guided-wave structures, millimeter-wave integrated circuits (MMICs), radar target scattering and antennas, and wireless communication applications; and (3) to demonstrate the advantages of the MRTD scheme over the conventional FDTD method for certain types of applications.

In order to fulfill the above objectives, the MRTD scheme is applied in conjunction with an adjustable multiple image technique (MIT) or an automatic image generator (AIG) to truncate the boundaries of the computational domain characterized by a perfect electric conductor (PEC) or a perfect magnetic conductor (PMC). An anisotropic perfectly matched layer (APML) is used to truncate open boundaries. Also, two-dimensional (2D) and three-dimensional (3D) versions of the MRTD plane-wave incidence and the near-to-far-zone field transforms are developed for scattering and radiation applications.

Before closing this section it may be useful to point out that the MRTD formulations presented herein still retain the concept of the leapfrog algorithm employed in the conventional FDTD method and inherit most of its advantages.

1.3 OVERVIEW

This monograph is designed to cover both the theoretical and practical aspects of the MRTD scheme and comprises twelve chapters and four appendixes, each of which focuses on a topic pertaining to the systematic development of the MRTD scheme and its applications. It covers the entire range of state-of-the-art topics that are pertinent to the current development of the MRTD, including the construction of the MRTD framework, choice of basis functions for field representations in the context of the MRTD, and applications to practical electromagnetic wave propagation problems.

Chapter 1 serves as an introduction, in which the background, objective, motivation, and an overview of the MRTD scheme are presented. In Chapter 2 the mathematical basis of the multiresolution analysis (MRA) is discussed, including the concepts of signal space and scaling and wavelet functions. This chapter also addresses the topic of generalized field expansions in terms of scaling and wavelet functions, which serves as the foundation of the MRTD development.

Chapter 3 mainly focuses on the following types of wavelet bases that are commonly used in the MRTD analysis: cubic spline Battle–Lemarié, Daubechies, biorthogonal Cohen–Daubechies–Feauveau (CDF), and biorthogonal interpolating wavelet bases. The construction—as well as the properties—of the scaling and wavelet functions is discussed and the three-dimensional multiresolution time domain update algorithms are developed in a homogeneous uniform space for all the basis function families.

Next, in Chapter 4, we introduce the fundamentals of the MRTD algorithm. This chapter provides a simple explanation of the MRTD kernel and develops a coherent relationship among the FDTD, MoM, and MRTD in a discrete computational space. In addition, the generalized MRTD update equations for dealing with a hybrid computational space that involves both conductors and dielectric materials are presented. The stability of the MRTD scheme and the orthogonal and integral relations in the implementation of the scheme are summarized as well.

In Chapter 5, we examine the issue of mesh truncations for a structure enclosed with either PEC or PMC walls. Also, we introduce the concept of the MIT—which is a flexible tool for determining the number of images on each side of a PEC-shielded structure, such as a shielded cavity, in the computational domain. In addition, we introduce the principle of the AIG, a software package that can easily generate images for a (fully or partially) shielded microwave structure. The MRTD-MIT method is then validated by analyzing a number of representative structures shielded by a PEC, including an empty rectangular cavity and a partially filled resonant cavity. The MRTD results are compared with those derived from the FDTD scheme as well as from analytical solutions. We show that even though the MRTD scheme requires only a small fraction of the computer memory and CPU time in comparison to those required by the conventional FDTD methods, its results compare very favorably with the latter in terms of accuracy.

Chapter 6 focuses on the applications of the unsplit APML s for mesh truncation of open boundaries in the context of the MRTD scheme. After introducing the APML medium, the governing equations inside the APML regions are developed. Following this, we discuss various aspects of implementing the APMLs, including the treatment of different types of APML regions, for example, APML faces, edges, and corners, choice of APML parameters, and the derivation of APML update equations.

Next, in Chapter 7, we describe the application of a one-dimensional (1D) MRTD scheme for electromagnetic wave propagation in a multilayer dielectric system, using scaling and high-level multiresolution wavelet functions to illustrate the structure and construction of the MRTD scheme in a systematic manner. We also examine the computational resources required by the MRTD scheme and compare the MRTD results with those derived from the FDTD. Additionally, we validate the results by comparing them against analytical solutions.

In Chapter 8, we go on to develop a two-dimensional (2D) MRTD model in conjunction with the unsplit APML for truncation of open regions in guided-wave structures. This chapter consists of two parts. The first presents a 2D-MRTD scheme for the analysis of printed millimeter-wave transmission lines by utilizing the transverse resonance property of guided-wave structures, while the second deals with the problem of TE_z and TM_z wave propagation in a parallel-plate transmission line. We also demonstrate that the MRTD scheme generates results that are in excellent agreement with those obtained by using the FDTD approach, the MoM, and analytical methods.

Chapter 9 discusses the development of a three-dimensional (3D) MRTD scheme, which is applied in conjunction with the APML, MIT, or AIG for the mesh truncation of open regions as well as PEC-shielded boundaries. Applications of the MRTD-APML, MRTD-MIT, and MRTD-AIG schemes are illustrated by analyzing a number of representative microwave structures. In particular, by utilizing only the field quantities defined in the original structure, a systematic algorithm for dealing with inhomogeneous media inside a microwave structure is developed in order to construct the constitutive relations and update equations in the MRTD transform domain. The characteristics of microstrip lines and inhomogeneously filled dielectric cavities are investigated and the MRTD results are validated once again.

In Chapter 10, the MRTD scheme is applied to the problem of monolithic millimeter-wave integrated circuit (MMIC) structures, including microstrip lines, microstrip patch antennas, and microstrip low-pass and band-pass filters. It is shown that good agreement between the MoM, FDTD, and MRTD techniques is achieved, despite the fact that the MRTD technique requires only around 15% of computational space and about 25% or less of the computational CPU time, as compared to the FDTD method.

In Chapters 11 and 12, we concentrate on scattering problems involving both two- and three-dimensional structures by utilizing the "scattered field" formulation. We also develop the 2D and 3D versions of the plane-wave incidence and the near-to-far-zone field transformation in the context of the MRTD to calculate either the scattering width (SW) or the radar cross section (RCS). We then employ these schemes to analyze a variety of PEC and dielectric targets, and we validate the MRTD results by comparing them with those derived from the FDTD as well as the MoM.

In addition to the twelve chapters, we also include four appendixes, which provide supplementary materials that help clarify the theoretical developments and show the derivation of various mathematical relations and update equations. In Appendix A, we introduce the concept of the generalized differential matrix operators (GDMOs), for formulation of engineering electromagnetic problems, and discuss the applications of the GDMOs in the formulation of practical electromagnetic problems. In Appendix B, we summarize most of orthogonal and integral relations that are frequently employed in the cubic spline Battle–Lemarié scaling and wavelet based MRTD scheme. Appendix C focuses on the development of the MRTD-APML algorithm for all APML faces, edges, and corners and presents a two-step approach for deriving the APML update equations. Finally, in Appendix D, we summarize the properties of the scaling and wavelet functions for the convenience of readers. In particular, this appendix provides various expressions of the cubic-spline Battle–Lemarie scaling and wavelet functions, as well as their mother functions, and also presents some of their important properties that are not commonly found in the literature.

REFERENCES

[1] K. S. Yee, "Numerical solution of initial boundary value problem involving Maxwell's equations in isotropic media," *IEEE Trans. Antennas Propag.*, vol. AP-14, no. 3, pp. 302–307, May 1966.

[2] K. S. Kunz and R. J. Luebbers, *The Finite Difference Time Domain Method for Electromagnetics*, CRC Press, Boca Raton, FL, 1993.

[3] A. Taflove, *Computational Electrodynamics: The Finite-Difference Time-Domain Method*, Artech House, Norwood, MA, 1995.

[4] J. Jin, *The Finite Element Method in Electromagnetics*, Wiley, Hoboken, NJ, 1993.

[5] J. A. Buck, *Fundamentals of Optical Fibers*, Wiley, Hoboken, NJ, 1996.

[6] G. Einarsson, *Principles of Lightwave Communications*, Wiley, Hoboken, NJ, 1996.

[7] J. C. Strikwerda, *Finite Difference Schemes and Partial Differential Equations*, Wadsworth & Brooks, Belmont, CA, 1989.

[8] X. Zhang, J. Fang, K. K. Mei, and Y. Liu, "Calculations of the dispersive characteristics of microstrips by the time-domain finite difference method," *IEEE Trans. Microwave Theory Tech.*, vol. MTT-36, no. 2, pp. 253–267, Feb. 1989.

[9] G. Liang, Y. Liu, and K. K. Mei, "Full wave analysis of coplanar waveguide and slotline using the time-domain finite-difference method," *IEEE Trans. Microwave Theory Tech.*, vol. MTT-37, no. 12, pp. 1949–1957, Dec. 1989.

[10] P. H. Aoyagi, J. F. Lee, and R. Mittra, "A hybrid pee algorithm/scalar wave equation approach," *IEEE Trans. Microwave Theory Tech.*, vol. MTT-41, no. 9, pp. 1993–1600, Sept. 1993.

[11] M. Mrozowski, "A hybrid PEE-FDTDalgorithm for accelerated time domain analysis of electromagnetic waves in shielded structure," *IEEE Microwave Guided Wave Lett.*, vol. 4, no. 10, pp. 323–325, Oct. 1994.

[12] L. Lapidus and G. F. Pinder, *Numerical Solution of Partial Differential Equations in Science and Engineering*, Wiley, Hoboken, NJ, 1982, pp. 171–179.

[13] J. P. Berenger, "A perfectly matched layer for the absorption of electromagnetic wave," *J. Comput. Phys.*, vol. 114, pp. 185–200, May 1994.

[14] M. Krumpholz and L. P. B. Katehi, "New prospects for time domain analysis," *IEEE Microwave Guided Wave Lett.*, vol. 5, no. 11, pp. 382–384, Nov. 1995.

[15] M. Krumpholz and L. P. B. Katehi, "MRTD: New time domain schemes based on multiresolution analysis," *IEEE Trans. Microwave Theory Tech.*, vol. 44, no. 4, pp. 555–571, Apr. 1996.

[16] R. L. Roberson, E. M. Tentzeris, M. Krumpholz, and L. P. B. Katehi, "Modeling of dielectric cavity structures using multiresolution time domain analysis," *Int. J. Numerical Modeling: Electronic Networks, Devices Fields*, vol. 11, pp. 55–68, Apr. 1998.

[17] R. L. Roberson, E. M. Tentzeris, M. Krumpholz, and L. P. B. Katehi, "MRTDanalysis of dielectric cavity structures," *Proc. IEEE MTT-S Dig.*, pp. 1861–1864, 1996.

[18] C. D. Sarris, L. P. B. Katehi, and J. F. Harvey, "Application of multiresolution analysis to the modelling of microwave and optical structures," *Opt. Quant. Electron.*, vol. 32, pp. 657–679, Aug. 2000.

[19] M. Fujii and W. J. R. Hoefer, "A three-dimensional Haar-wavelet-based multiresolution analysis similar to the FDTDmethod-derivation and application," *IEEE Trans. Microwave Theory Tech.*, vol. 46, no. 12, pp. 2463–2475, Dec. 1998.

[20] M. Fujii and W. J. R. Hoefer, "Field-singularity correction in 2-D time-domain Haar-wavelet modelling of waveguide components," *IEEE Trans. Microwave Theory Tech.*, vol. 49, no. 4, pp. 685–691, Apr. 2001.

[21] Y. W. Cheong, Y. M. Lee, K. H. Ra, J. G. Kang, and C. C. Shin, "Wavelet–Galerkin scheme of time-dependent inhomogeneous electromagnetic problem," *IEEE Microwave Guided Wave Lett.*, vol. 9, no. 8, pp. 297–299, Aug. 1999.

[22] M. Fujii and W. J. R. Hoefer, "Application of biorthogonal interpolating wavelet to the Galerkin scheme of time dependent Maxwell's equations," *IEEE Microwave Wireless Components Lett.*, vol. 11, no. 1, Jan. 2001.

[23] T. Dogaru and L. Carin, "Multiresolution time-domain using CDFbiothogonal wavelets," *IEEE Trans. Microwave Theory Tech.*, vol. 49, no. 5, pp. 902–912, May 2001.

[24] Q. Cao and Y. Chen, "Scaling-function based multiresolution time domain analysis for planar printed millimetre-wave integrated circuits," *IEE Proc. Microwaves Antennas Propag.*, vol. 148, no. 3, pp. 179–187, June 2001.

[25] E. M. Tentzeris, A. Cangellaris, L. P. B. Katehi, and J. Harvey, "Multiresolution time-domain (MRTD) adaptive schemes using arbitrary resolutions of wavelet," *IEEE Trans. Microwave Theory Tech.*, vol. 50, no. 2, pp. 501–516, Feb. 2002.

[26] Q. Cao, Y. Chen, and R. Mittra, "Multiple image technique (MIT) and anisotropic perfectly matched layer (APML) in implementation of MRTDscheme for boundary truncations of microwave structures," *IEEE Trans. Microwave Theory Tech.*, vol. 50, no. 6, pp. 1578–1589, June 2002.

CHAPTER TWO

Introduction to the Multiresolution Analysis

2.1 INTRODUCTION

Wavelet theory has become one of the most interesting areas of multidisciplinary research among mathematicians, physicists, and engineers. This new category of functions, called wavelets, has many obvious advantages over traditional systems of orthogonal functions. One of the best ways to introduce wavelets is through the multiresolution analysis (MRA), in which a single scaling function and a single wavelet are used to construct complete sets of bases for systems of function spaces. As we will see, the MRA has now been widely used in many electromagnetic wave applications.

Multiresolution refers to the simultaneous presence of various resolutions to approximate a field signal. By using a sequence of scaling function expansions with wavelets of successively higher resolutions, we can realize a desired resolution for representing an arbitrary signal, which is usually referred to as an electromagnetic field in this book. Wavelet and scaling function based applications of the MRA form a natural framework for highly efficient numerical analysis. In particular, the multiresolution time domain (MRTD) scheme, as an application of the MRA, is well suited for generalized electromagnetic engineering applications.

Our view of signals (or fields) defines their composition as a smooth background with fluctuations of details or noises on top of it. A quickly changing signal can only be discerned by high-resolution components of the signal, whereas its smoother parts can be recognized by the low-resolution expansion of the signal. With a given resolution level, a signal can be approximated by ignoring all

This chapter is coauthored with Xian Wu, Department of Mathematics, University of South Carolina, Columbia, SC.

Multiresolution Time Domain Scheme for Electromagnetic Engineering
By Yinchao Chen, Qunsheng Cao, and Raj Mittra
ISBN 0-471-27230-2

those fluctuations above this level, which in many cases just represent noise. With progressively increasing resolution levels, we add finer and finer resolution details to the signal. In other words, we provide successively better approximations of the original signal. For an extreme case, when the resolution level goes to infinity, we then completely recover the signal in its exact original form.

Unlike classical Fourier bases, wavelets and scaling functions have excellent localization properties for approximating a signal field in a compact form. In an analysis of electromagnetic fields using MRA theory, we can expand a field in terms of scaling basis functions at a lower resolution level in those regions where the electromagnetic field varies smoothly and slowly, and add in more wavelets in places where the field changes quickly and dramatically. Such wavelet and scaling function based techniques have proved very efficient in the analysis and prediction for a large category of electromagnetic wave phenomena.

In this chapter, we briefly discuss some fundamental concepts of signal space and the MRA without going into too much mathematical profundity. The goal is to build a solid foundation without being overly concerned with technical details. We intend to make a practical but still mathematically sound presentation that is suitable for engineers and scientists. For those who are interested in full theoretical development, we suggest consulting the many available references, such as those listed at the end of this chapter [1–12].

2.2 VECTORS AND SIGNAL SPACE

Linear analysis is essential in many applied electromagnetic wave boundary-value problems. Linear analysis can be represented elegantly in an algebraic structure called vector space, or signal space, which is frequently used in modern digital communication systems.

Vector Spaces

A vector space (over a scalar field F) is a set V of elements (called vectors) equipped with operations of addition and multiplication by scalars with the following familiar properties:

Addition. For any vectors $\vec{v} \in V$ and $\vec{u} \in V$, a vector $\vec{v} + \vec{u} \in V$ is defined as the sum of $\vec{v}$ and $\vec{u}$, such that:

- It is commutative; that is, $\vec{v} + \vec{u} = \vec{u} + \vec{v}$, for all $\vec{v}, \vec{u} \in V$.
- It is associative; that is, $\vec{v} + (\vec{u} + \vec{w}) = (\vec{v} + \vec{u}) + \vec{w}$, for all $\vec{v}, \vec{u}, \vec{w} \in V$.
- There is a zero vector $\vec{0} \in V$, such that $\vec{v} + \vec{0} = \vec{v}$ for all $\vec{v} \in V$.
- For any $\vec{v} \in V$, there is a $-\vec{v} \in V$ such that $\vec{v} + (-\vec{v}) = \vec{0}$.

Scalar Multiplication. For any scalar $\alpha \in F$ and vector $\vec{v} \in V$, there is a vector $\alpha\vec{v} \in V$ called the product of α and $\vec{v}$, such that:

- It is associative; that is, $\alpha(\beta\vec{v}) = (\alpha\beta)\vec{v}$ for all scalars, $\alpha, \beta \in F$, and $\vec{v} \in V$.
- For the unit scalar 1 in F, $1\vec{v} = \vec{v}$ for all $\vec{v} \in V$.

Distributive Laws

- For any scalar α and vectors $\vec{v}, \vec{u} \in V, \alpha(\vec{v}+\vec{u}) = \alpha\vec{v}+\alpha\vec{u}$.
- For any scalars $\alpha, \beta \in F$, and vector $\vec{v} \in V$, $(\alpha+\beta)\vec{v} = \alpha\vec{v}+\beta\vec{v}$.

We are particularly interested in complex and real vector spaces, that is, when the scalar field is C or R, the set of complex numbers or the set of real numbers, respectively.

Inner Product in Vector Spaces

An inner product in a complex vector space V is a complex valued function of two vectors $\vec{v}, \vec{u} \in V$, denoted by $\langle \vec{v}, \vec{u} \rangle$, which has the following properties:

- *Hermitian Symmetry*: $\langle \vec{v}, \vec{u} \rangle = \langle \vec{u}, \vec{v} \rangle^*$, where $*$ indicates the operation of the complex conjugate, in particular, $\langle \vec{v}, \vec{v} \rangle \in R$.
- *Hermitian Bilinearity*: $\langle \alpha\vec{v}+\beta\vec{u}, \vec{w} \rangle = \alpha\langle \vec{v}, \vec{w} \rangle + \beta\langle \vec{u}, \vec{w} \rangle$.
- *Strict Positivity*: $\langle \vec{v}, \vec{v} \rangle \geq 0$, with equality if and only if $\vec{v} = \vec{0}$.

In a similar way, an inner product can also be defined for real spaces.

A vector space equipped with an inner product is called an inner product space. The norm $||\vec{v}||$, or the length of a vector $\vec{v}$, is defined as

$$||\vec{v}|| = \sqrt{\langle \vec{v}, \vec{v} \rangle} \tag{2.1}$$

Orthogonality

Orthogonality is a very important concept in signal analysis as well as in the MRTD analysis. Two nonzero vectors $\vec{v}$ and $\vec{u}$ are said to be orthogonal if

$$\langle \vec{v}, \vec{u} \rangle = 0 \tag{2.2}$$

Inner Product Space R^n

For the vector space R^n of real n-tuples, that is, $\vec{v} = (v_1, \ldots, v_n)$, $v_j \in R$, the inner product of vectors $\vec{v} = (v_1, \ldots, v_n)$ and $\vec{u} = (u_1, \ldots, u_n)$ is usually defined as

$$\langle \vec{v}, \vec{u} \rangle = \sum_{j=1}^{n} v_j u_j \tag{2.3}$$

Especially, in R^3, a vector $\vec{v} = (v_1, v_2, v_3)$ represents a special point that is the intersection of three planes parallel to the coordinate planes. In this case, the inner product determines geometric quantities such as the length and the relative orientation of vectors. For example, the distance from the origin o to $\vec{u}$ is

$$||\vec{u}|| = \sqrt{\langle \vec{u}, \vec{u} \rangle} \tag{2.4}$$

In general, the distance from $\vec{v}$ to $\vec{u}$ (or from $\vec{u}$ to $\vec{v}$) is $||\vec{v} - \vec{u}||$, and the angle between $\vec{u}$ and $\vec{v}$ can be found by

$$\cos(\angle(\vec{v} - \vec{u})) = \frac{\langle \vec{v}, \vec{u} \rangle}{||\vec{v}||\ ||\vec{u}||} \tag{2.5}$$

Subspaces

A subspace S of a vector space V is a subset of vectors in V, which itself forms a vector space over the same scalar field as used by V.

For example, let us consider the vector space R^3, that is, the vector space of real 3-tuples. In this case, we treat the space as a volume since there exist three orthogonal coordinate axes, where the components u_1, u_2, u_3 define the length of $\vec{u}$ in each of these directions. If we define three unit vectors $\hat{e}_1$, $\hat{e}_2$, $\hat{e}_3$ in these three coordinate directions, respectively, then any vector $\vec{u}$ can be expressed as

$$\vec{u} = u_1\hat{e}_1 + u_2\hat{e}_2 + u_3\hat{e}_3 \tag{2.6}$$

An example of a subspace of R^3 is the set of all vectors that can be expressed as $\vec{u} = u_1\hat{e}_1 + u_2\hat{e}_2$. We may think of this subspace as R^2 in R^3.

In general, the set of all vectors that can be expressed as

$$\vec{u} = u_1\vec{v}_1 + u_2\vec{v}_2 + \cdots + u_n\vec{v}_n, \quad u_j \in F \tag{2.7}$$

is called the subspace spanned by vectors $\vec{v}_1, \vec{v}_2, \ldots, \vec{v}_n$ and is denoted by $Sp\{\vec{v}_1, \vec{v}_2, \ldots, \vec{v}_n\}$.

Linear Independence, Bases, and Dimensions

A set of vectors, $\vec{v}_1, \vec{v}_2, \vec{v}_3, \ldots, \vec{v}_n$, in a vector space V is linearly dependent if

$$\sum_{j=1}^{n} \alpha_j \vec{v}_j = 0 \tag{2.8}$$

for a set of scalars α_j that are not all equal to 0. Equivalently, the set is linearly dependent if one of those vectors, say, $\vec{v}_n$, is a linear combination of others, that is, if for some set of scalars α_j

$$\vec{v}_n = \sum_{j=1}^{n-1} \alpha_j \vec{v}_j \tag{2.9}$$

A set of vectors is linearly independent if any of its finite subsets is not linearly dependent.

A subset of a vector space V is called a basis of V if it is linearly independent and it spans the whole V.

It can be proved that any two bases of a vector space must contain the same number of vectors. We define this number as the dimension of the vector space.

In other words, the dimension of a subspace is the number of vectors in any largest linearly independent set in the subspace. For example, the dimension of R^3 is 3 and dimensions of its subspaces can be 0, 1, or 2.

Orthonormal Bases, Projections, and Riesz Bases

In an inner product space, a set of vectors $\vec{\phi}_1, \vec{\phi}_2, \ldots$ is orthonormal if

$$\langle \vec{\phi}_j, \vec{\phi}_k \rangle = \delta_{j,k} = \begin{cases} 0 & \text{for } j \neq k \\ 1 & \text{for } j = k \end{cases} \tag{2.10}$$

In other words, an orthonormal set is a set of orthogonal vectors where each vector is normalized in the sense of having unit length. If a set of vectors $\vec{v}_1, \vec{v}_2, \ldots$ is orthogonal, then the set

$$\vec{\phi}_j = \frac{\vec{v}_j}{||\vec{v}_j||} \tag{2.11}$$

is orthonormal. Note that an orthonormal set is necessarily a linearly independent set.

If a vector $\vec{u}$ is projected into a normalized vector $\vec{\phi}$, then the projection is expressed as

$$\vec{u}_| = \langle \vec{u}, \vec{\phi} \rangle \vec{\phi} \tag{2.12}$$

In general, we can write an arbitrary vector $\vec{u} \in V$ as $\vec{u} = \vec{u}_| + \vec{u}_\perp$, where $\vec{u}_| \in Sp\{\vec{\phi}\}$ and $\vec{u}_\perp$ is in the orthogonal subspace $Sp\{\vec{\phi}\}_\perp$, which consists of all vectors that are orthogonal to $\vec{\phi}$.

In the following, we generalize (2.12) to the projection of a vector $\vec{u} \in V$ into a finite dimensional subspace S of V.

Projection Theorem. Let S be an n-dimensional subspace in an inner product space V and assume that the set $\{\vec{\phi}_1, \vec{\phi}_2, \ldots, \vec{\phi}_n\}$ is an orthonormal basis for S. Then any vector $\vec{u} \in V$ may be decomposed as

$$\vec{u} = \vec{u}_{|S} + \vec{u}_{\perp S} \tag{2.13}$$

where $\vec{u}_{|S} \in S$ and $\langle \vec{u}_{\perp S}, \vec{s} \rangle = 0$ for all $\vec{s} \in S$. Furthermore, $\vec{u}_{|S}$ is uniquely determined by

$$\vec{u}_{|S} = \sum_{j=1}^{n} \langle \vec{u}, \vec{\phi}_j \rangle \vec{\phi}_j \tag{2.14}$$

Note that in the theorem we assume that there exist orthonormal bases of S.

Dealing with infinite-dimensional inner spaces is more involved. We first need to generalize our definition to include infinite sums as follows:

$$\vec{u} = \sum_{i=1}^{\infty} \alpha_i \vec{\phi}_i \quad \text{if} \quad \lim_{n \to \infty} \left\| \vec{u} - \sum_{i=1}^{n} \alpha_i \vec{\phi}_i \right\| = 0 \tag{2.15}$$

Riesz Bases

A sequence $\{\vec{\phi}_i\}$ is called a Riesz basis of an inner space V if there exist two constants $0 < A \leq B < \infty$, such that any vector $\vec{u} \in V$ can be expressed as $\vec{u} = \sum_{i=1}^{\infty} \alpha_i \vec{\phi}_i$ with

$$A \sum_{i=1}^{\infty} |\alpha_i|^2 \leq \left\| \sum_{i=1}^{\infty} \alpha_i \vec{\phi}_i \right\|^2 \leq B \sum_{i=1}^{\infty} |\alpha_i|^2 \tag{2.16}$$

Note that this condition automatically implies that $\{\vec{\phi}_i\}$ is linearly independent. This seemingly technical Riesz condition plays an important role in many ways. We may intuitively think of a Riesz basis as a "close-to-orthonormal" or "stable" basis. In particular, any orthonormal basis is a Riesz basis as we can take $A = B = 1$. Given these definitions, we can formally generalize the above Projection Theorem to infinite-dimensional cases.

Orthonormal Expansion in $L^2(R)$

The space $L^2[a, b]$ consists of all square-integrable functions defined over the region $[a, b]$ with the inner product and the norm defined as

$$\langle f_1, f_2 \rangle = \int_a^b f_1(t) f_2^*(t)\, dt \tag{2.17a}$$

$$\|f\| = \sqrt{\int_a^b |f(t)|^2\, dt} \tag{2.17b}$$

Given a subspace $S \subseteq L^2$ with an orthonormal basis $\{\phi_i\}$, we can apply the Projection Theorem to find the best approximation of any function $f \in L^2$ in terms of orthogonal expansions, $f \approx \sum_i \alpha_i \phi_i$. The best approximation is achieved if the expansion coefficients minimize the norm $\left\|f - \sum_i \alpha_i \phi_i\right\|$. Since $f = f_{|S} + f_{\perp S}$ by the Projection Theorem and $\langle g, f_{\perp S} \rangle = 0$ for all g in S, we have

$$\begin{aligned} \left\| f - \sum_i \alpha_i \phi_i \right\|^2 &= \left\| f_{|S} - \sum_i \alpha_i \phi_i + f_{\perp S} \right\|^2 \\ &= \left\langle f_{|S} - \sum_i \alpha_i \phi_i + f_{\perp S}, f_{|S} - \sum_i \alpha_i \phi_i + f_{\perp S} \right\rangle \\ &= \left\| f_{|S} - \sum_i \alpha_i \phi_i \right\|^2 + \|f_{\perp S}\|^2 \end{aligned} \tag{2.18}$$

Therefore, the best approximation is obtained if and only if $\sum_i \alpha_i \phi_i = f_{|S}$; that is, the projection into S gives rise to the best approximation with respect to S and the expansion coefficients are $\alpha_i = \langle f, \phi_i \rangle$.

If a basis $\{\phi_i\}$ of L^2 is given, then any $f(t) \in L^2$ can be expressed as an expansion series

$$f(t) = \sum_{i=1}^{\infty} \alpha_i \phi_i(t) \tag{2.19}$$

We must be careful here since the expansion may not be pointwise convergent. However, truncated series in (2.19) will give any appropriate L^2-norm approximation of $f(t)$.

Since we often consider an inner space as a signal space in this book, the norm of f is therefore referred to as the energy of the signal. For an orthonormal basis $\{\phi_i\}$, we can easily evaluate the norm (more generally, the inner product) in terms of expansion coefficients, namely,

$$\langle f, f \rangle = \int_a^b |f|^2\, dt = \int_a^b |(\alpha_1\phi_1 + \alpha_2\phi_2 + \cdots)|^2\, dt = \sum_{i=1}^{\infty} |\alpha_i|^2$$

The most well-known orthogonal system is the system of the trigonometric functions in $L^2[\tau, \tau + T]$. For example, if we define $\phi_k(t) = T^{-1/2} e^{j2\pi kt/T}$, then any $f(t) \in L^2[\tau, \tau + T]$ will have a Fourier series expansion

$$f(t) = \sum_{k=-\infty}^{\infty} \alpha_k \phi_k(t) \tag{2.20a}$$

where

$$\alpha_k = \int_{\tau}^{\tau+T} f(t)\phi_k^*(t)\, dt \tag{2.20b}$$

or

$$\alpha_k = \langle f, \phi_k \rangle \tag{2.20c}$$

Although the infinite sum frequently takes place in theory, in practice we usually truncate the series to a partial sum:

$$f \approx f^{(n)} = \sum_{k=-n}^{n} \langle f, \phi_k \rangle \phi_k \tag{2.21}$$

This gives the best L^2-approximation at the given level, as the summation is just the projection of f into the subspace S_n spanned by $\{\phi_k\}$, $-n \le k \le n$. For f as a signal, the meaning of approximations may depend on the basis used. In the case of the Fourier expansion, the approximation is purely frequency based. With wavelets, we will have much more flexibility. For this, we turn to the next section.

2.3 MULTIRESOLUTION ANALYSIS

Now we are in the position to carry out the task of presenting a brief introduction of MRA. We start with some heuristic explanations followed by the formal definition of the MRA. This leads to the definition of wavelets and two-scaling relations in Section 2.4. We then explore scaling functions and wavelets in the frequency domain, which is an integrated part of the MRA theory (Section 2.5). The canonical examples of the Haar and linear spline MRA are given in Section 2.6. We also give brief discussions of biorthogonal wavelets and multidimensional wavelets in Sections 2.7 and 2.8, respectively. Finally, we give a description of the basic setup of MRA in a MRTD scheme in Section 2.9.

To explain the basic idea behind the definition of the MRA for our intended readers, we start with an electromagnetic field signal distribution. In general, such a signal can be viewed as a composite of a smooth background and fluctuation noise or details on top of it. The distinction between the smooth part and the details is determined by the level of resolution. Therefore, at a given resolution level, a signal can be approximated by ignoring all effects of fluctuations above the level. To get better approximations, we can progressively increase resolution levels and successively add them to the coarser description in finer and finer detail. Eventually, when the resolution goes to infinity, we obtain the exact signal without any approximation error.

We can make the above intuitive description more precise as follows. Where the resolution level is labeled by an integer s and the corresponding scale is set to 2^{-s}, let us denote the approximation of a signal $f(t)$ at level s by $f_s(t)$. When we increase the level to $s+1$, we need to add the details, denoted by $d_s(t)$, to the signal at this new level in order to obtain the new approximation, that is, $f_{s+1}(t) = f_s(t) + d_s(t)$. By continuing this procedure, we can recover the original signal $f(t)$ at infinity, that is,

$$f(t) = f_s(t) + \sum_{k=s}^{\infty} d_k(t) \tag{2.22}$$

From a global point of view of the signal space, (2.22) means that we need to consider two sequences of subspaces $\{W_s\}$ and $\{V_s\}$, such that every $f(t)$ can be approximated at the resolution level $s+1$ with $f_s(t)$ in the scale space V_s and $d_s(t)$ in the detail space W_s. Those subspaces are the basic objects in our definition.

Definition of the Multiresolution Analysis

Our signal space is now considered to be the space of square-integrable functions. Here we define a sequence of the resolution levels labeled by integers such that all details of signals on scales smaller than 2^{-s} will be ignored at level s. Also, we define V_s as the subspace of the functions that contain the signal information down to the scale 2^{-s}. Formally, a MRA is just a sequence of V_s, which satisfies certain mathematical conditions. Instead of directly giving a formal definition of MRA, we try first to explain the meanings behind those conditions. In the following, we clarify the five fundamental properties of the MRA. Although most of those requirements appear naturally, it may take more time to fully understand some subtle MRA points.

(1) *Subspace Structure.* The subspaces V_s should be nested. This is easy to see since information at the resolution level s is necessarily included in the information at a higher resolution level; therefore, $V_s \subset V_{s+1}$ for all s. The difference between $f_{s+1}(t)$ and $f_s(t)$, denoted by $d_s(t) = f_{s+1}(t) - f_s(t)$, is the additional information about the detail on scales between 2^{-s} and $2^{-(s+1)}$. Since the best approximation in terms of V_s is given by projecting the function into V_s, we may expect that $d_s(t) \perp f_s(t)$. Accordingly, we can decompose the subspace V_{s+1} as $V_{s+1} = V_s \oplus W_s$, where W_s is called the detail space at the resolution level s and it should be orthogonal to V_s, that is, $W_s \perp V_s$. Repeating the decomposition, we obtain

$$\begin{aligned} V_{s+1} &= W_s \oplus V_s = W_s \oplus W_{s-1} \oplus V_{s-1} = \cdots \\ &= W_s \oplus W_{s-1} \oplus W_{s-2} \oplus \cdots \oplus W_{s-S} \oplus V_{s-S} \end{aligned} \tag{2.23}$$

It is easy to see that these subspaces are mutually orthogonal. In fact, since $W_s \perp V_s$, W_s should be orthogonal to any subspace of V_s. Since both $V_{s'}$ and $W_{s'}$ are contained in V_s for $s' < s$, we thus have $W_s \perp V_{s'}$ for all $s > s'$, and $W_s \perp W_{s'}$ when $s \neq s'$. In other words, any two detail spaces at different resolutions are orthogonal, and the detail space W_s is orthogonal to an approximation space $V_{s'}$ only when the detail space is at a higher resolution level.

(2) *Resolution of Functions.* In the MRA, we describe all square-integrable functions at the finest resolution and only the zero function at the coarsest level. More precisely, as the resolution increases, we include more and more details. As the resolution goes to infinity, the approximation should approach any signal in the entire initial space $L^2(R)$. In other words, $\lim_{s\to\infty} V_s = \cup_s V_s$ should be dense in $L^2(R)$. On the other hand, as the resolution gets coarser, we remove more and more details of the signal. Thus, when $s \to -\infty$, only constant functions survive. Since functions have to be square integrable, the zero function will be the only one left, that is, $\lim_{s\to-\infty} V_s = \cap_s V_s = \{0\}$.

(3) *Dilation of Function.* This scale or dilation invariance of the space V_s is an important feature of the MRA. More precisely, all V_s should be scaled versions of the central space V_0. In fact, if a function $f(t)$ is in the space V_s, then $f(t)$ contains no details or fluctuations of scale smaller than 2^{-s}. Since $f(2t)$ is a function obtained from $f(t)$ with a squeezing factor of 2,

it contains no details at scales smaller than $2^{-(s+1)}$. Therefore, $f(2t)$ is in V_{s+1}. Conversely, given that $f(2t)$ is in V_{s+1}, similar reasoning implies that $f(t)$ is in the space V_s. Therefore, $f(t) \in V_s, \Leftrightarrow f(2t) \in V_{s+1}$.

(4) *Translation of Function.* The translation or shift invariance of the space V_s is another important feature of the MRA. If we have a function $f(t) \in V_0$ and define $T_k f(t) = f(t-k)$, where $k \in Z$, then it is clear that the resolution level of $T_k f(t)$ should not change at all. We hence require that V_0 be the integral translation-invariant; that is, $T_k(V_0) \subset V_0$ for every integer k.

Readers may wonder why we require only integral translation-invariance of V_0. Translation-invariance is needed to construct a basis out of a scaling function. Since a basis is discrete, the above requirement will be enough. A continuous translation-invariant will put too strong a limitation on the MRA for it to work. In any case, it follows from the definition that V_s will only be guaranteed to be s-dyadic translation-invariant ($2^{-s}k$ translation-invariant, $k \in Z$).

Note that the third and fourth properties yield the following result: if $f(t) \in V_0$, then $f(2^j t - k) \in V_j$ for all $k \in Z$.

(5) *Scaling Basis Functions.* The final and most important property that we require is that there exists a function ϕ such that the set of its integral-translates forms a Riesz basis of the space V_0. The scale-invariance condition then implies that $\{\phi(2t-k)\}_{k \in Z}$ forms a Riesz basis of V_1. More generally, if we define $\phi_{s,k}(t) = 2^{s/2}\phi(2^s t - k)$, then $\{\phi_{s,k}(t)\}_{k \in Z}$ forms a Riesz basis of V_s. The function ϕ, which generates basis functions for all V_s, is called a scaling function of the MRA. This key feature is similar to what we have in Fourier analysis, where $e^{j\omega t}$ is the only function needed to generate an orthonormal basis.

Based on the above discussion, we now give the formal definition.

Formal Definition of the Multiresolution Analysis

A multiresolution analysis in $L^2(R)$ is defined as a set of closed subspaces V_s with $s \in Z$ such that

(1) $\cdots \subset V_{-1} \subset V_0 \subset V_1 \subset \cdots \subset L^2(R)$, that is, $V_s \subset V_{s+1}$ for all s in Z;
(2) $\bigcap_{s=-\infty}^{+\infty} V_s = \{0\}$ and $\bigcup_{s=-\infty}^{+\infty} V_s$ is dense in $L^2(R)$;
(3) $f(t) \in V_s$ if and only if $f(2t) \in V_{s+1}$;
(4) if $f(t) \in V_0$, then $f(t-k) \in V_0$;
(5) there exists a function $\phi(t) \in V_0$, called a scaling function, such that the set $\{\phi(t-k) | k \in Z\}$ is a Riesz basis of V_0.

If a scaling function actually generates an orthonormal basis, then we call it an orthonormal scaling function. In some references, this is required as a part of the definition of the MRA and the MRA as defined above is then called a Riesz MRA. While orthonormal bases are desirable, the orthogonality puts a strong limitation on possible scaling functions that one can choose in practice. Thus, it is more flexible to require only a "generalized orthonormal" basis in the definition. On the other hand, as we will see in Section 2.5, we can always use an orthogonalization procedure to produce an orthonormal scaling function from a Riesz scaling function such that they generate the same V_s. So, in this sense, we did not lose anything with the above "relaxed" definition.

2.4 SCALING FUNCTIONS AND WAVELETS

The function $\phi(t)$ in the definition of the MRA is also called a father scaling function, because it generates bases for all V_s. In fact, it is easy to see from properties (3), (4), and (5) of the MRA that the sequence $\{\phi_{s,l}(t) = 2^{s/2}\phi(2^s t - l) | l \in Z\}$ forms a Riesz basis of V_s for all s. It is clear from the definition that the scaling function plays a key role. In fact, such a function is often the starting point in practice.

Since V_0 is contained in V_1, $\phi(t)$ can be represented as a linear combination of the Riesz basis of V_1; that is,

$$\phi(t) = \sqrt{2} \sum_l h_l \phi(2t - l) \tag{2.24}$$

This is often called the refinement equation or dilation equation. It plays a very important role in both MRA theory and applications. For example, as we will see in Chapter 3, it provides a way to construct fast algorithms to evaluate a scaling function at dyadic points. This is often enough in many applications and is particularly useful when we do not know the scaling function itself explicitly.

On the other hand, when an orthonormal scaling function $\phi(t)$ is given, the values of h_l can easily be found as follows:

$$\begin{aligned}
\sqrt{2} \int_{-\infty}^{+\infty} \phi(t)\phi^*(2t - n)\, dt &= 2 \sum_l \int_{-\infty}^{+\infty} h_l \phi(2t - l)\phi^*(2t - n)\, dt \\
&= 2 \int_{-\infty}^{+\infty} h_n \phi(2t - n)\phi^*(2t - n)\, dt \qquad (2.25) \\
&= \int_{-\infty}^{+\infty} h_n \phi(x)\phi^*(x)\, dx \\
&= h_n
\end{aligned}$$

Since $V_s \subset V_{s+1}$, we may decompose V_{s+1} by $V_{s+1} = V_s \oplus W_s$. Recursive applications of this relation lead to

$$V_0 \oplus \sum_{s=0}^{n} W_s = \bigoplus_{s=-\infty}^{n} W_s = V_{n+1} \quad \text{and} \quad \bigoplus_{s=-\infty}^{+\infty} W_s = L^2 \tag{2.26}$$

Note that such W_s is not unique unless we require that W_s be the orthogonal complement space of V_s in V_{s+1}. We assume in this book that this is the case; that is, $W_s \perp V_s$, unless stated otherwise. This is referred to as an orthogonal MRA in some references because they want to deal with more general W_s, although the terminology may be misleading as it does not require or mean its scaling function generates orthogonal bases. However, it does imply that all W_s are mutually orthogonal as discussed before.

A function $\psi(t) \in W_0$ is called a wavelet (or semi-orthogonal wavelet as it called in some references) of the MRA if the sequence $\{\psi_l(t) = \psi_l(t-l) | l \in Z\}$ forms a Riesz basis of W_0. It is called an orthonormal wavelet if the wavelet actually generates an orthonormal basis. Similar to $\phi(t)$, the function $\psi(t)$ is also called a mother wavelet, as $\psi(t)$ generates $\{\psi_{s,l} | l \in Z\}$ to form Riesz bases for all W_s, where $\psi_{s,l}(t) = 2^{s/2}\psi(2^s t - l)$. Furthermore, since all W_s are mutually orthogonal, the union of their Riesz bases, namely, $\{\psi_{s,l} | s, l \in Z\}$, forms a Riesz basis of the whole space L^2. In fact, it is often called a semi-orthogonal basis since $\psi_{s,l}(t) \perp \psi_{r,k}(t)$ for $s \neq r$. Once again, note that in general it does not require (or imply) that the associated scaling function and wavelet are orthonormal, only that $W_s \perp V_s$ and all W_s are mutually orthogonal.

Since W_0 is contained in V_1, $\psi(t)$ can also be represented as a linear combination of the Riesz basis of V_1; that is,

$$\psi(t) = \sqrt{2} \sum_l g_l \phi(2t - l) \in V_1 \tag{2.27}$$

We call this the wavelet equation. Again, this equation plays a key role in our study, especially when it is used together with the refinement equation (2.24). Together they are often called the two-scale relations.

At this point we may ask: How can we know such a mother wavelet exists for a given MRA? More importantly, how do we find it when a MRA is given? Note that the existence of $\psi(t)$ is neither required nor guaranteed by the definition of the MRA. However, as we will see in the next section, there is a procedure to produce an associated wavelet for a given $\phi(t)$. For example, if $\phi(t)$ is orthonormal, then $\psi(t)$ can be constructed using (2.27), because we can take $g_l = (-1)^l\, h^*{}_{1-l}$ in this case, where h_k are the coefficients in the refinement equation (2.24). On the other hand, $\psi(t)$ is certainly not unique. In fact, flexibility is one of the important features of wavelet theory, but it does come at a cost of having more things to sort out.

2.5 MRA IN THE FREQUENCY DOMAIN

As mentioned in the previous section, there is an orthogonalization procedure for scaling functions. To fully understand this, we need to study MRAs in the frequency domain or Fourier space. This will also enable us to construct associated wavelets as stated at the end of the last section.

Since it is expected that most functions used in practice should be reasonably well behaved, we assume, for simplicity, that functions dealt with here satisfy the needed technical conditions without further specifications. For instance, we now assume that our scaling functions and wavelets are also in $L^1(R)$. Another example is that when a formula is stated to be true, sometimes it really means that the formula is true "almost everywhere" technically. This should not be a problem for us, as we deal almost exclusively with continuous functions.

For any $f(t) \in L^1(R)$, recall that its Fourier transform is defined by

$$\widetilde{f}(\omega) = \int_{-\infty}^{\infty} f(t)e^{-j\omega t}\, dt \tag{2.28}$$

Parseval's identity then states that, for any $f, g \in L^1(R) \cap L^2(R)$,

$$\langle f, g\rangle = \frac{1}{2\pi}\langle \widetilde{f}, \widetilde{g}\rangle \tag{2.29a}$$

that is,

$$\int_{-\infty}^{\infty} f(t)g^*(t)\, dt = \frac{1}{2\pi}\int_{-\infty}^{\infty} \widetilde{f}(\omega)\widetilde{g}^*(\omega)\, d\omega \tag{2.29b}$$

Therefore, let $c_k = \langle \phi, \phi_k\rangle = \int_{-\infty}^{\infty} \phi(t)\phi^*(t-k)\, dt$. We then have

$$\begin{aligned} c_k &= \int_{-\infty}^{\infty} \phi(t)\phi^*(t-k)\, dt \\ &= \frac{1}{2\pi}\int_{-\infty}^{\infty} \widetilde{\phi}(\omega)\widetilde{\phi}^*(\omega)e^{j\omega k}\, d\omega \\ &= \frac{1}{2\pi}\int_{0}^{2\pi} \sum_{l=-\infty}^{\infty} \left|\widetilde{\phi}(\omega + 2l\pi)\right|^2 e^{j\omega k}\, d\omega \end{aligned} \tag{2.30}$$

Note that if we define $F(\omega) = \sum_{l=-\infty}^{\infty} |\widetilde{\phi}(\omega + 2l\pi)|^2$, then the last expression in (2.30) means that c_k are exactly the coefficients in the Fourier series expansion of this 2π-periodic function; that is,

$$F(\omega) = \sum_{l=-\infty}^{\infty} \left|\widetilde{\phi}(\omega + 2l\pi)\right|^2 = \sum_{k=-\infty}^{\infty} c_k e^{-jk\omega} \tag{2.31}$$

Since $\{\phi_k = \phi(t-k)|k \in Z\}$ is orthonormal if and only if $c_k = \delta_{k,0}$, we then obtain a necessary and sufficient condition in the frequency domain for an orthonormal scaling function as

$$F(\omega) = c_0 = 1, \quad \text{that is,} \quad \sum_{l=-\infty}^{\infty} |\widetilde{\phi}(\omega + 2l\pi)|^2 = 1 \tag{2.32}$$

Of course, the above can also be applied to other functions as well, so a necessary and sufficient condition for an orthonormal wavelet can be expressed in the frequency domain as

$$L(\omega) = \sum_{l=-\infty}^{\infty} |\widetilde{\psi}(\omega + 2l\pi)|^2 = 1 \tag{2.33}$$

Moreover, if $\phi^*(t-k)$ is replaced by $\psi^*(t-k)$ in (2.30), then $V_0 \perp W_0$ if and only if all $c_k = 0$. Therefore, we obtain the following necessary and sufficient orthogonal condition between a scaling function and a wavelet in the frequency domain:

$$\sum_{l=-\infty}^{\infty} \widetilde{\phi}(\omega + 2l\pi)\widetilde{\psi}^*(\omega + 2l\pi) = 0 \tag{2.34}$$

We are now ready to carry out the following orthogonalization procedure as mentioned earlier. First, let us define

$$\widetilde{\phi}_{\text{orth}}(\omega) = \frac{\widetilde{\phi}(\omega)}{\sqrt{F(\omega)}} \tag{2.35}$$

This is justified since the Riesz condition guarantees that $F(\omega) > 0$. Using the fact that $F(\omega)$ is 2π-periodic, we then have

$$\begin{aligned}\sum_{l=-\infty}^{\infty} |\widetilde{\phi}_{\text{orth}}(\omega + 2l\pi)|^2 &= \sum_{l=-\infty}^{\infty} \left|\widetilde{\phi}(\omega + 2l\pi)/\sqrt{F(\omega + 2l\pi)}\right|^2 \\ &= \sum_{l=-\infty}^{\infty} \left|\widetilde{\phi}(\omega + 2l\pi)/\sqrt{F(\omega)}\right|^2 \\ &= \left(\sum_{l=-\infty}^{\infty} |\widetilde{\phi}(\omega + 2l\pi)|^2\right) \Big/ F(\omega) \\ &= 1\end{aligned} \tag{2.36}$$

Therefore, $\widetilde{\phi}_{\text{orth}}(\omega)$ satisfies the orthonormal condition (2.32). Hence, the corresponding $\phi_{\text{orth}}(t)$ is an orthonormal scaling function. In fact, $\{\phi_{\text{orth}}(t-k)|k \in Z\}$ is not only orthonormal but it actually generates the same subspace V_0. To see all of those, note once again that $F(\omega)$ is a 2π-periodic function. Hence, there exist $\{a_k\}$ and $\{b_k\}$ such that

$$(F(\omega))^{1/2} = \sum_{k=-\infty}^{\infty} a_k e^{-jk\omega} \tag{2.37a}$$

$$(F(\omega))^{-1/2} = \sum_{k=-\infty}^{\infty} b_k e^{-jk\omega} \tag{2.37b}$$

In other words, (2.37a) and (2.37b) are basically the Fourier series expansions of $(F(\omega))^{1/2}$ and $(F(\omega))^{-1/2}$, respectively. We claim that

$$\phi(t) = \sum_{k=-\infty}^{\infty} a_k \phi_{\text{orth}}(t-k) \tag{2.38a}$$

$$\phi_{\text{orth}}(t) = \sum_{k=-\infty}^{\infty} b_k \phi(t-k) \tag{2.38b}$$

We are going to verify one of the above formulas, (2.38b), since the same argument works for the other formula as well. To do that, it is enough to show that the Fourier transform of the summation on the left-hand side of (2.38b) is indeed equal to $\widetilde{\phi}_{\text{orth}}(\omega)$. Hence, applying the Fourier transform to the term, we then have

$$\begin{aligned}
\frac{1}{2\pi}\int_{-\infty}^{\infty}\sum_{k=-\infty}^{\infty} b_k\phi(t-k)e^{-j\omega t}\,dt &= \frac{1}{2\pi}\int_{-\infty}^{\infty}\sum_{k=-\infty}^{\infty} b_k\phi(x)e^{-j\omega(x+k)}\,dx \\
&= \sum_{k=-\infty}^{\infty} b_k e^{-j\omega k}\frac{1}{2\pi}\int_{-\infty}^{\infty}\phi(x)e^{-j\omega x}\,dx \quad (2.39)\\
&= (F(\omega))^{-1/2}\widetilde{\phi}(\omega)
\end{aligned}$$

This completes our verification of the claim (2.38), as the last term in the above equation is just the definition of $\widetilde{\phi}_{\text{orth}}(\omega)$ as given in (2.35).

Since $\phi_{\text{orth}}(t)$ generates $\phi(t)$ by (2.38a), it generates the same original MRA as claimed. Furthermore, by (2.36) and (2.38b), it is indeed an orthonormal scaling function.

The above discussion also provides a practical procedure to actually construct the orthonormal scaling function $\phi_{\text{orth}}(t)$ from a given Riesz scaling function $\phi(t)$. By (2.38b), we need mainly to find coefficients in the Fourier series expansion of the 2π-periodic function $(F(\omega))^{-1/2}$. We will study such examples in the next chapter. On the other hand, it should be pointed out that $\phi_{\text{orth}}(t)$ can be very different from $\phi(t)$. For example, in general it will not be compactly supported even if the original $\phi(t)$ is.

To construct an associated wavelet from a given scaling function, we need to look at two-scale relations in the frequency domain. Let us first consider the refinement equation. Multiplying both sides of (2.24) by $e^{-j\omega t}$ and integrating, we have

$$\begin{aligned}\tilde{\phi}(\omega) &= \int_{-\infty}^{\infty} \phi(t)e^{-j\omega t}\,dt \\ &= \sum_{l=-\infty}^{\infty} \sqrt{2}h_l \int_{-\infty}^{\infty} \phi(2t-l)e^{-j\omega t}\,dt \\ &= \sum_{l=-\infty}^{\infty} \frac{h_l}{\sqrt{2}} e^{-jl\omega/2} \int_{-\infty}^{\infty} \phi(2t-l)e^{-j(2t-l)\omega/2} d(2t-l)\end{aligned} \tag{2.40}$$

Therefore, the refinement equation in the frequency domain becomes

$$\tilde{\phi}(\omega) = H(\omega/2)\tilde{\phi}(\omega/2) \tag{2.41}$$

where $H(\omega) = \sum_{l=-\infty}^{\infty}(h_l/\sqrt{2})e^{-jl\omega}$.

Following the same argument, the wavelet equation (2.27) becomes

$$\tilde{\psi}(\omega) = G(\omega/2)\tilde{\phi}(\omega/2) \tag{2.42}$$

where $G(\omega) = \sum_{l=-\infty}^{\infty}(g_l/\sqrt{2})e^{-jl\omega}$.

We now need to reexamine the orthonormal conditions. Applying (2.41) to the definition of $F(2\omega)$, we have

$$F(2\omega) = \sum_{l=-\infty}^{\infty} \left|\tilde{\phi}(2\omega + 2l\pi)\right|^2 = \sum_{l=-\infty}^{\infty} \left|H(\omega + l\pi)\right|^2 \left|\tilde{\phi}(\omega + l\pi)\right|^2 \tag{2.43}$$

To further simplify (2.43), we split the last summation into even ($l = 2k$) and odd ($l = 2k + 1$) terms. Using the fact that $H(\omega)$ is 2π-periodic, we have

$$\begin{aligned}F(2\omega) &= \sum_{k=-\infty}^{\infty} \left|H(\omega + 2k\pi)\right|^2 \left|\tilde{\phi}(\omega + 2k\pi)\right|^2 \\ &\quad + \sum_{k=-\infty}^{\infty} \left|H(\omega + \pi + 2k\pi)\right|^2 \left|\tilde{\phi}(\omega + \pi + 2k\pi)\right|^2 \\ &= \left|H(\omega)\right|^2 \sum_{k=-\infty}^{\infty} \left|\tilde{\phi}(\omega + 2k\pi)\right|^2 + \left|H(\omega + \pi)\right|^2 \sum_{k=-\infty}^{\infty} \left|\tilde{\phi}(\omega + \pi + 2k\pi)\right|^2 \\ &= \left|H(\omega)\right|^2 F(\omega) + \left|H(\omega + \pi)\right|^2 F(\omega + \pi)\end{aligned} \tag{2.44}$$

In particular, if the scaling function is orthonormal, then $F(\alpha) \equiv 1$ for all α by (2.32). We therefore obtain the following equation in h_k when the given scaling function is orthonormal:

$$|H(\omega)|^2 + |H(\omega+\pi)|^2 = 1 \tag{2.45}$$

Note that, unlike (2.32), the above equation only provides a necessary orthogonal condition for the scaling function.

Applying the same argument to $L(2\omega)$ yields the following equation:

$$L(2\omega) = |G(\omega)|^2 F(\omega) + |G(\omega+\pi)|^2 F(\omega+\pi) \tag{2.46}$$

Therefore, assuming (2.32) ($F(\alpha) = 1$), (2.33) is equivalent to the following:

$$|G(\omega)|^2 + |G(\omega+\pi)|^2 = 1 \tag{2.47}$$

Similarly, if $F(\alpha) = 1$ holds, then the following equation is equivalent to (2.34):

$$H(\omega)G^*(\omega) + H(\omega+\pi)G^*(\omega+\pi) = 0 \tag{2.48}$$

Equation (2.47) gives a necessary and sufficient condition for the coefficients g_k in the wavelet equation for a wavelet being orthonormal if an orthonormal scaling function is already given. Equation (2.48) is more interesting, as it relates the coefficients g_k in the wavelet equation to h_k in the refinement equation and provides a necessary and sufficient condition that guarantees $V_0 \perp W_0$ when an orthonormal scaling function is given. This paves the way for constructing a wavelet from an orthonormal scaling function. To see this, note that (2.48) is satisfied with $G(\omega) = A(\omega)H^*(\omega+\pi)$ if we choose $A(\omega)$ such that $A(\omega) = -A(\omega+\pi)$. Such $A(\omega)$ must be 2π-periodic since both $H(\omega)$ and $G(\omega)$ are. Moreover, (2.47) implies that we will get an orthonormal wavelet if and only if $|A(\omega)| = 1$, since $F(\alpha) = 1$ as we start with an orthonormal scaling function. There are many such $A(\omega)$, but the simplest form would be $A(\omega) = ce^{-j(2k+1)\omega}$ with $|c| = 1$ for any $k \in Z$. Here we will use the standard choice $A(\omega) = -e^{-j\omega}$. It should be borne in mind that any $A(\omega)$ as discussed above should work as well. For example, another standard choice used in many references and computational software is $A(\omega) = e^{-j\omega}$. Therefore, readers should be aware that, depending on the software used, the associated wavelet constructed below could be off by a sign.

Finally, writing out $H(\omega)$ explicitly in $G(\omega) = -e^{-j\omega}H^*(\omega+\pi)$, we have

$$\begin{aligned} G(\omega) &= -e^{-j\omega} \sum_{l=-\infty}^{\infty} \frac{h_l^*}{\sqrt{2}} e^{jl(\omega+\pi)} \\ &= \sum_{l=-\infty}^{\infty} \frac{h_l^*}{\sqrt{2}} e^{-j(1-l)\omega}(-1)^{1-l} = \sum_{k=-\infty}^{\infty} \frac{(-1)^k h_{1-k}^*}{\sqrt{2}} e^{-jk\omega} \end{aligned} \tag{2.49}$$

This yields a very simple formula:

$$g_k = (-1)^k h_{1-k}^* \tag{2.50}$$

We thus define

$$\psi(t) = \sqrt{2} \sum_l (-1)^l h_{1-l}^* \phi(2t - l) \in V_1 \tag{2.51}$$

and let W_0 be the subspace generated by $\{\psi(t-k)|k \in Z\}$. By our construction, $\{\psi(t-k)|k \in Z\}$ forms an orthonormal basis of W_0 and we also have $W_0 \perp V_0$. We therefore obtain an associated orthonormal wavelet $\psi(t)$. In fact, the only thing left to show is that $V_1 = V_0 \oplus W_0$. This is not too hard to do and we refer interested readers to Appendix 2.1 of [2] for details of an elementary proof.

In summary, we can construct an associated orthonormal wavelet $\psi(t)$ from a given orthonormal scaling function $\phi(t)$ as follows. First, find the coefficients h_l in the refinement equation by calculating the inner products of $\phi(t)$ and $\phi_{1,l}(t)$. The wavelet $\psi(t)$ can then be constructed by (2.51) based on the wavelet equation and the formula (2.50). Note that the wavelet constructed should have features that mirror the given scaling function. For example, it will have a compact support if and only if the scaling function is compactly supported. In the case where the original scaling function is not orthonormal, we first need to apply the orthogonalization procedure outlined in the first part of this section.

2.6 EXAMPLES

In this section we illustrate our discussion of the MRA with two classical examples. More comprehensive and sophisticated MRA families will be discussed in Chapter 3.

Box Function and Haar MRA

Our first example is the Haar MRA generated by the box function.

Let V_0 be the space of functions in $L^2(R)$ that are constant over intervals $[k, k+1)$ for $k \in Z$. More generally, let V_s be the space of functions that are constant over intervals $[k/2^s, (k+1)/2^s)$. It is not difficult to verify that this collection of spaces forms a MRA of $L^2(R)$. In fact, besides finding a scaling function, the only other condition that needs to be verified is that V_s will tend to L^2. This verification can be performed by repeating the definition of integrals by Riemann sums.

With the above defined V_s, it is easy to see that the characteristic function ϕ on the interval $[0, 1)$, often called the box function, as shown in Figure 2.1, can be chosen as the Haar scaling function:

$$\phi(t) = \begin{cases} 1 & \text{for } 0 \le t < 1 \\ 0 & \text{otherwise} \end{cases} \tag{2.52}$$

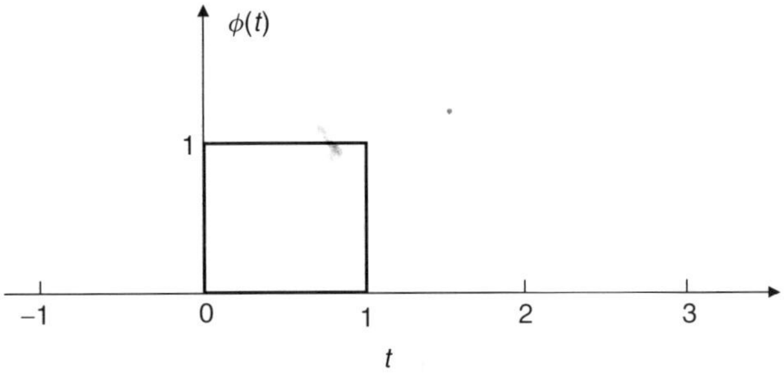

FIGURE 2.1 Haar scaling function $\phi(t)$.

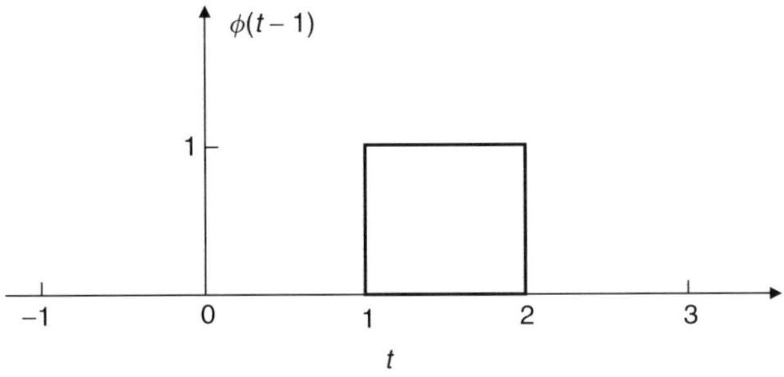

FIGURE 2.2 Translated version of the Haar scaling function $\phi(t-1)$.

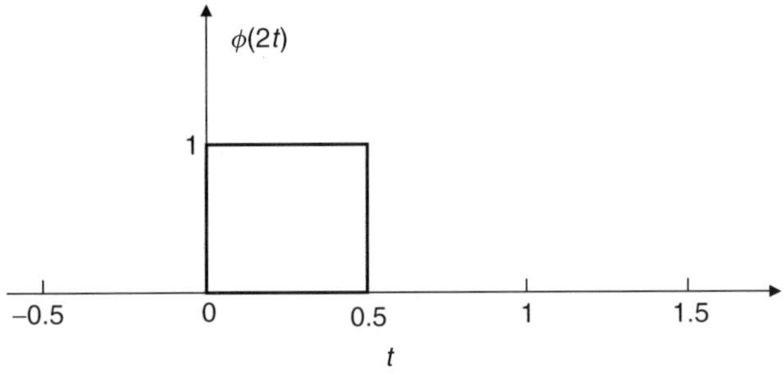

FIGURE 2.3 Dilated version of the Haar scaling function $\phi(2t)$.

In Figure 2.2, we show the translated version of the Haar scaling function $\phi(t-1)$, which shifts one unit along the x-axis.

Since $\phi(t)$ and $\phi(t-1)$ are supported on disjoined sets, they are clearly orthogonal. In fact, it is easy to see that $\phi(t)$ is an orthonormal scaling function.

Next, we demonstrate how to construct the associated wavelet $\psi(t)$. The graph of the function $\phi(2t)$ is just that of $\phi(t)$ squeezed into half the width (Figure 2.3). The graph of $\phi(2t-1) = \phi(2(t-\frac{1}{2}))$ is the graph of $\phi(2t)$ translated to the right by the amount $\frac{1}{2}$ (Figure 2.4).

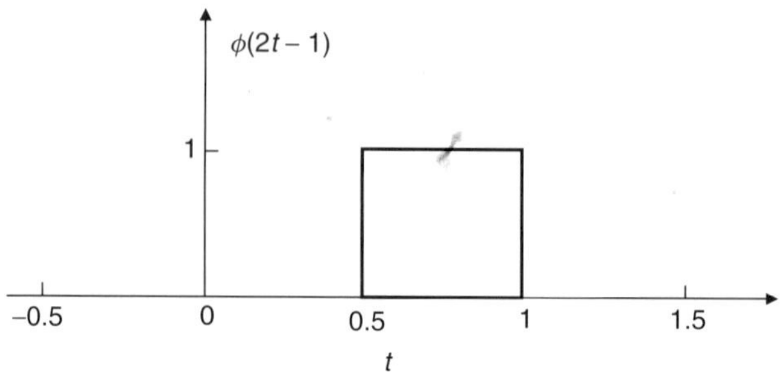

FIGURE 2.4 Dilated/translated version of the Haar scaling function $\phi(2t-1)$.

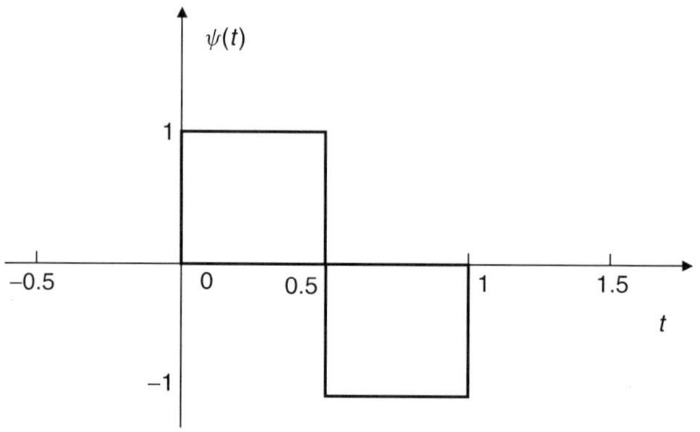

FIGURE 2.5 Haar wavelet function $\psi(t)$.

Therefore, it is obvious that the refinement equation is

$$\phi(t) = \phi(2t) + \phi(2t-1) \tag{2.53}$$

Hence, $h_0 = h_1 = 1/\sqrt{2}$ and all other coefficients are zero. Those coefficients can also be obtained easily by calculating the inner product directly. Therefore, following the construction method from Section 2.5, we represent the Haar wavelet as follows:

$$\psi(t) = \sqrt{2}\sum_l g_l \phi(2t-l) = \sqrt{2}\sum_l (-1)^l h_{1-l}^* \phi(2t-l) = \phi(2t) - \phi(2t-1) \tag{2.54}$$

The graph of the Haar wavelet is shown in Figure 2.5:

$$\psi(t) = \begin{cases} 1, & 0 \le t < \frac{1}{2} \\ -1, & \frac{1}{2} \le t < 1 \\ 0, & \text{elsewhere} \end{cases} \tag{2.55}$$

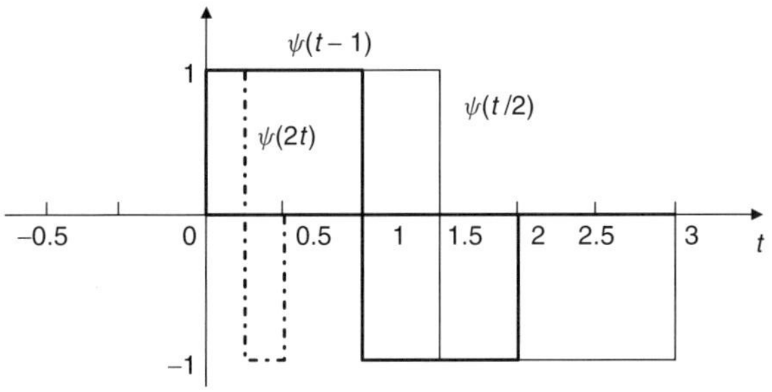

FIGURE 2.6 Haar wavelet functions $\psi(t)$, $\psi(t-1)$, and $\psi(t/2)$.

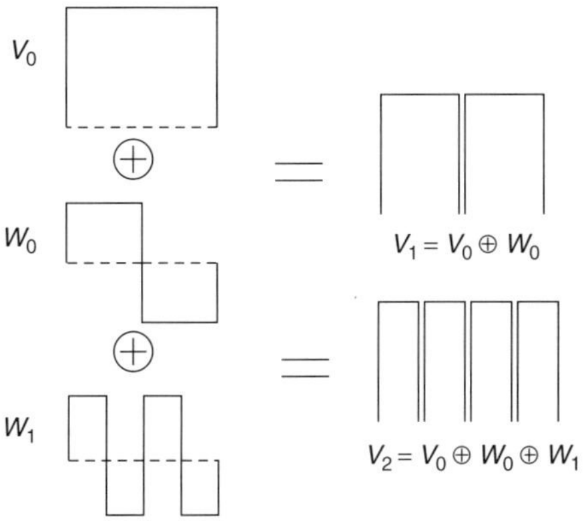

FIGURE 2.7 Demonstration of the Haar MRA for the first two levels.

Our construction guarantees that the Haar wavelets are orthonormal, and indeed this fact can easily be seen from the graphs of the wavelets. The graphs of $\psi(2t)$, $\psi(t-1)$, and $\psi(t/2)$ are represented in Figure 2.6.

The first two levels of the Haar MRA are shown in Figure 2.7. Note that here both the scaling function and the wavelet are compactly supported.

We can readily present the frequency domain expressions for the Haar scaling function and wavelet as

$$\widetilde{\phi}(\omega) = \int_{-\infty}^{+\infty} \phi(x) e^{-j\omega x}\, dx = \int_{0}^{1} e^{-j\omega x}\, dx = \left(\frac{1 - e^{-j\omega}}{j\omega}\right)$$

$$= e^{-j\omega/2} \frac{\sin(\omega/2)}{(\omega/2)} \tag{2.56}$$

$$\widetilde{\psi}(\omega) = \int_{0}^{1/2} e^{-j\omega x}\, dx - \int_{1/2}^{1} e^{-j\omega x}\, dx = \left(\frac{1 - 2e^{-j\omega/2} + e^{-j\omega}}{j\omega}\right)$$

$$= j e^{-j\omega/2} \frac{\sin^2(\omega/4)}{(\omega/4)} \tag{2.57}$$

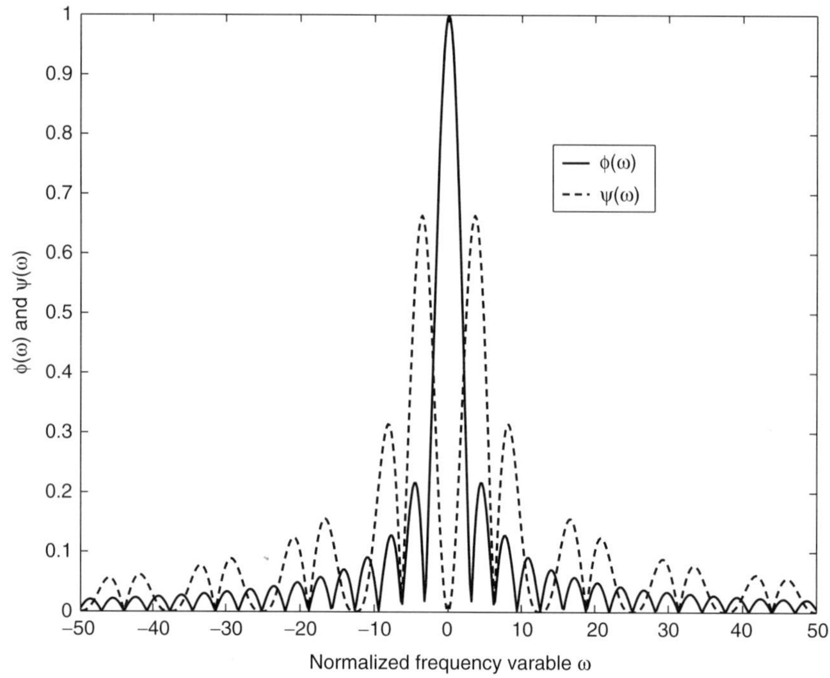

FIGURE 2.8 Magnitude of the Haar scaling function and wavelet in the frequency domain.

In Figure 2.8, we have displayed the magnitude distribution of the Haar scaling function and wavelet in the frequency domain. As seen from either formulas or graphs, the Haar basis functions exhibit good localization in the space domain but poor localization in the frequency domain.

Hat Function and Linear Spline MRA

As another example, we review the linear spline MRA generated by the hat function.

Let V_s be the space of continuous functions in $L^2(R)$ that are linear over intervals $[k/2^s, (k+1)/2^s]$, $k \in Z$. It is not hard to verify that $\{V_s\}$ defines a MRA with the following hat function (Figure 2.9) as its scaling function:

$$\beta_1(t) = \begin{cases} 1 - |t| & \text{for } 0 \le t \le 2 \\ 0 & \text{otherwise} \end{cases} \tag{2.58}$$

For example, we see that the integral of any continuous function in L^2 can be approximated by the integral of a sum using functions in V_s. Since the space of continuous functions is dense in L^2, the union of all V_s must be dense in L^2 as well. To see that the scaling function generates a basis of V_0, we write an

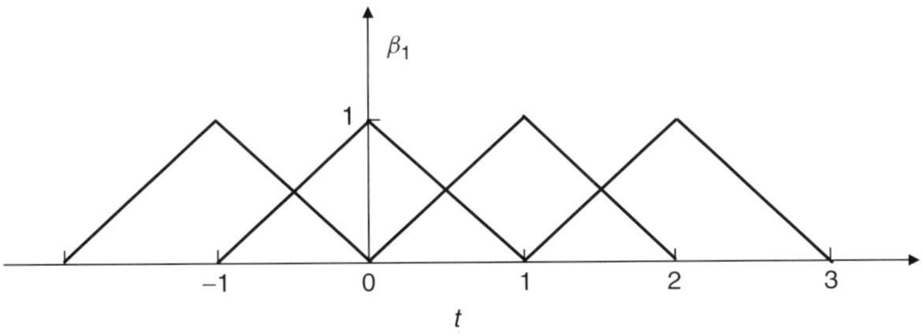

FIGURE 2.9 Hat function $\beta_1(t)$ and three of its translations.

arbitrary function $f(t)$ in V_0 as the superposition of the basis functions by

$$f(t) = \sum_l f(l)\beta_1(t - l) \tag{2.59}$$

In fact, since both sides of (2.59) are linear over an arbitrary interval $[k, k + 1]$, we need only to verify that they agree at two points, say, k and $k + 1$ in each interval. This is easy to see, as $\beta_1(k - l) = \delta_{k,l}, k, l \in Z$.

We want to point out that the hat function does not generate an orthonormal basis of V_0, but the basis is very close to orthonormal. In fact, we see that $\langle \beta_1(t - k), \beta_1(t - l) \rangle = 0$ unless $k = l$ or $|k - l| = 1$. Anyway, it is easy to get the refinement equation here, since it is again finite:

$$\beta_1(t) = 2^{-1}\beta_1(2t + 1) + \beta_1(2t) + 2^{-1}\beta_1(2t - 1) \tag{2.60}$$

To construct an associated wavelet following the method of Section 2.5, we first need to apply the orthogonalization procedure to the hat scaling function. We will show this in a more general setting in Chapter 3.

The above two examples are the first two cases of a family of the so-called B-spline MRAs. Those and other MRA families will be described in more detail in the next chapter.

2.7 BIORTHOGONAL MRA AND WAVELETS

Since orthogonality puts a strong limitation on a MRA, we often have to deal with nonorthogonal bases. In such cases, biorthogonal dual systems are very useful.

We would like to use an orthonormal basis because we can easily evaluate expansion coefficients; for example, if $\{\phi_i\}$ is an orthonormal basis, then

$$f = \sum \alpha_i \phi_i \tag{2.61a}$$

if and only if

$$\alpha_i = \langle f, \phi_i \rangle \tag{2.61b}$$

However, the above can also be achieved in another way. If we have another set $\{\hat{\phi}_i\}$ such that $\langle \phi_i, \hat{\phi}_j \rangle = \delta_{i,j}$, then we will have

$$f = \sum \alpha_i \phi_i \tag{2.62a}$$

with

$$\alpha_i = \langle f, \hat{\phi}_i \rangle \tag{2.62b}$$

Note that, at least from the computational point of view, any such $\{\hat{\phi}_i\}$ will work. For example, $\{\phi_i\}$ and $\{\hat{\phi}_i\}$ do not have to generate the same subspace.

With the above motivation, we now define the dual MRA. We make use of the exact same notations for the new dual MRA except we add "$\wedge$" to mark them as the dual. We say that this new MRA is a dual of a given MRA if

$$\langle \phi_{s,l'}, \hat{\phi}_{s,k} \rangle = \delta_{l,k'} \tag{2.63a}$$

$$\langle \psi_{s,l}, \hat{\psi}_{s',l'} \rangle = \delta_{s,s'}\delta_{l,l'}, \tag{2.63b}$$

$$\langle \phi_{s,l}, \hat{\psi}_{s,k} \rangle = 0 \tag{2.63c}$$

$$\langle \psi_{s,l}, \hat{\phi}_{s,l'} \rangle = 0 \tag{2.63d}$$

The above are often referred to as the biorthogonality conditions. Some of the above biorthogonality conditions can be expressed in terms of the subspaces as

$$V_s \perp \hat{W}_s,\ \hat{V}_s \perp W_s, \quad \text{and} \quad W_s \perp \hat{W}_{s'} \text{ for } s \neq s' \tag{2.64}$$

A MRA is called a biorthogonal MRA if its dual MRA exists and in this case ψ (or together with $\hat{\psi}$) is called a biorthogonal wavelet.

In general, the dual MRA often defines different subspaces from the primary MRA. Unlike the case of orthogonal MRAs, we do not require that W_s be orthogonal to V_s in a biorthogonal MRA, since such a requirement will put an unnecessary constraint on the existence of biorthogonal MRAs. However, if all W_s are orthogonal to V_s, then it implies $W_s = \hat{W}_s$ and $V_s = \hat{V}_s$ for all s. In this case, corresponding bases are real dual bases to each other and a dual scaling function $\hat{\phi}(t)$ can be constructed by setting

$$\widetilde{\hat{\phi}}(\omega) = \frac{\widetilde{\phi}(\omega)}{F(\omega)} \tag{2.65}$$

where $F(\omega) = \sum_{l=-\infty}^{\infty} \left|\widetilde{\phi}(\omega + 2l\pi)\right|^2$ is the same 2π-periodic function as defined in (2.31). In fact, with the above defined $\hat{\phi}(t)$, it is easy to see that

$$\sum_{l=-\infty}^{\infty} \widetilde{\phi}(\omega + 2l\pi)\widetilde{\hat{\phi}}^*(\omega + 2l\pi) = 1 \tag{2.66}$$

Note that (2.66) is the dual version of the orthonormal condition (2.32) in the frequency domain. Therefore, (2.63a) is indeed satisfied. Similarly, if the primary wavelet $\psi(t)$ is given, then a dual wavelet $\hat{\psi}(t)$ can be constructed by setting

$$\tilde{\hat{\psi}}(\omega) = \frac{\tilde{\psi}(\omega)}{L(\omega)} \tag{2.67}$$

where $L(\omega) = \sum_{l=-\infty}^{\infty} |\tilde{\psi}(\omega + 2l\pi)|^2$ is the same 2π-periodic function as defined in (2.33).

We will not get into more details about the existence and general construction procedures of a dual MRA at this point, as a detailed construction procedure will be given in Chapter 3. For now, we list some equations that will be needed later. Note that (2.45) in h_k now becomes one in h_k and $\hat{h}_k$ as follows:

$$H(\omega)\hat{H}^*(\omega) + H(\omega + \pi)\hat{H}^*(\omega + \pi) = 1 \tag{2.68}$$

This is a necessary orthogonal condition between the primary and the dual scaling functions so it has to be satisfied. Other orthogonal conditions, such as (2.47) and (2.48), can also be converted in a similar fashion. However, once the scaling functions are known, those conditions are satisfied automatically, as we often define the dual and primary wavelets by

$$\hat{\psi}(x) = \sqrt{2}\sum_{l=-\infty}^{+\infty} \hat{g}_l \hat{\phi}(2x - l) = \sqrt{2}\sum_{l=-\infty}^{+\infty} (-1)^l h_{1-l}^* \hat{\phi}(2x - l) \tag{2.69}$$

$$\psi(x) = \sqrt{2}\sum_{l=-\infty}^{+\infty} g_l \phi(2x - l) = \sqrt{2}\sum_{l=-\infty}^{+\infty} (-1)^l \hat{h}_{1-l}^* \phi(2x - l) \tag{2.70}$$

As a result, we will not list other orthogonal conditions here.

2.8 MULTIDIMENSIONAL WAVELETS

Many applications occur in higher dimensions. Although the theory of multidimensional MRA is more involved, there is a straightforward way to generalize the one-dimensional MRA defined above to handle higher-dimensional cases, which is simply to take the tensor product. For example, we may define a two-dimensional scaling function in $L^2(R^2)$ as

$$\phi(x, y) = \phi(x)\phi(y) \tag{2.71}$$

The set of its two-dimensional integral translations, $\{\phi_{k,l}(x, y) = \phi(x - k, y - l) | (k, l) \in Z^2\}$, then generates the subspace $V_0 = \{f(x, y) = \sum_{k,l} c_{k,l}\phi_{k,l}(x, y)\}$. The dyadic dilations of V_0 further generate a sequence of subspaces to form a MRA of $L^2(R^2)$. The rest of the MRA can also be generalized in a similar fashion. For example, the refinement equation becomes

$$\phi(x, y) = 2\sum_{k,l} h_{k,l}\phi(2x - k, 2y - l) \tag{2.72}$$

Since $\phi(x)$ and $\phi(y)$ both satisfy the refinement dilation equation, namely,

$$\phi(\alpha) = \sqrt{2}\sum_{k} h_k\phi(2\alpha - k) \quad (\alpha = x, y) \tag{2.73}$$

we then have $h_{k,l} = h_k h_l$. On the other hand, we must be careful about constructing two-dimensional wavelets. In fact, one basic wavelet is not enough to generate the needed detail spaces, since

$$\begin{aligned} V_{s+1}(x, y) &= V_{s+1}(x) \otimes V_{s+1}(y) \\ &= (V_s(x) \oplus W_s(x)) \otimes (V_s(y) \oplus W_s(y)) \\ &= V_s(x) \otimes V_s(x) \oplus \{V_s(x) \otimes W_s(y) \oplus W_s(x) \otimes V_s(y) \oplus W_s(x) \otimes W_s(y)\} \\ &= V_s(x, y) \oplus W_s(x, y) \end{aligned} \tag{2.74}$$

Therefore, we need the following three basic wavelets to do the job:

$$\psi^{\mathrm{I}}(x, y) = \phi(x)\psi(y) \tag{2.75a}$$

$$\psi^{\mathrm{II}}(x, y) = \psi(x)\phi(y) \tag{2.75b}$$

$$\psi^{\mathrm{III}}(x, y) = \psi(x)\psi(y) \tag{2.75c}$$

2.9 FIELD EXPANSIONS IN THE MRTD ANALYSIS

As indicated in the introduction, our main focus is on the MRA based MRTD scheme. We now give a brief description of the basic setup—field expansions in terms of scaling and wavelet functions. In principle, because scaling functions are low-pass functions and wavelet functions are band-pass functions, we can expand an arbitrary field component $f(x)$, where x can represent either time or space, as a summation of wavelets and scaling basis functions in a hierarchical decomposition manner:

$$f(x) = \sum_{i=-\infty}^{\infty} c_{s,i}\phi_{s,i}(x) + \sum_{k=s}^{\infty}\sum_{l=-\infty}^{\infty} d_{k,l}\psi_{k,l}(x) \tag{2.76}$$

The first summation in (2.76) presents the projection of $f(x)$ into the scaling or approximation subspace V_s, which corresponds to an approximation of f at a preselected "coarse" resolution level s or low spatial frequency information of a signal. The second summation consists of projections of f into the wavelet subspaces W_k, and each of them should be orthogonal to V_k as $k \geq s$.

In practice, we need to truncate (2.76) into a finite sum to approximate f. This can be done naturally for the first summation, since f is in general defined over a finite interval—or it at least has a fast decay in practice. Therefore, a scaling function with a compact support or at least a very fast decay is preferred. The same principle applies to the inside terms in the second summation, as it also involves translations only. The dilation part in the second summation corresponds to the details, so we can control resolution levels by adjusting this part of the summation. Note that we have a lot of flexibility here. For example, we may pick up an initial base level s so that it satisfies the coarse level for most values of x, then add more wavelets only at places where they are needed.

In order to compute expansion coefficients efficiently, orthonormal bases are most desirable. Otherwise, we would welcome a biorthogonal system. In any case, it can be shown that the total number of coefficients in an appropriate expansion is the same as or less than what is given in a similar Fourier (or any orthogonal) expansion.

As an example, we use the Haar MRA to approximate $f(x) = \exp(-x/4)$ over the interval [0, 16]. Here, we use the base level $s = 0$ and we thus have

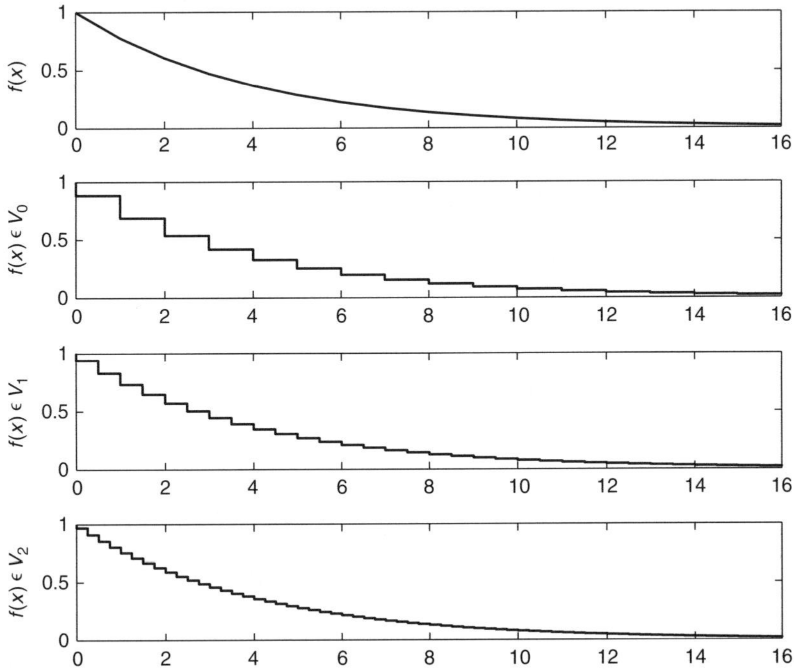

FIGURE 2.10 Projections of the function $f(x) = \exp(-x/4)$ into three approximation spaces V_0, V_1, and V_2, respectively.

the following:

$$f(x) \approx \sum_{i=0}^{15} c_{0,i}\phi_{0,i}(x) + \sum_{k=0}^{L-1} \sum_{l=0}^{2^L 16-1} d_{k,l}\psi_{k,l}(x) \tag{2.77}$$

where L is the additional detail level desired. All coefficients in (2.77) can easily be evaluated by the inner product of f and basis functions, as the Haar MRA is an orthonormal system; that is,

$$c_{0,i} = \langle f, \phi_{0,i} \rangle \tag{2.78a}$$

$$d_{k,l} = \langle f, \psi_{k,l} \rangle \tag{2.78b}$$

Figure 2.10 shows graphs of $f(x)$ and three of its approximations for detail level $L = 0$, 1, and 2, respectively. As we can see, we have a lot of flexibility and control here; for example, we may drop some $d_{k,l}$ terms for large l and still achieve a good approximation.

REFERENCES

[1] G. Bachmam, L. Narici, and E. Beckenstein, *Fourier and Wavelet Analysis*, Springer, New York, 2000.

[2] A. Boggess and F. J. Narcowich, *A First Course in Wavelets with Fourier Analysis*, Prentice-Hall, New York, 2001.

[3] R. N. Bracewell, *The Fourier Transform and Its Applications*, McGraw-Hill Book Company, New York, 1965.

[4] E. O. Brigham, *The Fast Fourier Transform and Its Applications*, Prentice-Hall, New York, 1988.

[5] C. K. Chui, *An Introduction to Wavelets*, Academic Press, San Diego, 1992.

[6] A. Cohen, "Biorthogonal wavelets," in *Wavelets: A Tutorial in Theory and Applications*, (ed. C. K. Chui), Academic Press, San Diego, pp. 123–152, 1992.

[7] I. Daubechies, *Ten Lectures on Wavelets*, SIAM, Philadelphia, 1992.

[8] B. Jawerth and W. Sweldens, "An overview of wavelet based multiresolution analysis," *SIAM Rev.*, vol. 36, no. 3, pp. 377–412, 1994.

[9] B. P. Lathi, *Linear Systems and Signals*, Berkeley-Cambridge Press, Carmichael, CA, 1992.

[10] S. G. Mallat, "A theory for multiresolution signal decomposition: the wavelet representation," *IEEE Trans. Pattern Anal. Machine Intell.*, vol. 11, no. 7, pp. 674–693, 1989.

[11] S. G. Mallat, "Multifrequency channel decompositions of images and wavelet model," *IEEE Trans. Acoust. Speech Signal Process.*, vol. 37, no. 12, pp, 2091–2110, 1989.

[12] Y. Meyer, *Ondelettes et Opérateurs*, I: *Ondelettes*, II: *Opérateurs de Calderón Zygmund*, III: (with R. Coifman), *Opérateurs multilinéaires*. Hermann, Paris, 1990. English translation of first volume, *Wavelets and Operators*, is published by Cambridge University Press, New York, 1993.

CHAPTER THREE

MRA Families in MRTD Analysis

3.1 INTRODUCTION

Having discussed some fundamental concepts of the multiresolution analysis (MRA) in Chapter 2, we present in this chapter some basic families of scaling functions and wavelets that are commonly used in the current multiresolution time domain (MRTD) applications. Here we pay great attention to practical matters that are essential in the implementation of the MRTD. In particular, we present various useful detailed formulations, graphs, and numerical coefficients throughout the chapter. For those readers who are really interested in full theoretical development or other related topics, we suggest they check the many available references such as those listed at the end of this chapter [1–29].

In implementing the MRTD analysis for electrical engineering applications, we usually use a MRA to expand a field as a summation of scaling functions and wavelets. Unlike what we have seen in the formal definition, in practice, a MRA often starts with a scaling function, since we care much less about the subspace structures in numerical applications. Although many different families of scaling functions and wavelets are available, there are some fundamental requirements that may restrict our selection in practice. The following are some important properties of scaling functions and wavelets that we should take into consideration:

(1) *Smoothness and Differentiability*. This is particularly important in our MRTD applications since derivatives are involved in the MRTD formulation and electromagnetic field representation should be as close as possible to a realistic field distribution.

This chapter is coauthored with Xian Wu, Department of Mathematics, University of South Carolina, Columbia, SC.

Multiresolution Time Domain Scheme for Electromagnetic Engineering
By Yinchao Chen, Qunsheng Cao, and Raj Mittra
ISBN 0-471-27230-2

(2) *Orthogonality.* The orthogonality not only simplifies the MRTD formulation in the evaluation of inner products but also stabilizes numerical computations; that is, initial errors are less likely to grow with sequential computations.
(3) *Compact Support.* With this property, we can reduce the number of terms in field expansions, easily realize localizations, and carry out computations accurately in finite summations. If a compact support is not possible, then a fast decay may serve the same purpose with reasonable approximations in practice.
(4) *Symmetry or Asymmetry.* This is critical in setting up field update leapfrogging patterns and signal processing applications. Phase distortions are more likely to occur without symmetry.
(5) *Explicit Analytic Expression.* This may not be essential, as are the other properties, in numerical computations, but it is certainly very helpful in formulating the MRTD scheme when complex materials are involved.

The chapter is organized as follows: In Section 3.2 we introduce the basic spline (B-spline) family—a generalization of two examples (the Haar and the linear scaling functions) given in Chapter 2. Since the basis functions in B-spline families are not orthogonal except at order 0, we apply the orthogonal procedure to obtain the Battle–Lemarié (orthonormal) spline family in Section 3.3. Section 3.4 follows with a detailed example of the cubic spline Battle–Lemarié MRA, one of the most frequently used MRAs in MRTD applications. Section 3.5 is devoted to Daubechies' procedure, an important way of constructing MRA families using preselected filters. Examples of Daubechies' original orthogonal families with compact support are given in the next section. In Section 3.7 we present the Coiflet family, a variation of the Daubechies family with better symmetric properties. The final two sections are on biorthogonal MRAs. A general construct procedure and the biorthogonal Cohen–Daubechies–Feauveau spline family are given in Sections 3.8 and 3.9, respectively.

3.2 BASIC SPLINE MRA FAMILY

We begin with the construction of the nth order B-spline functions, as they generate the basic spline MRA. Let $B_0(x)$ be the Haar scaling function (the box function) as given in (2.52). The nth order B-spline function $B_n(x)$ is defined by a repeated convolution of $B_0(x)$, namely,

$$B_n(x) = \underbrace{B_0 * B_0 * \cdots * B_0(x)}_{(n+1)\text{ times}} = B_0 * B_{n-1}(x) \tag{3.1}$$

where the sign $*$ denotes the convolution operation defined by

$$y_1 * y_2(x) = \int_{-\infty}^{+\infty} y_1(\lambda) y_2(x - \lambda)\, d\lambda \tag{3.2}$$

It is easy to express the B-spline functions explicitly. For example, we can see that the first-order B-spline function $B_1(x)$ is just the linear scaling function (the hat function) as given in (2.58) shifted one unit to the right. In fact,

$$B_1(x) = B_0 * B_0(x) = \int_{-\infty}^{+\infty} B_0(t) B_0(x-t)\, dt$$

$$= \int_0^1 B_0(x-t)\, dt = \begin{cases} x, & 0 \le x \le 1 \\ 2-x, & 1 \le x \le 2 \\ 0, & \text{otherwise} \end{cases} \tag{3.3}$$

Therefore, B-spline functions are extensions of two examples (the Haar and the linear scaling functions) given in Chapter 2. More generally, we have

$$B_n(x) = B_0 * B_{n-1}(x) = \int_{-\infty}^{+\infty} B_0(t) B_{n-1}(x-t)\, dt$$

$$= \int_0^1 B_{n-1}(x-t)\, dt = \int_{x-1}^{x} B_{n-1}(s)\, ds \tag{3.4}$$

Taking the derivative of (3.4), we then have

$$B_n'(x) = B_{n-1}(x) - B_{n-1}(x-1) \tag{3.5}$$

Using the above formulas inductively, we can generalize some properties of the first two B-spline functions as follows:

(1) $B_n(x)$ is compactly supported on the interval $[0, n+1]$; that is, we have $B_n(x) = 0$ outside the interval. (This follows immediately from (3.4).)
(2) $B_n(x)$ is symmetric about the center of its support interval; that is, $B_n((n+1)/2 + x) = B_n((n+1)/2 - x)$. (This is an easy exercise using (3.4).)
(3) $B_n(x)$ is $(n-1)$-times continuously differentiable; that is, it has the $(n-1)$th continuous derivative. (This follows immediately from (3.5).)
(4) $B_n(x)$ is a polynomial of degree n over each of the intervals $[k,\ k+1], 0 \le k \le n$ and is $(n-1)$-times continuously differentiable at the overlaps. (This follows again from (3.4).) Moreover, it is easy to see that explicit analytic expressions can be obtained readily.
(5) $B_n(x)$ is a scaling function; in other words, it generates a MRA. (This is an easy generalization of the first two cases already proved in Chapter 2.)

However, $B_n(x)$ is not orthogonal unless $n = 0$. This posts a major obstruction to applying them directly in the MRTD scheme.

Since any integer-translation of a scaling function is still a scaling function, in practice, we often use a shifted version $\beta_n(x)$ of $B_n(x)$ that moves the center of the support interval close to the origin:

$$\beta_n(x) = \begin{cases} B_n\left(x + \dfrac{n}{2}\right) & \text{for } n \text{ even} \\ B_n\left(x + \dfrac{n+1}{2}\right) & \text{for } n \text{ odd} \end{cases} \tag{3.6}$$

Thus, for odd n, the spline function $\beta_n(x)$ is supported on $[-(n+1)/2, (n+1)/2]$ and is symmetric about $x = 0$; and for even n, $\beta_n(x)$ is supported on $[-n/2, n/2+1]$ and is symmetric about $x = \frac{1}{2}$. These are often called the centralized B-spline functions. An explicit formula for $\beta_n(x)$ is given below for the reader's convenience [11, 12]:

$$\beta_n(x) = \begin{cases} \dfrac{1}{n!}\displaystyle\sum_{k=0}^{n+1}(-1)^k \binom{n+1}{k} [x + n/2 - k]_+^n & \text{for } n \text{ even} \\ \dfrac{1}{n!}\displaystyle\sum_{k=0}^{n+1}(-1)^k \binom{n+1}{k} [x + (n+1)/2 - k]_+^n & \text{for } n \text{ odd} \end{cases} \tag{3.7}$$

where

$$[x]_+^n = \begin{cases} 0 & \text{if } x \le 0 \\ x^n & \text{if } x > 0 \end{cases} \tag{3.8}$$

For example, if $n = 2$, (3.7) becomes

$$\beta_2(x) = \begin{cases} \dfrac{(x+1)^2}{2}, & -1 \le x \le 0 \\ \dfrac{(x+1)^2 - 3x^2}{2}, & 0 \le x \le 1 \\ \dfrac{(x+1)^2 - 3x^2 + 3(x-1)^2}{2}, & 1 \le x \le 2 \\ 0, & \text{otherwise} \end{cases} \tag{3.9}$$

Note that the relation

$$\frac{(x+1)^2 - 3x^2 + 3(x-1)^2 - (x-2)^2}{2} = 0$$

is used in deriving the above equation.

The B-spline functions have a much simpler explicit form in the frequency domain. We start with the Fourier transform of the B-spline box function $B_0(x)$. As we have seen in Chapter 2,

$$\begin{aligned} \widetilde{B}_0(\omega) &= \int_{-\infty}^{+\infty} B_0(x)e^{-j\omega x}\,dx = \int_0^1 e^{-j\omega x}\,dx \\ &= \left(\frac{1}{j\omega}\right)(1 - e^{-j\omega}) = e^{-j\omega/2}\frac{\sin(\omega/2)}{\omega/2} \end{aligned} \tag{3.10}$$

Since $B_n(x)$ is the $(n+1)$-fold convolution of $B_0(x)$, its Fourier transform is the $(n+1)$th power of $\widetilde{B}_0(\omega)$; that is,

$$\widetilde{B}_n(\omega) = (\widetilde{B}_0(\omega))^{n+1} = e^{-j((n+1)/2)\omega}\left(\frac{\sin(\omega/2)}{\omega/2}\right)^{n+1} \tag{3.11}$$

Since the centralized B-spline function $\beta_n(x)$ is obtained from $B_n(x)$ by shifting, $\widetilde{\beta}_n(\omega)$ is obtained from $\widetilde{B}_n(\omega)$ by changing the phase factor $e^{-j((n+1)/2)\omega}$ to $e^{-jp\omega}$; that is,

$$\widetilde{\beta}_n(\omega) = e^{-jp\omega}\left(\frac{\sin(\omega/2)}{\omega/2}\right)^{n+1} \tag{3.12}$$

where the phase factor $e^{-jp\omega}$ is determined by the location of the center, which depends on the choice of the shifting. In particular, in our case, $p = 0$ for odd n and $p = \frac{1}{2}$ for even n.

Next, we proceed to find the refinement equation for $\beta_n(x)$. Recall (2.41) in Chapter 2. The refinement equation in the frequency domain for $\widetilde{\beta}_n(\omega)$ has the following form:

$$\widetilde{\beta}_n(\omega) = H(\omega/2)\widetilde{\beta}_n(\omega/2) \tag{3.13}$$

where

$$H(\omega) = \sum_{l=-\infty}^{\infty} \frac{h_l}{\sqrt{2}} e^{-jl\omega} \tag{3.14}$$

Therefore, we have

$$\begin{aligned}
H(\omega) &= \widetilde{\beta}_n(2\omega)/\widetilde{\beta}_n(\omega) \\
&= \frac{e^{-j2p\omega}}{e^{-jp\omega}}\left(\frac{\sin\omega}{\omega}\,\frac{\omega/2}{\sin(\omega/2)}\right)^{n+1} \\
&= e^{-jp\omega}\cos^{n+1}(\omega/2) \\
&= e^{-jp\omega}\left(\frac{e^{-j\omega/2}+e^{j\omega/2}}{2}\right)^{n+1} \\
&= \frac{1}{2^n\sqrt{2}}\sum_{k=0}^{n+1}\binom{n+1}{k}\frac{e^{-j(p-(n+1)/2+k)\omega}}{\sqrt{2}}
\end{aligned} \tag{3.15}$$

Comparing the last expression in (3.15) with (3.14) and the definition of the refinement equation (2.24) in Chapter 2, we finally get the following explicit refinement equation for $\beta_n(x)$ in the space domain:

$$\beta_n(x) = \begin{cases} \dfrac{1}{2^n}\displaystyle\sum_{k=0}^{n+1}\binom{n+1}{k}\beta_n(2x+n/2-k) & \text{for } n \text{ even} \\ \dfrac{1}{2^n}\displaystyle\sum_{k=0}^{n+1}\binom{n+1}{k}\beta_n(2x+(n+1)/2-k) & \text{for } n \text{ odd} \end{cases} \tag{3.16}$$

The graphs of $\beta_2(x)$ and $\beta_3(x)$ are shown in Figure 3.1a. As we can see, they are almost seamlessly connected at the overlaps. The graphs of $|\widetilde{\beta}_2(\omega)|$ and $|\widetilde{\beta}_3(\omega)|$ are shown in Figure 3.1b.

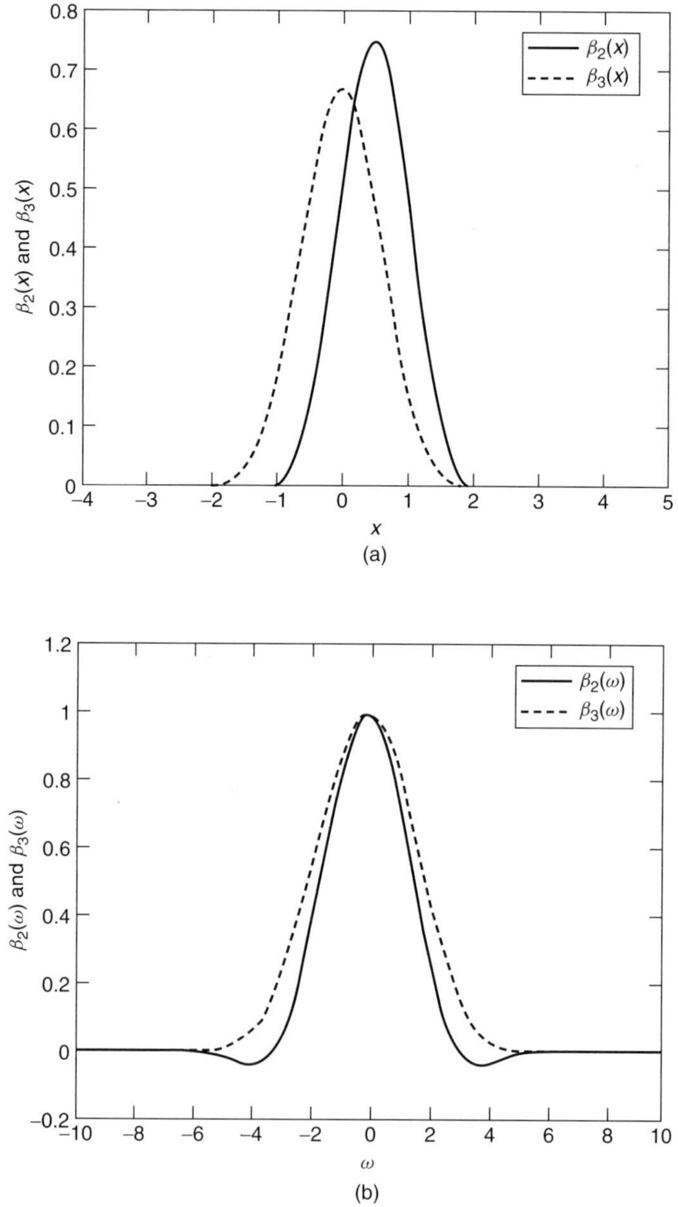

FIGURE 3.1 (a) Centralized B-spline functions in the space domain, $\beta_2(x)$ and $\beta_3(x)$. (b) B-spline functions in the frequency domain, the magnitude of $\widetilde{\beta}_2(\omega)$ and $\widetilde{\beta}_3(\omega)$.

3.3 BATTLE–LEMARIÉ SPLINE MRA FAMILY

Since $B_n(x)$ is not orthogonal except for $n = 0$, in this section we apply the orthogonalization procedure as described in Chapter 2. The resulting orthogonalized spline functions are called the Battle–Lemarié scaling functions, so named because they were introduced by Battle [15] and Lemarié [16].

To orthogonalize $B_n(x)$, we first calculate $F(\omega) = \sum_l |\widetilde{B}_n(\omega + 2\pi l)|^2$ as defined in Section 2.5.

By (3.11), we have

$$F(\omega) = \sum_l |\widetilde{B}_n(\omega + 2\pi l)|^2 = \sum_l \frac{(2\sin(\omega/2))^{2(n+1)}}{(\omega + 2\pi l)^{2(n+1)}}$$
$$= (2\sin(\omega/2))^{2(n+1)} f_{2(n+1)}(\omega) \tag{3.17}$$

where, for $n > 1$,

$$f_n(\omega) = \sum_l \frac{1}{(\omega + 2\pi l)^n} = \frac{(-1)^{n-2}}{(n-1)!} \frac{d^{n-2}(f_2(\omega))}{d\omega^{n-2}} \tag{3.18}$$

To get a more explicit expression, we claim that

$$f_2(\omega) = \frac{1}{4\sin^2(\omega/2)} \tag{3.19}$$

Note that $B_0(x)$ is orthonormal and therefore, by the orthonormal condition (2.32),

$$1 = \sum_l |\widetilde{B}_0(\omega + 2\pi l)|^2 = \sum_l \frac{(2\sin(\omega/2))^2}{(\omega + 2\pi l)^2} = (2\sin(\omega/2))^2 f_2(\omega) \tag{3.20}$$

thus proving the claim.

Putting the above all together, we then obtain

$$F(\omega) = (2\sin(\omega/2))^{2(n+1)} \frac{1}{(2n+1)!} \frac{d^{2n}((2\sin(\omega/2))^{-2})}{d\omega^{2n}} \tag{3.21}$$

Substituting (3.21) into (2.35) of Chapter 2, we finally obtain an explicit expression for the orthogonalized spline function in the frequency domain as follows:

$$\widetilde{\phi}_{\text{orth}}(\omega) = \frac{\widetilde{B}_n(\omega)}{\sqrt{F(\omega)}}$$
$$= e^{-j((n+1)/2)\omega} \left(\frac{1}{\omega}\right)^{n+1} \left(\frac{1}{(2n+1)!} \frac{d^{2n}((2\sin(\omega/2))^{-2})}{d\omega^{2n}}\right)^{-1/2} \tag{3.22}$$

Using the inverse Fourier transformation, we can then obtain the Battle–Lemarié spline scaling function $\phi_{\text{orth}}(x)$ in the space domain from $\widetilde{\phi}_{\text{orth}}(\omega)$ in the frequency domain. In practice, as shown in Chapter 2, we often use (2.37b) and (2.38b) to express the Battle–Lemarié scaling functions $\phi_{\text{orth}}(x)$ in terms of the original B-spline functions; in other words, we calculate the Fourier series,

$$(F(\omega))^{-1/2} = \sum_k b_k e^{-jk\omega} \tag{3.23}$$

and use $\{b_k\}$ to construct $\phi_{\text{orth}}(x)$ by

$$\phi_{\text{orth}}(x) = \sum_k b_k B_n(x-k) \tag{3.24}$$

Note that, from its definition, $F(\omega)$ is independent of the shifting of scaling functions. Therefore, we obtain the same $\{b_k\}$ if we start with $\beta_n(x)$ instead. In particular, we get the centralized Battle–Lemarié scaling functions with

$$\phi(x) = \sum_k b_k \beta_n(x-k) \tag{3.25}$$

This is the version often used in practice. Moreover, since $F(\omega)$ is an even function here, we always have

$$b_k = b_{-k} \tag{3.26}$$

In particular, the orthogonalized scaling functions will always have the same symmetry as the original ones. On the other hand, the Battle–Lemarié scaling functions are no longer compactly supported since there are infinitely many nonzero terms of b_k . In practice, since the infinite sequence $\{b_k\}$ has exponential decay [15], we can truncate the series into a finite sum to satisfy the required accuracy. In any case, we can easily calculate the needed values of Fourier coefficient b_k using MatLab or other standard software. A detailed example will be worked out in the next section.

The refinement equation for the Battle–Lemarié scaling functions can be obtained in many different ways. In particular, we can proceed as follows:

$$\widetilde{\phi}(\omega) = \frac{\widetilde{\beta}_n(\omega)}{\sqrt{F(\omega)}} = \frac{H(\omega/2)\widetilde{\beta}_n(\omega/2)}{\sqrt{F(\omega)}} = \frac{H(\omega/2)\sqrt{F(\omega/2)}\widetilde{\phi}(\omega/2)}{\sqrt{F(\omega)}} \tag{3.27}$$

Comparing (3.27) with the refinement equation for $\widetilde{\phi}(\omega)$ in the frequency domain listed below,

$$\widetilde{\phi}(\omega) = H_\phi(\omega/2)\widetilde{\phi}(\omega/2) \tag{3.28}$$

we then obtain

$$H_\phi(\omega) = \frac{H(\omega)\sqrt{F(\omega)}}{\sqrt{F(2\omega)}} \tag{3.29}$$

Using explicit formulas for $H(\omega)$ in (3.15) and $F(\omega)$ in (3.21), we can proceed to calculate the desired coefficients by

$$_{\phi}h_k = \sqrt{2}\frac{1}{2\pi}\int_0^{2\pi} H_{\phi}(\omega)e^{jk\omega}d\omega \tag{3.30}$$

Our next step is to construct the associated wavelet. Since our Battle–Lemarié scaling function $\phi(x)$ is orthonormal, the associated $\psi(x)$ can be constructed by the standard procedure given in Chapter 2, namely, by the wavelet equation (2.27) and equation (2.50):

$$\psi(x) = \sqrt{2}\sum_{l=-\infty}^{+\infty} {}_{\phi}g_l\phi(2x-l) = \sqrt{2}\sum_{l=-\infty}^{+\infty}(-1)^l{}_{\phi}h_{1-l}^*\phi(2x-l) \tag{3.31}$$

where $_{\phi}h_k$ are the coefficients in the refinement equation of $\phi(x)$ as given in (3.30).

Finally, substituting (3.25) into the above equation, we can express the associated wavelet in terms of the original B-spline basis functions generated by $\beta_n(x)$; namely,

$$\begin{aligned}\psi(x) &= \sqrt{2}\sum_{l=-\infty}^{+\infty}(-1)^l{}_{\phi}h_{1-l}^*\phi(2x-l)\\ &= \sqrt{2}\sum_{l=-\infty}^{+\infty}\sum_{i=-\infty}^{+\infty}(-1)^l{}_{\phi}h_{1-l}^*b_i\beta_n(2x-l-i)\\ &= \sum_{k=-\infty}^{+\infty}c_k\beta_n(2x-k)\end{aligned} \tag{3.32}$$

where

$$c_k = \sqrt{2}\sum_{l=-\infty}^{+\infty}(-1)^l{}_{\phi}h_{1-l}^*b_{k-l} \tag{3.33}$$

We will work out an example to see how to actually carry out those steps in the next section.

3.4 CUBIC SPLINE BATTLE–LEMARIÉ MRA—AN EXAMPLE

The cubic ($n = 3$) spline Battle–Lemarié scaling function generates one of the most frequently used MRAs in the MRTD analysis and it has the following nice properties:

(1) It is twice continuously differentiable.

(2) It is symmetric (we often use the centralized version that is symmetric about $x = 0$).

(3) Although it is not compactly supported, it does have exponential decay.
(4) It is piecewise cubic with an explicit analytic expression.

In this section, we work out more computational detail for the cubic spline Battle–Lemarié MRA following the general steps and formulas developed in the last two sections.

As mentioned earlier, we use the centralized version here so that the origin is the center. When $n = 3$ in (3.7), it is straightforward to deduct a more explicit formula for the original cubic B-spline scaling function $\beta_3(x)$ (the centralized version) in the following symmetric form:

$$\beta_3(x) = \begin{cases} \dfrac{(x+2)^3}{6}, & -2 \le x \le -1 \\ \dfrac{-3x^3 - 6x^2 + 4}{6}, & -1 \le x \le 0 \\ \dfrac{3x^3 - 6x^2 + 4}{6}, & 0 \le x \le 1 \\ \dfrac{(2-x)^3}{6}, & 1 \le x \le 2 \\ 0, & \text{otherwise} \end{cases} \tag{3.34}$$

The above formula is useful since we will express everything in terms of basis functions generated by $\beta_3(x)$. In particular, using (3.25) and (3.26), we have

$$\phi(x) = \sum_{k=-\infty}^{+\infty} b_k \beta_3(x-k) = b_0\beta_3(x) + \sum_{k=1}^{+\infty} b_k(\beta_3(x-k) + \beta_3(x+k)) \tag{3.35}$$

where b_k can be computed by using the relations defined in (3.21) and (3.23).

We now outline some explicit computations as follows. First, for $n = 3$, (3.21) becomes

$$F(\omega) = (2\sin(\omega/2))^8 \frac{1}{7!}\frac{d^6((2\sin(\omega/2))^{-2})}{d\omega^6} \tag{3.36}$$

Taking derivatives six times and simplifying, we then obtain

$$\begin{aligned} F(\omega) &= \cos^6\left(\frac{\omega}{2}\right) + \frac{5}{3}\sin^2\left(\frac{\omega}{2}\right)\cos^4\left(\frac{\omega}{2}\right) \\ &\quad + \frac{11}{15}\sin^4\left(\frac{\omega}{2}\right)\cos^2\left(\frac{\omega}{2}\right) + \frac{17}{315}\sin^6\left(\frac{\omega}{2}\right) \\ &= 1 - \frac{4}{3}\sin^2\left(\frac{\omega}{2}\right) + \frac{2}{5}\sin^4\left(\frac{\omega}{2}\right) - \frac{4}{315}\sin^6\left(\frac{\omega}{2}\right) \end{aligned} \tag{3.37}$$

We can evaluate b_k in (3.35) by

$$b_k = \frac{1}{2\pi}\int_0^{2\pi} (F(\omega))^{-1/2} e^{jk\omega}\, d\omega \tag{3.38}$$

In practice, such calculations can be done with any standard software using either (3.37) or (3.36) directly. Because of the symmetry, we need only compute the b_k for $k \geq 0$. The first few values of b_k from our computations are listed in Table 3.1.

To construct the associated wavelet following steps given in the last section, we first note that (3.29) becomes

$$H_\phi(\omega) = \frac{H(\omega)\sqrt{F(\omega)}}{\sqrt{F(2\omega)}}$$
$$= \frac{\cos^4\left(\frac{\omega}{2}\right)\sqrt{1 - \frac{4}{3}\sin^2\left(\frac{\omega}{2}\right) + \frac{2}{5}\sin^4\left(\frac{\omega}{2}\right) - \frac{4}{315}\sin^6\left(\frac{\omega}{2}\right)}}{\sqrt{1 - \frac{4}{3}\sin^2\omega + \frac{2}{5}\sin^4\omega - \frac{4}{315}\sin^6\omega}} \tag{3.39}$$

Therefore, we can evaluate the coefficients in the refinement equation using (3.30):

$$_\phi h_k = \sqrt{2}\frac{1}{2\pi}\int_0^{2\pi} H_\phi(\omega)e^{jk\omega}d\omega \tag{3.40}$$

Finally, by (3.32) and (3.33), we obtain

$$\psi(x) = \sum_{k=-\infty}^{+\infty} c_k\beta_3(2x-k) \tag{3.41}$$

where

$$c_k = \sqrt{2}\sum_{l=-\infty}^{+\infty}(-1)^l {}_\phi h^*_{1-l} b_{k-l} \tag{3.42}$$

We want to point out that, since both $\{_\phi h_k\}$ and $\{b_k\}$ are symmetric with respect to $k = 0$ for odd n, the sequence $\{c_k\}$ is symmetric with respect to $k = 1$; that is,

TABLE 3.1 Expansion Coefficients b_k and c_k

k	b_k	c_k
0	1.96976168178803	$c_0 = c_2$
1	−0.67243048528362	−2.89173706361184
2	0.26870424357567	2.00521685190847
3	−0.11851994477803	−0.54228152595883
4	0.05519146442494	0.01207588451057
5	−0.02652033906672	−0.14411868190353
6	0.01299816593975	0.14597256636735
7	−0.00645749182905	−0.00309487337866
8	0.00323986310165	−0.02817338043451
9	−0.00163777604041	−0.01963582787186
10	0.00083283614954	0.02353256405256

$c_{1+k} = c_{1-k}$. Therefore, we need only compute the values of c_k for $k \geq 1$. The first few calculated values of c_k are listed together with those of b_k in Table 3.1.

Since $\beta_3(x)$ is symmetric about $x = 0$, it is easy to verify that the symmetry in $\{c_k\}$ then implies that the cubic spline Battle–Lemarié wavelet is symmetric about $x = \frac{1}{2}$. Obviously, the wavelet shares similar properties with the cubic spline Battle–Lemarié scaling function. In Figure 3.2a–d, we have plotted graphs of

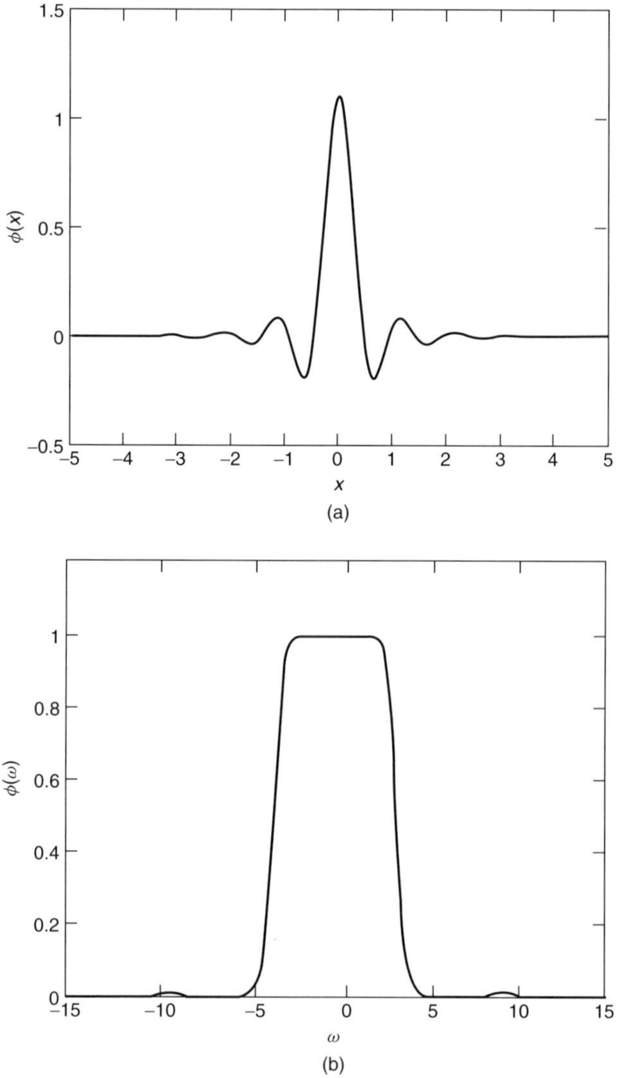

FIGURE 3.2 (a) Cubic spline Battle–Lemarié scaling function in the space domain. (b) Cubic spline Battle–Lemarié scaling function in the frequency domain. (c) Cubic spline Battle–Lemarié wavelets in the space domain. (d) Cubic spline Battle–Lemarié wavelets in the frequency domain.

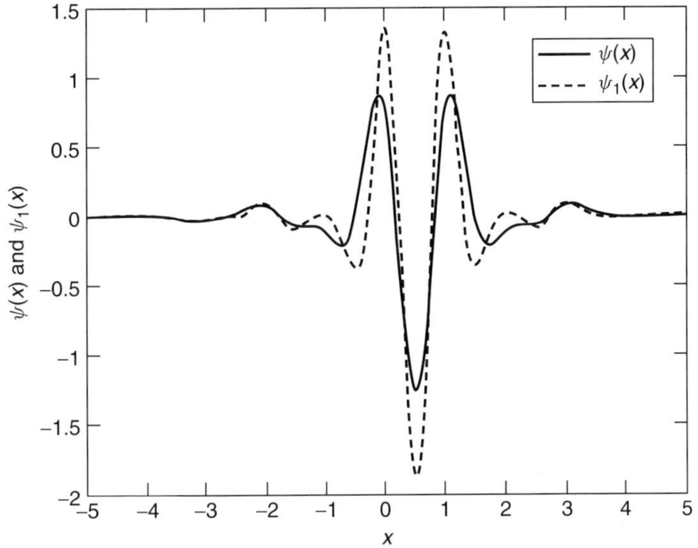

(c)

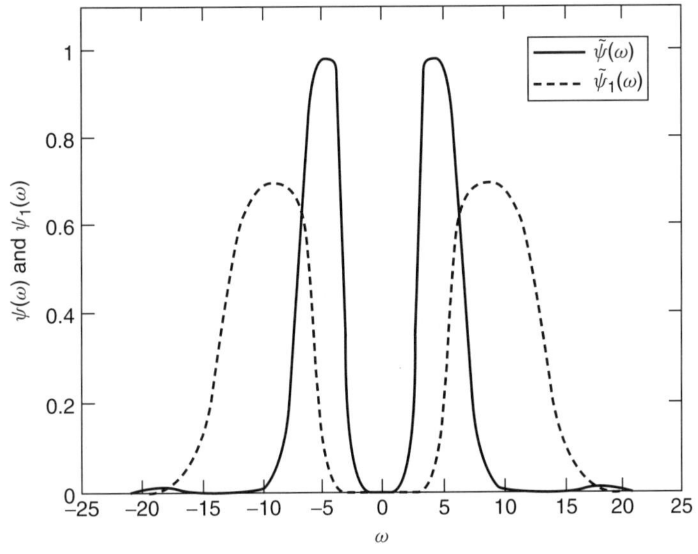

(d)

FIGURE 3.2 *(Continued)*

a few selected scaling functions and wavelets, namely, $\phi(x)$, $\psi(x)$, and $\psi_1(x)$; and $\widetilde{\phi}(\omega)$, $\widetilde{\psi}(\omega)$, and $\widetilde{\psi}_1(\omega)$, in the space and frequency domains, respectively. As seen from the figures, although they are not compactly supported, the cubic spline Battle–Lemarié scaling function and wavelet do have very fast decay and exhibit good localization properties in both the space and frequency domains.

3.5 DAUBECHIES' PROCEDURE OF MRA CONSTRUCTION

As seen in the last section, the orthogonalization procedure generally produces the new scaling functions and wavelets with the infinite support. In her 1988 paper [24], Daubechies constructs a family of orthonormal scaling functions and wavelets with compact support. The novel construction does not come without a price. In particular, this family does not have explicit analytic expressions and symmetry. For this reason, MRAs from this family may not appear to be ideal for use in MRTD applications. However, they can be computed numerically and do have some distinct properties. The principle of the construction is a very valuable one and can be applied in constructing more general types of families. In this section, we outline the basic procedure of the construction, which will be used to construct examples in Section 3.6.

In our study thus far, we have started with a scaling function $\phi(x)$ and then proceeded to deduct the rest of the MRA. For example, with a given $\phi(x)$, we often proceed to find its refinement equation

$$\phi(x) = \sqrt{2}\sum_{l} h_l \phi(2x - l) \tag{3.43}$$

or, equivalently, its corresponding equation in the frequency domain

$$\widetilde{\phi}(\omega) = H(\omega/2)\widetilde{\phi}(\omega/2) \tag{3.44}$$

with

$$H(\omega) = \frac{1}{\sqrt{2}}\sum_{l} h_l e^{-jl\omega} \tag{3.45}$$

Such $H(\omega)$ is often called a filter and plays a key role in many studies and applications. In fact, we can deduct the entire MRA from $H(\omega)$ itself. Indeed, applying (3.44) recursively for $\omega/2, \omega/4, \ldots$, we can recover the scaling function through

$$\widetilde{\phi}(\omega) = H(\omega/2)H(\omega/4)\widetilde{\phi}(\omega/4)\cdots = a\prod_{k=1}^{+\infty} H(2^{-k}\omega) \tag{3.46}$$

where

$$a = \widetilde{\phi}(0) = \int_{-\infty}^{+\infty} \phi(t)\, dt \tag{3.47}$$

In practice, we often require the scaling function to be normalized so $a = 1$. Moreover, as shown in Chapter 2, in the case where the scaling function is orthonormal, we can construct the associated wavelet using the wavelet equation

$$\psi(x) = \sqrt{2}\sum_{l} g_l \phi(2x - l) \tag{3.48}$$

or, equivalently, its corresponding equation in the frequency domain

$$\tilde{\psi}(\omega) = \frac{1}{\sqrt{2}} \sum_k g_k e^{-jk\omega/2} \tilde{\phi}(\omega/2) = G(\omega/2)\tilde{\phi}(\omega/2) \tag{3.49}$$

by selecting

$$G(\omega) = -e^{-j\omega} H^*(\omega + \pi) \tag{3.50}$$

Therefore, once the filter coefficients are known, the entire MRA can be constructed. This is an interesting new way to construct MRA families. Instead of starting with a scaling function, we may start with a suitable filter and define the associated scaling function and wavelet using equations (3.46), (3.48) or (3.49), and (3.50). Of course, the named filter must satisfy some necessary conditions. For example, it must be a 2π-periodic function that satisfies (2.45) in Chapter 2:

$$|H(\omega)|^2 + |H(\omega + \pi)|^2 = 1 \tag{3.51}$$

The key is to choose a suitable function for the filter so that it indeed generates a MRA with the desired properties. Since the full theory in this regard is beyond the scope of this book, our goal here is to outline basic steps and facts so that engineers can use them effectively.

The following facts should be kept in mind. First, it is easy to see that a filter yields compact support for the associated scaling function and wavelet if and only if it is a trigonometric polynomial. However, not every trigonometric polynomial that satisfies the needed conditions such as (3.51) will work, and there are many subtle restrictions imposed on their selections. Second, in spite of restrictions, there is flexibility within the framework in the choice of filters to satisfy specified properties leading to different families.

Let us first describe in more detail how to construct the associated scaling function in practice. Once the filter coefficients are known, we can directly compute the associated scaling function $\phi(x)$ numerically by using the refinement equation (3.43) recursively, as follows: The values of $\phi(x)$ at integers are solved from (3.43), which amounts to an eigenvalue problem (note that $\phi(x)$ is compactly supported). With the initial values at integers computed, (3.43) can then be applied recursively to yield values of $\phi(x)$ at dyadic points $x = \pm k2^{-s}$, $k, s \in N$, where N is the set of natural numbers. Since any $x \in R$ is a limit of the sequence of dyadic points, $\phi(x)$ can be approximated by its values at dyadic points to any given accuracy as long as $\phi(x)$ is continuous at x. The same approach can be used to compute derivatives of $\phi(x)$ numerically. For example, to calculate $\phi'(x)$, we formally take the derivative of both sides of (3.43) to get

$$\phi'(x) = 2\sqrt{2} \sum_l h_l \phi'(2x - l) \tag{3.52}$$

The same method can then be applied. Of course, once $\phi(x)$ and its derivatives are calculated, we can calculate the wavelet and its derivatives using the wavelet equation (3.48) and its differentiations.

Daubechies' construction is indexed by the vanishing moments N of the associated wavelet. A wavelet $\psi(x)$ is said to have N vanishing moments if

$$\int_{-\infty}^{\infty} x^l \psi(x)\,dx = 0, \quad l = 0, 1, \ldots, N-1 \tag{3.53}$$

As we will see, this is a desirable property to require. For example, as an immediate consequence, it implies that the scaling function $\phi(x)$ alone is enough to generate all polynomials of degree smaller than N.

The Daubechies family is so indexed also because the vanishing moment is a property that is reflected directly by the filter $H(\omega)$. In fact, it is easy to see that integral equations in (3.53) translate into the following derivative conditions:

$$H^{(l)}(\pi) = 0 \quad \text{for } l = 0, 1, \ldots, N-1 \tag{3.54}$$

As it turns out, the differentiability of $\psi(x)$ is also proportional to N. Therefore, this provides a way to choose a filter such that the resulting wavelet will have expected smoothness.

Since $H(\omega)$ is a trigonometric polynomial, to fulfill the moment conditions (3.54) it must assume the following form:

$$H(\omega) = \left(\frac{1+e^{-j\omega}}{2}\right)^N L(\omega) = e^{-jN\omega/2} \cos^N\left(\frac{\omega}{2}\right) L(\omega) \tag{3.55}$$

It can be shown [1] that, to satisfy the key condition (3.51), $L(\omega)$ must be subject to

$$|L(\omega)|^2 = \sum_{p=0}^{N-1} \binom{N-1+p}{p} \left(\sin^2\frac{\omega}{2}\right)^p + \left(\sin^2\frac{\omega}{2}\right)^N f(\cos\omega) \tag{3.56}$$

where $f(x)$ is an odd polynomial subject to the obvious condition that (3.56) be positive. Interested readers may check directly that (3.51) is indeed satisfied if we assume (3.56).

There are two kinds of choices that can and need to be made here. Obviously, there are infinitely many choices of different $f(x)$ in (3.56). The subtler kind is that, for a fixed $f(x)$, we still need to find an $L(\omega)$ that fulfills (3.56). This amounts to finding the "square roots" of the polynomial in (3.56) and this can be done numerically using the so-called spectral factorization procedure. Following the steps above, we can construct all possible filters that satisfy (3.51) and the moment conditions (3.54). However, since (3.51) is only a necessary orthonormal condition, as shown in Chapter 2, not all such $H(\omega)$ will lead to an orthogonal MRA. There are both necessary and sufficient conditions that guarantee a successful construction, and we refer interested readers to [1] for details. For now, we end this section and turn our attention to examples.

3.6 DAUBECHIES' ORIGINAL FAMILY

Daubechies' original family, constructed in [24], corresponds to the simplest case in the sense that $f(x) = 0$ is taken. The filters in the family are indexed by the vanishing moments N and will be denoted as D_N. Choosing $f(x) = 0$ means that D_N has only $2N$ nonzero filter coefficients, the least number possible for the given N. The phase of D_N is so chosen such that the first nonzero coefficient is indexed at zero (in practice, we may want to choose the "centralized" phase to index the first nonzero coefficient at $1 - N$). Therefore, we have

$$D_N(\omega) = \left(\frac{1+e^{-j\omega}}{2}\right)^N L(\omega) = \frac{1}{\sqrt{2}} \sum_{l=0}^{2N-1} h_l e^{-jl\omega} \tag{3.57}$$

As a direct consequence, the associated orthonormal scaling function ${}_N\phi(x)$ and wavelet ${}_N\psi(x)$ are compactly supported on intervals $[0, 2N - 1]$ and $[1 - N, N]$, respectively. It can also be shown that both are approximately $0.2N$-times differentiable. Except for $N = 1$, in which case we get the Haar MRA, ${}_N\phi(x)$ and ${}_N\psi(x)$ do not have explicit analytic expressions and they are far from being symmetric. The family is characterized as the one with extremal phase and with the shortest support length for the given number of vanishing moments.

For $N = 1$, we have from (3.55) and (3.56)

$$D_1(\omega) = \frac{1+e^{-j\omega}}{2} \tag{3.58}$$

By (3.46), the associated scaling function in the spectral domain can be expressed as

$${}_1\widetilde{\phi}(\omega) = \lim_{n\to\infty} \prod_{k=1}^{n} \left(\frac{1+e^{-2^{-k}j\omega}}{2}\right) \tag{3.59}$$

Since

$$\begin{aligned}
\prod_{k=1}^{n} \left(\frac{1+e^{-2^{-k}j\omega}}{2}\right) &= \frac{1-e^{-2^{-n}j\omega}}{1-e^{-2^{-n}j\omega}} \prod_{k=1}^{n} \left(\frac{1+e^{-2^{-k}j\omega}}{2}\right) \\
&= \frac{2^{-1}(1-e^{-2^{-(n-1)}j\omega})}{1-e^{-2^{-n}j\omega}} \prod_{k=1}^{n-1} \left(\frac{1+e^{-2^{-k}j\omega}}{2}\right) \\
&= \cdots \\
&= \frac{2^{-n}(1-e^{-j\omega})}{1-e^{-2^{-n}j\omega}}
\end{aligned} \tag{3.60}$$

we have

$${}_1\widetilde{\phi}(\omega) = \lim_{n\to\infty} \frac{2^{-n}(1-e^{-j\omega})}{1-e^{-2^{-n}j\omega}} = \frac{1-e^{-j\omega}}{j\omega} \tag{3.61}$$

Therefore, ${}_1\phi(x)$ is just the Haar scaling function; thus, the Daubechies D_1 MRA coincides with the Haar MRA.

For $N = 2$, by (3.56), we have the corresponding Daubechies D_2 filter that is subject to

$$|D_2(\omega)|^2 = \left|\left(\frac{1+e^{-j\omega}}{2}\right)^2\right|^2 \left(1 + 2\sin^2\left(\frac{\omega}{2}\right)\right) = \left|\left(\frac{1+e^{-j\omega}}{2}\right)^2\right|^2 (2 - \cos\omega) \tag{362}$$

Following the standard procedure we obtain $D_2(\omega)$ in the form of

$$D_2(\omega) = \left(\frac{1+e^{-j\omega}}{2}\right)^2 \frac{1+\sqrt{3}+(1-\sqrt{3})e^{-j\omega}}{2} = \frac{1}{\sqrt{2}}\sum_{l=0}^{3} h_k e^{-jl\omega} \tag{3.63}$$

It is easy to verify that (3.62) is indeed satisfied by the above $D_2(\omega)$. Therefore, the nonzero filter coefficients are

$$h_0 = \frac{1+\sqrt{3}}{4\sqrt{2}}; \quad h_1 = \frac{3+\sqrt{3}}{4\sqrt{2}}; \quad h_2 = \frac{3-\sqrt{3}}{4\sqrt{2}}; \quad h_3 = \frac{1-\sqrt{3}}{4\sqrt{2}} \tag{3.64}$$

For general N, we can use the built-in functions from any standard software to generate the needed filter coefficients. We want to point out that different references and the corresponding software may use slightly different indexing, and one way to confirm a given index is to check the total number of coefficients or the length of the support interval. We should also be aware that, due to different forms of the refinement equation used, the coefficients generated by different software may also be off by a factor. Table 3.2a lists values of the Daubechies D_N-filter coefficients for $N = 1$ to $N = 6$ generated by MatLab.

Once the values of the filter coefficients are known, the associated scaling function and wavelet can be computed numerically as described before. In Figure 3.3a–f, using MatLab, we have plotted graphs of the scaling functions and wavelets for the cases of $N = 2$, 3, and 4. Note that the wavelets shown have been shifted so they have the same support interval $[0, 2N - 1]$ instead of $[1 - N, N]$, as they should. As we can see, the smoothness increases with N and there is very marked asymmetry.

We want to point out that, even with the simplest $|L(\omega)|^2$ fixed, there are still different choices of "square roots" other than the one used to generate the above family. Such choices will lead to different families. A particular one worth mentioning is Daubechies' least asymmetric family. It is so named because it produces the least asymmetric scaling functions and wavelets of those possible when $f(x) = 0$. Table 3.2b lists values of Daubechies' least asymmetric filter coefficients for $N = 2$ to $N = 5$ generated by MatLab. In Figure 3.4a–f, graphs of the least asymmetric scaling functions and wavelets for cases of $N = 2$, 3, and 4 are plotted. As we can see, although they have less marked asymmetry, they are far from being symmetric. In fact, it is known that complete symmetry is not possible under Daubechies' construction. So the best we can do is to construct a near-symmetric family. This is possible when a different $f(x)$ is chosen. For this end, we will go to the next section.

TABLE 3.2a Filter Coefficients for Daubechies' Original Family

	n	$h_n/\sqrt{2}$		n	$h_n/\sqrt{2}$
D_1	0	0.50000000000000	D_2	0	0.34150635094622
	1	0.50000000000000		1	0.59150635094587
				2	0.15849364905378
				3	−0.09150635094587
D_3	0	0.23523360389270	D_4	0	0.16290171402562
	1	0.57055845791731		1	0.50547285754565
	2	0.32518250026371		2	0.44610006912319
	3	−0.09546720778426		3	−0.01978751311791
	4	−0.06041610415535		4	−0.13225358368437
	5	0.02490874986589		5	0.02180815023739
				6	0.02325180053556
				7	−0.00749349466513
D_5	0	0.11320949129173	D_6	0	0.07887121600143
	1	0.42697177135271		1	0.34975190703757
	2	0.51216347213016		2	0.53113187994121
	3	0.09788348067375		3	0.22291566146505
	4	−0.17132835769133		4	−0.15999329944587
	5	−0.02280056594205		5	−0.09175903203003
	6	0.05485132932108		6	0.06894404648720
	7	−0.00441340005433		7	0.01946160485396
	8	−0.00889593505093		8	−0.02233187416548
	9	0.00235871396920		9	0.00039162557603
				10	0.00337803118151
				11	−0.00076176690258

TABLE 3.2b Filter Coefficients for Daubechies' Least Asymmetric Family

	n	$h_n/2$		n	$h_n/2$
Sym2	0	0.34150635094622	Sym3	0	0.23523360389270
	1	0.59150635094587		1	0.57055845791731
	2	0.15849364905378		2	0.32518250026371
	3	−0.09150635094587		3	−0.09546720778426
				4	−0.06041610415535
				5	0.02490874986589
Sym4	0	0.02278517294800	Sym5	0	0.01381607647893
	1	−0.00891235072085		1	−0.01492124993438
	2	−0.07015881208950		2	−0.12397568130675
	3	0.21061726710200		3	0.01173946156807
	4	0.56832912170500		4	0.44829082419092
	5	0.35186953432800		5	0.51152648344605
	6	−0.02095548256255		6	0.14099534842729
	7	−0.05357445070900		7	−0.02767209305836
				8	0.02087343221079
				9	0.01932739797744

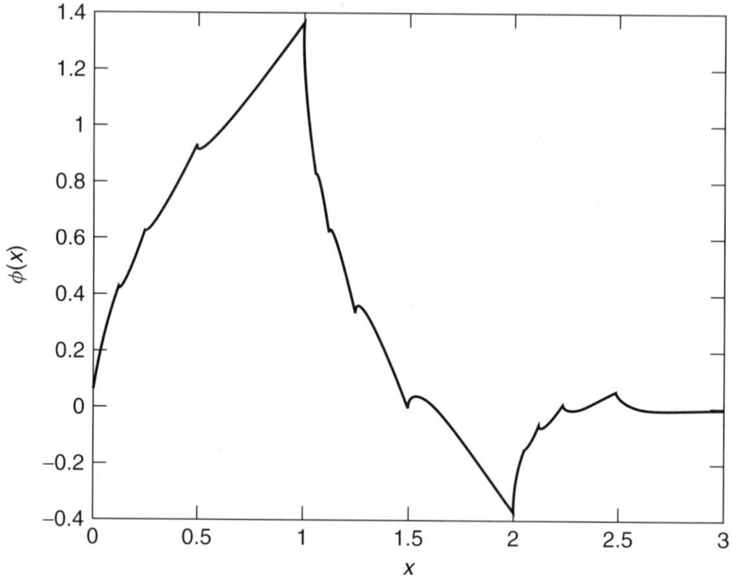

(a)

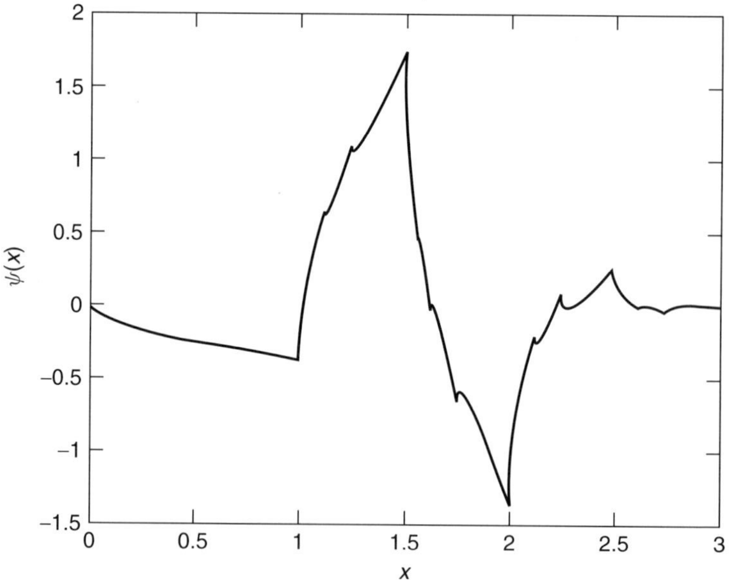

(b)

FIGURE 3.3 (a) Graph of the D_2 scaling function. (b) Graph of the D_2 wavelet. (c) Graph of the D_3 scaling function. (d) Graph of the D_3 wavelet. (e) Graph of the D_4 scaling function. (f) Graph of the D_4 wavelet.

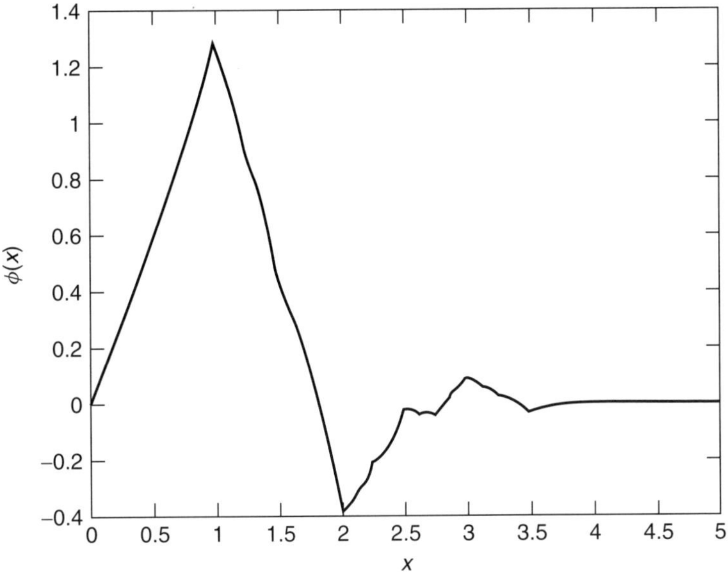

(c)

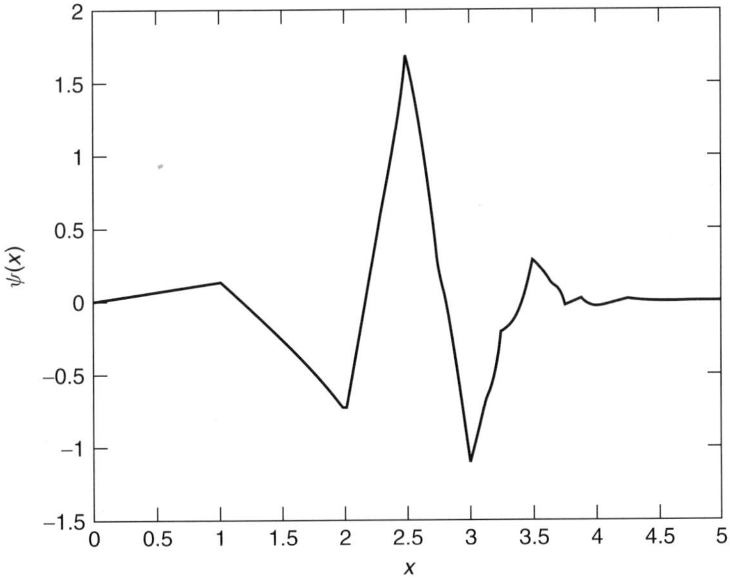

(d)

FIGURE 3.3 *(Continued)*

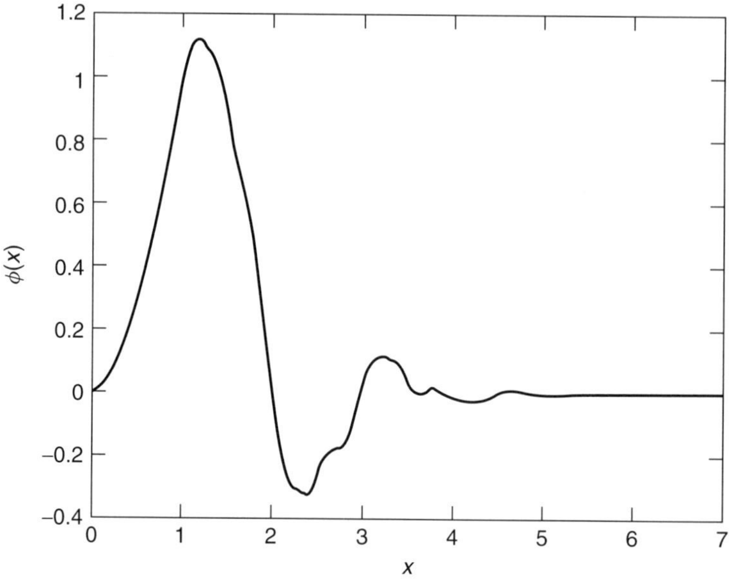

(e)

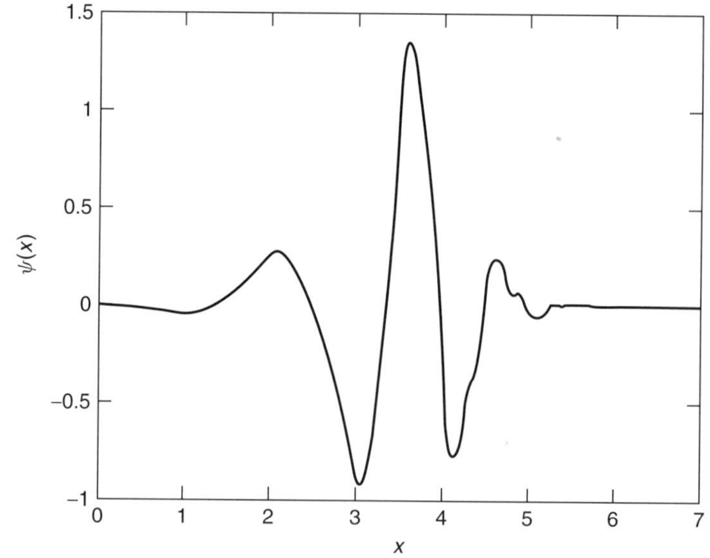

(f)

FIGURE 3.3 *(Continued)*

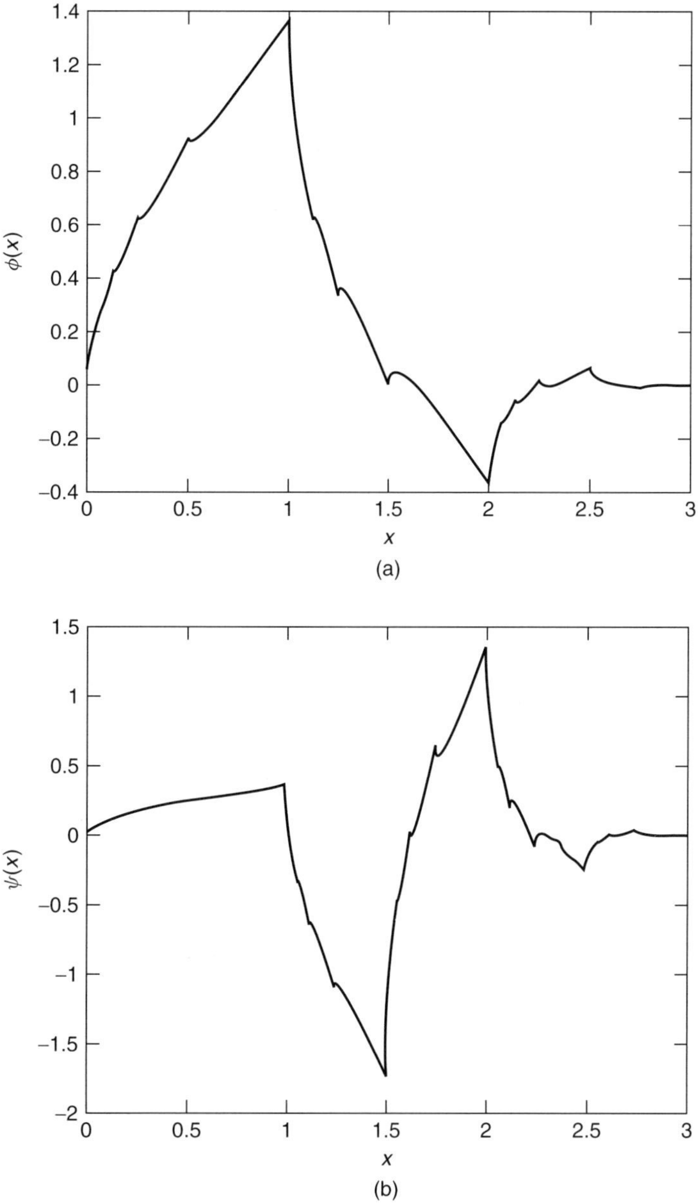

FIGURE 3.4 (a) Graph of Daubechies' least asymmetric scaling function for $N = 2$. (b) Graph of Daubechies' least asymmetric wavelet for $N = 2$. (c) Graph of Daubechies' least asymmetric scaling function for $N = 3$. (d) Graph of Daubechies' least asymmetric wavelet for $N = 3$. (e) Graph of Daubechies' least asymmetric scaling function for $N = 4$. (f) Graph of Daubechies' least asymmetric wavelet for $N = 4$.

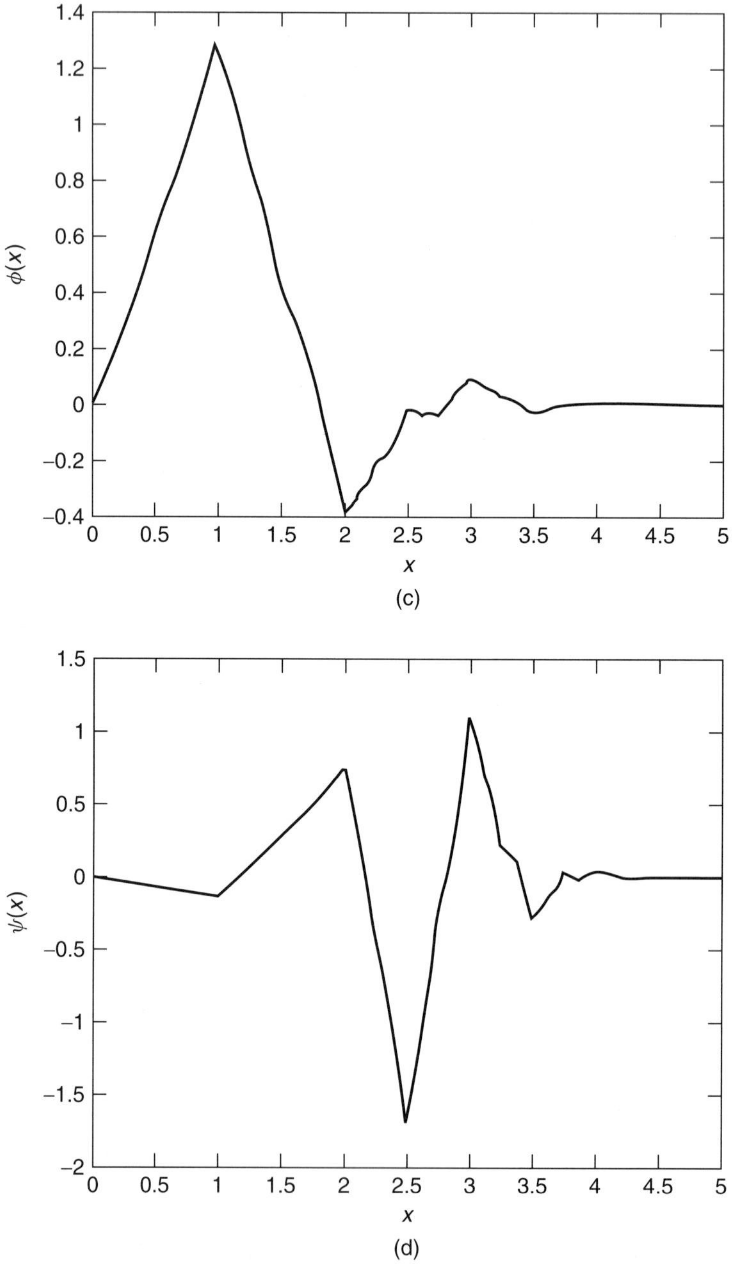

FIGURE 3.4 (*Continued*)

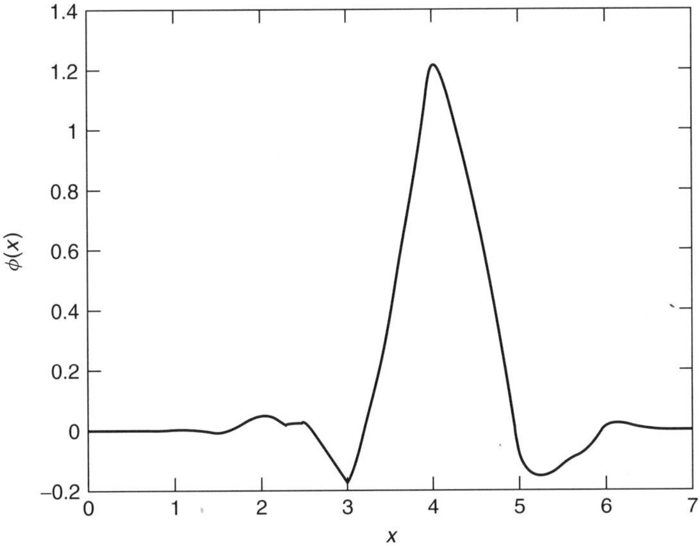

(e)

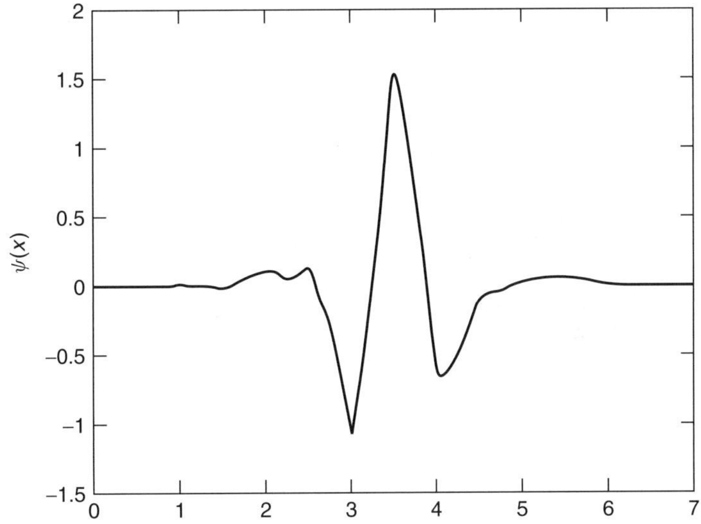

(f)

FIGURE 3.4 (*Continued*)

3.7 COIFLET FAMILY

The Coiflet family is a variation of the Daubechies family. It is so named because, in 1989, Coifman suggested that Daubechies consider the case when both scaling functions and wavelets have the required vanishing moments.

Since we must have

$$\int_{-\infty}^{\infty} \phi(x)dx = a \neq 0, \quad a = 1 \text{ if normalized} \tag{3.65}$$

the best we can do is to require

$$\int_{-\infty}^{\infty} x^l \phi(x)dx = 0 \quad \text{for } l = 1, \ldots, N-1 \tag{3.66}$$

The Coiflet family is constructed under the same framework of Daubechies' general procedure with those additional requirements. For the sake of simplicity, we often assume that $N = 2K$ is even. They are therefore indexed by K.

To construct the family, we first have to find the filter $H(\omega)$ under the extra constraint (3.66). Similar to the process that converts (3.53) to (3.54), the above vanishing conditions can easily be converted into the following derivative conditions:

$$H^{(l)}(0) = 0 \quad \text{for } l = 1, 2, \ldots, N-1 \tag{3.67}$$

Therefore, in addition to satisfying (3.55), reiterated below for convenience,

$$H(\omega) = \left(\frac{1 + e^{-j\omega}}{2}\right)^N L(\omega) = e^{-jN\omega/2} \cos^N\left(\frac{\omega}{2}\right) L(\omega) \tag{3.68a}$$

$H(\omega)$ must also have the following form (note that $H(0)$ must be equal to 1):

$$H(\omega) = 1 + \left(\frac{1 - e^{-j\omega}}{2}\right)^N \hat{L}(\omega) \tag{3.68b}$$

As it turns out, with this extra condition (3.68b), the same method used to obtain the required general form (3.56) of $|L(\omega)|^2$ can be used to obtain a required general form as listed below. The difference is that this time it is for $H(\omega)$ itself instead of the square of its norm:

$$H(\omega) = \left(\cos^2\frac{\omega}{2}\right)^K \left[\sum_{p=0}^{K-1} \binom{K-1+p}{p} \left(\sin^2\frac{\omega}{2}\right)^p + \left(\sin^2\frac{\omega}{2}\right)^K f(\omega)\right] \tag{3.69}$$

where $2K = N$ and $f(\omega)$ is an arbitrary trigonometric polynomial.

Of course, the above is only a necessary condition and we need at least to choose $f(\omega)$ such that orthonormal condition (3.51) is satisfied.

In her 1990 paper [28], Daubechies chose $f(\omega)$ in the form

$$f(\omega) = \sum_{n=0}^{N-1} f_n e^{-jn\omega} \tag{3.70}$$

and showed that the possible values of the coefficient f_n are determined by a system of quadratic equations. This provides a way to find filter coefficients numerically. For example, for $K = 1$, $(N = 2)$, the Coiflet filter coefficients are given by

$$\{h_n/\sqrt{2}\} = \{-0.05143, 0.23893, 0.60286, 0.27214, -0.05143, -0.01107\}.$$

In general, there are $6K - 1 = 3N - 1$ nonzero filter coefficients for the Coiflet of order K. We list coefficients for $K = 1$ to $K = 4$ in Table 3.3. Those

TABLE 3.3 Coiflet Filter Coefficients for K from 1 to 4

	n	$h_n/\sqrt{2}$		n	$h_n/\sqrt{2}$
$K = 1$	0	−0.05142972847100	$K = 2$	0	0.01158759673900
	1	0.23892972847100		1	−0.02932013798000
	2	0.60285945694200		2	−0.04763959031000
	3	0.27214054305800		3	0.27302104653500
	4	−0.05142972847100		4	0.57468239385700
	5	−0.01107027152900		5	0.29486719369600
				6	−0.05408560709200
				7	−0.04202648046100
				8	0.01674441016300
				9	0.00396788361300
				10	−0.00128920335600
				11	−0.00050950539900
$K = 3$	0	−0.00268241867100	$K = 4$	0	0.00063096104600
	1	0.00550312670900		1	−0.00115222485200
	2	0.01658356047900		2	−0.00519452402600
	3	−0.04650776447900		3	0.01136245924400
	4	−0.04322076356000		4	0.01886723537800
	5	0.28650333527400		5	−0.05746423442900
	6	0.56128525687000		6	−0.03965264851700
	7	0.30298357177300		7	0.29366739089500
	8	−0.05077014075500		8	0.55312645256200
	9	−0.05819625076200		9	0.30715732619800
	10	0.02443409432100		10	−0.04711273886500
	11	0.01122924096200		11	−0.06803812705100
	12	−0.00636960101100		12	0.02781364015300
	13	−0.00182045891600		13	0.01773583743800
	14	0.00079020510100		14	−0.01075631851700
	15	0.00032966517400		15	−0.00400101288600
	16	−0.00005019277500		16	0.00265266594600
	17	−0.00002446573400		17	0.00089559452900
				18	−0.00041650057100
				19	−0.00018382976900
				20	0.00004408035400
				21	0.00002208285700
				22	−0.00000230494200
				23	−0.00000126217500

coefficient indexes have been shifted so they start at 0. Therefore, the corresponding scaling functions are supported on $[0, 6K - 1]$ here.

A few graphs of the Coiflet scaling functions and wavelets ($K = 1, 2, 3,$ and 4) are plotted in Figure 3.5a–h. As we can see, they have much more symmetry than what the original Daubechies scaling functions and wavelets have displayed. Of course, we do not get this improved symmetry without paying a price: those functions are now supported on intervals of length $3N - 1$ instead of $2N - 1$. On the other hand, they seem to have very fast decay so the effective supporting interval may not be much larger after all.

We summarize some basic properties of the Coiflet scaling function and wavelet of order K as follows:

- They are orthonormal.
- They are compactly supported on intervals of length $6K - 1$.
- They are almost symmetric.
- They both have vanishing moments up to $2K$.
- They do not have explicit analytic expressions but can be evaluated numerically, including their derivatives.

Finally, we want to point out that it is possible to choose other types of $f(\omega)$. This will lead to other Coiflet families. For example, it is possible to construct a smoother family but the trade-off is less symmetry.

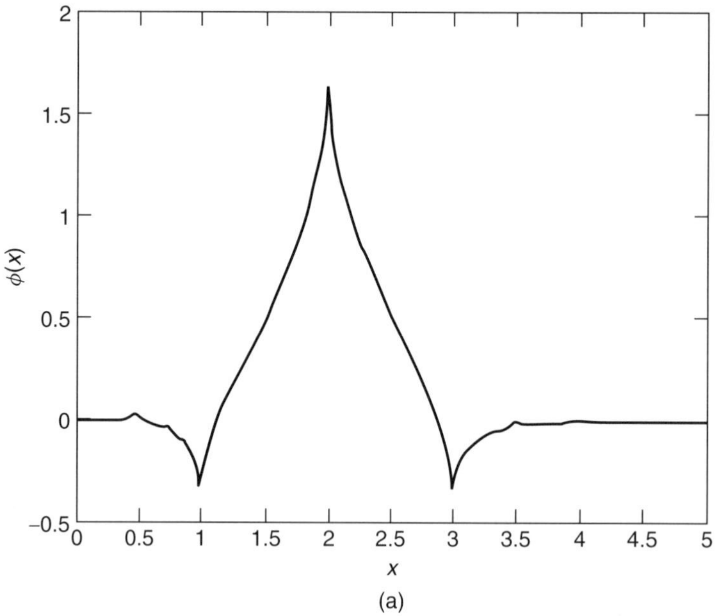

FIGURE 3.5 (a) Coiflet scaling function of order 1 ($K = 1$). (b) Coiflet wavelet of order 1 ($K = 1$). (c) Coiflet scaling function of order 2 ($K = 2$). (d) Coiflet wavelet of order 2 ($K = 2$). (e) Coiflet scaling function of order 3 ($K = 3$). (f) Coiflet wavelet of order 3 ($K = 3$). (g) Coiflet scaling function of order 4 ($K = 4$). (h) Coiflet wavelet of order 4 ($K = 4$).

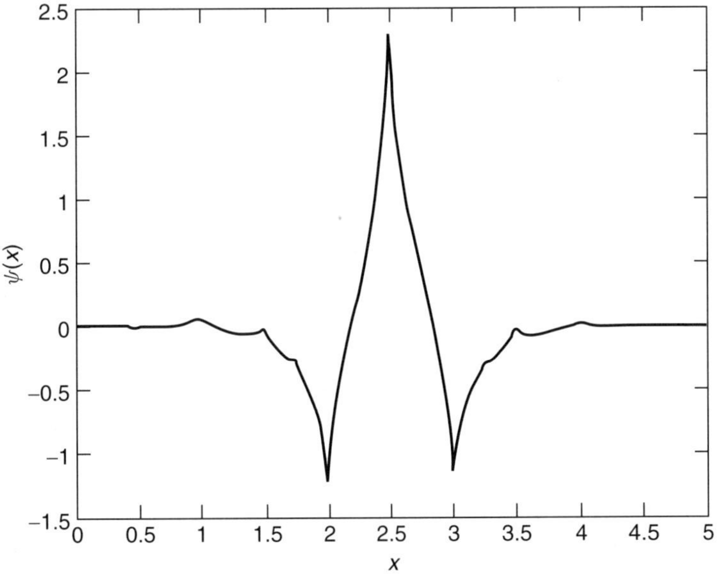

(b)

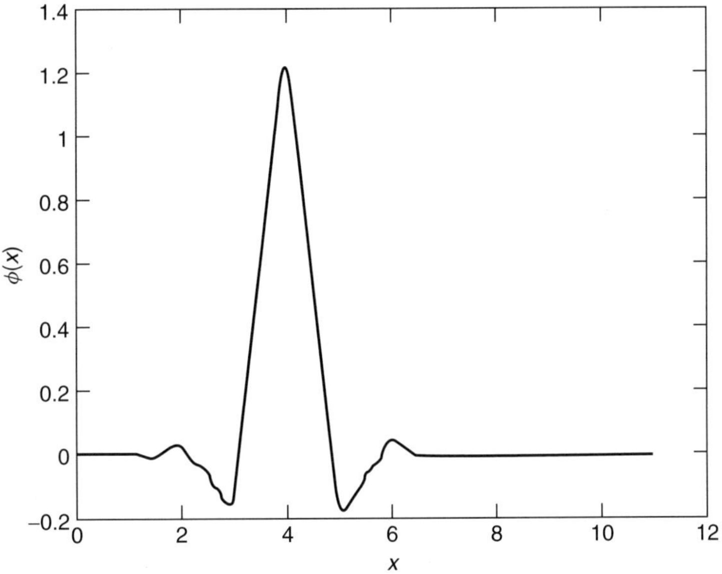

(c)

FIGURE 3.5 *(Continued)*

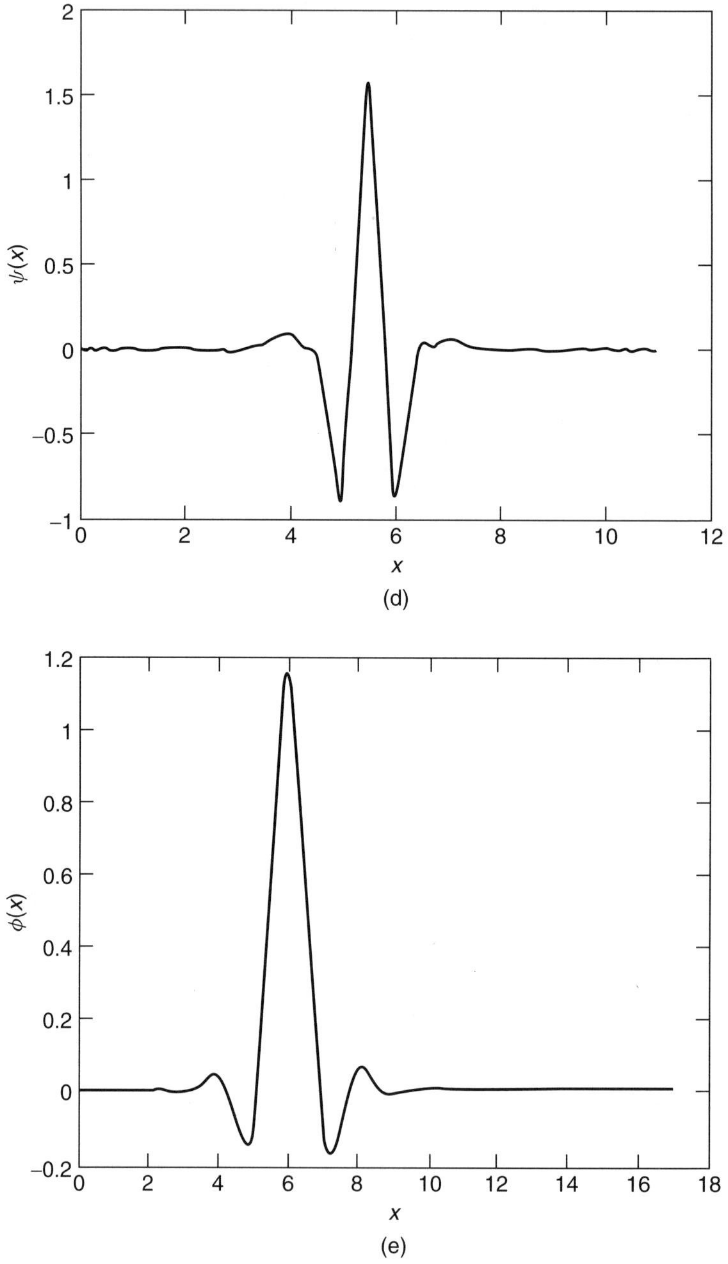

FIGURE 3.5 *(Continued)*

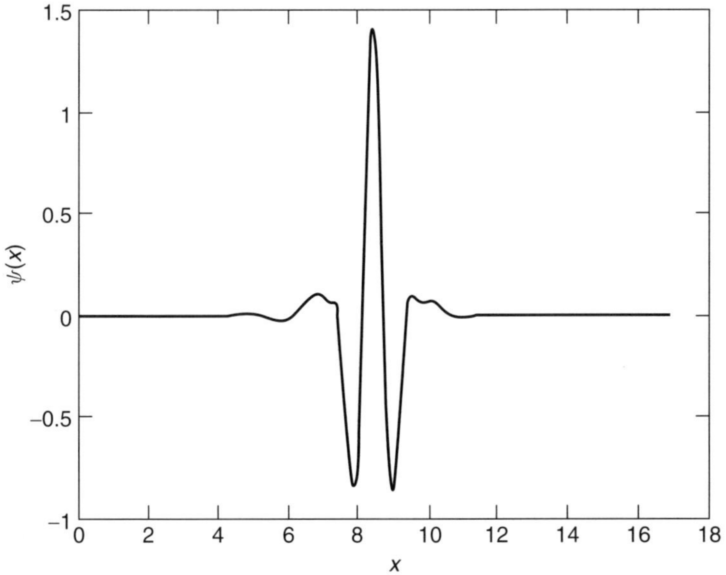

(f)

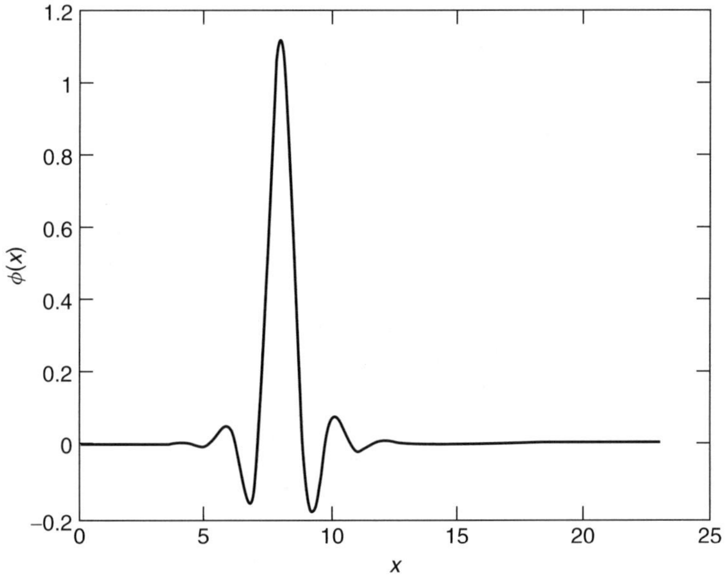

(g)

FIGURE 3.5 (*Continued*)

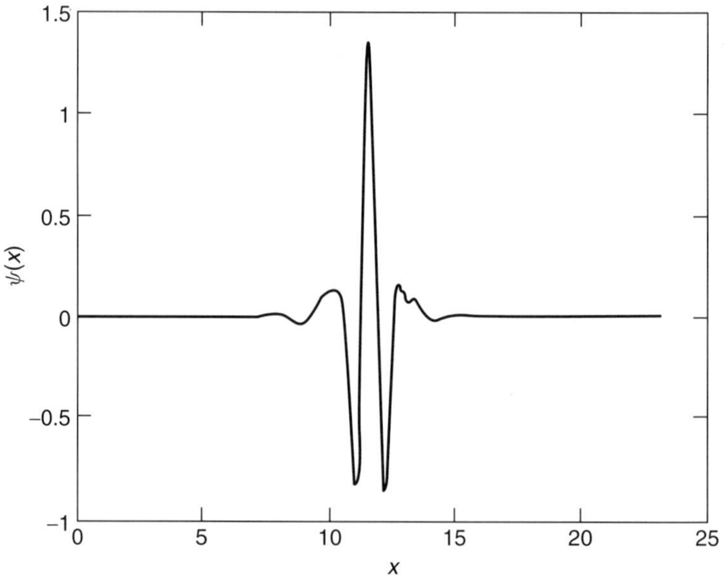

(h)

FIGURE 3.5 *(Continued)*

3.8 BIORTHOGONAL MRA

While Daubechies' construction will always produce orthonormal families with compact support, they are in general not symmetric. On the other hand, if we start with symmetric scaling functions with compact support, they are in general not orthonormal and will not be compactly supported anymore after the orthogonalization. This is not a coincidence, as it can be shown that the Haar MRA is the only MRA that is orthogonal, symmetric, and compactly supported. The biorthogonal MRA makes it possible to have all those properties at once. The trade-off here is that we need now to consider both the primary MRA and the dual MRA instead of just one MRA.

Let us first recall the basic biorthogonal setting from Chapter 2. By the definition, the basic setup consists of a primary scaling function $\phi(x)$, a dual scaling function $\hat{\phi}(x)$, a primary wavelet $\psi(x)$, and a dual wavelet $\hat{\psi}(x)$ such that

$$\langle \hat{\phi}(x), \psi(x-l) \rangle = 0 \quad \text{and} \quad \langle \phi(x), \hat{\psi}(x-l) \rangle = 0 \tag{3.71a}$$

$$\langle \phi(x), \hat{\phi}(x-l) \rangle = \delta_{l,0} \quad \text{and} \quad \langle \psi(x), \hat{\psi}(x-l) \rangle = \delta_{l,0} \tag{3.71b}$$

Obviously, the role of the primary and the dual can be interchanged.

Note that we do not require the wavelets to be orthogonal to their corresponding scaling functions (only to their duals). This is not a problem to be concerned with here, since we will in general expand formally in terms of the primary basis but apply the dual to compute coefficients. In fact, for an arbitrary function

$f \in L^2(R)$, we have

$$\begin{aligned} f(x) &= \sum_i c_{s_0,i}\phi_{s_0,i}(x) + \sum_{s=s_0}^{s_{\max}} \sum_l d_{s,l}\psi_{s,l}(x) \\ &= \sum_i \langle f, \hat{\phi}_{s_0,i}\rangle \phi_{s_0,i}(x) + \sum_{s=s_0}^{s_{\max}} \sum_l \langle f, \hat{\psi}_{s,l}\rangle \psi_{s,l}(x) \end{aligned} \tag{3.72}$$

Therefore, we care only that the primary scaling function and wavelet have such beneficial properties as symmetry, smoothness, and short supporting intervals. In some applications, we may not actually use the dual wavelet or the dual scaling function at all.

How do we construct a biorthogonal system? So far, we have used two very different approaches to construct a MRA: starting either from a scaling function or from a filter. These two different methods will be combined together to construct the biorthogonal families. In this section we outline a general procedure.

We first choose a scaling function with the desired properties as the primary scaling function (or as the dual scaling function as it is called in some references). The primary filter $H(\omega)$ can then be computed. Next, we proceed to select a dual filter $\hat{H}(\omega)$ subject to the corresponding orthonormal condition (2.68):

$$H(\omega)\hat{H}^*(\omega) + H(\omega+\pi)\hat{H}^*(\omega+\pi) = 1 \tag{3.73}$$

We may impose additional conditions such as vanishing moments on the dual scaling function and/or on the wavelets. At this point, we can borrow the similar techniques used in Daubechies' construction to find a suitable dual filter. As expected, it will be much easier this time since half of (3.73) is already known. For example, we no longer need to use the spectral factorization procedure. In any case, after a dual filter is found, the associated dual scaling function can be constructed in exactly the same way as given in Daubechies' procedure. Finally, the dual and primary wavelets can be constructed by the corresponding wavelet equations, namely,

$$\hat{\psi}(x) = \sqrt{2}\sum_{l=-\infty}^{+\infty} \hat{g}_l\hat{\phi}(2x-l) = \sqrt{2}\sum_{l=-\infty}^{+\infty} (-1)^l h_{1-l}^* \hat{\phi}(2x-l) \tag{3.74a}$$

$$\psi(x) = \sqrt{2}\sum_{l=-\infty}^{+\infty} g_l\phi(2x-l) = \sqrt{2}\sum_{l=-\infty}^{+\infty} (-1)^l \hat{h}_{1-l}^* \phi(2x-l) \tag{3.74b}$$

We will illustrate the above outlined procedure in the next section to construct the biorthogonal Cohen–Daubechies–Feauveau spline family. To end this section, we point out that, for a given scaling function, there are in general many different dual scaling functions and dual wavelets from which to choose. Since the primary wavelet is determined not only by the primary scaling function but also by the dual filter, there will be many different wavelets for a fixed scaling function as well.

Furthermore, we could also use a different procedure altogether. For example, we may start with a preselected primary filter instead of a primary scaling function.

3.9 BIORTHOGONAL COHEN–DAUBECHIES–FEAUVEAU FAMILY

As the name suggests, the biorthogonal Cohen–Daubechies–Feauveau (CDF) spline family can be thought of as a product of the marriage between the spline family and Daubechies' construction. For the first part, we simply select the centralized B-spline function $\beta_n(x)$ of order n as the primary scaling function $\phi(x)$. The selection of the dual filter is made under the assumption that the dual wavelet $\hat{\psi}(x)$ will have $\hat{n}$ vanishing moments, just like what we see in Daubechies' procedure. The CDF spline family is therefore indexed by the pair $(n, \hat{n})$ and denoted by CDF $(n, \hat{n})$.

Let us work out more details as follows. First, from (3.15), we have

$$H(\omega) = e^{-jp\omega}\cos^{n+1}(\omega/2) \tag{3.75}$$

where $p = 0$ for odd n and $p = \frac{1}{2}$ for even n.

Therefore, the equation (3.73) becomes

$$e^{-jp\omega}\cos^{n+1}(\omega/2)\hat{H}^*(\omega) + e^{-jp(\omega+\pi)}\cos^{n+1}(\omega/2 + \pi/2)\hat{H}^*(\omega+\pi) = 1 \tag{3.76}$$

From this equation, once again, the similar method used in Daubechies' procedure yields the following familiar form for the dual filter:

$$\hat{H}(\omega) = e^{-jp\omega}\left(\cos\frac{\omega}{2}\right)^{\hat{n}}\left[\sum_{k=0}^{K-1}\binom{K-1+k}{k}\left(\sin^2\frac{\omega}{2}\right)^k + \left(\sin^2\frac{\omega}{2}\right)^K f(\cos\omega)\right] \tag{3.77}$$

where $2K = \hat{n} + n + 1$, $f(x)$ is an odd polynomial, and $p = 0$ for odd n and $p = \frac{1}{2}$ for even n. Interested readers may check directly that (3.76) will be satisfied by such a dual filter in (3.77). Note that $\hat{n}$ and $n + 1$ must be both even or both odd.

Of course, we will take the simplest form by setting $f(x) = 0$. The rest of the procedure can then be followed routinely. We do want to point out that not all pairs $(n, \hat{n})$ with $2\ K = \hat{n} + n + 1$ will lead to a biorthogonal MRA. For example, (2, 1) does not lead to a square-integral dual-scaling function.

Let us consider two simple examples CDF(0, 1) and CDF(1, 2). Setting $n = 0$ and $\hat{n} = 1$ in (3.75) and (3.77), respectively, we obtain

$$H(\omega) = \hat{H}(\omega) = e^{-j\omega/2}\cos(\omega/2) \tag{3.78}$$

Therefore, in the case of CDF(0, 1), the primary and dual MRAs coincide and are equivalent to the Haar MRA.

For CDF(1, 2), the primary MRA is the linear spline MRA. Therefore, we have the primary filter coefficients

$$\{h_k\} = \left\{\frac{\sqrt{2}}{2}, \sqrt{2}, \frac{\sqrt{2}}{2}\right\} \tag{3.79}$$

By (3.77), we have

$$\hat{H}(\omega) = \left(\cos\frac{\omega}{2}\right)^2 \sum_{k=0}^{1} \binom{1+k}{k} \left(\sin^2\frac{\omega}{2}\right)^k$$
$$= \left(\frac{e^{j\omega/2}+e^{-j\omega/2}}{2}\right)^2 \left(1+2\left(\frac{e^{j\omega/2}-e^{-j\omega/2}}{2}\right)^2\right) \tag{3.80}$$

Expanding the product in (3.79), we then obtain the following dual filter coefficients:

$$\{\hat{h}_k\} = \left\{-\tfrac{\sqrt{2}}{8}, \tfrac{\sqrt{2}}{4}, \tfrac{3\sqrt{2}}{4}, \tfrac{\sqrt{2}}{4}, -\tfrac{\sqrt{2}}{8}\right\} \tag{3.81}$$

In general, the primary scaling function in CDF $(n, \hat{n})$ has a supporting interval of length $n+1$ (it is just the B-spline function of order n), and the dual primary scaling function in CDF $(n, \hat{n})$ has a supporting interval of length $(2\hat{n}+n-1)$. In Figure 3.6a–f, we have displayed graphs of the primary scaling functions, the dual scaling functions, the primary wavelets, and the dual wavelets for CDF(1, 2), CDF(1, 4), CDF(2, 3), and CDF(2, 5). We used MatLab to generate those graphs. Be advised that MatLab uses a slightly different index; that is, our CDF $(n, \hat{n})$ is denoted by CDF $(n+1, \hat{n})$ in MatLab.

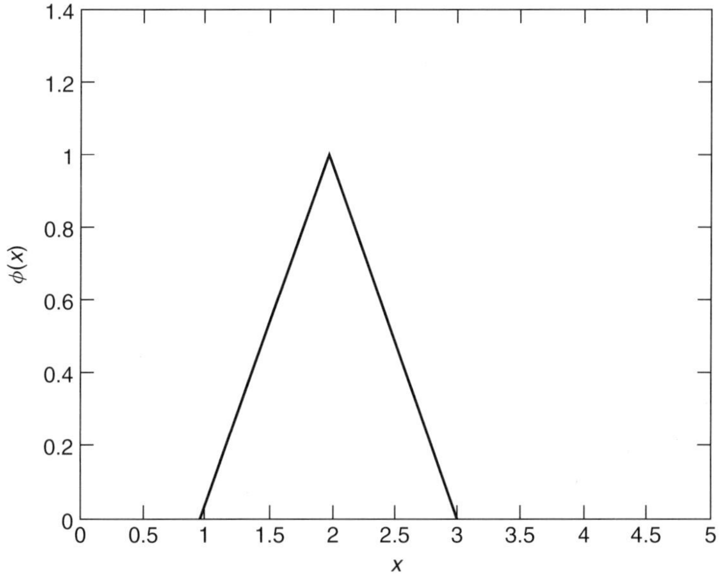

(a)

FIGURE 3.6 (a) Biorthogonal CDF(1, 2) primary scaling function. (b) Biorthogonal CDF(1, 2) primary wavelet. (c) Biorthogonal CDF(1, 2) dual scaling function. (d) Biorthogonal CDF(1, 2) dual wavelet. (e) Biorthogonal CDF(1, 4) primary wavelet. (f) Biorthogonal CDF(1, 4) dual scaling function. (g) Biorthogonal CDF(1, 4) dual wavelet. (h) Biorthogonal CDF(2, 3) primary scaling function. (i) Biorthogonal CDF(2, 3) primary wavelet. (j) Biorthogonal CDF(2, 3) dual scaling function. (k) Biorthogonal CDF(2, 3) dual wavelet. (l) Biorthogonal CDF(2, 5) primary wavelet. (m) Biorthogonal CDF(2, 5) dual scaling function. (n) Biorthogonal CDF(2, 5) dual wavelet.

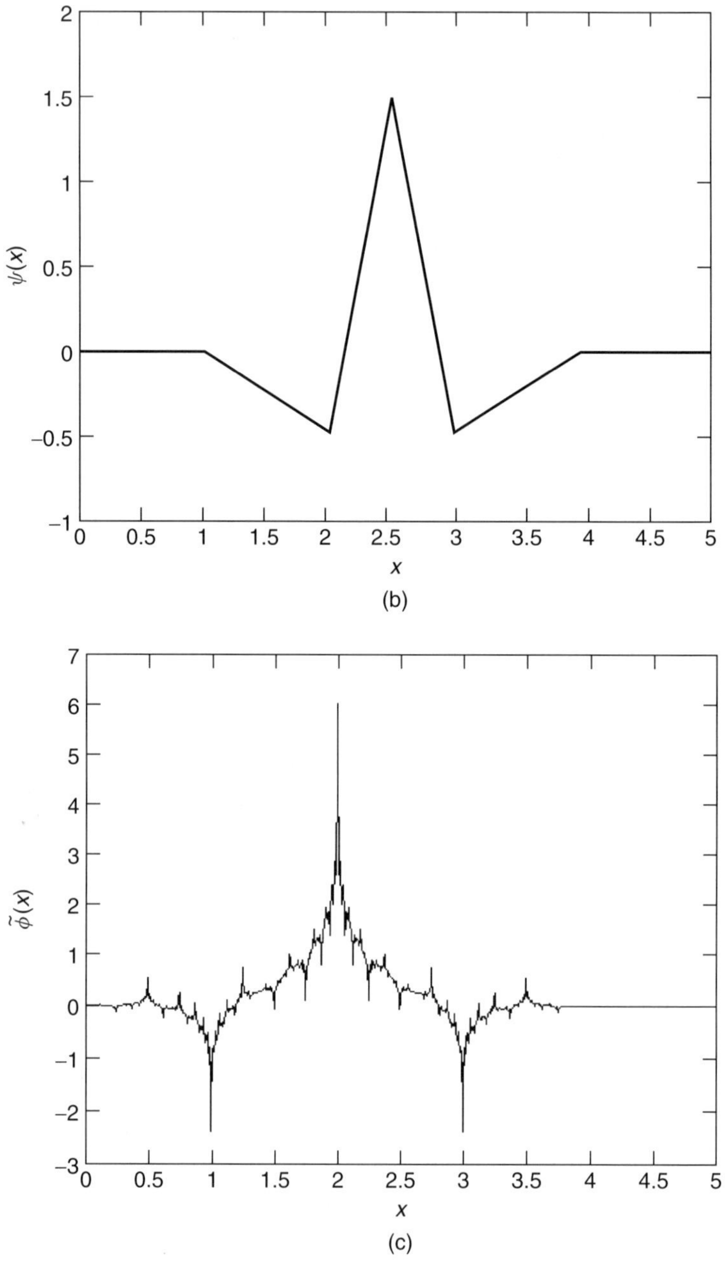

FIGURE 3.6 *(Continued)*

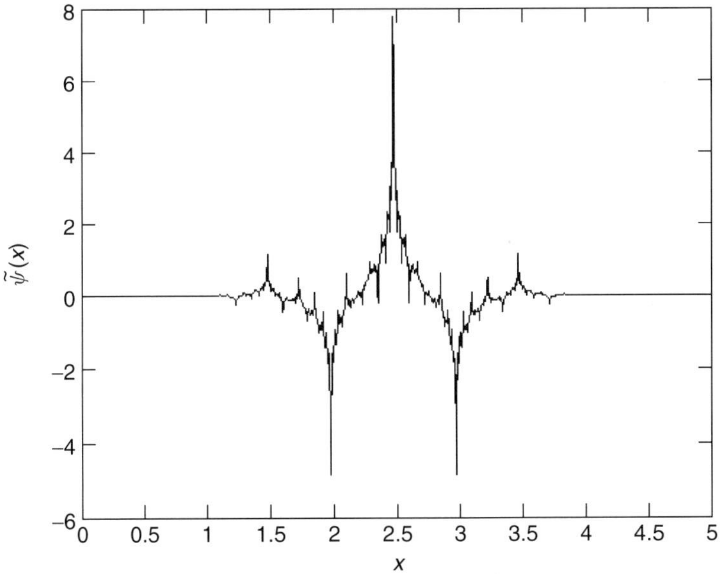

(d)

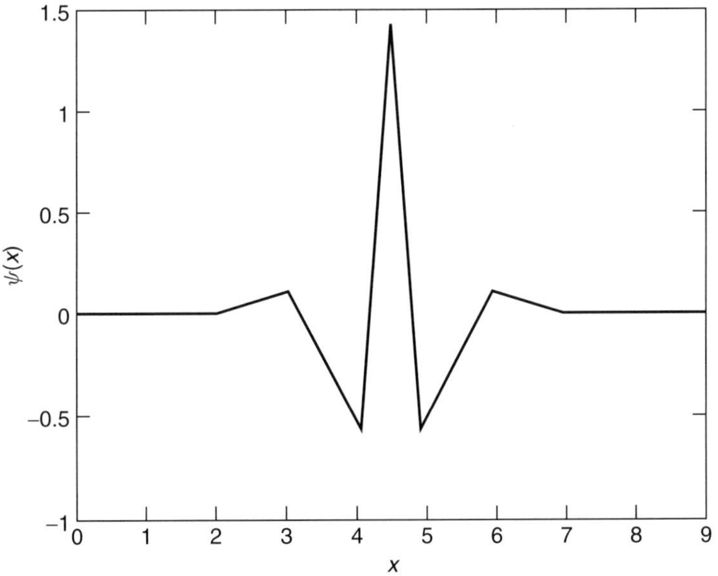

(e)

FIGURE 3.6 *(Continued)*

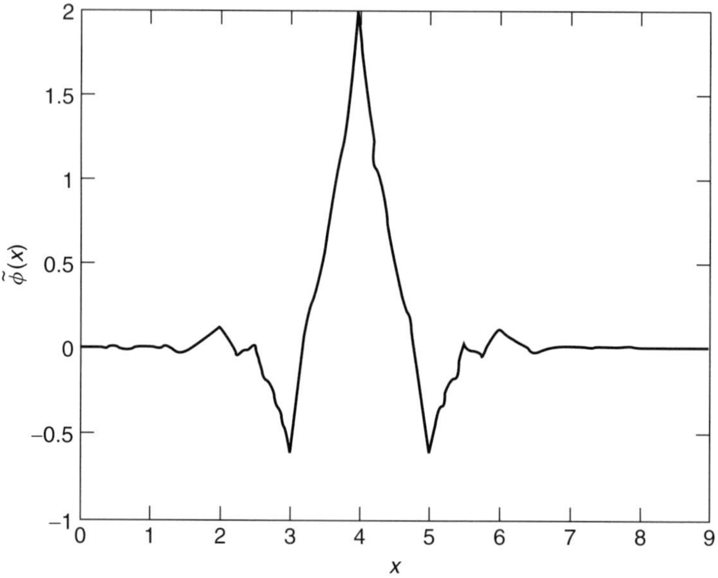

(f)

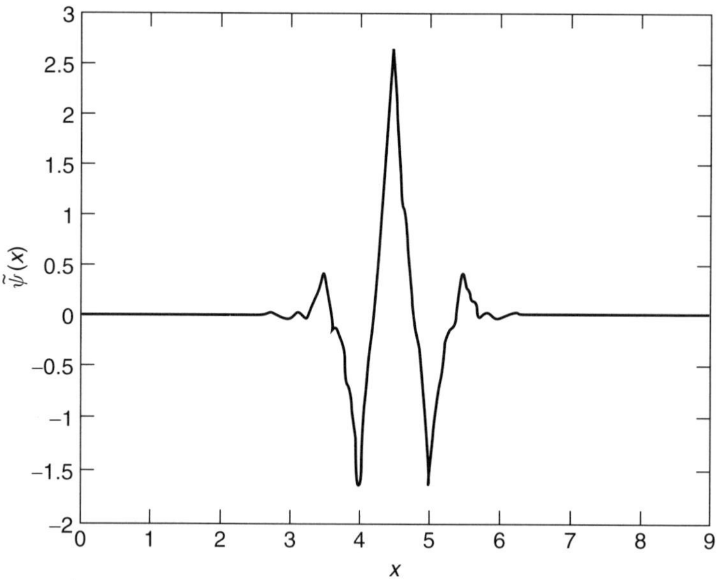

(g)

FIGURE 3.6 *(Continued)*

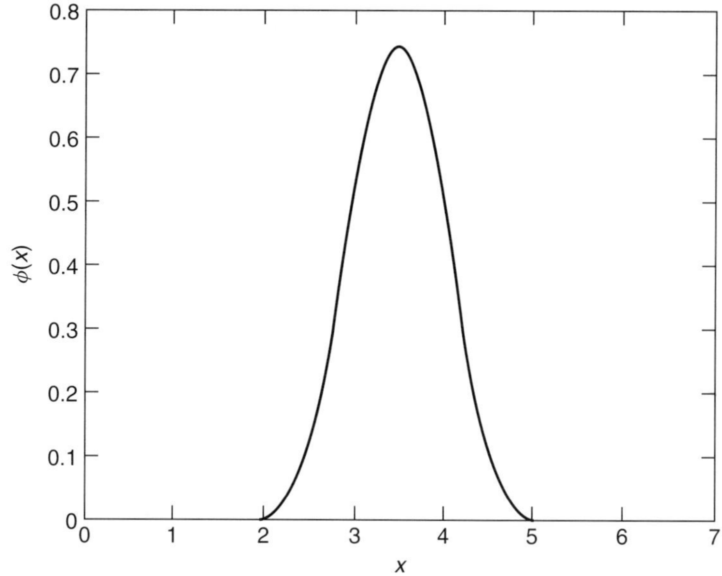

(h)

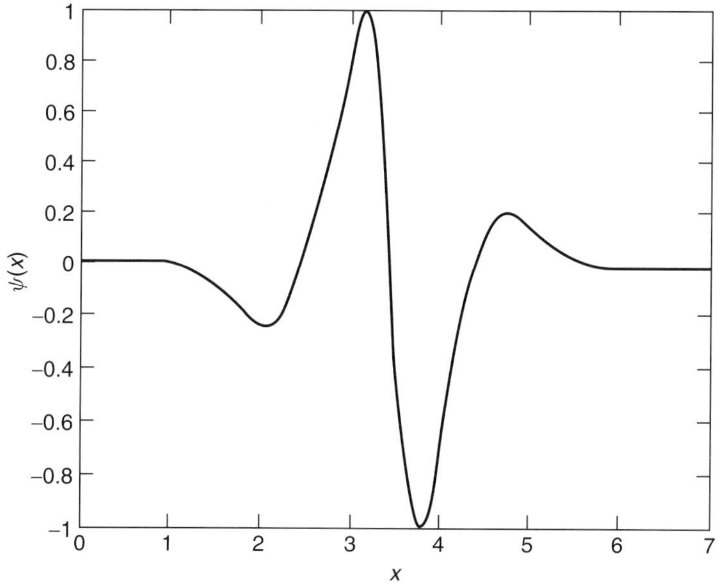

(i)

FIGURE 3.6 *(Continued)*

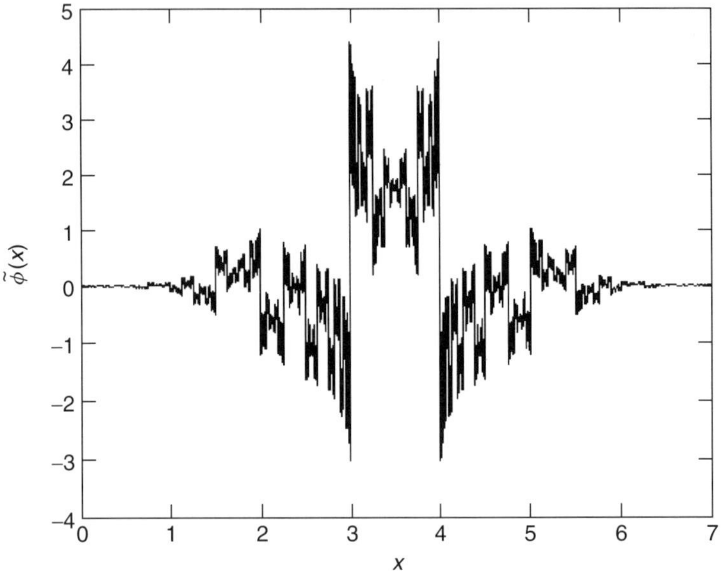

(j)

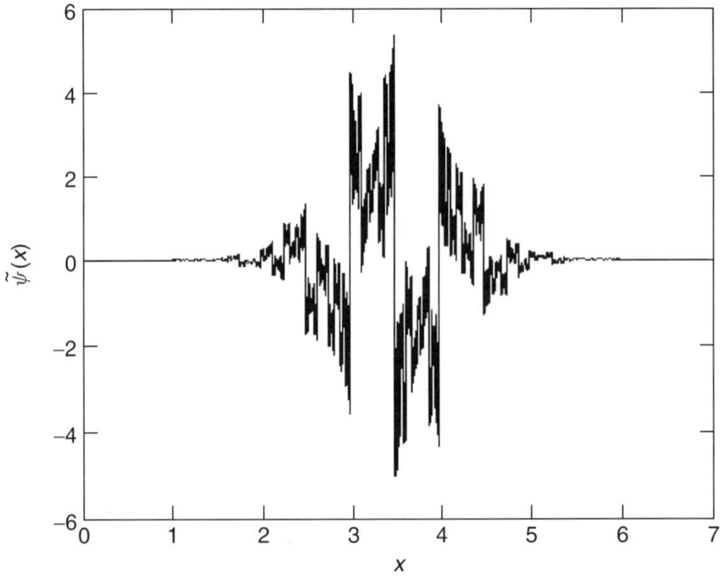

(k)

FIGURE 3.6 *(Continued)*

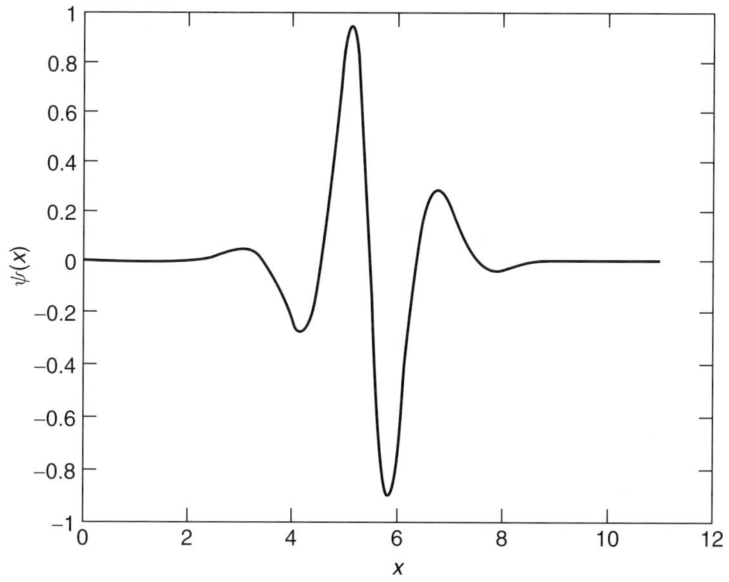

(l)

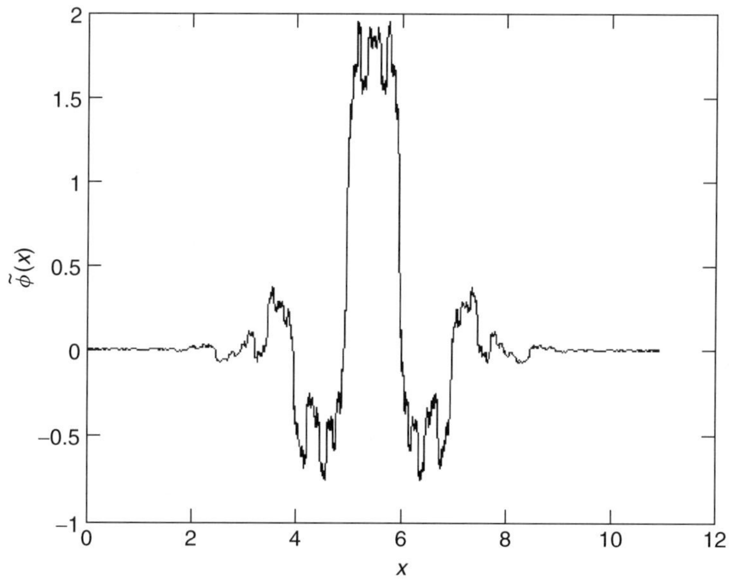

(m)

FIGURE 3.6 (*Continued*)

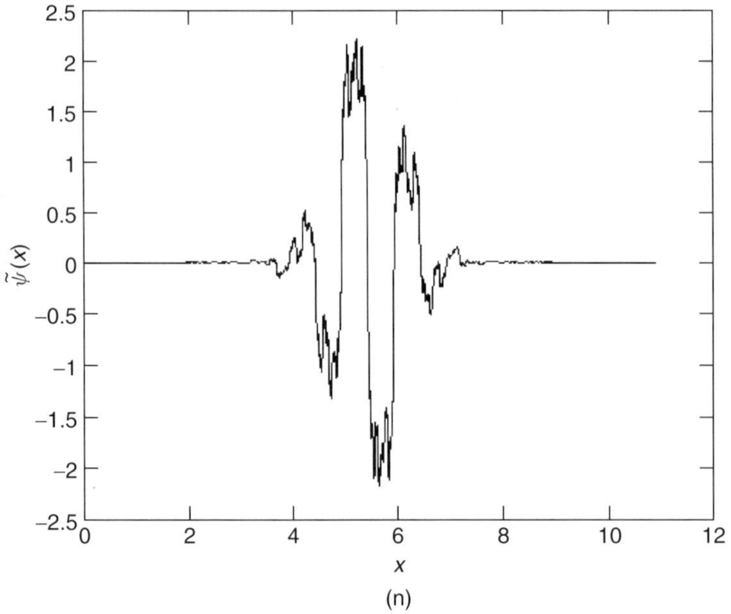

FIGURE 3.6 *(Continued)*

REFERENCES

[1] I. Daubechies, *Ten Lectures on Wavelets*, SIAM, Philadelphia, 1992.

[2] T. Dogaru and L. Carin, "Multiresolution time-domain using CDF biorthogonal wavelets," *IEEE Trans. Microwave Theory Tech.*, vol. 49, pp. 902–912, May 2001.

[3] J. J. H. Wang, *Generalized Moment Methods in Electromagnetics: Formulation and Computer Solution of Integral Equations*, Wiley-Interscience, Hoboken, NJ, 1991.

[4] M. Krumpholz and L. P. B. Katehi, "MRTD: New time domain schemes based on multiresolution analysis," *IEEE Trans. Microwave Theory Tech.*, vol. 44, no. 4, pp. 555–571, April 1996.

[5] M. Fujii and W. J. R. Hoefer, "A three-dimensional Haar wavelet-based multiresolution analysis similar to the 3-D FDTD method derivation and application," *IEEE Trans. Microwave Theory Tech.*, vol. 46, no. 12, pp. 2463–2475, Dec. 1998.

[6] X. Zhu and L. Carin, "Multiresolution time-domain analysis of plane-wave scattering from general three-dimensional surface and subsurface dielectric targets," *IEEE Trans. Microwave Antennas Propag.*, vol. 49, no. 11, pp. 1568–1578, Nov. 2001.

[7] T. Dogaru and L. Carin, "Application of Haar-wavelet-based multiresolution time-domain schemes to electromagnetic scattering problems," *IEEE Trans. Microwave Antennas Propag.*, vol. 50, no. 6, pp. 774–784, Nov. 2001.

[8] D. L Donoho, "Interpolating wavelets transforms," *Dept. Statistic, Stanford Univ., Stanford, CA, Tech. Rep.*, 1992.

[9] W. Sweldens and R. Piessens, "Wavelet sampling techniques," in *Proceedings of the 1993 Statistical Computing Section*, 1993, pp. 20–29.

[10] M. Fujii and W. J. R. Hoefer, "A wavelet formulation of finite-difference method: full-vector analysis of optical waveguide junctions," *IEEE J. Quantum Electron.*, vol. 37, no. 8, pp. 1015–1029, Aug. 2001.

[11] M. Unser "Ten good reasons for using spline wavelets," in *Proceedings of the SPIE*, vol. 3169, Wavelet Applications in Signal and Image Processing V, pp. 422–431, 1997.

[12] C. K. Chui and J. Z. Wang, "On compactly supported spline wavelets and a duality principle," *Trans. Am. Math. Soc.*, vol. 330, no. 2, pp. 903–915, 1992.

[13] C. K. Chui, *An Introduction to Wavelets*, Vol. 1, *Wavelet Analysis and Its Applications*, Academic Press, San Diego, CA, 1992.

[14] S. G. Mallat, "Multifrequency channel decompositions of images and wavelet model," *IEEE Trans. Acoust. Speech Signal Process.*, vol. 37, no. 12, pp. 2091–2110, 1989.

[15] G. Battle, "A block spin construction of ondeletts. Part I: Lemarié ," *Commun. Math. Phys.*, vol. 110, pp. 601–615, 1987.

[16] P. -G. Lemarié, "Ondelettes à localization exponentielles," *J. Math. Pures Appl.*, vol. 67, no. 3, pp. 227–236, 1988.

[17] Y. Meyer, *Ondelettes et Opérateurs,* I: *Ondelettes*, II: *Opérateurs de Calderón Zygmund*, III: (with R. Coifman), *Opérateurs multilinéaires*. Hermann, Paris, 1990. English translation of first volume, *Wavelets and Operators*, is published by Cambridge University Press, New York, 1993.

[18] A. Cohen, I. Daubechies, and J. Feauveau, "Bi-orthogonal bases of compactly supported wavelets," *Commun. Pure Appl. Math.*, vol. 45, pp. 485–560, 1992.

[19] B. Jawerth and W. Sweldens, "An overview of wavelet based multiresolution analysis," *SIAM Rev.*, vol. 36, no. 3, pp. 377–412, 1994.

[20] M. Unser, A. Aldroubi, and M. Eden, "On the asymptotic convergence of B-spline wavelets to Gabor functions," *IEEE Trans. Inform. Theory Machine Intell.*, vol. 38, no. 2, pp. 864–872, Mar. 1992.

[21] M. Unser, P. Thévenaz, and A. Aldroubi, "Shift-orthogonal wavelet bases using splines," *IEEE Signal Process. Lett.*, vol. 3, no. 3, pp. 85–88, Mar. 1996.

[22] G. Strang and T. Nguyen, *Wavelets and Filter Banks*, Wellesley-Cambridge, Wellesley, MA, 1996.

[23] A. Haar, "Zur Throie der orthogonalen Funktionnensysteme," *Math. Ann.*, vol. 69, pp. 331–71, 1910.

[24] I. Daubechies, "Orthogonal bases of compactly supported wavelets," *Commun. Pure Appl. Math.*, vol. XLI(41), pp. 909–996, 1988.

[25] I. Daubechies and J. G. Lagarlas, "Two-scale difference equation I. Existence and global regularity of solutions," *SIAM J. Math. Anal.*, vol 22, pp. 1388–1400, 1991.

[26] I. Daubechies, "Orthogonal bases of compactly supported wavelets II. Variations on a theme," *SIAM J. Math. Anal.*, vol. 24, no. 2, pp. 499–519, Mar. 1993.

[27] M. Antonini, M. Barlaud, P. Mathieu, and I. Daubechies, "Image coding using wavelet transform," *IEEE Trans. Image Process.*, vol. 1, no. 2, pp. 205–220, Apr. 1992.

[28] I. Daubechies, "The wavelet transformation, time-frequency localization and signal analysis," *IEEE Trans. Inform. Theory*, vol. 36, pp. 961–1005, 1990.

[29] A. Cohen, "Biorthogonal wavelets," in *Wavelets: A Tutorial in Theory and Applications* (ed. C. K. Chui), Academic Press, San Diego, 1992, pp. 123–152.

CHAPTER FOUR

Kernel of Multiresolution Time Domain Scheme

4.1 INTRODUCTION

In this chapter we apply the concept of multiresolution analysis (MRA) developed earlier, based on scaling and wavelet function expansions, as well as the application of the method of moments (MoM); then we apply the MRA to solve the Maxwell equations in the time domain. We show how the application of MRA leads to a new time domain numerical technique, referred to as the multiresolution time domain (MRTD) scheme [1, 2]. We then demonstrate that the MRTD scheme is superior to the conventional finite difference time domain (FDTD) method, both in terms of computer memory and computational time for a variety of electromagnetic application problems.

The MRTD scheme is a generalized version of Yee's FDTD method; hence, we can use a unified approach to develop both the MRTD and FDTD algorithms. On the basis of the MRA in the time domain, we will show that the conventional FDTD method [3] can be derived by applying the MoM to the Maxwell equations using a pulse basis function to expand the unknown fields [4]. The reason we are interested in developing the MRTD is that it promises to greatly improve computational efficiency and substantially reduce the computer resources required, in comparison to the FDTD method. In the MRTD applications, we apply the image principle to handle the boundary truncations of the perfect electric conductor (PEC) and perfect magnetic conductor (PMC) and to overcome the problems arising from the use of nonlocalized basis functions in the MRTD boundary modeling [2, 5–6]. To facilitate future comparisons of various schemes, it is convenient to introduce the following nomenclatures: (1) S-MRTD—the

Multiresolution Time Domain Scheme for Electromagnetic Engineering
By Yinchao Chen, Qunsheng Cao, and Raj Mittra
ISBN 0-471-27230-2

MRTD scheme with the electromagnetic field expansions only in terms of scaling functions (in all directions); (2) SW-MRTD—the MRTD scheme using both scaling and wavelet functions; and (3) SWx-MRTD, SWy-MRTD, or SWz-MRTD scheme—the MRTD scheme using scaling functions in all directions, and wavelet functions only in the x-, y-, or z-axis, respectively.

We first discuss the interrelationships among the FDTD, MoM, and MRTD methods. Next, we focus on developing a full version of the scaling and wavelet based SW-MRTD scheme and present both the field expansions and the derivation of the MRTD update equations. Third, we describe guidelines for the MRTD time update stability and discuss total field calculations. Finally, at the end of the chapter, we summarize a set of frequently used orthogonal and integral relationships that are very useful in developing the MRTD scheme.

4.2 RELATIONSHIPS AMONG THE FDTD, MoM, AND MRTD

The MRTD update equations are typically derived via an application of the MoM [4], which utilizes field expansions as summations of wavelets and scaling functions. In this section, we relate the FDTD, MoM, and MRTD approaches and show how use of the MoM and MRTD approaches leads to the FDTD update algorithm if the pulse functions are chosen as the basis functions.

To help the reader understand the relationship among the FDTD, MoM, and MRTD, we begin with a simple example, namely, the well-known transmission line equations in the time domain, which take the form

$$\frac{\partial V(x,t)}{\partial x} = -RI(x,t) - L\frac{\partial I(x,t)}{\partial t} \tag{4.1}$$

$$\frac{\partial I(x,t)}{\partial x} = -GV(x,t) - C\frac{\partial V(x,t)}{\partial t} \tag{4.2}$$

where $V(x,t)$ and $I(x,t)$ are voltage and current defined at (x,t) along a transmission line; R is the series resistance per unit length in Ω/m; L is the series inductance per unit length in H/m; G is the shunt conductance per unit length in S/m; and, C is the shunt capacitance per unit length in F/m, respectively.

FDTD Approach

Let us first derive the FDTD update equations from (4.1) and (4.2) by approximating the derivatives using differencing equations. Let $V(i\Delta x, n\Delta t) \equiv V_i^n$, $I(i\Delta x, n\Delta t) \equiv I_i^n$ be the discretized voltage and current at space–time point $(i\Delta x, n\Delta t)$. By using the central differencing technique [3, 4, 7], we can approximate the partial derivatives as

$$\begin{aligned}\frac{\partial V(i\Delta x, n\Delta t)}{\partial x} &= \frac{V((i+1)\Delta x, n\Delta t) - V((i-1)\Delta x, n\Delta t)}{2\Delta x} + O(\Delta x^2)\\ &\approx \frac{V_{i+1}^n - V_{i-1}^n}{2\Delta x}\end{aligned} \tag{4.3}$$

$$\begin{aligned}\frac{\partial I(i\Delta x, n\Delta t)}{\partial x} &= \frac{I((i+1)\Delta x, n\Delta t) - I((i-1)\Delta x, n\Delta t)}{2\Delta x} + O(\Delta x^2) \\ &\approx \frac{I_{i+1}^n - I_{i-1}^n}{2\Delta x} \end{aligned} \tag{4.4}$$

$$\begin{aligned}\frac{\partial V(i\Delta x, n\Delta t)}{\partial t} &= \frac{V(i\Delta x, (n+1)\Delta t) - V(i\Delta x, (n-1)\Delta t)}{2\Delta t} + O(\Delta t^2) \\ &\approx \frac{V_i^{n+1} - V_i^{n-1}}{2\Delta t} \end{aligned} \tag{4.5}$$

$$\begin{aligned}\frac{\partial I(i\Delta x, n\Delta t)}{\partial t} &= \frac{I(i\Delta x, (n+1)\Delta t) - I(i\Delta x, (n-1)\Delta t)}{2\Delta t} + O(\Delta t^2) \\ &\approx \frac{I_i^{n+1} - I_i^{n-1}}{2\Delta t} \end{aligned} \tag{4.6}$$

and the time average as

$$V_i^n = \tfrac{1}{2}(V_i^{n+1} + V_i^{n-1}) \tag{4.7}$$

$$I_i^n = \tfrac{1}{2}(I_i^{n+1} + I_i^{n-1}) \tag{4.8}$$

Using the above differencing relations, we can now derive the following FDTD update equations:

$$I_i^{n+1} = \left(\frac{L - R\Delta t}{L + R\Delta t}\right) I_i^{n-1} - \frac{1}{L + R\Delta t}(V_{i+1}^n - V_{i-1}^n)\frac{\Delta t}{\Delta x} \tag{4.9}$$

$$V_i^{n+1} = \left(\frac{C - G\Delta t}{C + G\Delta t}\right) V_i^{n-1} - \frac{1}{C + G\Delta t}(I_{i+1}^n - I_{i-1}^n)\frac{\Delta t}{\Delta x} \tag{4.10}$$

We can use the above two equations to compute the temporal signatures of the FDTD voltage and current for a transmission line.

MoM Method

Next, we move on to the MoM technique, in which the voltage and current are expanded in terms of the rectangular pulse basis functions $\{h_i(x)\}$ and $\{h_n(t)\}$ in both space and time as follows:

$$V(x, t) = \sum_{n,i} V_i^n h_i(x) h_n(t) \tag{4.11}$$

$$I(x, t) = \sum_{n,i} I_i^n h_i(x) h_n(t) \tag{4.12}$$

where the rectangular pulse is defined as

$$h_i(x) = \begin{cases} 1, & i\Delta x \le x < (i+1)\Delta x \\ 0, & \text{otherwise} \end{cases} \tag{4.13}$$

Substitution of (4.11), (4.12), and (4.13) into (4.1) and (4.2) yields

$$\sum_{n,i} V_i^n \frac{dh_i(x)}{dx} h_n(t) = -R \sum_{n,i} I_i^n h_i(x) h_n(t) - L \sum_{i,n} I_i^n h_i(x) \frac{dh_n(t)}{dt} \tag{4.14}$$

$$\sum_{n,i} I_i^n \frac{dh_i(x)}{dx} h_n(t) = -G \sum_{n,i} V_i^n h_i(x) h_n(t) - C \sum_{i,n} V_i^n h_i(x) \frac{dh_n(t)}{dt} \tag{4.15}$$

Our next step is to apply the Galerkin method [5], in which we use the complex conjugate of the basis functions as the testing functions to sample the equations (4.14) and (4.15) in space and time, respectively. For instance, testing the left-hand side of (4.14), we get

$$\begin{aligned} &\int_{-\infty}^{+\infty} \int_{-\infty}^{+\infty} h_i(x) h_n(t) \sum_{n',i'} V_{i'}^{n'} h_{n'}(t) \frac{dh_{i'}(x)}{dx} \, dx \, dt \\ &= \sum_{n',i'} V_{i'}^{n'} \int_{-\infty}^{+\infty} h_i(x) \frac{dh_{i'}(x)}{dx} \, dx \int_{-\infty}^{+\infty} h_n(t) h_{n'}(t) \, dt \\ &= \sum_{n',i'} V_{i'}^{n'} \frac{1}{2} (\delta_{i',i+1} - \delta_{i',i-1}) \delta_{n'n} \, \Delta t = (V_{i+1}^n - V_{i-1}^n) \, \frac{\Delta t}{2} \end{aligned} \tag{4.16}$$

where we have used the following integral relations:

$$\int_{-\infty}^{+\infty} h_i(x) \frac{dh_{i'}(x)}{dx} \, dx = \frac{1}{2} (\delta_{i',i+1} - \delta_{i',i-1}) \tag{4.17}$$

$$\int_{-\infty}^{+\infty} h_n(t) h_{n'}(t) \, dt = \delta_{n'n} \, \Delta t = \begin{cases} \Delta t, & \text{if } n' = n \\ 0, & \text{if } n' \neq n \end{cases} \tag{4.18}$$

and $\delta_{n,n'}$ is the Kronecker symbol: if $n = n'$, $\delta_{n,n'} = 1$; otherwise, $\delta_{n,n'} = 0$. Note that in evaluating an inner product of two functions, we use the same notation to represent a function and its complex conjugate when they are both real.

Similarly, by sampling and testing the first term of the right-hand side (RHS) of (4.14) we obtain

$$\begin{aligned} &- R \int_{-\infty}^{+\infty} \int_{-\infty}^{+\infty} h_i(x) h_n(t) \sum_{n',i'} I_{i'}^{n'} h_{i'}(x) h_{n'}(t) \, dx \, dt \\ &= -R \sum_{n',i'} I_{i'}^{n'} \int_{-\infty}^{+\infty} h_{i'}(x) h_i(x) \, dx \int_{-\infty}^{+\infty} h_{n'}(t) h_n(t) \, dt \end{aligned}$$

$$= -R \sum_{n',i'} I_{i'}^{n'} \delta_{i',i} \delta_{n,n'} \; \Delta x \; \Delta t \tag{4.19}$$
$$= -R I_i^n \Delta x \; \Delta t$$
$$= -\frac{R}{2}(I_i^{n+1} + I_i^{n-1}) \; \Delta x \; \Delta t$$

In a similar manner, we can simplify the second term of the RHS of (4.14) to obtain

$$- \int_{-\infty}^{+\infty} \int_{-\infty}^{+\infty} h_i(x) h_n(t) L \sum_{n',i'} I_{i'}^{n'} h_{i'}(x) \frac{dh_{n'}(t)}{dt} \, dx \, dt$$
$$= -L \sum_{n',i'} I_{i'}^{n'} \int_{-\infty}^{+\infty} h_i(x) h_{i'}(x) \, dx \int_{-\infty}^{+\infty} h_n(t) \frac{dh_{n'}(t)}{dt} \, dt \tag{4.20}$$
$$= -L \sum_{n',i'} I_{i'}^{n'} \delta_{i',i} \; \Delta x \; \frac{1}{2}(\delta_{n',n+1} - \delta_{n',n-1}) = -L(I_i^{n+1} - I_i^{n-1}) \; \frac{\Delta x}{2}$$

By combining (4.16), (4.19), and (4.20), we obtain

$$(V_{i+1}^n - V_{i-1}^n) \; \frac{\Delta t}{2} = -\frac{R}{2}(I_i^{n+1} + I_i^{n-1}) \; \Delta x \; \Delta t - L(I_i^{n+1} - I_i^{n-1}) \; \frac{\Delta x}{2} \tag{4.21}$$

which can be solved by using the leapfrog update algorithm. For instance, we can update the current I in time by using

$$I_i^{n+1} = \left(\frac{L - R \; \Delta t}{L + R \; \Delta t} \right) I_i^{n-1} - \frac{1}{L + R \; \Delta t}(V_{i+1}^n - V_{i-1}^n) \; \frac{\Delta t}{\Delta x} \tag{4.22}$$

Note that (4.22), derived using the MoM, is identical to (4.9), which was derived using the typical FDTD approach. Similarly, we can derive the update equation for the voltage V and find that it is identical to that in (4.10). Thus, for this particular example, we have demonstrated that the FDTD can be treated as a special case of the MoM when the pulse function is used for the spatial and temporal basis functions.

MRTD Scheme

We now turn to the MRTD scheme, which shares the same basic philosophy as that embodied in the MoM. In the MRTD scheme, the fields are first expanded as summations of the scaling and wavelet basis functions. In practice, the basis functions are frequently designated as the cubic spline Battle–Lemarié, Daubechies, and biorthogonal Cohen–Daubechies–Feauveau (CDF) and the biorthogonal scaling and wavelet functions. The MRTD update equations can also be obtained by applying Galerkin's method to the Maxwell equations. In the MRTD scheme, the fields (voltage and current in our example) are expanded in terms of a summation of scaling and wavelet functions as follows:

$$V(x,t) = \sum_{n=-\infty}^{+\infty} \left[\sum_{i=-\infty}^{+\infty} {}_{\phi}V_i^n \phi_i(x) + \sum_{s=0}^{s_{\max}} \sum_{l=-\infty}^{+\infty} {}_{\psi}V_{s,l}^n \psi_{s,l}(x) \right] h_n(t) \quad (4.23)$$

$$I(x,t) = \sum_{n=-\infty}^{+\infty} \left[\sum_{i=-\infty}^{+\infty} {}_{\phi}I_i^n \phi_i(x) + \sum_{s=0}^{s_{\max}} \sum_{l=-\infty}^{+\infty} {}_{\psi}I_{s,l}^n \psi_{s,l}(x) \right] h_n(t) \quad (4.24)$$

where s is the resolution level, which may range from 0 to $s_{\max}$ (the maximum wavelet resolution level), and $\phi(x)$ and $\psi(x)$ represent the scaling and mother wavelet functions, respectively. On a dyadic scale of 2^s (i.e., at the sth scale level, with $s \geq 0$), $\phi_{s,n}(x)$ and $\psi_{s,n}(x)$ are defined as

$$\phi_{s,n}(x) = 2^{s/2}\phi(2^s x - n) \quad (4.25)$$

$$\psi_{s,n}(x) = 2^{s/2}\psi(2^s x - n) \quad (4.26)$$

Note that the central symmetry index of the scaling function ϕ_i is located at i, and that the wavelet with two indices (s, l) is localized with the central position grid index $x = 2^{-s}(l + \frac{1}{2})$ in the space domain. For example if $s = s_{\max} = 1$, then $l = 0, 1$, and the corresponding wavelet centers are located at $x = \frac{1}{4}$ and $x = \frac{3}{4}$; if $s = s_{\max} = 2$, then $l = 0, 1, 2, 3$, and the corresponding wavelet centers are located at $x = \frac{1}{8}$, $x = \frac{3}{8}$, $x = \frac{5}{8}$, and $x = \frac{7}{8}$, respectively. In the field expansions, it is much more convenient to relate the wavelet location index (s, l) to the scaling function index i. Thus, a set of newly defined indices (i, s, l) with reference to i is introduced at the location of $i + 2^{-s}(l + \frac{1}{2})$ as the central symmetrical location index. Table 4.1 lists the central symmetrical coordinates of wavelet basis functions with different values of (i, s, l), and Figure 4.1 shows wavelet central positions for different values of (i, s, l).

Using this modified reference to the central symmetrical location index i of the scaling function, $V(x, t)$ and $I(x, t)$ can be rewritten as follows:

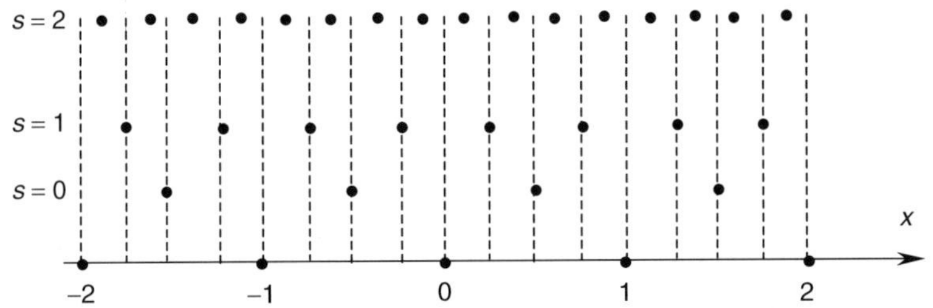

FIGURE 4.1 Location of the wavelet central symmetry index in the space domain.

TABLE 4.1 Central Symmetrical Coordinates of Wavelet Basis

i		-2	-1	0	1
$s=0$	$l=0$	$-\frac{3}{2}$	$-\frac{1}{2}$	$\frac{1}{2}$	$\frac{3}{2}$
$s=1$	$l=0$	$-\frac{7}{4}$	$-\frac{3}{4}$	$\frac{1}{4}$	$\frac{5}{4}$
	$l=1$	$-\frac{5}{4}$	$-\frac{1}{4}$	$\frac{3}{4}$	$\frac{7}{4}$
$s=2$	$l=0$	$-\frac{15}{4}$	$-\frac{7}{8}$	$\frac{1}{8}$	$\frac{9}{8}$
	$l=1$	$-\frac{13}{8}$	$-\frac{5}{8}$	$\frac{3}{8}$	$\frac{11}{8}$
	$l=2$	$-\frac{11}{8}$	$-\frac{3}{8}$	$\frac{5}{8}$	$\frac{13}{8}$
	$l=3$	$-\frac{9}{8}$	$-\frac{1}{8}$	$\frac{7}{8}$	$\frac{15}{8}$

$$V(x,t) = \sum_{n,i} \left[{}_{\phi}V_i^n \phi_i(x) + \sum_{s=0}^{s_{\max}} \sum_{l=0}^{2^s-1} {}_{\psi}^{s,l}V_{i+2^{-s}(1/2+l)}^n \psi_{s,i+2^{-s}(1/2+l)}(x) \right] h_n(t) \tag{4.27}$$

$$I(x,t) = \sum_{n,i} \left[{}_{\phi}I_i^n \phi_i(x) + \sum_{s=0}^{s_{\max}} \sum_{l=0}^{2^s-1} {}_{\psi}^{s,l}I_{i+2^{-S}(1/2+l)}^n \psi_{s,i+2^{-s}(1/2+l)}(x) \right] h_n(t) \tag{4.28}$$

with

$$\psi_{s,i+2^{-s}(l+1/2)}(x) = \psi_s\left(\frac{x}{\Delta x} - i - 2^{-s}l\right) = 2^{s/2}\psi\left[2^s\left(\frac{x}{\Delta x} - i\right) - l\right] \tag{4.29}$$

where $s = 0, 1, \ldots, s_{\max}$, and $l = 0, 1, 2, \ldots, 2^s - 1$ for each value of s. Note that in the present notation, the central symmetrical position of each wavelet specified by (i, s, l) in (4.27) or (4.28) is defined at $i + 2^{-s}(l + \frac{1}{2})$, although the index l does not appear in the right-hand side of the equation. It is also useful to point out that all the wavelets with identical values of i and s (l can be different) have identical shapes, and only their locations are translated.

Next, substitution of (4.27) and (4.28) into the transmission line equation in (4.1) leads to

$$\begin{aligned}
&\sum_{n,i} \left[{}_{\phi}V_i^n \frac{\partial \phi_i(x)}{\partial x} + \sum_{s=0}^{s_{\max}} \sum_{l=0}^{2^S-1} {}_{\psi}^{s,l}V_{i+2^{-S}(1/2+l)}^n \frac{\partial \psi_{s,i+2^{-S}(1/2+l)}(x)}{\partial x} \right] h_n(t) \\
&= -R \sum_{n,i} \left[{}_{\phi}I_i^n \phi_i(x) + \sum_{s=0}^{s_{\max}} \sum_{l=0}^{2^S-1} {}_{\psi}^{s,l}I_{i+2^{-S}(1/2+l)}^n \psi_{s,i+2^{-S}(1/2+l)}(x) \right] h_n(t) \\
&\quad - L \sum_{n,i} \left[{}_{\phi}I_i^n \phi_i(x) + \sum_{s=0}^{s_{\max}} \sum_{l=0}^{2^S-1} {}_{\psi}^{s,l}I_{i+2^{-S}(1/2+l)}^n \psi_{s,i+2^{-S}(1/2+l)}(x) \right] \frac{\partial h_n(t)}{\partial t}
\end{aligned} \tag{4.30}$$

We now apply Galerkin's method, choose the complex conjugate of the basis functions $\phi_i(x)h_n(t)$ and $\psi_{s,i+2^{-S}(l+1/2)}(x)h_n(t)$ as testing functions, sample (4.30) in space and time, and apply the orthogonal relationships (see Appendix B). The left-hand side of (4.30) then becomes

$$
\begin{aligned}
&\int_{-\infty}^{+\infty}\int_{-\infty}^{+\infty} \phi_i(x)h_n(t)\sum_{i',n'}\left({}_{\phi}V_{i'}^{n'}\frac{\partial\phi_{i'}(x)}{\partial x}+\sum_{s=0}^{s_{\max}}\sum_{l=0}^{2^S-1}{}^{s,l}_{\psi}V^{n'}_{i'+2^{-S}(1/2+l)}\right.\\
&\qquad\qquad\left.\times\frac{\partial\psi_{s,i'+2^{-S}(l+1/2)}(x)}{\partial x}\right)h_{n'}(t)\,dx\,dt\\
&=\sum_{i',n'}{}_{\phi}V_{i'}^{n'}\int_{-\infty}^{+\infty}\phi_i(x)\frac{\partial\phi_{i'}(x)}{\partial x}\,dx\int_{-\infty}^{+\infty}h_n(t)h_{n'}(t)\,dt\\
&+\sum_{i',n'}\sum_{s=0}^{s_{\max}}\sum_{l=0}^{2^S-1}{}^{s,l}_{\psi}V^{n'}_{i'+2^{-S}(1/2+l)}\int_{-\infty}^{+\infty}\phi_i(x)\frac{\partial\psi_{s,i'+2^{-S}(l+1/2)}(x)}{\partial x}\,dx\int_{-\infty}^{+\infty}h_n(t)h_{n'}(t)\,dt\\
&=\sum_{i'}\sum_{\nu=-\infty}^{+\infty}{}_{\phi}V_{i'}^{n}a(\nu)\delta_{i+\nu,i'}\;\Delta t+\sum_{i'}\sum_{\nu=-\infty}^{+\infty}\sum_{s=0}^{s_{\max}}\\
&\quad\times\sum_{l=0}^{2^S-1}{}^{s,l}_{\psi}V^{n}_{i'+2^{-S}(1/2+l)}d_{s,l}(\nu)\delta_{i+\nu-1/2,i'}\;\Delta t\\
&=\sum_{\nu=-\infty}^{+\infty}{}_{\phi}V_{i+\nu}^{n}a(\nu)\;\Delta t+\sum_{\nu=-\infty}^{+\infty}\sum_{s=0}^{s_{\max}}\sum_{l=0}^{2^S-1}{}^{s,l}_{\psi}V^{n}_{i+\nu-1/2+2^{-S}(1/2+l)}d_{s,l}(\nu)\;\Delta t
\end{aligned}
\tag{4.31a}
$$

$$
\begin{aligned}
&\int_{-\infty}^{+\infty}\int_{-\infty}^{+\infty}\psi_{s,i+2^{-s}(l+1/2)}(x)h_n(t)\sum_{i',n'}\left({}_{\phi}V_{i'}^{n'}\frac{\partial\phi_{i'}(x)}{\partial x}+\sum_{s'=0}^{s_{\max}}\sum_{l'=0}^{2^s-1}{}^{s',l'}_{\psi}V^{n'}_{i'+2^{-s'}(1/2+l')}\right.\\
&\qquad\qquad\left.\times\frac{\partial\psi_{s',i'+2^{-s}(l'+1/2)}(x)}{\partial x}\right)h_{n'}(t)\,dx\,dt\\
&=\sum_{i'}\left[\begin{array}{l}{}_{\phi}V_{i'}^{n}\int_{-\infty}^{+\infty}\psi_{s,i+2^{-s}(l+1/2)}(x)\dfrac{\partial\phi_{i'}(x)}{\partial x}\,dx+\displaystyle\sum_{s'=0}^{s_{\max}}\sum_{l'=0}^{2^S-1}{}^{s',l'}_{\psi}V^{n'}_{i'+2^{-s'}(1/2+l')}\\
\times\int_{-\infty}^{+\infty}\psi_{s,i+2^{-s}(l+1/2)}(x)\dfrac{\partial\psi_{s',i'+2^{-s}(l'+1/2)}(x)}{\partial x}\,dx\end{array}\right]\Delta t
\end{aligned}
$$

$$= \sum_{i'} \left[\begin{array}{l} \sum\limits_{\nu=-\infty}^{+\infty} {}_{\phi}V_{i'}^{n} c_{s,l}(\nu) \delta_{i+\nu+1/2,i'} \Delta t \\ + \sum\limits_{\nu=-\infty}^{+\infty} \sum\limits_{s'=0}^{s_{\max}} \sum\limits_{l'=0}^{2^{s}-1} {}^{s',l'}_{\psi}V^{n'}_{i'+2^{-s'}(1/2+l')} b^{s,s'}_{l,l'}(\nu) \delta_{i+\nu-1/2,i'} \end{array} \right] \Delta t$$

$$= \sum_{\nu=-\infty}^{+\infty} {}_{\phi}V^{n}_{i+\nu+1/2} c_{s,l}(\nu)\ \Delta t$$

$$+ \sum_{\nu=-\infty}^{+\infty} \sum_{s'=0}^{s_{\max}} \sum_{l'=0}^{2^{s}-1} {}^{s',l'}_{\psi}V^{n'}_{i+\nu-1/2+2^{-s'}(1/2+l')} b^{s,s'}_{l,l'}(\nu)\ \Delta t \tag{4.31b}$$

Following a similar approach and using the orthogonal relationships (see Appendix B), the first term in the RHS of (4.30) can be represented as

$$- R \int_{-\infty}^{+\infty} \int_{-\infty}^{+\infty} \phi_i(x) h_n(t)$$

$$\times \left({}_{\phi}I^{n'}_{i'} \phi_{i'}(x) + \sum_{s=0}^{s_{\max}} \sum_{l=0}^{2^{s}-1} {}^{s,l}_{\psi}I^{n'}_{i'+2^{-s}(1/2+l)} \psi_{s,i'+2^{-s}(1/2+l)}(x) \right) h_{n'}(t)\, dx\, dt$$

$$= -R_{\phi}I^{n'}_{i'} \delta_{i,i'}\ \Delta x \delta_{n,n'}\ \Delta t$$

$$= -RI^{n}_{i} \Delta x\ \Delta t$$

$$= -\frac{R}{2}(I^{n+1}_{i} + I^{n-1}_{i})\ \Delta x\ \Delta t \tag{4.32a}$$

$$- R \int_{-\infty}^{+\infty} \int_{-\infty}^{+\infty} \psi_{s,i+2^{-s}(1/2+l)}(x) h_n(t)$$

$$\times \left({}_{\phi}I^{n'}_{i'} \phi_{i'}(x) + \sum_{s=0}^{s_{\max}} \sum_{l=0}^{2^{s}-1} {}^{s,l}_{\psi}I^{n'}_{i'+2^{-s}(l+1/2)} \psi_{s,i'+2^{-s}(l+1/2)}(x) \right) h_{n'}(t)\, dx\, dt$$

$$= -R \sum_{s'=0}^{S_{\max}} \sum_{l'=0}^{2^{s}-1} {}^{s,l}_{\psi}I^{n'}_{i'+2^{-s}(1/2+l')} \delta_{i,i'}\ \Delta x \delta_{n,n'}\ \Delta t$$

$$= -R\, {}^{s,l}_{\psi}I^{n}_{i+2^{-s}(1/2+l)}\ \Delta x\ \Delta t$$

$$= -\frac{R}{2} \left({}^{s,l}_{\psi}I^{n+1}_{i+2^{-s}(1/2+l)} + {}^{s,l}_{\psi}I^{n-1}_{i+2^{-s}(1/2+l)} \right)\ \Delta x\ \Delta t \tag{4.32b}$$

Similarly, the second term of the RHS of (4.30) becomes

$$
\begin{aligned}
&- L \int_{-\infty}^{+\infty}\int_{-\infty}^{+\infty} \phi_i(x) h_n(t) \\
&\quad \times \left({}_{\phi}I_{i'}^{n'} \phi_{i'}(x) + \sum_{s=0}^{s_{\max}} \sum_{l=0}^{2^s-1} {}_{\psi}^{s,l}I_{i'+2^{-s}(1/2+l)}^{n'} \psi_{s,i'+2^{-s}(1/2+l)}(x) \right) \frac{\partial h_{n'}(t)}{\partial t}\, dx\, dt \\
&= -L_{\phi} I_{i'}^{n'} \delta_{i,i'}\ \Delta x \int_{-\infty}^{+\infty} h_n(x) \frac{\partial h_{n'}(x)}{\partial t}\, dt \\
&= -\frac{L}{2} ({}_{\phi}I_i^{n+1} - {}_{\phi}I_i^{n-1})\ \Delta x
\end{aligned}
\tag{4.33a}
$$

$$
\begin{aligned}
&- L \int_{-\infty}^{+\infty}\int_{-\infty}^{+\infty} \psi_{s,i+2^{-s}(1/2+l)}(x) h_n(t) \\
&\quad \times \left({}_{\phi}I_{i'}^{n'} \phi_{i'}(x) + \sum_{s=0}^{s_{\max}} \sum_{l=0}^{2^s-1} {}_{\psi}^{s,l}I_{i'+2^{-s}(l+1/2)}^{n'} \psi_{s,i'+2^{-s}(l+1/2)}(x) \right) \frac{\partial h_{n'}(t)}{\partial t}\, dx\, dt \\
&= -L \sum_{s'=0}^{s_{\max}} \sum_{l'=0}^{2^s-1} {}_{\psi}^{s,l}I_{i'+2^{-s}(1/2+l')}^{n'} \delta_{i,i'}\ \Delta x \int_{-\infty}^{+\infty} h_n(x) \frac{\partial h_{n'}(x)}{\partial t}\, dt \\
&= -\frac{L}{2} \left({}_{\psi}^{s,l}I_{i+2^{-s}(1/2+l)}^{n+1} - {}_{\psi}^{s,l}I_{i+2^{-s}(1/2+l)}^{n-1} \right)\ \Delta x
\end{aligned}
\tag{4.33b}
$$

where the coefficients $a(\nu)$, $b_{l,l'}^{s,s'}(\nu)$, $c_{s,l}(\nu)$, and $d_{s,l}(\nu)$ are defined as follows:

$$
a(\nu) = \frac{1}{\pi} \int_0^{+\infty} |\widetilde{\phi}(\omega)|^2 \omega \sin[\omega(\nu + 1/2)]\, d\omega \tag{4.34a}
$$

$$
\begin{aligned}
b_{l,l'}^{s,s'}(\nu) &= \frac{1}{\pi} \int_0^{\infty} |\psi_s(\omega)||\psi_{s'}(\omega)| \omega \sin \\
&\quad \times \left[\omega \left(\nu + \frac{1}{2} + \frac{l'}{2^{s'}} - \frac{l}{2^s} - \frac{1}{2^{s'+1}} + \frac{1}{2^{s+1}} \right) \right] d\omega
\end{aligned}
\tag{4.34b}
$$

$$
c_{s,l}(\nu) = \frac{1}{\pi} \int_0^{\infty} \widetilde{\phi}(\omega|\widetilde{\psi}_s(\omega)| \omega \sin \left[\omega \left(\nu + \frac{1}{2} - \frac{l}{2^s} - \frac{1}{2^{s+1}} \right) \right] d\omega \tag{4.34c}
$$

$$d_{s,l}(\nu) = \frac{1}{\pi} \int_0^\infty \widetilde{\phi}(\omega|\widetilde{\psi}_s(\omega)|\omega \sin\left[\omega\left(\nu - \frac{1}{2} + \frac{l}{2^s} + \frac{1}{2^{s+1}}\right)\right] d\omega \quad (4.34d)$$

The numerical values of these coefficients are given in Tables 4.2a–4.2c (for cases of $s = 0$ and $s = 1$).

We next substitute (4.31) through (4.33) into (4.28) to derive the update equations for the current I:

$$\begin{aligned} {}_{\phi}I_i^{n+1} &= \left(\frac{L - R\,\Delta t}{L + R\,\Delta t}\right) {}_{\phi}I_i^{n-1} - \frac{1}{L + R\,\Delta t} \\ &\quad \times \sum_{\nu=-\infty}^{+\infty} \left({}_{\phi}V_{i+\nu}^{n} a(\nu) + \sum_{s=0}^{s_{\max}} \sum_{l=0}^{2^s-1} {}_{\psi}^{s,l}V_{i+\nu-1/2+2^{-s}(1/2+l)}^{n} d_{s,l}(\nu) \right) \frac{\Delta t}{\Delta x} \end{aligned} \quad (4.35a)$$

$$\begin{aligned} {}_{\psi}^{s,l}I_{i+2^{-s}(1/2+l)}^{n+1} &= \left(\frac{L - R\,\Delta t}{L + R\,\Delta t}\right) {}_{\psi}^{s,l}I_{i+2^{-s}(1/2+l)}^{n-1} \\ &\quad - \frac{1}{L + R\,\Delta t} \left[\sum_{\nu=-\infty}^{+\infty} \left({}_{\phi}V_{i+\nu+1/2}^{n} c_{s,l}(\nu) \right.\right. \\ &\quad \left.\left. + \sum_{s'=0}^{s_{\max}} \sum_{l'=0}^{2^s-1} {}_{\psi}^{s',l'}V_{i+\nu-1/2+2^{-s'}(1/2+l')}^{n} b_{l,l'}^{s,s'}(\nu) \right) \right] \frac{\Delta t}{\Delta x} \end{aligned} \quad (4.35b)$$

TABLE 4.2a Integral Coefficients for $a(\nu)$, $b_{l,l'}^{s,s'}(\nu)$, and $c_{s,l}(\nu)$ ($s = 0$)

ν	$a(\nu)$	$b_{0,0}^{0,0}(\nu)$	$c_{0,0}(\nu)$
0	1.291846	2.472539	0
1	−0.156076	0.956228	−0.046608
2	0.059639	0.166059	0.054565
3	−0.029310	0.093924	−0.037051
4	0.015372	0.003141	0.020673
5	−0.008189	0.013494	−0.011337
6	0.004379	−0.002859	0.006321
7	−0.002343	0.002779	−0.003846
8	0.001254	−0.001129	0.001714
9	−0.000671	0.000707	−0.000918

Note: $a(-\nu - 1) = -a(\nu)$ and $c_{s,l}(-\nu) = -c_{s,l}(\nu)$.

TABLE 4.2b Integral Coefficients for $d_{s,l}(\nu)$ and $c_{s,l}(\nu)$ $(s = 1)$

ν	$c_{1,0}(\nu)$	$c_{1,1}(\nu)$	$d_{1,0}(\nu)$	$d_{1,1}(\nu)$
−8	0.000077791	−0.000107917	−0.000107917	0.000077791
−7	−0.000154260	0.000216187	0.000216187	−0.000154260
−6	0.000313336	−0.000440914	−0.000440914	0.000313336
−5	−0.000625694	0.000906723	0.000906723	−0.000625694
−4	0.001348953	−0.001959836	−0.001959836	0.001348953
−3	−0.002586165	0.004273722	0.004273722	−0.002586165
−2	0.007532622	−0.011329640	−0.011329640	0.007532622
−1	−0.020265614	0.024321565	0.024321565	−0.020265614
0	0.030300911	−0.030300911	−0.030300911	0.030300911
1	−0.024321565	0.020265614	0.020265614	−0.024321565
2	0.011329640	−0.007532622	−0.007532622	0.011329640
3	−0.004273722	0.002586165	0.002586165	−0.004273722
4	0.001959836	−0.001348953	−0.001348953	0.001959836
5	−0.000906723	0.00088486	0.00088486	−0.000906723
6	0.000440914	−0.000313336	−0.000313336	0.00062356
7	−0.000216187	0.000154260	0.000154260	−0.000216187
8	0.000107917	−0.000077791	−0.000077791	0.000107917

Note: $c_{s,l}(-\nu) = -c_{s,l}(\nu)$ and $d_{s,l}(-\nu) = -d_{s,l}(\nu)$.

TABLE 4.2c Integral Coefficients for $b_{l,l'}^{s,s'}(\nu)$ $(s = 1)$

ν	$b_{0,0}^{0,1}(\nu)$	$b_{0,1}^{0,1}(\nu)$	$b_{0,0}^{1,0}(\nu)$	$b_{0,0}^{1,1}(\nu)$	$b_{0,1}^{1,1}(\nu)$
−9	0.000068126	−0.000157979	−0.000157979	0.000062393	−0.000116529
−8	0.000143468	−0.000352128	−0.000352128	0.000217810	−0.000406512
−7	0.001101551	−0.001591902	−0.001591902	0.000760823	−0.001416527
−6	0.002106587	−0.004918699	−0.004918699	0.002663166	−0.004916515
−5	0.010876491	−0.018197365	−0.018197365	0.009390025	−0.016826645
−4	0.030774267	−0.062187652	−0.062187652	0.033934368	−0.054686440
−3	0.121900158	−0.216773548	−0.216773548	0.132668009	−0.142270550
−2	0.394611424	−0.739278550	−0.739278550	0.641536536	0.056428700
−1	1.225829337	−1.609301081	−1.609301081	5.039549683	0.000000000
0	1.609301081	−1.225829337	−1.225829337	−5.039549683	−0.056428700
1	0.739278549	−0.394611424	−0.394611424	−0.641536536	0.142270550
2	0.216773549	−0.121900158	−0.121900158	−0.132668009	0.054686440
3	0.062187653	−0.030774267	−0.030774267	−0.033934368	0.016826645
4	0.018197366	−0.010876491	−0.010876491	−0.009390025	0.004916515
5	0.004918699	−0.002106587	−0.002106587	−0.002663166	0.001416527
6	0.001591902	−0.001101551	−0.001101551	−0.000760823	0.000406512
7	0.0003521281	−0.000143468	−0.000143468	−0.000217810	0.000116529
8	0.0001579794	−0.000068126	−0.000068126	−0.000062393	0.000003339

Note: $b_{0,0}^{1,1}(-\nu) = -b_{0,0}^{1,1}(\nu - 1)$, if $\nu < 0$; $b_{0,1}^{1,1}(-\nu) = -b_{0,1}^{1,1}(\nu - 2)$, if $\nu < -1$; $b_{1,0}^{1,0}(\nu) = b_{0,0}^{0,1}(\nu)$; $b_{1,0}^{1,1}(\nu) = b_{0,1}^{1,1}(\nu)$; and $b_{1,1}^{1,1}(\nu) = b_{0,0}^{1,1}(\nu)$.

Following the same procedure that we used to derive the update equations for the current, we can obtain the update equations for the voltage V:

$$
\begin{aligned}
{}_{\phi}V_i^{n+1} &= \left(\frac{C - G\,\Delta t}{C + G\,\Delta t}\right) {}_{\phi}V_i^{n-1} - \frac{1}{C + G\,\Delta t} \\
&\quad \times \sum_{\nu=-\infty}^{+\infty} \left({}_{\phi}I_{i+\nu}^{n} a(\nu) + \sum_{s=0}^{s_{\max}} \sum_{l=0}^{2^s-1} {}_{\psi}^{s,l}I_{i+\nu-1/2+2^{-s}(1/2+l)}^{n} d_{s,l}(\nu) \right) \frac{\Delta t}{\Delta x}
\end{aligned} \tag{4.36a}
$$

$$
\begin{aligned}
{}_{\psi}^{s,l}V_{i+2^{-s}(1/2+l)}^{n+1} &= \left(\frac{C - G\,\Delta t}{C + G\,\Delta t}\right) {}_{\psi}^{s,l}V_{i+2^{-s}(1/2+l)}^{n} \\
&\quad - \frac{1}{C + G\,\Delta t} \left[\sum_{\nu=-\infty}^{+\infty} \left({}_{\phi}I_{i+\nu+1/2}^{n} c_{s,l}(\nu) \right.\right. \\
&\quad \left.\left. + \sum_{s'=0}^{s_{\max}} \sum_{l'=0}^{2^{s}-1} {}_{\psi}^{s',l'}I_{i+\nu-1/2+2^{-s'}(1/2+l')}^{n} b_{l,l'}^{s,s'}(\nu) \right) \right] \frac{\Delta t}{\Delta x}
\end{aligned} \tag{4.36b}
$$

We see from (4.35a) and (4.35b) that if we choose the order of resolution $s_{\max} = 0$, the update equations are simplified to

$$
\begin{aligned}
{}_{\phi}I_i^{n+1} &= \left(\frac{L - R\,\Delta t}{L + R\,\Delta t}\right) {}_{\phi}I_i^{n-1} \\
&\quad - \frac{1}{L + R\,\Delta t} \sum_{\nu=-\infty}^{+\infty} \left({}_{\phi}V_{i+\nu}^{n} a(\nu) + {}_{\psi}^{0,0}I_{i+\nu}^{n} d_{0,0}(\nu) \right) \frac{\Delta t}{\Delta x}
\end{aligned} \tag{4.37a}
$$

$$
\begin{aligned}
{}_{\psi}^{0,0}I_i^{n+1} &= \left(\frac{L - R\,\Delta t}{L + R\,\Delta t}\right) {}_{\psi}^{0,0}I_i^{n-1} - \frac{1}{L + R\,\Delta t} \\
&\quad \times \left[\sum_{\nu=-\infty}^{+\infty} \left({}_{\phi}V_{i+\nu+1/2}^{n} c_{0,0}(\nu) + {}_{\psi}^{0,0}I_{i+\nu}^{n} b_{0,0}^{0,0}(\nu) \right) \right] \frac{\Delta t}{\Delta x}
\end{aligned} \tag{4.37b}
$$

Moreover, if we only use the scaling functions in the dominant update equations and neglect the wavelet contributions, we obtain

$$
{}_{\phi}I_i^{n+1} = \left(\frac{L - R\,\Delta t}{L + R\,\Delta t}\right) {}_{\phi}I_i^{n-1} - \frac{1}{L + R\,\Delta t} \sum_{\nu=-\infty}^{+\infty} {}_{\phi}V_{i+\nu}^{n} a(\nu) \frac{\Delta t}{\Delta x} \tag{4.38}
$$

Using a similar method and considering only the scaling functions, we can obtain the update equation for the voltage V, which reads

$$
{}_{\phi}V_i^{n+1} = \left(\frac{C - G\,\Delta t}{C + G\,\Delta t}\right) {}_{\phi}V_i^{n-1} - \frac{1}{C + G\,\Delta t} \sum_{\nu=-\infty}^{+\infty} {}_{\phi}I_{i+\nu}^{n} a(\nu) \frac{\Delta t}{\Delta x} \tag{4.39}
$$

We note that (4.38) and (4.39) are quite similar to (4.9) and (4.10), which were derived previously using the FDTD and MoM methods, respectively. In fact, when we designate the coefficients $\alpha(\nu)$ in (4.38) and (4.39) as -1 and 1, (4.38) and (4.39) become identical to (4.9) and (4.10).

In general, as seen above, the MRTD scheme is a superset of the MoM and FDTD methods. When the basis functions are chosen as pulse functions, both the MoM and MRTD approaches reduce to the FDTD update algorithm.

4.3 THE MRTD SCHEME

As discussed earlier, in the application of the MRTD we need to expand an arbitrary function $f(x)$ in the frame of the MRA. As seen in the previous example, we find that the key step in the application of the MRTD scheme is to select the basis functions, including both the scaling and wavelet functions. This section develops a general MRTD treatment for solving the Maxwell equations and for deriving the generalized MRTD update equations with an arbitrary multiresolution level s in a region containing both dielectrics and conductors. For simplicity, we expand only the wavelet components along one direction, the y-direction. Here we assume that readers possess basic knowledge of the FDTD method and the MoM. This background is needed to follow the discussion presented in the rest of this section.

Let us begin with the Maxwell curl equations:

$$\nabla \times \vec{E} = -\mu_0 \overline{\overline{\mu}}_r \cdot \frac{\partial \vec{H}}{\partial t} - \overline{\overline{\sigma}}_m \cdot \vec{H} \tag{4.40}$$

$$\nabla \times \vec{H} = \varepsilon_0 \overline{\overline{\varepsilon}}_r \cdot \frac{\partial \vec{E}}{\partial t} + \overline{\overline{\sigma}}_e \cdot \vec{E} \tag{4.41}$$

where $\overline{\overline{\varepsilon}}_r, \overline{\overline{\mu}}_r, \overline{\overline{\sigma}}_e$, and $\overline{\overline{\sigma}}_m$ are material parameters defined in diagonal matrix forms in Cartesian coordinates. For instance,

$$\overline{\overline{a}} = [a] = \begin{bmatrix} a_{xx} & 0 & 0 \\ 0 & a_{xx} & 0 \\ 0 & 0 & a_{xx} \end{bmatrix}, \quad a = \varepsilon_r, \mu_{\mathrm{r}}, \sigma_e, \text{ or } \sigma_m \tag{4.42}$$

Our next step is to use the generalized differential matrix operators (GDMOs) [8] (see Appendix A) to express the Maxwell equations as

$$\begin{bmatrix} 0 & -\partial_z & \partial_y \\ \partial_z & 0 & -\partial_x \\ -\partial_y & \partial_x & 0 \end{bmatrix} \begin{bmatrix} H_x \\ H_y \\ H_z \end{bmatrix} = \varepsilon_0 \begin{bmatrix} \varepsilon_{xx} & 0 & 0 \\ 0 & \varepsilon_{yy} & 0 \\ 0 & 0 & \varepsilon_{zz} \end{bmatrix} \frac{\partial}{\partial t} \begin{bmatrix} E_x \\ E_y \\ E_z \end{bmatrix} + \begin{bmatrix} \sigma_{exx} & 0 & 0 \\ 0 & \sigma_{eyy} & 0 \\ 0 & 0 & \sigma_{ezz} \end{bmatrix} \begin{bmatrix} E_x \\ E_y \\ E_z \end{bmatrix} \tag{4.43}$$

$$\begin{bmatrix} 0 & -\partial_z & \partial_y \\ \partial_z & 0 & -\partial_x \\ -\partial_y & \partial_x & 0 \end{bmatrix} \begin{bmatrix} E_x \\ E_y \\ E_z \end{bmatrix} = -\mu_0 \begin{bmatrix} \mu_{xx} & 0 & 0 \\ 0 & \mu_{yy} & 0 \\ 0 & 0 & \mu_{zz} \end{bmatrix} \frac{\partial}{\partial t} \begin{bmatrix} H_x \\ H_y \\ H_z \end{bmatrix} - \begin{bmatrix} \sigma_{mxx} & 0 & 0 \\ 0 & \sigma_{myy} & 0 \\ 0 & 0 & \sigma_{mzz} \end{bmatrix} \begin{bmatrix} E_x \\ E_y \\ E_z \end{bmatrix} \tag{4.44}$$

All the field quantities in the SWy-MRTD scheme can then be expanded as

$$E_x(\vec{r},t) = \sum_{n,i,j,k=-\infty}^{+\infty} \left[{}_{\phi x}E^n_{i+1/2,j,k}\phi_j(y) + \sum_{s=0}^{s_{\max}} \sum_{l=0}^{2^s-1} {}^{s,l}_{\psi x}E^n_{i+1/2j+2^{-s}(l+1/2),k}\psi_{s,j+2^{-s}(l+1/2)}(y) \right] \times \phi_{i+1/2}(x)\phi_k(z)h_n(t) \tag{4.45a}$$

$$E_y(\vec{r},t) = \sum_{n,i,j,k=-\infty}^{+\infty} \left[{}_{\phi y}E^n_{i,j+1/2,k}\phi_{j+1/2}(y) + \sum_{s=0}^{s_{\max}} \sum_{l=0}^{2^s-1} {}^{s,l}_{\psi y}E^n_{i,j+2^{-s}(l+1/2),k}\psi_{s,j+2^{-s}(l+1/2)}(y) \right] \times \phi_i(x)\phi_k(z)h_n(t) \tag{4.45b}$$

$$E_z(\vec{r},t) = \sum_{n,i,j,k=-\infty}^{+\infty} \left[{}_{\phi z}E^n_{i,j,k+1/2}\phi_j(y) + \sum_{s=0}^{s_{\max}} \sum_{l=0}^{2^s-1} {}^{s,l}_{\psi z}E^n_{i,j+2^{-s}(l+1/2),k+1/2}\psi_{s,j+2^{-s}(l+1/2)}(y) \right] \times \phi_i(x)\phi_{k+1/2}(z)h_n(t) \tag{4.45.c}$$

$$H_x(\vec{r},t) = \sum_{n,i,j,k=-\infty}^{+\infty} \left[{}_{\phi x}H^{n+1/2}_{i,j+1/2,k+1/2}\phi_{j+1/2}(y) + \sum_{s=0}^{s_{\max}} \sum_{l=0}^{2^s-1} {}^{s,l}_{\psi x}H^{n+1/2}_{i,j-1/2+2^{-s}(l+1/2),k+1/2}\psi_{s,j-1/2+2^{-s}(l+1/2)}(y) \right] \times \phi_i(x)\phi_{k+1/2}(z)h_{n+1/2}(t) \tag{4.46a}$$

$$H_y(\vec{r},t) = \sum_{n,i,j,k=-\infty}^{+\infty} \left[{}_{\phi y}H^{n+1/2}_{i+1/2,j,k+1/2}\phi_j(y) + \sum_{s=0}^{s_{\max}} \sum_{l=0}^{2^s-1} {}^{s,l}_{\psi y}H^{n+1/2}_{i+1/2,j+2^{-s}(l+1/2),k+1/2}\psi_{s,j+2^{-s}(l+1/2)}(y) \right] \times \phi_{i+1/2}(x)\phi_{k+1/2}(z)h_{n+1/2}(t) \tag{4.46b}$$

$$H_z(\vec{r},t) = \sum_{n,i,j,k=-\infty}^{+\infty} \left[{}_{\phi z}H^{n+1/2}_{i+1/2,j+1/2,k}\phi_{j+1/2}(y) + \sum_{s=0}^{s_{\max}} \sum_{l=0}^{2^s-1} {}^{s,l}_{\psi z}H^{n+1/2}_{i+1/2,j-1/2+2^{-s}(l+1/2),k}\psi_{s,j-1/2+2^{-s}(l+1/2)}(y) \right] \times \phi_{i+1/2}(x)\phi_k(z)h_{n+1/2}(t) \tag{4.46c}$$

where the basis functions are chosen to be the cubic spline Battle–Lemarié scaling and the sth-order resolution level wavelet functions; and $({}_{\phi\kappa}E^{n}_{i,j,k}, {}_{\phi\kappa}H^{n+1/2}_{i,j,k})$ and $({}^{s,l}_{\psi\kappa}E^{n}_{i,j+2^{-s}l,k}, {}^{s,l}_{\psi\kappa}H^{n+1/2}_{i,j+2^{-s}l,k})$ with $\kappa = x, y, z$ are the expansion coefficients, respectively. The indices i, j, k, and n are discrete space and time indices related to the space and time coordinates via $x = i\ \Delta x$, $y = j\ \Delta y$, $z = k\ \Delta z$, and $t = n\ \Delta t$, where Δx, Δy, Δz, and Δt are the space and time discretization intervals in $x-$, $y-$, $z-$, and t-directions.

Figures 4.2 and 4.3 show the scaling function based S-MRTD scheme (without including wavelet expansions) and the SWy-MRTD at $s = 0$, respectively, which, in some degree, are similar to the conventional Yee FDTD lattice [3]. Note that

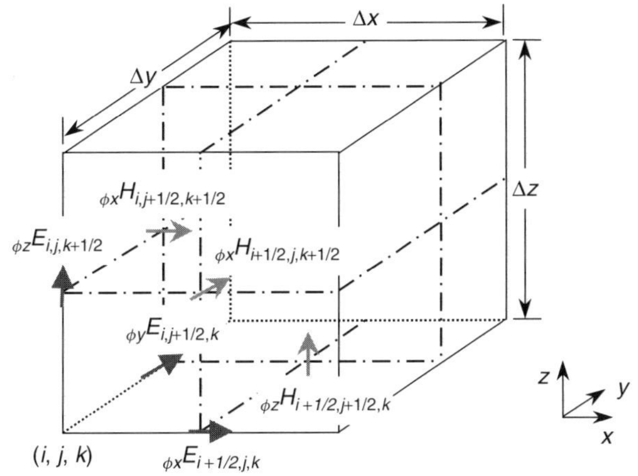

FIGURE 4.2 S-MRTD lattice.

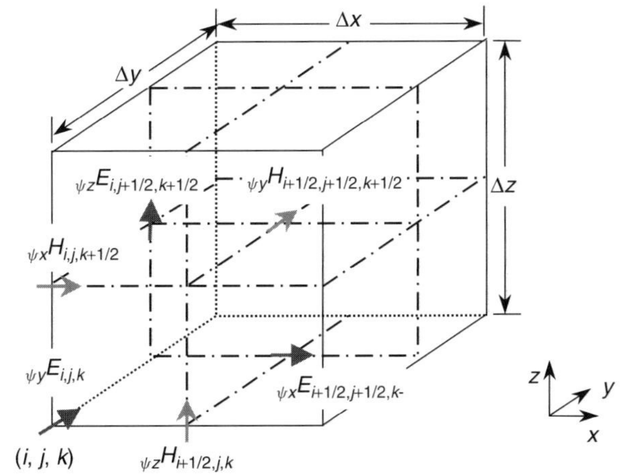

FIGURE 4.3 SW-MRTD ($s = 0$) lattice.

the field components shown in the above lattices are not total fields, which will be discussed later in this chapter.

The procedure for deriving all six scalar equations in (4.45) and (4.46) is quite similar. Hence, for the sake of illustration, we derive only one component of the update equations for the electric and magnetic fields, for instance, their x-components. We assume $\sigma_m = 0$, $\mu_r = 1$ and define ε_r and σ_e as follows:

$$\varepsilon_{\alpha\alpha} = \varepsilon_r(\kappa), \quad \alpha_1(\kappa) \le \alpha \le \alpha_2(\kappa), \quad (\alpha = x, y, z) \tag{4.47}$$

$$\sigma_{e\alpha\alpha} = \begin{cases} \sigma_e(\kappa), & \alpha_1(\kappa) \le \alpha \le \alpha_2(\kappa) \\ 0, & \text{otherwise} \end{cases} \quad (\alpha = x, y, z) \tag{4.48}$$

where $\kappa = 0, 1, 2, \ldots, \kappa_{\max}$ is the number of rectangular objects in the computational volume.

Let us consider the two x-components of H_x and E_x:

$$\frac{\partial E_z}{\partial y} - \frac{\partial E_y}{\partial z} = -\mu_0 \frac{\partial H_x}{\partial t} \tag{4.49}$$

$$\frac{\partial H_z}{\partial y} - \frac{\partial H_y}{\partial z} = \varepsilon_0 \varepsilon_{xx} \frac{\partial E_x}{\partial t} + \sigma_{exx} E_x \tag{4.50}$$

Substituting (4.45) and (4.46) into (4.49), and sampling and testing the equation with the functions $\phi_i(x)\phi_{j+1/2}(y)\phi_{k+1/2}(z)h_n(t)$ and $\phi_i(x)\psi_{s,j-1/2+2^{-s}(l+1/2)}(y)\phi_{k+1/2}(z)h_n(t)$, respectively, we can derive the following update equations of ${}_{\phi x}H$ and ${}^{s,l}_{\psi x}H$ as

$$\begin{aligned} {}_{\phi x}H^{n+1/2}_{i,j+1/2,k+1/2} &= {}_{\phi x}H^{n-1/2}_{i,j+1/2,k+1/2} + \frac{1}{\mu_0} \sum_{v=-\infty}^{+\infty} a(v)\, {}_{\phi y}E^{n}_{i,j+1/2,k+v+1} \frac{\Delta t}{\Delta z} \\ &\quad - \frac{1}{\mu_0} \left(\sum_{v=-\infty}^{+\infty} a(v)\, {}_{\phi z}E^{n}_{i,j+v+1,k+1/2} \right. \\ &\quad \left. + \sum_{v=-\infty}^{+\infty} \sum_{s=0}^{s_{\max}} \sum_{l=0}^{2^s-1} d_{s,l}(v)\, {}^{s,l}_{\psi z}E^{n}_{i,j+v+2^{-s}(l+1/2),k+1/2} \right) \frac{\Delta t}{\Delta y} \end{aligned} \tag{4.51}$$

$$\sum_{s=0}^{s_{\max}} \sum_{l=0}^{2^s-1} {}^{s,l}_{\psi x}H^{n+1/2}_{i,j-1/2+2^{-s}(l+1/2),k}$$

$$= \sum_{s=0}^{s_{\max}} \sum_{l=0}^{2^s-1} {}^{s,l}_{\psi x}H^{n-1/2}_{i,j-1/2+2^{-s}(l+1/2),k} + \frac{1}{\mu_0} \sum_{v=-\infty}^{+\infty} a(v)\, {}^{s,l}_{\psi y}E^{n}_{i,j+2^{-s}(l+1/2),k+v+1} \frac{\Delta t}{\Delta z}$$

$$
-\frac{1}{\mu_0}\left(\sum_{\nu=-\infty}^{+\infty} c_{s,l}(\nu)_{\phi z} E^{n}_{i,j+\nu,k+1/2}\right.
$$

$$
\left.-\sum_{\nu=-\infty}^{+\infty}\sum_{s'=0}^{s_{\max}}\sum_{l=0}^{2^{s}-1} b^{s,s'}_{l,l'}(\nu)^{s',l'}_{\psi z} E^{n}_{i,j+\nu+2^{-s'}(l'+1/2),k+1/2}\right)\frac{\Delta t}{\Delta y} \tag{4.52}
$$

where the coefficients $d_{s,l}(\nu)$ and $c_{s,l}(\nu)$ are defined as follows:

$$
\int_{-\infty}^{+\infty} \phi_{j+1/2}(y)\frac{\partial \psi_{s,j'+2^{-s}(l+1/2)}(y)}{\partial y}\,dy = \sum_{\nu=-\infty}^{+\infty} d_{s,l}(\nu)\delta_{j',j+\nu} \tag{4.53}
$$

$$
\int_{-\infty}^{+\infty} \psi_{s,j-1/2+2^{-s}(l+1/2)}(y)\frac{\partial \phi_{j'}(y)}{\partial y}\,dy = \sum_{\nu=-\infty}^{\nu=+\infty} c_{s,l}(\nu)\delta_{j',j+\nu} \tag{4.54}
$$

$$
\int_{+\infty}^{+\infty} \psi_{s,i-1/2+2^{-s}(l+1/2)}(x)\frac{\partial \psi_{s,i'+2^{-s'}(l'+1/2)}(x)}{\partial x}\,dx = \sum_{\nu=-\infty}^{+\infty} b^{s,s'}_{l,l'}(\nu)\delta_{i',i+\nu} \tag{4.55}
$$

The expressions in (4.51) and (4.52) are a group of equations that relate to the sth-order resolution of wavelet functions with different values of $l(l = 0, 1, \ldots, 2^s - 1)$.

We now consider the case of $s_{\max} = 0$, in (4.51), and derive the following update equations of the scaling and the lowest level wavelet components:

$$
{}_{\phi x}H^{n+1/2}_{i,j+1/2,k+1/2} = {}_{\phi x}H^{n-1/2}_{i,j+1/2,k+1/2} + \frac{1}{\mu_0}\sum_{\nu=-\infty}^{+\infty} a(\nu)_{\phi y} E^{n}_{i,j+1/2,k+\nu+1}\frac{\Delta t}{\Delta z}
$$

$$
-\frac{1}{\mu_0}\left(\sum_{\nu=-\infty}^{+\infty} a(\nu)_{\phi z} E^{n}_{i,j+\nu+1,k+1/2}\right.
$$

$$
\left.+\sum_{\nu=-\infty}^{+\infty} d_{0,0}(\nu)^{0,0}_{\psi z} E^{n}_{i,j+\nu+1/2}\right)\frac{\Delta t}{\Delta y} \tag{4.56}
$$

$$
{}^{0,0}_{\psi x}H^{n+1/2}_{i,j,k} = {}^{0,0}_{\psi x}H^{n-1/2}_{i,j,k} + \frac{1}{\mu_0}\sum_{\nu=-\infty}^{+\infty} a(\nu)^{0,0}_{\psi y} E^{n}_{i,j+1/2,k+\nu+1}\frac{\Delta t}{\Delta z}
$$

$$
-\frac{1}{\mu_0}\left(\sum_{\nu=-\infty}^{+\infty} c_{0,0}(\nu)_{\phi z} E^{n}_{i,j+\nu,k+1/2}\right.
$$

$$
\left.-\sum_{\nu=-\infty}^{+\infty} b^{0,0}_{0,0}(\nu)^{0,0}_{\psi z} E^{n}_{i,j+\nu+1/2,k+1/2}\right)\frac{\Delta t}{\Delta y} \tag{4.57}
$$

If $s_{\max} = 1$ in (4.51) and (4.52), the index s in the summation can take values of both $s = 0$ and $s = 1$, and the corresponding values of l in the equations are $l = 0$ and $l = 0, 1$, respectively. Derived from (4.51), the H-field update equation for the scaling function components is

$$
\begin{aligned}
{}_{\phi x}H^{n+1/2}_{i,j+1/2,k+1/2} &= {}_{\phi x}H^{n-1/2}_{i,j+1/2,k+1/2} + \frac{1}{\mu_0}\sum_{\nu=-\infty}^{+\infty} a(\nu)\,{}_{\phi y}E^{n}_{i,j+1/2,k+\nu+1}\frac{\Delta t}{\Delta z} \\
&\quad - \frac{1}{\mu_0}\left(\sum_{\nu=-\infty}^{+\infty} a(\nu)\,{}_{\phi z}E^{n}_{i,j+\nu+1,k+1/2}\right. \\
&\quad \left. + \sum_{\nu=-\infty}^{+\infty}\sum_{s=0}^{1}\sum_{l=0}^{1} d_{s,l}(\nu)\,{}^{s,l}_{\psi z}E^{n}_{i,j+\nu+2^{-s}(l+1/2),k+1/2}\right)\frac{\Delta t}{\Delta y}
\end{aligned}
\tag{4.58}
$$

In contrast to the scaling components, the H-field update equation for the wavelet components with $s = 1$ splits into four update equations corresponding to different values of l as shown in (4.52).

For the case of $s = 0$ and $l = 0$, (4.57) results in

$$
\begin{aligned}
{}^{0,0}_{\psi x}H^{n+1/2}_{i,j,k} &= {}^{0,0}_{\psi x}H^{n-1/2}_{i,j,k} + \frac{1}{\mu_0}\sum_{\nu=-\infty}^{+\infty} a(\nu)\,{}^{0,0}_{\psi y}E^{n}_{i,j+1/2,k+\nu+1}\frac{\Delta t}{\Delta z} \\
&\quad - \frac{1}{\mu_0}\left(\sum_{\nu=-\infty}^{+\infty} c_{0,\,0}(\nu)\,{}_{\phi z}E^{n}_{i,j+\nu,k+1/2}\right. \\
&\quad \left. - \sum_{\nu=-\infty}^{+\infty}\sum_{s'=0}^{1}\sum_{l=0}^{1} b^{0,s'}_{0,l'}(\nu)\,{}^{s',l'}_{\psi z}E^{n}_{i,j+\nu+2^{-s'}(l'+1/2),k+1/2}\right)\frac{\Delta t}{\Delta y}
\end{aligned}
\tag{4.59}
$$

For $s = 1$, $l = 0$,

$$
\begin{aligned}
{}^{1,0}_{\psi x}H^{n+1/2}_{i,j-1/4,k} &= {}^{1,0}_{\psi x}H^{n-1/2}_{i,j-1/4,k} + \frac{1}{\mu_0}\sum_{\nu=-\infty}^{+\infty} a(\nu)\,{}^{1,0}_{\psi y}E^{n}_{i,j+1/4,k+\nu+1}\frac{\Delta t}{\Delta z} \\
&\quad - \frac{1}{\mu_0}\left(\sum_{\nu=-\infty}^{+\infty} c_{1,\,0}(\nu)\,{}_{\phi z}E^{n}_{i,j+\nu,k+1/2}\right. \\
&\quad \left. - \sum_{\nu=-\infty}^{+\infty}\sum_{s'=0}^{1}\sum_{l=0}^{1} b^{1,s'}_{0,l'}(\nu)\,{}^{s',l'}_{\psi z}E^{n}_{i,j+\nu+2^{-s'}(l'+1/2),k+1/2}\right)\frac{\Delta t}{\Delta y}
\end{aligned}
\tag{4.60}
$$

For $s = 1$, $l = 1$,

$$
{}^{1,1}_{\psi x}H^{n+1/2}_{i,j+1/4,k} = {}^{1,1}_{\psi x}H^{n-1/2}_{i,j+1/4,k} + \frac{1}{\mu_0}\sum_{\nu=-\infty}^{+\infty} a(\nu)\,{}^{1,1}_{\psi y}E^{n}_{i,j+3/4,k+\nu+1}\frac{\Delta t}{\Delta z}
$$

$$
-\frac{1}{\mu_0}\left(\sum_{\nu=-\infty}^{+\infty} c_{1,1}(\nu)_{\phi z}E^n_{i,j+\nu,k+1/2}\right.
$$

$$
\left.-\sum_{\nu=-\infty}^{+\infty}\sum_{s'=0}^{1}\sum_{l=0}^{1} b^{1,s'}_{1,l'}(\nu)^{s',l'}_{\psi z}E^n_{i,j+\nu+2^{-s'}(l'+1/2),k+1/2}\right)\frac{\Delta t}{\Delta y} \tag{4.61}
$$

Similarly, substituting (4.45) and (4.46) into (4.50), sampling and testing it with the functions $\phi_{i+1/2}(x)\phi_j(y)\phi_k(z)h_{n+1/2}(t)$ and $\phi_{i+1/2}(x)\psi_{s,j+2^{-s}(l+1/2)}(y)$ $\phi_k(z)h_{n+1/2}(t)$, respectively, we can obtain the update equations of ${}_{\phi x}E$ and ${}^{s,l}_{\psi x}E$, which read

$$
{}_{\phi x}E^{n+1}_{i+1/2,j,k} = \frac{A^-_\phi}{A^+_\phi}{}_{\phi x}E^n_{i+1/2,j,k}
$$

$$
= \frac{1}{A^+_\phi\varepsilon_0}\left(\sum_{\nu=-\infty}^{+\infty} a(\nu)_{\phi z}H^{n+1/2}_{i+1/2,j+\nu+1/2,k}\right.
$$

$$
\left.+\sum_{\nu=-\infty}^{+\infty}\sum_{s=0}^{s_{\max}}\sum_{l=0}^{2^S-1} c_s(\nu)^{s,l}_{\psi z}E^n_{i,j+\nu-1/2+2^{-S}(l+1/2),k}\right)\frac{\Delta t}{\Delta y}
$$

$$
-\frac{1}{A^+_\phi\varepsilon_0}\sum_{\nu=-\infty}^{+\infty} a(\nu)_{\phi y}H^{n+1/2}_{i+1/2,j,k+\nu+1/2}\frac{\Delta t}{\Delta z} \tag{4.62}
$$

where the coefficients $A^\pm$ are defined as follows:

$$
A^+_\phi = 1-\sum_{\kappa=0}^{\kappa_{\max}}\left[(\varepsilon_r(\kappa)-1)+\frac{\sigma_e(\kappa)\Delta t}{2\varepsilon_0}\right]\alpha^{\phi\phi}_{i+1/2,i+1/2}(\kappa)\alpha^{\phi\phi}_{j,j}(\kappa)\alpha^{\phi\phi}_{k,k}(\kappa) \tag{4.63a}
$$

$$
A^-_\phi = 1-\sum_{\kappa=0}^{\kappa_{\max}}\left[(\varepsilon_r(\kappa)-1)-\frac{\sigma_e(\kappa)\Delta t}{2\varepsilon_0}\right]\alpha^{\phi\phi}_{i+1/2,i+1/2}(\kappa)\alpha^{\phi\phi}_{j,j}(\kappa)\alpha^{\phi\phi}_{k,k}(\kappa) \tag{4.63b}
$$

with

$$
\alpha^{\phi\phi}_{i+1/2,i+1/2}(\kappa) = \frac{1}{\Delta x}\int_{x_1(\kappa)}^{x_2(\kappa)} \phi_{i+1/2}(x)\phi_{i+1/2}(x)\,dx \tag{4.64a}
$$

$$
\alpha^{\phi\phi}_{j,j}(\kappa) = \frac{1}{\Delta y}\int_{y_1(\kappa)}^{y_2(\kappa)} \phi_j(y)\phi_j(y)\,dy \tag{4.64b}
$$

Note that the diagonal approximation has been used in the above derivation. Specifically, $\alpha^{\phi\phi}_{i+1/2,i'+1/2}(\kappa)\alpha^{\phi\phi}_{j,j'}(\kappa)\alpha^{\phi\phi}_{k,k'}(\kappa)$ has been approximated to $\alpha^{\phi\phi}_{i+1/2,i'+1/2}$

TABLE 4.3 Coefficients $\alpha^{\phi\phi}_{i,i+m}(x_1 = 0, x_2 = 5\,\Delta x)$

m / $\alpha_{i,i+m}$ / i	−2	−1	0	1	2
0	−0.00054	0.00137	0.99787	0.00137	−0.00054
1	−0.00043	0.00137	0.99777	0.00164	−0.00116
2	−0.00054	0.00164	0.99716	0.00299	−0.00433
3	−0.00116	0.00299	0.99418	0.01005	−0.04105
4	−0.00433	0.01005	0.97718	0.09609	−0.00017
5	−0.04105	0.09609	0.49896	−0.00948	−0.04071

TABLE 4.4 Coefficients $\alpha^{\phi\phi}_{i+1/2,i+1/2+m}$ $(x_1 = 0, x_2 = 5\,\Delta x)$

m / $\alpha_{i+1/2,\,i+1/2+m}$ / i	−2	−1	0	1	2
0	−0.00042	0.00132	0.99788	0.00139	−0.00061
1	−0.00042	0.00194	0.99769	0.00183	−0.00165
2	−0.00061	0.00183	0.99673	0.00416	−0.00876
3	−0.00165	0.00416	0.99098	0.02252	−0.02178
4	−0.00876	0.02252	0.92732	0.00063	0.00021

$(\kappa)\alpha^{\phi\phi}_{j,j'}(\kappa)\alpha^{\phi\phi}_{k,k'}(\kappa)\delta_{i,i'}\delta_{j,j'}\delta_{k,k'}$ by considering the localization of scaling and wavelet functions. The expansion coefficients are summarized in Tables 4.3 and 4.4.

From the calculated coefficients $\alpha^{\phi\phi}_{i,i+m}$, $\alpha^{\phi\phi}_{i,i'}$, $\alpha^{\phi\phi}_{i+1/2,i+1/2+m}$, and $\alpha^{\phi\phi}_{i+1/2,i'+1/2}$, we find primarily that the diagonal elements, which are the elements with the same two subscripts, have values that are two orders of magnitude higher than the off-diagonal elements. Thus, the off-diagonal elements contribute little to the updated fields and, in practice, it is convenient to use the diagonal approximation to simplify the update equations for regions containing both dielectric and conducting materials. We have

$$\sum_{s=0}^{s_{\max}}\sum_{l=0}^{2^s-1} {}_{\psi x}^{s,l}E^{n+1}_{i+1/2,j+2^{-s}(l+1/2),k}$$

$$= \frac{A^-_\psi}{A^+_\psi}\sum_{s=0}^{s_{\max}}\sum_{l=0}^{2^s-1} {}_{\psi x}^{s,l}E^{n}_{i+1/2,j+2^{-s}(l+1/2),k} + \frac{1}{A^+_\psi\varepsilon_0}\left(\sum_{\nu=-\infty}^{+\infty} c_s(\nu)_{\phi z}H^{n+1/2}_{i+1/2,j+\nu+1/2,k}\right.$$

$$+ \sum_{\nu=-\infty}^{+\infty} \sum_{s'=0}^{s'_{\max}} \sum_{l'=0}^{2^{s'}-1} b_{l,l'}^{s,s'}(\nu)_{\psi z}^{s',l'} H^{n}_{i,j+\nu-1/2+2^{-s'}(l'+1/2),k} \Bigg) \frac{\Delta t}{\Delta y}$$

$$- \frac{1}{A_{\psi}^{+}\varepsilon_0} \sum_{\nu=-\infty}^{+\infty} a(\nu)_{\psi y}^{s,l} H^{n+1/2}_{i+1/2,j+2^{-s}(l+1/2),k+\nu+1/2} \frac{\Delta t}{\Delta z} \tag{4.65}$$

where the coefficients $A^{\pm}$ are defined as

$$A_{\psi}^{+} = 1 - \sum_{\kappa=0}^{\kappa_{\max}} \left[(\varepsilon_r(\kappa) - 1) + \frac{\sigma_e(\kappa)\ \Delta t}{2\varepsilon_0} \right] \times \alpha^{\phi\phi}_{i+1/2,i+1/2}(\kappa) \alpha^{\psi\psi}_{j+2^{-s}(l+1/2),j+2^{-s}(l+1/2)}(\kappa) \alpha^{\phi\phi}_{k,k}(\kappa) \tag{4.66a}$$

$$A_{\psi}^{-} = 1 - \sum_{\kappa=0}^{\kappa_{\max}} \left[(\varepsilon_r(\kappa) - 1) - \frac{\sigma_e(\kappa)\ \Delta t}{2\varepsilon_0} \right] \times \alpha^{\phi\phi}_{i+1/2,i+1/2}(\kappa) \alpha^{\psi\psi}_{j+2^{-s}(l+1/2),j+2^{-s}(l+1/2)}(\kappa) \alpha^{\phi\phi}_{k,k}(\kappa) \tag{4.66b}$$

with

$$\alpha^{\psi\psi}_{j+2^{-s}(l+1/2),j+2^{-s}(l+1/2)}(\kappa) = \frac{1}{\Delta y} \int_{y_1(\kappa)}^{y_2(\kappa)} \psi_{s,j+2^{-s}(l+1/2)}(y) \psi_{s,j+2^{-s}(l+1/2)}(y)\, dy \tag{4.67}$$

Similarly, for the same reasons given above, $\alpha^{\phi\phi}_{i+1/2,i'+1/2}(\kappa)$ $\alpha^{\psi\psi}_{j+2^{-s}(l+1/2),j'+2^{-s'}(l'+1/2)}(\kappa)\alpha^{\phi\phi}_{k,k'}(\kappa)$ is approximated by $\alpha^{\phi\phi}_{i+1/2,i'+1/2}(\kappa)$ $\alpha^{\psi\psi}_{j+2^{-s}(l+1/2),j'+2^{-s'}(l'+1/2)}(\kappa)\alpha^{\phi\phi}_{k,k'}\delta_{i,i'}\delta_{j,j'}\delta_{k,k'}\delta_{s,s'}\delta_{l,l'}$.

The E-field update equations in (4.62) and (4.65) also define a set of equations for a specified s and different values of l, given by $l = 0, 1, \ldots, 2^s$. For example, if $s_{\max} = 0$, the s value is set to 0 and the corresponding $l = 0$. Then (4.62) and (4.65) read

$$_{\phi x}E^{n+1}_{i+1/2,j,k} = \frac{A_{\phi}^{-}}{A_{\phi}^{+}} {}_{\phi x}E^{n}_{i+1/2,j,k}$$

$$= \frac{1}{A_{\phi}^{+}\varepsilon_0} \Bigg(\sum_{\nu=-\infty}^{+\infty} a(\nu)_{\phi z} H^{n+1/2}_{i+1/2,j+\nu+1/2,k}$$

$$+ \sum_{\nu=-\infty}^{+\infty} c_{0,0}(\nu)_{\psi z}^{0,0} E^{n}_{i,j+\nu,k} \Bigg) \frac{\Delta t}{\Delta y}$$

$$- \frac{1}{A_{\phi}^{+}\varepsilon_0} \sum_{\nu=-\infty}^{+\infty} a(\nu)_{\phi y} H^{n+1/2}_{i+1/2,j,k+\nu+1/2} \frac{\Delta t}{\Delta z} \tag{4.68}$$

$$
\begin{aligned}
{}^{0,0}_{\psi x}E^{n+1}_{i+1/2,j+1/2,k} = {} & \frac{A^-_\psi}{A^+_\psi}\,{}^{0,0}_{\psi x}E^{n}_{i+1/2,j+1/2,k} \\
& + \frac{1}{A^+_\psi \varepsilon_0}\left(\sum_{\nu=-\infty}^{+\infty} c_0(\nu)_{\phi z} H^{n+1/2}_{i+1/2,j+\nu+1/2,k}\right. \\
& \left. + \sum_{\nu=-\infty}^{+\infty} b^{0,0}_{0,0}(\nu)\,{}^{0,0}_{\psi z}H^{n}_{i,j+\nu,k}\right)\frac{\Delta t}{\Delta y} \\
& - \frac{1}{A^+_\psi \varepsilon_0}\sum_{\nu=-\infty}^{+\infty} a(\nu)\,{}^{0,0}_{\psi y}H^{n+1/2}_{i+1/2,j+1/2,k+\nu+1/2}\frac{\Delta t}{\Delta z}
\end{aligned}
\tag{4.69}
$$

If $s_{\max} = 1$, the s values can be $s = 0$ and $s = 1$. Equation (4.62) then reads

$$
\begin{aligned}
{}_{\phi x}E^{n+1}_{i+1/2,j,k} = {} & \frac{A^-_\phi}{A^+_\phi}\,{}_{\phi x}E^{n}_{i+1/2,j,k} \\
= {} & \frac{1}{A^+_\phi \varepsilon_0}\left(\sum_{\nu=-\infty}^{+\infty} a(\nu)_{\phi z} H^{n+1/2}_{i+1/2,j+\nu+1/2,k}\right. \\
& \left. + \sum_{\nu=-\infty}^{+\infty}\sum_{s=0}^{1}\sum_{l=0}^{1} c_{s,l}(\nu)\,{}^{s,l}_{\psi z}E^{n}_{i,j+\nu-1/2+2^{-s}(l+1/2),k}\right)\frac{\Delta t}{\Delta y} \\
& - \frac{1}{A^+_\phi \varepsilon_0}\sum_{\nu=-\infty}^{+\infty} a(\nu)_{\phi y} H^{n+1/2}_{i+1/2,j,k+\nu+1/2}\frac{\Delta t}{\Delta z}
\end{aligned}
\tag{4.70}
$$

Correspondingly, (4.65) then splits into three equations with ($s = 0$, $l = 0$), ($s = 1$, $l = 0$), and ($s = 1$, $l = 1$) with three update equations for the wavelet components:

For $s = 0, l = 0$:

$$
\begin{aligned}
{}^{0,0}_{\psi x}E^{n+1}_{i+1/2,j+1/2,k} = {} & \frac{A^-_\psi}{A^+_\psi}\,{}^{0,0}_{\psi x}E^{n}_{i+1/2,j+1/2,k} \\
& + \frac{1}{A^+_\psi \varepsilon_0}\left(\sum_{\nu=-\infty}^{+\infty} c_0(\nu)_{\phi z} H^{n+1/2}_{i+1/2,j+\nu+1/2,k}\right. \\
& \left. + \sum_{\nu=-\infty}^{+\infty}\sum_{s'=0}^{1}\sum_{l'=0}^{1} b^{0,s'}_{0,l'}(\nu)\,{}^{s',l'}_{\psi z}H^{n}_{i,j+\nu-1/2+2^{-s'}(l'+1/2),k}\right)\frac{\Delta t}{\Delta y} \\
& - \frac{1}{A^+_\psi \varepsilon_0}\sum_{\nu=-\infty}^{+\infty} a(\nu)\,{}^{0,0}_{\psi y}H^{n+1/2}_{i+1/2,j+1/2,k+\nu+1/2}\frac{\Delta t}{\Delta z}
\end{aligned}
\tag{4.71}
$$

For $s = 1, l = 0$:

$$
\begin{aligned}
{}^{1,0}_{\psi x}E^{n+1}_{i+1/2,j+1/4,k} &= \frac{A^{-}_{\psi}}{A^{+}_{\psi}}\,{}^{1,0}_{\psi x}E^{n}_{i+1/2,j+1/4,k} \\
&\quad + \frac{1}{A^{+}_{\psi}\varepsilon_0}\left(\sum_{\nu=-\infty}^{+\infty} c_{1,0}(\nu)_{\phi z} H^{n+1/2}_{i+1/2,j+\nu+1/2,k}\right. \\
&\quad \left. + \sum_{\nu=-\infty}^{+\infty}\sum_{s'=0}^{1}\sum_{l'=0}^{1} b^{1,s'}_{0,l'}(\nu)^{s',l'}_{\psi z} H^{n}_{i,j+\nu-1/2+2^{-s'}(l'+1/2),k}\right)\frac{\Delta t}{\Delta y} \\
&\quad - \frac{1}{A^{+}_{\psi}\varepsilon_0}\sum_{\nu=-\infty}^{+\infty} a(\nu)^{1,0}_{\psi y} H^{n+1/2}_{i+1/2,j+1/4,k+\nu+1/2}\frac{\Delta t}{\Delta z}
\end{aligned} \tag{4.72}
$$

For $s = 1, l = 1$:

$$
\begin{aligned}
{}^{1,1}_{\psi x}E^{n+1}_{i+1/2,j+3/4,k} &= \frac{A^{-}_{\psi}}{A^{+}_{\psi}}\,{}^{1,1}_{\psi x}E^{n}_{i+1/2,j+3/4,k} \\
&\quad + \frac{1}{A^{+}_{\psi}\varepsilon_0}\left(\sum_{\nu=-\infty}^{+\infty} c_{1,1}(\nu)_{\phi z} H^{n+1/2}_{i+1/2,j+\nu+1/2,k}\right. \\
&\quad \left. + \sum_{\nu=-\infty}^{+\infty}\sum_{s'=0}^{1}\sum_{l'=0}^{1} b^{1,s'}_{1,l'}(\nu)^{s',l'}_{\psi z} H^{n}_{i,j+\nu-1/2+2^{-s'}(l'+1/2),k}\right)\frac{\Delta t}{\Delta y} \\
&\quad - \frac{1}{A^{+}_{\psi}\varepsilon_0}\sum_{\nu=-\infty}^{+\infty} a(\nu)^{1,1}_{\psi y} H^{n+1/2}_{i+1/2,j+3/4,k+\nu+1/2}\frac{\Delta t}{\Delta z}
\end{aligned} \tag{4.73}
$$

As a result, for $s_{\max} = 1$ we also have four update equations with different values of l.

Until now, we have derived the MRTD update equations with an arbitrary resolution level for both the E- and H-fields. If the field expansions include only the scaling functions, then there are only two sets of update equations for the E- and H-fields; however, if the expansions contain both the scaling and wavelet functions with $s_{\max} = 0$, then the number of equations is restricted to two sets. If $s_{\max} = 1$, the number of equations increases to four, and if the multiresolution level $s_{\max} = 2$, then the number of equations increases to eight. It is apparent that both the mathematical complexity and computational burden increase dramatically as $s_{\max}$ is increased. On the other hand, a large value of $s_{\max}$ yields better resolution and greater accuracy, which is usually needed for geometries with fine features. In practice, we attempt to strike a balance between accuracy and efficiency when using the MRTD scheme.

4.4 STABILITY CRITERIA

In the FDTD method, it is necessary to impose the Courant condition, which dictates [9] the time step Δt, to ensure a stable solution. The condition is expressed as

$$\Delta t_{\text{FDTD}} \leq \frac{1}{v_{\max}\sqrt{\dfrac{1}{(\Delta x)^2} + \dfrac{1}{(\Delta y)^2} + \dfrac{1}{(\Delta z)^2}}} \tag{4.74}$$

where $v_{\max}$ represents the maximum phase velocity that exists in the lattice. For the case of $\Delta x = \Delta y = \Delta z = \Delta$, it simplifies to

$$\Delta t_{\text{FDTD}} \leq \frac{\Delta}{v_{\max}\sqrt{3}} \tag{4.75}$$

The stability analysis of the MRTD scheme is described in [1], and it is based on the solution of an eigenvalue problem in time and space. The numerical dispersion relation is given by

$$\left[\frac{1}{c\,\Delta t}\sin(\omega\,\Delta t/2)\right]^2 = \sum_{i=1}^{3}\left[\frac{1}{\Delta i}\sum_{\nu=0}^{+10} a(\nu)\sin(k_i\,\Delta i(\nu+\tfrac{1}{2}))\right]^2, \qquad i = x, y, x \tag{4.76}$$

or

$$\begin{aligned}\left[\frac{1}{c\,\Delta t}\sin(\omega\Delta t/2)\right]^2 &= \left[\frac{1}{\Delta x}\sum_{\nu=0}^{+10} a(\nu)\sin\left(k_x\Delta x\left(\nu+\tfrac{1}{2}\right)\right)\right]^2 \\ &+ \left[\frac{1}{\Delta y}\sum_{\nu=0}^{+10} a(\nu)\sin\left(k_y\Delta y\left(\nu+\tfrac{1}{2}\right)\right)\right]^2 \\ &+ \left[\frac{1}{\Delta z}\sum_{\nu=0}^{+10} a(\nu)\sin\left(k_z\Delta z\left(\nu+\tfrac{1}{2}\right)\right)\right]^2\end{aligned} \tag{4.77}$$

where $a(\nu)$ is defined in (4.34a), which satisfies the symmetrical relation $a(-1-\nu) = -a(\nu)$. Solution of the eigenvalue problem of (4.77) leads to the satiability condition for the time step Δt of the S-MRTD scheme [10, 11]:

$$\Delta t_{\text{S-MRTD}} \leq \frac{1}{v_{\max}\sum_{\nu=0}^{+10}|a(\nu)|\sqrt{\dfrac{1}{(\Delta x)^2} + \dfrac{1}{(\Delta y)^2} + \dfrac{1}{(\Delta z)^2}}} \tag{4.78}$$

When $\Delta x = \Delta y = \Delta z = \Delta$, (4.78) reduces to

$$\begin{aligned}\Delta t_{\text{S-MRTD}} &\leq \frac{\Delta}{v_{\max}\sum_{\nu=0}^{+10} a(\nu)\sqrt{3}} \\ &\approx \frac{\Delta}{1.5684\sqrt{3}v_{\max}} \\ &= 0.368112\frac{\Delta}{v_{\max}}\end{aligned} \tag{4.79}$$

For the SW-MRTD scheme (for the resolution level $s = 0$), the stability condition is given by [11]

$$\Delta t_{\text{SW-MRTD}} \leq \chi \frac{1}{v_{\max}\sqrt{\dfrac{1}{(\Delta x)^2} + \dfrac{1}{(\Delta y)^2} + \dfrac{1}{(\Delta z)^2}}} \tag{4.80}$$

where the coefficient χ is defined as

$$\chi = \frac{2}{\sqrt{3(A)^2 + (B)^2 + 2(C)^2 + (AB^+)\sqrt{(AB^-)^2 + 4(C)^2}}} \tag{4.81}$$

and

$$A = \sum_{\nu} |a(\nu)| \tag{4.82a}$$

$$B = \sum_{\nu} |b_{0,0}(\nu)| \tag{4.82b}$$

$$C = \sum_{\nu} |c_0(\nu)| \tag{4.82c}$$

$$AB^+ = \sum_{\nu} |a(\nu) + b_{0,0}(\nu)| \tag{4.82d}$$

$$AB^- = \sum_{\nu} |a(\nu) - b_{0,0}(\nu)| \tag{4.82e}$$

If $\Delta x = \Delta y = \Delta z = \Delta$, (4.80) becomes

$$\begin{aligned}\Delta t_{\text{SW-MRTD}} &\leq \chi \frac{\Delta}{v_{\max}\sqrt{3}} \\ &= 0.3433\frac{\Delta}{v_{\max}\sqrt{3}} \\ &= 0.19821\frac{\Delta}{v_{\max}}\end{aligned} \tag{4.83}$$

For the case of $s > 0$, we may follow the same procedure described in [11] to solve for the stability condition for the SW-MRTD scheme with an arbitrary level of s; in this case we expect a more complicated relation to define the stability condition.

4.5 COMPUTATION OF TOTAL FIELDS

The only quantities that require time-updating in the field expansions (4.45) and (4.46) are the field expansion coefficients. For the S-MRTD and SW-MRTD schemes, the total field at a particular point in space must be calculated from the field expansion formulations. For example, we may calculate the x-component of the total electric field $E_x(\vec{r}_0, t_0) = E_x(x_0, y_0, z_0, t_0)$ at an arbitrary spatial location $\vec{r}_0 = (i_0\ \Delta x, j_0\ \Delta x, k_0\ \Delta x)$ and time point $t_0 = n_0\ \Delta t$ with $|t/\Delta t - t_0| < \frac{1}{2}(s = 0)$:

$$
\begin{aligned}
E_x(\vec{r}_0, t_0) &= \int_{-\infty}^{+\infty} E_x(\vec{r}, t)\delta(x - x_0)\delta(y - y_0)\delta(z - z_0)\delta(t - t_0)\, dx\, dy\, dz\, dt \\
&= \sum_{i',j',k',=-\infty}^{+\infty} [{}_{\phi x}E^n_{i'+1/2,j',k'}\phi_{j'}(y_0) + {}_{\psi x}E^n_{i'+1/2,j'+1/2,k'}\psi_{j'+1/2}(y_0)] \\
&\quad \times \phi_{i'+1/2}(x_0)\phi_{k'}(z_0) \\
&= \sum_{i',j',k',=-\infty}^{+\infty} [{}_{\phi x}E^n_{i'+1/2,j',k'}\phi(j_0 - j') + {}_{\psi x}E^n_{i'+1/2,j'+1/2,k'}\psi(j_0 - j')] \\
&\quad \times \phi\left(i_0 - i' - \tfrac{1}{2}\right)\phi(k_0 - k') \\
&= \sum_{i,j,k,=-\infty}^{+\infty} \left[{}_{\phi x}E^n_{i_0-i,j_0-j,k_0-k}\phi(j) + {}_{\psi x}E^n_{i_0-i,j_0-j+1/2,k_0-k}\psi(j)\right]\phi(i)\phi(k)
\end{aligned}
\tag{4.84}
$$

where

$$
\int_{-\infty}^{+\infty} h_n(t)\delta(t - t_0)\, dt = \int_{-\infty}^{+\infty} h\left(\frac{t}{\Delta t} - n\right)\delta(t - t_0)\, dt = h\left(\frac{t_0}{\Delta t} - n\right) = 1
\tag{4.85}
$$

Because of the exponentially decaying nature of the cubic spline Battle–Lemarié scaling and wavelet functions, we may need only to consider a few terms in the summations for the values of i, j, and k. The values of $\phi(i)$ and $\psi(i)$ for the first nine integer location indices are listed in Table 4.5, and they can be used in (4.84) for calculating the total fields.

TABLE 4.5 Value of $\phi(i)$ and $\psi(i)$

i	$\phi(i)$	$\psi(i)$
0	1.089031	1.259419
1	−0.075209	0.025306
2	0.047311	0.069728
3	−0.025031	−0.017583
4	0.012621	0.004724
5	−0.006315	0.000000
6	0.003169	0.000000
7	−0.001599	0.000000
8	0.000811	0.000000
9	−0.000413	0.000000

4.6 ORTHOGONAL AND INTEGRAL RELATIONS

In the process of deriving the MRTD scheme update equations, we have to employ Galerkin's method. The derivation requires a number of orthogonal and integral relations satisfied by the cubic spline Battle–Lemarié scaling and wavelet functions. We have summarized the most frequently used relationships in Appendix B for the convenience of the readers and present a few representative formulas in the following.

Orthogonal Relations

$$\int_{-\infty}^{+\infty} \phi_i(\varsigma)\phi_{i'}(\varsigma)\,dx = \delta_{i',i}\Delta\varsigma \tag{4.86}$$

$$\int_{-\infty}^{+\infty} \psi_{s,i+2^{-s}(l+1/2)}(\varsigma)\psi_{s',i'+2^{-s'}(l'+1/2)}(\varsigma)\,d\varsigma = \delta_{s,s'}\delta_{l,l'}\delta_{i,i'}\Delta\varsigma \tag{4.87}$$

$$\int_{-\infty}^{+\infty} \phi_i(\varsigma)\psi_{i'+2^{-s}(l+1/2)}(\varsigma)\,d\varsigma = 0 \tag{4.88}$$

$$\int_{-\infty}^{+\infty} h_{n'}(t)h_n(t)\,dt = \delta_{n',n}\Delta t \tag{4.89}$$

Integral Relations

$$\int_{-\infty}^{+\infty} h_{n'}(t)\frac{\partial}{\partial t}h_{n+1/2}(t)\,dt = \delta_{n',n} - \delta_{n',n+1} \tag{4.90}$$

$$\int_{-\infty}^{\infty} \phi_i(\varsigma)\frac{\partial \phi_{i'}(\varsigma)}{\partial \varsigma} d\varsigma = \sum_{\nu=-\infty}^{+\infty} a(\nu)\delta_{i',i+\nu} \tag{4.91}$$

$$\int_{-\infty}^{\infty} \phi_{i+1/2}(\varsigma)\frac{\partial \psi_{s,i'+2^{-s}(l+1/2)}(\varsigma)}{\partial \varsigma} d\varsigma = \sum_{\nu=-\infty}^{+\infty} d_{s,l}(\nu)\delta_{i',i+\nu} \tag{4.92}$$

$$\int_{-\infty}^{\infty} \psi_{s,i-1/2+2^{-s}(l+1/2)}(\varsigma)\frac{\partial \phi_{i'}(\varsigma)}{\partial \varsigma} d\varsigma = \sum_{\nu=-\infty}^{+\infty} c_{s,l}(\nu)\delta_{i',i+\nu+1/2} \tag{4.93}$$

$$\int_{-\infty}^{\infty} \psi_{s,i+2^{-s}(l+1/2)}(\varsigma)\frac{\partial \psi_{s',i'+2^{-s'}(l'+1/2)}(\varsigma)}{\partial \varsigma} d\varsigma = \sum_{\nu=-\infty}^{+\infty} b_{l,l'}^{s,s'}(\nu)\delta_{i',i+\nu-1/2} \tag{4.94}$$

$$\int_{-\infty}^{\infty} \psi_{s,i-1/2+2^{-s}(l+1/2)}(x)\frac{\partial \psi_{s',i'+2^{-s'}(l'+1/2)}(x)}{\partial x} dx = \sum_{\nu=-\infty}^{+\infty} b_{l,l'}^{s,s'}(\nu)\delta_{i',i+\nu} \tag{4.95}$$

REFERENCES

[1] M. Krumpholz and L. P. B. Katehi, "New prospects for time domain analysis," *IEEE Microwave Guided Wave Lett.*, vol. 5, no. 11, pp. 382–384, Nov. 1995.

[2] M. Krumpholz and L. P. B. Katehi, "MRTD: New time domain schemes based on multiresolution analysis," *IEEE Trans. Microwave Theory Tech.*, vol. 44, no. 4, pp. 555–571, Apr. 1996.

[3] K. S. Yee, "Numerical solution of initial boundary value problem involving Maxwell's equations in isotropic media," *IEEE Trans. Antennas Propag.*, vol. AP-14, no. 3, pp. 302–307, May 1966.

[4] R. F. Harrington, *Field Computation by Moment Methods*, Macmillan, New York, 1968.

[5] Q. Cao, Y. Chen, and R. Mittra, "Multiple image technique (MIT) and anistropic perfectly matched layer (APML) in implementation of MRTD scheme for boundary truncations of microwave structures," *IEEE Trans. Microwave Theory Tech.*, vol. 50, no. 6, pp. 1578–1589, June 2002.

[6] C. D. Sarris, L. P. B. Katehi, and J. F. Harvey, "Application of multiresolution analysis to the modeling of microwave and optical structures," *Opt. Quantum Electron.*, (invited paper), vol. 32, pp. 657–679, Aug. 2000.

[7] J. C. Strikwerda, *Finite Difference Schemes and Partial Differential Equations*, Wadsworth, Belmont, CA, 1989.

[8] Y. Chen, K. Sun, B. Berker, and R. Mittra, "Unified matrix presentation of Maxwell's and wave equations using generalized different differential matrix operators," *IEEE Trans. Educ.*, vol. 41, pp. 61–69, Feb. 1998.

[9] A. Taflove, *Computational Electrodynamics*, Artech House, Norwood, MA, 1995.

[10] I. Hong, N. Yoon, and H. Park, "Numerical dispersive characteristics and stability condition of the multi-resolution time-domain (MRTD) method," *Electromagnetic Compatibility Proceedings, International Symposium of 1997*, pp. 455–458, 1997.

[11] E. M. Tentzeris, R. L. Robertson, J. F. Harvey, and L. P. B. Ketehi, "Stability and dispersion analysis of Battle–Lemarié-based MRTD schemes," *IEEE Trans. Microwave Theory Tech.*, vol. 47, no. 7, July 1999.

CHAPTER FIVE

PEC Boundary Truncations

5.1 INTRODUCTION

Since the MRTD scheme uses an expanded basis function, it cannot be applied directly to a localized boundary condition. For example, we cannot directly enforce tangential electric fields to be zero at a perfectly conducting boundary. An image technique, introduced by Krumpholz and Katehi [1, 2], has been developed to model a perfect electric conductor (PEC) or perfect magnetic conductor (PMC) boundary truncation technique in the context of the MRTD scheme. The procedure for generating the images on different sides of a structure is neither obvious nor straightforward, particularly when complex materials and multiple images are involved.

A systematic MRTD approach for PEC-shielded boundary truncation is presented in this chapter, referred to herein as the MRTD multiple image technique (MRTD-MIT) [3]. The MRTD approach described in this chapter retains the philosophy of the leapfrog algorithm employed in the conventional FDTD; also, the MRTD scheme is placed in the context of the FDTD. However, the MRTD-MIT approach is innovative in its application of image theory in this MRTD scheme for boundary truncation of PEC-shielded structures. Based on the MIT, an automatic image generator (AIG) is further introduced to generate images in a systematic manner and to reduce unnecessary computer processing [4]. The primary motivation behind introducing the MIT is to provide a flexible tool for varying the number of side images and thereby to balance accuracy and computer resources. An obvious advantage is that additional images do not require additional memory in the MRTD-MIT implementation since only the fields defined in the original structures are necessary to generate them.

In this chapter, an application of the MIT is illustrated by analyzing a dielectric-loaded cavity and by developing a systematic formulation for constructing the

Multiresolution Time Domain Scheme for Electromagnetic Engineering
By Yinchao Chen, Qunsheng Cao, and Raj Mittra
ISBN 0-471-27230-2

constitutive relations and update equations in the MRTD transform domain with the use of the field quantities that are defined in the real structures only. Furthermore, a criterion for determining the number of images needed along each side of the structure is presented, which is based on the localization properties of the scaling basis functions. Although, in principle, we can construct as many images as we wish, it is highly desirable to limit this number, from the point of view of balancing the numerical efficiency and accuracy.

5.2 METHOD OF ANALYSIS

MRTD Update Equations

The Maxwell curl equations can be written as

$$\nabla \times \vec{E} = -\mu_0 \frac{\partial \vec{H}}{\partial t} \tag{5.1}$$

$$\nabla \times \vec{H} = \frac{\partial \vec{D}}{\partial t} \tag{5.2}$$

with the constitutive relation

$$\vec{D} = \varepsilon(\vec{r})\vec{E} \tag{5.3}$$

In the MRTD-transformed domain, the above local relationship between the $\vec{D}$- and $\vec{E}$- fields is no longer valid; hence, we have to work simultaneously with both the $\vec{D}$- and $\vec{E}$-fields.

For simplicity, we consider only the S-MRTD scheme. In this case, all the field quantities (D- and H-fields) are expanded in terms of the scaling function $\phi_i(x)$ in space and pulse function $h_n(t)$ in time:

$$D_x(\vec{r},t) = \sum_{i,j,k,n=-\infty}^{\infty} {}_{\phi x}D^{n}_{i+1/2,j,k}\phi_{i+1/2}(x)\phi_j(y)\phi_k(z)h_n(t) \tag{5.4a}$$

$$D_y(\vec{r},t) = \sum_{i,j,k,n=-\infty}^{\infty} {}_{\phi y}D^{n}_{i,j+1/2,k}\phi_i(x)\phi_{j+1/2}(y)\phi_k(z)h_n(t) \tag{5.4b}$$

$$D_z(\vec{r},t) = \sum_{i,j,k,n=-\infty}^{\infty} {}_{\phi z}D^{n}_{i,j,k+1/2}\phi_i(x)\phi_j(y)\phi_{k+1/2}(z)h_n(t) \tag{5.4c}$$

$$H_x(\vec{r},t) = \sum_{i,j,k,n=-\infty}^{\infty} {}_{\phi x}H^{n+1/2}_{i,j+1/2,k+1/2}\phi_i(x)\phi_{j+1/2}(y)\phi_{k+1/2}(z)h_{n+1/2}(t) \tag{5.5a}$$

$$H_y(\vec{r},t) = \sum_{i,j,k,n=-\infty}^{\infty} {}_{\phi y}H^{n+1/2}_{i+1/2,j,k+1/2}\phi_{i+1/2}(x)\phi_j(y)\phi_{k+1/2}(z)h_{n+1/2}(t) \tag{5.5b}$$

$$H_z(\vec{r},t) = \sum_{i,j,k,n=-\infty}^{\infty} {}_{\phi z}H^{n+1/2}_{i+1/2,j+1/2,k}\phi_{i+1/2}(x)\phi_{j+1/2}(y)\phi_k(z)h_{n+1/2}(t) \tag{5.5c}$$

Substitution of the above field expansions into the Maxwell equations, followed by the application of Galerkin's method [5], leads to the following set of update equations:

$$\phi_x D^{n+1}_{i+1/2,j,k} = \phi_x D^{n}_{i+1/2,j,k} + \sum_{\nu} a(\nu)\left[\phi_z H^{n+1/2}_{i+1/2,j+\nu+1/2,k}\frac{\Delta t}{\Delta y} - \phi_y H^{n+1/2}_{i+1/2,j,k+\nu+1/2}\frac{\Delta t}{\Delta z}\right] \tag{5.6a}$$

$$\phi_y D^{n+1}_{i,j+1/2,k} = \phi_x D^{n}_{i,j+1/2,k} + \sum_{\nu} a(\nu)\left[\phi_x H^{n+1/2}_{i,j+1/2,k+\nu+1/2}\frac{\Delta t}{\Delta z} - \phi_z H^{n+1/2}_{i+\nu+1/2,j+1/2,k}\frac{\Delta t}{\Delta x}\right] \tag{5.6b}$$

$$\phi_z D^{n+1}_{i,j,k+1/2} = \phi_z D^{n}_{i,j,k+1/2} + \sum_{\nu} a(\nu)\left[\phi_y H^{n+1/2}_{i+\nu+1/2,j,k+1/2}\frac{\Delta t}{\Delta x} - \phi_x H^{n+1/2}_{i,j+\nu+1/2,k+1/2}\frac{\Delta t}{\Delta y}\right] \tag{5.6c}$$

$$\phi_x H^{n+1/2}_{i,j+1/2,k+1/2} = \phi_x H^{n-1/2}_{i,j+1/2,k+1/2} + \frac{1}{\mu_0}\sum_{\nu} a(\nu)\left[\phi_y E^{n}_{i,j+1/2,k+\nu+1}\frac{\Delta t}{\Delta z} - \phi_z E^{n}_{i,j+\nu+1,k+1/2}\frac{\Delta t}{\Delta y}\right] \tag{5.7a}$$

$$\phi_y H^{n+1/2}_{i+1/2,j,k+1/2} = \phi_y H^{n-1/2}_{i+1/2,j,k+1/2} + \frac{1}{\mu_0}\sum_{\nu} a(\nu)\left[\phi_z E^{n}_{i+\nu+1,j,k+1/2}\frac{\Delta t}{\Delta x} - \phi_x E^{n}_{i+1/2,j,k+\nu+1}\frac{\Delta t}{\Delta z}\right] \tag{5.7b}$$

$$\phi_z H^{n+1/2}_{i+1/2,j+1/2,k} = \phi_z H^{n-1/2}_{i+1/2,j+1/2,k} + \frac{1}{\mu_0}\sum_{\nu} a(\nu)\left[\phi_x E^{n}_{i+1/2,j+\nu+1,k}\frac{\Delta t}{\Delta y} - \phi_y E^{n}_{i+\nu+1,j+1/2,k}\frac{\Delta t}{\Delta x}\right] \tag{5.7c}$$

where the quantities being updated are the expansion coefficients in the field expansions; their corresponding superscripts and subscripts represent the discretized time and space positions. The coefficient $a(\nu)$, which appears in (5.6) and (5.7), can be derived by utilizing an integral relationship [1]. Although the summation index ν in (5.6) and (5.7) spans from positive to negative infinities, it is usually sufficient to truncate it at 10, by taking advantage of the localized nature of the cubic spline Battle–Lemarié scaling function. Here, we would like to point out that if the basis functions $\phi(x)$ in the expansion functions of D

and H components are not chosen to be the cubic spline Battle–Lemarié scaling functions, we have to use a different coefficient $a(\nu)$ and different truncation number N. For example, if we use Daubechies' or the CDFwavelet basis functions, the index ν in the summations in the above equations will be truncated with a different integer N. At the same time, we have to reevaluate the coefficient $a(\nu)$ by using corresponding orthogonal relations of the wavelet functions.

MRTD-MIT Constitutive Relationship

The constitutive relationship in the transformed domain of the MRTD is considerably more involved than its counterpart in the spatial domain presented in (5.3). This is because a MRTD-transformed D-field at a particular point is determined not only by the E-field at the same location, but also by the distribution of the E-fields in its neighborhood. To solve for the constitutive relation in the MRTD domain, we once again represent the E-fields as expansions in terms of the same basis functions as in (5.4):

$$E_x(\vec{r}, t) = \sum_{i,j,k,n=-\infty}^{\infty} {}_{\phi x}E^n_{i+1/2,j,k}\phi_{i+1/2}(x)\phi_j(y)\phi_k(z)h_n(t) \tag{5.8a}$$

$$E_y(\vec{r}, t) = \sum_{i,j,k,n=-\infty}^{\infty} {}_{\phi y}E^n_{i,j+1/2,k}\phi_i(x)\phi_{j+1/2}(y)\phi_k(z)h_n(t) \tag{5.8b}$$

$$E_z(\vec{r}, t) = \sum_{i,j,k,n=-\infty}^{\infty} {}_{\phi z}E^n_{i,j,k+1/2}\phi_i(x)\phi_j(y)\phi_{k+1/2}(z)h_n(t) \tag{5.8c}$$

Then we can substitute the above expansions into (5.3). For example, application of Galerkin's method to the x-component of (5.3) leads to the constitutive relationships in the MRTD domain:

$$ {}_{\phi x}D^n_{i+1/2,j,k} = \varepsilon_0 \sum_{i',j',k'=-\infty}^{+\infty} (\varepsilon_{xx})_{ii',jj',kk'} \, {}_{\phi x}E^n_{i'+1/2,j',k'} \tag{5.9}$$

where

$$\begin{aligned}(\varepsilon_{xx})_{ii',jj',kk'} &= \int_{-\infty}^{+\infty}\int_{-\infty}^{+\infty}\int_{-\infty}^{+\infty} \phi_{i+1/2}(x)\phi_j(y)\phi_k(z)\varepsilon_{xx} \\ &\quad \times (\vec{r})\phi_{i'+1/2}(x)\phi_{j'}(y)\phi_k(z)\frac{dx\,dy\,dz}{\Delta x\ \Delta y\ \Delta z} \qquad (5.10) \\ &= \delta_{i,i'}\delta_{j,j'}\delta_{k,k'} + \sum_{\kappa=0}^{\kappa_{\max}}[\varepsilon_{xx}(\kappa) - 1]\alpha^{\phi\phi}_{i+1/2,i'+1/2}(\kappa)\alpha^{\phi\phi}_{j,j'}(\kappa)\alpha^{\phi\phi}_{k,k'}(\kappa)\end{aligned}$$

$$\alpha_{i+1/2,i'+1/2}^{\phi\phi}(\kappa) = \int_{x_1(\kappa)}^{x_2(\kappa)} \phi_{i+1/2}(x)\phi_{i'+1/2}(x)\frac{dx}{\Delta x} \tag{5.11a}$$

$$\alpha_{j,j'}^{\phi\phi}(\kappa) = \int_{y_1(\kappa)}^{y_2(\kappa)} \phi_j(y)\phi_{j'}(y)\frac{dy}{\Delta y} \tag{5.11b}$$

$$\alpha_{k,k'}^{\phi\phi}(\kappa) = \int_{z_1(\kappa)}^{z_2(\kappa)} \phi_k(z)\phi_{k'}(z)\frac{dz}{\Delta z} \tag{5.11c}$$

where $\kappa (\kappa = 1, 2, \ldots, \kappa_{\max})$ is the current number of dielectric objects and all dielectric objects are assumed to be rectangular in shape.

Now we illustrate a derivation for the constitutive relationship for a dielectric-loaded cavity shown in Figure 5.1. The relative permittivity inside the cavity is described as follows:

$$\varepsilon_{\alpha\alpha}(x, y, z) = \varepsilon_{\alpha\alpha}(y) = \begin{cases} \varepsilon_r, & y_1(1) \leq y < y_2(1) \\ 1, & y_1(2) \leq y \leq y_2(2) \end{cases} \tag{5.12}$$

where $\alpha = x$, y, and z. Then the x-component of the constitutive relation in the MRTD domain is derived as

$$_{\phi x}D_{i+1/2,j,k}^{n} = \varepsilon_0 \sum_{j'=j-M_0}^{j+M_0} (\varepsilon_{xx})_{j,j'\ \phi x}E_{i+1/2,j',k}^{n} \tag{5.13}$$

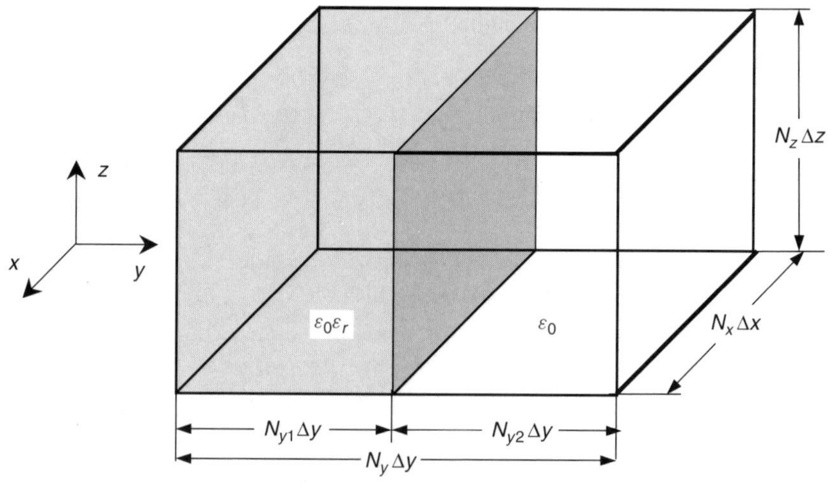

FIGURE 5.1 Geometry of a half-filled dielectric cavity.

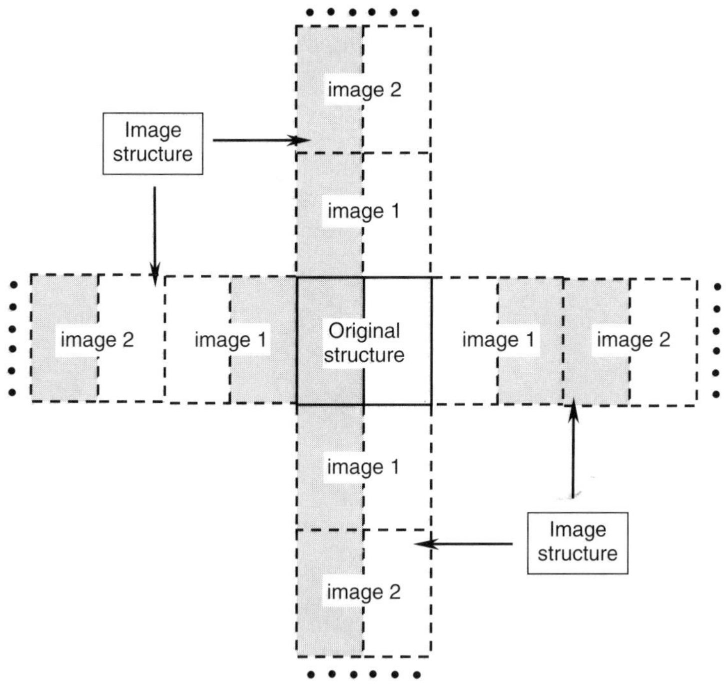

FIGURE 5.2 Original structure with multiple images in the y-z plane.

with

$$(\varepsilon_{xx})_{j,j'} = \delta_{j,j'} + \sum_{\kappa=1}^{2}[\varepsilon_r(\kappa) - 1] \int_{y_1(\kappa)}^{y_2(\kappa)} \phi_j(y)\phi_{j'}(y)\frac{dy}{\Delta y} \tag{5.14}$$

where M_0 is the number of sampling points near the location j in the MRTD domain and κ is the summation index for the total number of dielectric objects. M_0 can be chosen as a value depending on the localization of the scaling function and is usually set equal to a number slightly greater than 9 when the cubic spline Battle–Lemarié scaling function is applied.

It is very important to note that the summations in (5.13) cover not only the real structure region but also all image regions. Now let us look at a half-filled dielectric-loaded cavity to illustrate the MIT. Figure 5.1 shows the geometrical configuration of the cavity structure, and Figure 5.2 displays the original structure and its multiple images in the y-z plane.

For a PEC-shielded structure, all field quantities in the image regions can be expressed in terms of ones inside the original cavity structure. Toward this end, we employ the following principle:

The tangential components of the image E-fields (parallel to the PEC mirror wall) and the normal components of the image H-fields (normal to the PEC mirror wall) are always odd-symmetric relative to the original fields, while the normal components of the image E-fields and the tangential components of the image H-fields are even-symmetric.

Using the guidelines above, we can express all image field quantities in terms of the fields in the original region, and this leads to a degeneration of the summation index in (5.13) to N_y, the grid number along the y-direction, defined only inside the original cavity structure. Consequently, (5.13) can be simplified to read

$$ {}_{\phi x}D^n_{i+1/2,j,k} = \varepsilon_0 \sum_{j'=0}^{N_y} (\hat{\varepsilon}_{xx})_{j,j'} {}_{\phi x}E^n_{i+1/2,j',k} \tag{5.15}$$

where we define the equivalent permittivity matrix in the MRTD domain as

$$ \sum_{j'=0}^{N_y} (\hat{\varepsilon}_{xx})_{j,j} = \sum_{j'=0}^{N_y} (\varepsilon_{xx})_{j,j'} - \sum_{j'=-1}^{-N_y} (\varepsilon_{xx})_{j,j'} - \sum_{j'=N_y+1}^{2N_y} (\varepsilon_{xx})_{j,j'} - \cdots + (-1)^m \sum_{j'=-(m-1)N_y-1}^{j-M_0} (\varepsilon_{xx})_{j,j'} + (-1)^m \sum_{j'=mN_y+1}^{j+M_0} (\varepsilon_{xx})_{j,j'} \tag{5.16a}$$

or in a matrix form as

$$ [\hat{\varepsilon}_{xx}] = [\varepsilon_{xx}]_{\text{Orig}} - [\varepsilon_{xx}]_{\text{Img}(-1)} - [\varepsilon_{xx}]_{\text{Img}(+1)} - \cdots + (-1)^m [\varepsilon_{xx}]_{\text{Img}(-m)} + (-1)^m [\varepsilon_{xx}]_{\text{Img}(+m)} \tag{5.16b}$$

where m is the number of images and $m > 0$. The first term in (5.16a) or (5.16b) corresponds to the original structure, and the rest are associated with the image regions

Similarly, we can derive the y- and z-components of the update equations:

$$ {}_{\phi y}D^n_{i,j+1/2,k} = \varepsilon_0 \sum_{j'=0}^{N_y} (\hat{\varepsilon}_{yy})_{j,j'} {}_{\phi y}E^n_{i,j'+1/2,k} \tag{5.17}$$

$$ \sum_{j'=0}^{N_y} (\hat{\varepsilon}_{yy})_{j,j} = \sum_{j'=0}^{N_y-1} (\varepsilon_{yy})_{j,j'} - \sum_{j'=-1}^{-N_y} (\varepsilon_{yy})_{j,j'} - \sum_{j'=N_y}^{2N_y-1} (\varepsilon_{yy})_{j,j'} - \cdots + (-1)^m \sum_{j'=-(m-1)N_y}^{j-M_0} (\varepsilon_{yy})_{j,j'} + (-1)^m \sum_{j'=mN_y}^{j+M_0} (\varepsilon_{yy})_{j,j'} \tag{5.18a}$$

$$ [\hat{\varepsilon}_{yy}] = [\varepsilon_{yy}]_{\text{Orig}} - [\varepsilon_{yy}]_{\text{Img}(-1)} - [\varepsilon_{yy}]_{\text{Img}(+1)} - \cdots + (-1)^m [\varepsilon_{yy}]_{\text{Img}(-m)} + (-1)^m [\varepsilon_{yy}]_{\text{Img}(+m)} \tag{5.18b}$$

$$
{}_{\phi z}D^{n}_{i,j,k+1/2} = \varepsilon_0 \sum_{j'=0}^{N_y} (\hat{\varepsilon}_{zz})_{j,j'} \, {}_{\phi z}E^{n}_{i,j',k+1/2} \tag{5.19}
$$

$$
\begin{aligned}
\sum_{j'=0}^{N_y} (\hat{\varepsilon}_{zz})_{j,j} &= \sum_{j'=0}^{N_y} (\varepsilon_{zz})_{j,j'} - \sum_{j'=-1}^{-N_y} (\varepsilon_{zz})_{j,j'} - \sum_{j'=N_y+1}^{2N_y} (\varepsilon_{zz})_{j,j'} \\
&\quad - \cdots + (-1)^m \sum_{j'=-(m-1)N_y-1}^{j-M_0} (\varepsilon_{zz})_{j,j'} + (-1)^m \sum_{j'=mN_y+1}^{j+M_0} (\varepsilon_{zz})_{j,j'}
\end{aligned} \tag{5.20a}
$$

$$
\begin{aligned}
[\hat{\varepsilon}_{zz}] &= [\varepsilon_{zz}]_{\text{Orig}} - [\varepsilon_{zz}]_{\text{Img}(-1)} - [\varepsilon_{zz}]_{\text{Img}(+1)} \\
&\quad - \cdots + (-1)^m [\varepsilon_{zz}]_{\text{Img}(-m)} + (-1)^m [\varepsilon_{zz}]_{\text{Img}(+m)}
\end{aligned} \tag{5.20b}
$$

Theoretically, parallel PEC plates should have an infinite number of images. However, in the application of the MRTD-MIT technique, we only need to use a few images since the scaling functions are localized. In the following, we present a criterion for determining m, the number of images, along the $+y$ or $-y$ directions. The criterion is derived from the effective range of the basis functions at the PEC boundary locations:

$$
m = \begin{cases} INT\left[\dfrac{M_0}{N_y}\right] + 1 & \text{if } \dfrac{M_0}{N_y} \text{ is not an integer} \\ \dfrac{M_0}{N_y} & \text{if } \dfrac{M_0}{N_y} \text{ is an integer} \end{cases}, \tag{5.21}
$$

where INT is the integer-converting function that truncates the decimal part of a number. For example, if $M_0 = 10$, $N_y = 4$, $m = INT(2.5) + 1 = 3$, we may need to add three images along both sides of the $+y$ and $-y$ directions.

Now, let us illustrate the procedure for deriving the equivalent permittivity matrix in the MRTD domain. Here we consider a cavity with the discretization grid number $N_y = 6$ in the y-direction. For the sake of simplifying the notation, we omit the subscript x and write $(\varepsilon_{xx})_{j,j}$ simply as $\varepsilon_{j,j'}$. In this example, we choose the values $M_0 = 9$ and $m = 2$ for the equation in (5.16b), yielding

$$
\begin{bmatrix} {}_{\phi x}D^{n}_{i+1/2,0,k} \\ {}_{\phi x}D^{n}_{i+1/2,1,k} \\ {}_{\phi x}D^{n}_{i+1/2,2,k} \\ {}_{\phi x}D^{n}_{i+1/2,3,k} \\ {}_{\phi x}D^{n}_{i+1/2,4,k} \\ {}_{\phi x}D^{n}_{i+1/2,5,k} \\ {}_{\phi x}D^{n}_{i+1/2,6,k} \end{bmatrix} = \varepsilon_0 \begin{bmatrix} \hat{\varepsilon}_{0,0} & \hat{\varepsilon}_{0,1} & \hat{\varepsilon}_{0,2} & \hat{\varepsilon}_{0,3} & \hat{\varepsilon}_{0,4} & \hat{\varepsilon}_{0,5} & \hat{\varepsilon}_{0,6} \\ \hat{\varepsilon}_{1,0} & \hat{\varepsilon}_{1,1} & \hat{\varepsilon}_{1,2} & \hat{\varepsilon}_{1,3} & \hat{\varepsilon}_{1,4} & \hat{\varepsilon}_{1,5} & \hat{\varepsilon}_{1,6} \\ \hat{\varepsilon}_{2,0} & \hat{\varepsilon}_{2,1} & \hat{\varepsilon}_{2,2} & \hat{\varepsilon}_{2,3} & \hat{\varepsilon}_{2,4} & \hat{\varepsilon}_{2,5} & \hat{\varepsilon}_{2,6} \\ \hat{\varepsilon}_{3,0} & \hat{\varepsilon}_{3,1} & \hat{\varepsilon}_{3,2} & \hat{\varepsilon}_{3,3} & \hat{\varepsilon}_{3,4} & \hat{\varepsilon}_{3,5} & \hat{\varepsilon}_{3,6} \\ \hat{\varepsilon}_{4,0} & \hat{\varepsilon}_{4,1} & \hat{\varepsilon}_{4,2} & \hat{\varepsilon}_{4,3} & \hat{\varepsilon}_{4,4} & \hat{\varepsilon}_{4,5} & \hat{\varepsilon}_{4,6} \\ \hat{\varepsilon}_{5,0} & \hat{\varepsilon}_{5,1} & \hat{\varepsilon}_{5,2} & \hat{\varepsilon}_{5,3} & \hat{\varepsilon}_{5,4} & \hat{\varepsilon}_{5,5} & \hat{\varepsilon}_{5,6} \\ \hat{\varepsilon}_{6,0} & \hat{\varepsilon}_{6,1} & \hat{\varepsilon}_{6,2} & \hat{\varepsilon}_{6,3} & \hat{\varepsilon}_{6,4} & \hat{\varepsilon}_{6,5} & \hat{\varepsilon}_{6,6} \end{bmatrix} \begin{bmatrix} {}_{\phi x}E^{n}_{i+1/2,0,k} \\ {}_{\phi x}E^{n}_{i+1/2,1,k} \\ {}_{\phi x}E^{n}_{i+1/2,2,k} \\ {}_{\phi x}E^{n}_{i+1/2,3,k} \\ {}_{\phi x}E^{n}_{i+1/2,4,k} \\ {}_{\phi x}E^{n}_{i+1/2,5,k} \\ {}_{\phi x}E^{n}_{i+1/2,6,k} \end{bmatrix} \tag{5.22}
$$

where

$$[\varepsilon_{xx}]_{\text{Orig}} = \begin{bmatrix} \varepsilon_{0,0} & \varepsilon_{0,1} & \varepsilon_{0,2} & \varepsilon_{0,3} & \varepsilon_{0,4} & \varepsilon_{0,5} & \varepsilon_{0,6} \\ \varepsilon_{1,0} & \varepsilon_{1,1} & \varepsilon_{1,2} & \varepsilon_{1,3} & \varepsilon_{1,4} & \varepsilon_{1,5} & \varepsilon_{1,6} \\ \varepsilon_{2,0} & \varepsilon_{2,1} & \varepsilon_{2,2} & \varepsilon_{2,3} & \varepsilon_{2,4} & \varepsilon_{2,5} & \varepsilon_{2,6} \\ \varepsilon_{3,0} & \varepsilon_{3,1} & \varepsilon_{3,2} & \varepsilon_{3,3} & \varepsilon_{3,4} & \varepsilon_{3,5} & \varepsilon_{3,6} \\ \varepsilon_{4,0} & \varepsilon_{4,1} & \varepsilon_{4,2} & \varepsilon_{4,3} & \varepsilon_{4,4} & \varepsilon_{4,5} & \varepsilon_{4,6} \\ \varepsilon_{5,0} & \varepsilon_{5,1} & \varepsilon_{5,2} & \varepsilon_{5,3} & \varepsilon_{5,4} & \varepsilon_{5,5} & \varepsilon_{5,6} \\ \varepsilon_{6,0} & \varepsilon_{6,1} & \varepsilon_{6,2} & \varepsilon_{6,3} & \varepsilon_{6,4} & \varepsilon_{6,5} & \varepsilon_{6,6} \end{bmatrix} \tag{5.23a}$$

$$[\varepsilon_{xx}]_{\text{Img}(-1)} = \begin{bmatrix} 0 & \varepsilon_{0,-1} & \varepsilon_{0,-2} & \varepsilon_{0,-3} & \varepsilon_{0,-4} & \varepsilon_{0,-5} & \varepsilon_{0,-6} \\ 0 & \varepsilon_{1,-1} & \varepsilon_{1,-2} & \varepsilon_{1,-3} & \varepsilon_{1,-4} & \varepsilon_{1,-5} & \varepsilon_{1,-6} \\ 0 & \varepsilon_{2,-1} & \varepsilon_{2,-2} & \varepsilon_{2,-3} & \varepsilon_{2,-4} & \varepsilon_{2,-5} & \varepsilon_{2,-6} \\ 0 & \varepsilon_{3,-1} & \varepsilon_{3,-2} & \varepsilon_{3,-3} & \varepsilon_{3,-4} & \varepsilon_{3,-5} & \varepsilon_{3,-6} \\ 0 & \varepsilon_{4,-1} & \varepsilon_{4,-2} & \varepsilon_{4,-3} & \varepsilon_{4,-4} & \varepsilon_{4,-5} & 0 \\ 0 & \varepsilon_{5,-1} & \varepsilon_{5,-2} & \varepsilon_{5,-3} & \varepsilon_{5,-4} & 0 & 0 \\ 0 & \varepsilon_{6,-1} & \varepsilon_{6,-2} & \varepsilon_{6,-3} & 0 & 0 & 0 \end{bmatrix} \tag{5.23b}$$

$$[\varepsilon_{xx}]_{\text{Img}(+1)} = \begin{bmatrix} 0 & 0 & 0 & \varepsilon_{0,9} & \varepsilon_{0,8} & \varepsilon_{0,7} & 0 \\ 0 & 0 & \varepsilon_{1,10} & \varepsilon_{1,9} & \varepsilon_{1,8} & \varepsilon_{1,7} & 0 \\ 0 & \varepsilon_{2,11} & \varepsilon_{2,10} & \varepsilon_{2,9} & \varepsilon_{2,8} & \varepsilon_{2,7} & 0 \\ \varepsilon_{3,12} & \varepsilon_{3,11} & \varepsilon_{3,10} & \varepsilon_{3,9} & \varepsilon_{3,8} & \varepsilon_{3,7} & 0 \\ \varepsilon_{4,12} & \varepsilon_{4,11} & \varepsilon_{4,10} & \varepsilon_{4,9} & \varepsilon_{4,8} & \varepsilon_{4,7} & 0 \\ \varepsilon_{5,12} & \varepsilon_{5,11} & \varepsilon_{5,10} & \varepsilon_{5,9} & \varepsilon_{5,8} & \varepsilon_{5,7} & 0 \\ \varepsilon_{6,12} & \varepsilon_{6,11} & \varepsilon_{6,10} & \varepsilon_{6,9} & \varepsilon_{6,8} & \varepsilon_{6,7} & 0 \end{bmatrix} \tag{5.23c}$$

$$[\varepsilon_{xx}]_{\text{Img}(-2)} = \begin{bmatrix} 0 & 0 & 0 & \varepsilon_{0,-9} & \varepsilon_{0,-8} & \varepsilon_{0,-7} & 0 \\ 0 & 0 & 0 & 0 & \varepsilon_{1,-8} & \varepsilon_{1,-7} & 0 \\ 0 & 0 & 0 & 0 & 0 & \varepsilon_{2,-7} & 0 \\ 0 & 0 & 0 & 0 & 0 & 0 & 0 \\ 0 & 0 & 0 & 0 & 0 & 0 & 0 \\ 0 & 0 & 0 & 0 & 0 & 0 & 0 \\ 0 & 0 & 0 & 0 & 0 & 0 & 0 \end{bmatrix} \tag{5.23d}$$

$$[\varepsilon_{x}]_{\text{Img}(+2)} = \begin{bmatrix} 0 & 0 & 0 & 0 & 0 & 0 & 0 \\ 0 & 0 & 0 & 0 & 0 & 0 & 0 \\ 0 & 0 & 0 & 0 & 0 & 0 & 0 \\ 0 & 0 & 0 & 0 & 0 & 0 & 0 \\ 0 & \varepsilon_{4,13} & 0 & 0 & 0 & 0 & 0 \\ 0 & \varepsilon_{5,13} & \varepsilon_{5,14} & 0 & 0 & 0 & 0 \\ 0 & \varepsilon_{6,13} & \varepsilon_{6,14} & \varepsilon_{6,15} & 0 & 0 & 0 \end{bmatrix} \tag{5.23e}$$

where (5.23a) is associated with the original structure, while (5.23b) through (5.23e) are contributed by the image regions. As seen from the above, the latter contributions to the equivalent permittivity matrix decrease dramatically with the increase in image region number (Img). Since the elements of the equivalent permittivity matrix depend only on the basis functions and the material properties of the structure, both the $[\hat{\varepsilon}_{xx}]$ and its inverse $[\hat{\varepsilon}_{xx}]^{-1}$ can be computed in advance and saved in the MRTD Maxwell solver for the updated equations. Consequently, the E-field can be updated as follows:

$$_{\phi x}E^{n}_{i+1/2,j,k} = \frac{1}{\varepsilon_0}\sum_{j'=0}^{N_y}([\hat{\varepsilon}_{xx}]^{-1})_{j,j'\,\phi x}D^{n}_{i+1/2,j',k} \tag{5.24}$$

where, apparently, we need only to compute the product of two small matrices, whose dimensions are typically 20 or less (depending on the number of mesh discretizations along the expansion direction) to update the E-fields at each time step.

AIG Technique

Based on the MIT, we develop an automatic image generator (AIG), a software tool, to truncate PEC boundaries and to reduce memory requirements for image fields. In principle, to maintain the simplicity of update equations and minimize memory requirements, for a generalized PEC-shielded structure, we define (1) all updated field quantities existing only in the original structure, whose image contributions can be expressed in a symmetric or asymmetric manner; and (2) $(\varepsilon_{\alpha\alpha})_{ii',jj',kk'}$ located in both original and image regions. In fact, with this application, we can significantly reduce memory requirements of the image fields. Based on the MIT presented, for example, all image fields along the x-direction for a PEC wall can be expressed as

$$\begin{aligned} &E_t[-i] = -E_t[i], && 1 \le i \le N_x \\ &E_t[k\cdot 2N_x + i] = E_t[i], && -N_x \le i < N_x \\ &E_n[-i+\tfrac{1}{2}] = E_n[i-\tfrac{1}{2}], && 1 \le i \le N_x \\ &E_n[k\cdot 2N_x + i + \tfrac{1}{2}] = E_n[i+\tfrac{1}{2}], && -N_x \le i < N_x \end{aligned} \tag{5.25}$$

$$\begin{aligned} &H_n[-i] = -H_n[i], && 1 \le i \le N_x \\ &H_n[k\cdot 2n_x + i] = H_n[i], && N_x \le i < N_x \\ &H_t[-i+\tfrac{1}{2}] = H_t[i-\tfrac{1}{2}], && 1 \le i \le N_x \\ &H_t[k\cdot 2N_x + i + \tfrac{1}{2}] = H_t[i+\tfrac{1}{2}], && -N_x \le i < N_x \end{aligned} \tag{5.26}$$

where k stands for the serial number of the image and $k = 0, \pm 1, \pm 2, \ldots, N_x$—the total number of cells for the structure along the x-direction—and $E_t(H_t)$ and $E_n(H_n)$ are E-field (H-field) tangential and normal components, respectively.

As an example of the AIG application, we consider an arbitrarily partially PEC-shielded structure, whose cross section is shown in Figure 5.3 (dark black lines indicate PEC walls and three rectangular areas with various gray shading represent different materials). Based on the structure described in Figure 5.3, we use the AIG to automatically generate an entire computational domain, including

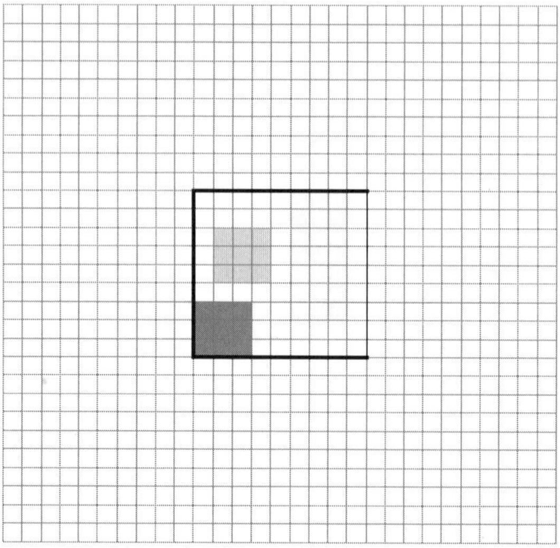

FIGURE 5.3 An arbitrarily partially PEC-shielded structure.

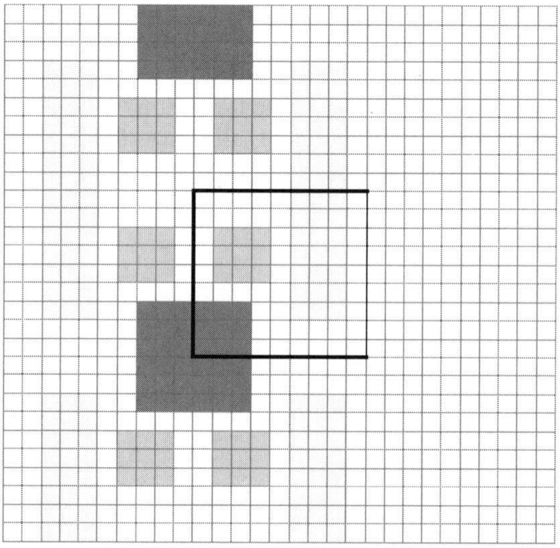

FIGURE 5.4 The image structures automatically created by the developed AIG.

both the original structure and its images. The computational domain is displayed in Figure 5.4, where we use only one image on each side.

MIT in Update Equations

Although (5.6) and (5.7) can, in principle, be employed to update the fields, they are numerically inefficient and somewhat impractical for coding and computation, because the summations in these update expressions include both the original and image regions. In this section, we truncate the summations in these update equations by excluding the image field quantities. This is accomplished by carrying out the following two steps: (1) expressing all field quantities in terms of those defined in the original structure by the MIT technique described in the previous section; and (2) maintaining the values and positions of the coefficients $a(\nu)$ unchanged for all terms, but expressing them only by using the grid indices inside the structure. Following this procedure, we can rewrite the update equations for the x-components using the MIT as follows:

$$
\begin{aligned}
{}_{\phi x}D^{n+1}_{i+1/2,j,k} &= {}_{\phi x}D^{n}_{i+1/2,j,k} + \left\{ \sum_{\nu=0}^{N_y-1} a(\nu - j)_{\phi z}H^{n+1/2}_{i+1/2,\nu+1/2,k} \right. \\
&\quad + \left[\sum_{\text{Img}=\text{even}} \left(\sum_{\nu=0}^{N_y-1} a(-(\text{Img}-2)\cdot N_y - \nu - j - 1) \right.\right. \\
&\qquad \left.\left. + \sum_{\nu=0}^{N_y-1} a(\text{Img}\cdot N_y + \nu - j) \right) {}_{\phi z}H^{n+1/2}_{i+1/2,\nu+1/2,k} \right] \\
&\quad + \left[\sum_{\text{Img}=\text{odd}} \left(\sum_{\nu=0}^{N_y-1} a(-\text{Img}\cdot N_y - \nu - j - 1) \right.\right. \\
&\qquad \left.\left.\left. + \sum_{m=0}^{N_y-1} a(\text{Img}\cdot N_y + \nu - j) \right) {}_{\phi z}H^{n+1/2}_{i+1/2,N_y-\nu-1+1/2,k} \right] \right\} \frac{\Delta t}{\Delta y} \\
&\quad + \left\{ \sum_{\nu=0}^{N_z-1} a(\nu - k)_{\phi y}H^{n+1/2}_{i+1/2,j,\nu+1/2} \right. \\
&\quad + \left[\sum_{\text{Img}=\text{even}} \left(\sum_{\nu=0}^{N_z-1} a(-(\text{Img}-2)\cdot N_z - \nu - k - 1) \right.\right. \\
&\qquad \left.\left. + \sum_{\nu=0}^{N_z-1} a(\text{Img}\cdot N_z + \nu - k) \right) {}_{\phi y}H^{n+1/2}_{i+1/2,j,\nu+1/2} \right] \\
&\quad + \left[\sum_{\text{Img}=\text{odd}} \left(\sum_{\nu=0}^{N_z-1} a(-\text{Img}\cdot N_z - \nu - k - 1) \right.\right. \\
&\qquad \left.\left.\left. + \sum_{\nu=0}^{N_z-1} a(\text{Img}\cdot N_z + \nu - k) \right) {}_{\phi y}H^{n+1/2}_{i+1/2,j,N_z-\nu-1+1/2} \right] \right\} \frac{\Delta t}{\Delta z}
\end{aligned}
\tag{5.27}
$$

$$
\begin{aligned}
{}_{\phi x}H^{n+1/2}_{i,j+1/2,k+1/2} = {}_{\phi x}H^{n-1/2}_{i,j+1/2,k+1/2} &+ \frac{1}{\mu_0}\left\{\sum_{\nu=0}^{N_z} a(\nu - k - 1)_{\phi y}E^{n}_{i,j+1/2,\nu}\right. \\
&+ \left[\sum_{\mathrm{Img=even}}\left(-\sum_{\nu=0}^{N_z-1} a(-(\mathrm{Img}-2)\cdot N_z - \nu - k - 2)\right.\right. \\
&\quad \left.\left. + \sum_{m=0}^{Nz-1} a(\mathrm{Img}\cdot N_z + \nu - k)\right)_{\phi y}E^{n}_{i,j+1/2,\nu+1}\right] \\
&+ \left[\sum_{\mathrm{Img=odd}}\left(\sum_{\nu=0}^{N_z-1} a(-\mathrm{Img}\cdot N_z - \nu - k - 2)\right.\right. \\
&\quad \left.\left.\left. - \sum_{\nu=0}^{N_z-1} a(\mathrm{Img}\cdot N_z + \nu - k)\right)_{\phi z}E^{n+1/2}_{i,j+1/2,N_z-\nu-1}\right]\right\}\frac{\Delta t}{\Delta z} \\
&- \frac{1}{\mu_0}\left\{\sum_{\nu=0}^{N_y} a(\nu - j - 1)_{\phi z}E^{n}_{i,\nu,k+1/2}\right. \\
&+ \left[\sum_{\mathrm{Img=even}}\left(-\sum_{\nu=0}^{N_y-1} a(-(\mathrm{Img}-2)\cdot N_y - \nu - j - 2)\right.\right. \\
&\quad \left.\left. + \sum_{\nu=0}^{N_y-1} a(\mathrm{Img}\cdot N_y + \nu - j)\right)_{\phi z}E^{n}_{i,\nu+1,k+1/2}\right] \\
&+ \left[\sum_{\mathrm{Img=odd}}\left(\sum_{\nu=0}^{N_y-1} a(-\mathrm{Img}\cdot N_y - \nu - j - 2)\right.\right. \\
&\quad \left.\left.\left. - \sum_{\nu=0}^{N_y-1} a(\mathrm{Img}\cdot N_y + \nu - j)\right)_{\phi z}E^{n}_{i,N_y-\nu-1,k+1/2}\right]\right\}\frac{\Delta t}{\Delta y}
\end{aligned}
\tag{5.28}
$$

where N_y and N_z are the number of cells along the y- and z-directions inside the structure, respectively, and the summation indices, odd (1, 3, ...) and even (2, 4, ...), denote the contributions to the updated field quantities from the specified odd or even image regions, respectively.

Before closing this section, it would be worthwhile to point out that although (5.27) and (5.28) are quite lengthy, they are actually quite efficient from the point of view of coding and computation. All the field quantities usually have relatively small dimensions, because the coefficient $a(\nu)$ can be computed in advance and saved in the MRTD solver for later use.

5.3 NUMERICAL RESULTS

In this section, the MRTD scheme is applied in conjunction with the MIT to analyze an empty cavity and a half-filled dielectric cavity. In all these cases, the dimensions of the cavity are set to $1 \times 2 \times 1.5$ m^3, in the x-, y-, and z-directions, respectively.

We begin with an empty cavity that is discretized with $2 \times 4 \times 3$ cells in the x-, y-, and z-directions. The frequency spectrum of E_x sampled at the grid point [1, 3, 1] is displayed in Figure 5.5, and the extracted resonant frequencies for the dominant and higher-order modes are summarized in Table 5.1 along with the discretization parameters. It is evident that the results obtained with the MRTD scheme show good agreement, with both the analytical results and those derived by using the FDTD. However, the MRTD scheme demands only 0.8% of the computational resources needed in the conventional FDTD technique. In addition, the required CPU time for conventional MRTD and the MRTD-MIT schemes has been investigated. For a total time step of $N_t = 10^4$, the CPU time for the conventional MRTD is about 19.27 seconds, while that of the MRTD-MIT requires about 27.08 seconds on a 500-MHz Alpha digital Pentium III workstation. Obviously, this increment in the CPU time is due to the use of additional images in the process of implementing the MRTD-MIT technique. Specifically, we use 5 images along the positive and negative x-directions, and 3 images along the rest. By comparison, the conventional MRTD uses only one image in each of

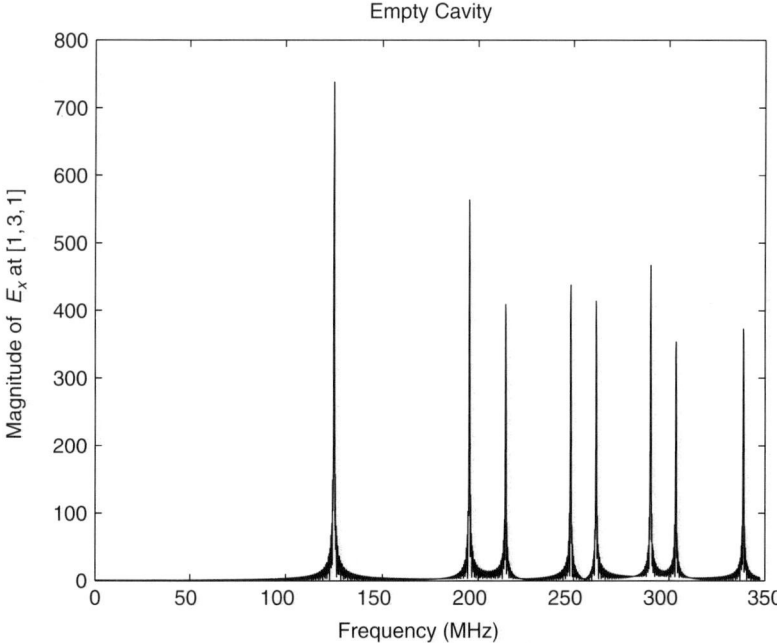

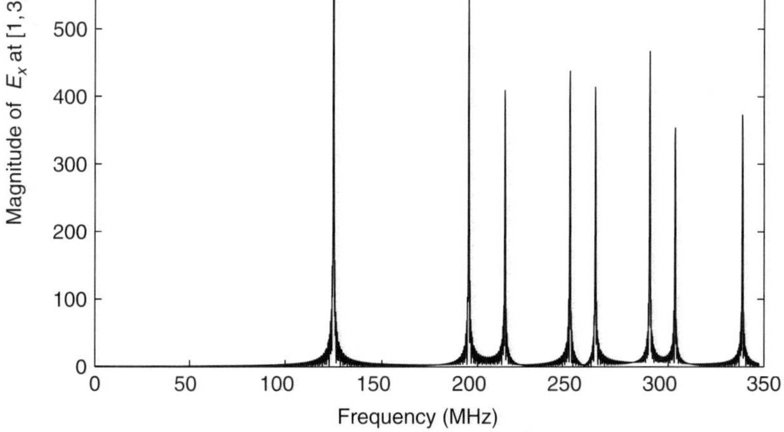

FIGURE 5.5 Frequency spectrum of E_x at the location [1, 3, 1] for an empty cavity.

the directions. Note that the introduction of the MIT in the MRTD scheme and the use of a greater number of images lead to more accurate results.

The MRTD analysis of the half-filled dielectric cavity, with $\varepsilon_r = 64$, also yields excellent results, as is evident from Figure 5.6, and Table 5.2. Once again, the MRTD scheme is found to be highly efficient, when compared to the conventional FDTD method, as it requires only 0.8% of the CPU memory using the FDTD. The accuracy of the MRTD technique is again improved when it is combined with the MIT.

TABLE 5.1 Resonant Frequencies in MHz for an Empty Cavity ($V = 1 \times 2 \times 1.5$ m^3; $\Delta x = \Delta y = \Delta z = \frac{1}{2}$ m)

Analytic [6]	FDTD [1] (10 × 20 × 15)	S-MRTD [1] (2 × 4 × 3)	This Work (2 × 4 × 3)	Percentage Difference
125.00	124.85	125.10	124.95	−0.040%
180.27	179.75	180.50	180.39	0.067%
213.60	212.40	214.60	214.30	0.328%
246.22	244.50	248.70	248.55	0.946%
250.00	248.70	251.00	250.75	0.300%
301.04	298.95	303.90	303.65	0.580%
336.34	334.35	339.20	338.91	0.764%

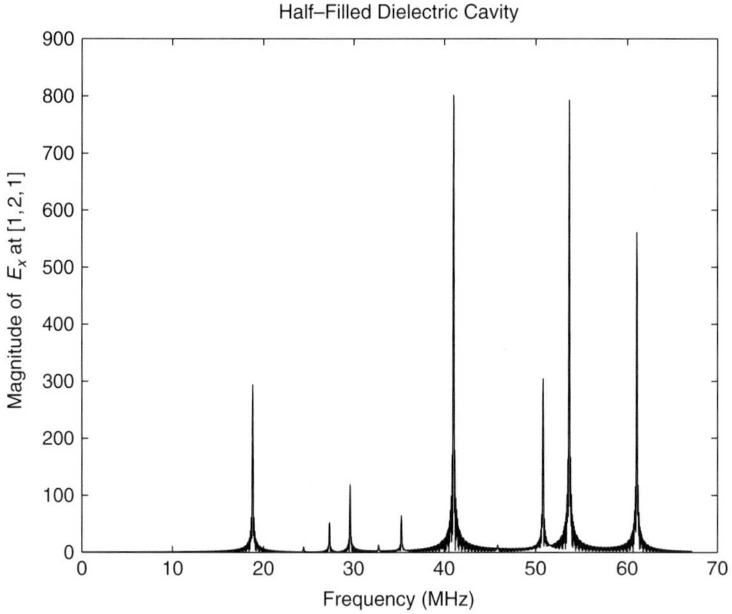

FIGURE 5.6 Frequency spectrum of E_x at the location [1, 2, 1] for a half-filled dielectric cavity.

TABLE 5.2 Resonant Frequencies in MHz for a Half-Filled Dielectric Cavity ($\varepsilon_r = 64$; $N_{y1} = N_{y2} = 2.0$; $V = 1 \times 2 \times 1.5\ m^3$; $\Delta x = \Delta y = \Delta z = 1/2\ m$)

Analytic [6]	FDTD [1] (10 × 20 × 15)	S-MRTD [1] (2 × 4 × 3)	This Work (2 × 4 × 3)	Percentage Difference
18.627	18.615	18.715	18.692	0.349%
27.172	27.140	27.350	27.313	0.519%
29.375	29.215	29.580	29.526	0.514%
35.069	34.970	35.280	35.247	0.507%

5.4 CONCLUSIONS

In this chapter, an adjustable MIT and AIG have been developed, which can be incorporated in the MRTD scheme for the boundary truncation of PEC-shielded structures. The techniques can easily be extended for boundary truncation of a PMC-shielded structure. We have also applied this systematic technique for constructing multiple images, extracting the constitutive relations, deriving the update equations, and determining the number of images in the transformed domain of the MRTD. The present MRTD-MIT algorithm slightly improves the accuracy of the conventional MRTD scheme and yields results with high accuracy and efficiency when compared with traditional FDTD techniques.

REFERENCES

[1] M. Krumpholz and L. P. B. Katehi, "MRTD: New time domain schemes based on multiresolution analysis," *IEEE Trans. Microwave Theory Tech.*, vol. 44, no. 4, pp. 555–571, Apr. 1996.

[2] R. Robertson, E. Tentzeris, M. Krumpholz, and L. P. B. Katehi, "MRTD analysis of dielectric cavity structures," *Proc. IEEE MTT-S Dig.*, pp. 1861–1864, 1996.

[3] Q. Cao, Y. Chen, and R. Mittra, "Multiple image technique (MIT) in implementation of multiresolution time domain scheme for boundary truncation of PEC-shielded structures," *IEEE Trans. Microwave Theory Tech.*, vol. 50, no. 6, pp. 1578–1589, June 2002.

[4] M. Yang, Q. Cao, and Y. Chen, "A systematic and memory-saving scaling function based multiresolution scheme," in *2000 IEEE International Symposium on AP-S and USNC/URSI National Radio Science Meeting*, pp. 761–766, Salt Lake City, USA, July 2000.

[5] R. F. Harrington, *Field Computation by Moment Methods*, Krieger, Melbourne FL, 1982.

[6] C. A. Balanis, *Advanced Engineering Electromagnetics*, Wiley, Hoboken, 1989.

CHAPTER SIX

Open Boundary Truncation

6.1 INTRODUCTION

Although the conventional finite difference time domain (FDTD) method has provided, in principle, a generalized Maxwell solver for complicated electromagnetic problems [1], it usually requires an extremely large computational capacity and very long computation time. Various efforts have been made to improve the efficiency of the FDTD technique by using high-order expansion techniques. Two representative techniques in this area are the pseudospectral time domain (PSTD) technique and the multiresolution time domain (MRTD) scheme [2–6]. One of greatest challenges in applying these high-order expansion techniques is the boundary truncation for both open and shielded structures, because of the high-order expansions and field coupling between cells. An image technique has been introduced for the boundary truncation of perfectly electric conductor (PEC) shielded structures [3], and a multiple image technique (MIT) has been presented in the previous chapter in an implementation of the MRTD scheme [5]. Recently, a perfectly matched layer (PML) absorber has been presented as a superabsorbing boundary condition for open structures to truncate the FDTD lattice [6–8]. It has been extended by both split-field and unsplit-field PML ABCs for open boundary truncation in the MRTD analysis for fundamental guided-wave structures and various open microwave circuits [9].

This chapter presents an anisotropic perfectly matched layer (APML) [7, 8] for the analysis of practical structures, in particular, for planer printed millimeter-wave integrated structures. The primary reason that we choose the APML as presented in [8] is because this APML approach is consistent with the Maxwell equations, requires no splitting of fields, and is based on a widely accepted theory. Considering the difficulty of the MRTD application in a complicated material region, this chapter begins with the derivation of the MRTD algorithm for

Multiresolution Time Domain Scheme for Electromagnetic Engineering
By Yinchao Chen, Qunsheng Cao, and Raj Mittra
ISBN 0-471-27230-2

a hybrid dielectric–conductor, planar microstrip-type structure. Various aspects of the APML applications, including the treatment of different APML regions, selection of APML parameters, and derivation of update APML update equations, will be addressed herein [10].

6.2 MRTD UPDATE EQUATIONS IN APML REGIONS

Governing Equations in APML Regions

Beginning with an anisotropic perfectly matched layer formulation [4–6], a MRTD update scheme is developed in all APML regions, including the APML faces, edges, and corners as shown in Figure 6.1.

In general, the Maxwell curl equations in the APML region in the frequency domain can be written as

$$\begin{bmatrix} 0 & -\partial_z & \partial_y \\ \partial_z & 0 & -\partial_x \\ -\partial_y & \partial_x & 0 \end{bmatrix} \begin{bmatrix} H_x \\ H_y \\ H_z \end{bmatrix} = j\omega\varepsilon_0 \begin{bmatrix} S_{xx} & 0 & 0 \\ 0 & S_{yy} & 0 \\ 0 & 0 & S_{zz} \end{bmatrix} \begin{bmatrix} E_x \\ E_y \\ E_z \end{bmatrix} \tag{6.1a}$$

$$\begin{bmatrix} 0 & -\partial_z & \partial_y \\ \partial_z & 0 & -\partial_x \\ -\partial_y & \partial_x & 0 \end{bmatrix} \begin{bmatrix} E_x \\ E_y \\ E_z \end{bmatrix} = -j\omega\mu_0 \begin{bmatrix} S_{xx} & 0 & 0 \\ 0 & S_{yy} & 0 \\ 0 & 0 & S_{zz} \end{bmatrix} \begin{bmatrix} H_x \\ H_y \\ H_z \end{bmatrix} \tag{6.1b}$$

where the relative permeability and permittivity are set to $\mu_r = 1$ and $\varepsilon_r = 1$. The treatment for a generalized $\varepsilon_r(\vec{r})$ will be discussed in the following chapters.

The APML parameter matrix $[S]$ is the product of three submatrices [8]:

$$[S] = [S_x][S_y][S_z] = \begin{bmatrix} s_x^{-1} & 0 & 0 \\ 0 & s_x & 0 \\ 0 & 0 & s_x \end{bmatrix} \begin{bmatrix} s_y & 0 & 0 \\ 0 & s_y^{-1} & 0 \\ 0 & 0 & s_y \end{bmatrix} \begin{bmatrix} s_z & 0 & 0 \\ 0 & s_z & 0 \\ 0 & 0 & s_z^{-1} \end{bmatrix} \tag{6.2}$$

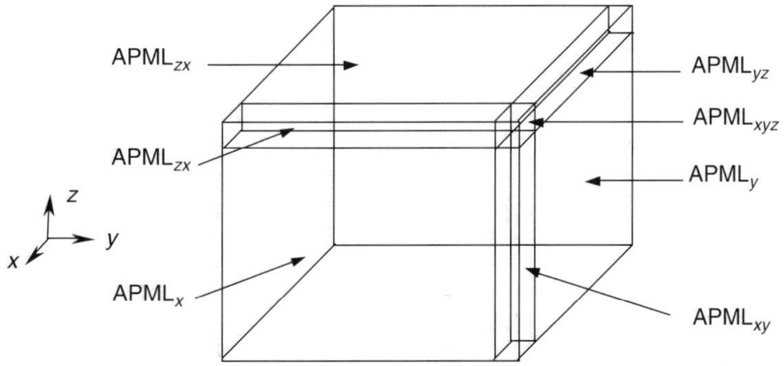

FIGURE 6.1 Rectangular computational volume with different APML orientations.

with

$$s_\alpha = 1 + \frac{\sigma_\alpha}{j\omega\varepsilon_0} \quad (\alpha = x, y, \text{ or } z) \tag{6.3}$$

where σ_α is spatially varying conductivity along the absorption axis. Note that $[S_x]$, $[S_y]$, and $[S_z]$, defined as APML constitutive submatrices, are associated with wave absorption in the x, y, and z directions, respectively. As shown in Figure 6.1, when an APML is applied on a typical rectangular computational volume, the APML regions consist of three different boundary orientations, namely, face-APMLs, edge-APMLs, and corner-APMLs.

If we investigate the S-MRTD case, all field quantities are expanded in terms of the scaling functions in space and pulse functions in time; that is,

$$E_x(\vec{r}, t) = \sum_{i,j,k,n=-\infty}^{\infty} {}_{\phi x}E^{n}_{i+1/2,j,k}\phi_{i+1/2}(x)\phi_j(y)\phi_k(z)h_n(t) \tag{6.4a}$$

$$E_y(\vec{r}, t) = \sum_{i,j,k,n=-\infty}^{\infty} {}_{\phi y}E^{n}_{i,j+1/2,k}\phi_i(x)\phi_{j+1/2}(y)\phi_k(z)h_n(t) \tag{6.4b}$$

$$E_z(\vec{r}, t) = \sum_{i,j,k,n=-\infty}^{\infty} {}_{\phi z}E^{n}_{i,j,k+1/2}\phi_i(x)\phi_j(y)\phi_{k+1/2}(z)h_n(t) \tag{6.4c}$$

$$H_x(\vec{r}, t) = \sum_{i,j,k,n=-\infty}^{\infty} {}_{\phi x}H^{n+1/2}_{i,j+1/2,k+1/2}\phi_i(x)\phi_{j+1/2}(y)\phi_{k+1/2}(z)h_{n+1/2}(t) \tag{6.5a}$$

$$H_y(\vec{r}, t) = \sum_{i,j,k,n=-\infty}^{\infty} {}_{\phi y}H^{n+1/2}_{i+1/2,j,k+1/2}\phi_{i+1/2}(x)\phi_j(y)\phi_{k+1/2}(z)h_{n+1/2}(t) \tag{6.5b}$$

$$H_z(\vec{r}, t) = \sum_{i,j,k,n=-\infty}^{\infty} {}_{\phi z}H^{n+1/2}_{i+1/2,j+1/2,k}\phi_{i+1/2}(x)\phi_{j+1/2}(y)\phi_k(z)h_{n+1/2}(t) \tag{6.5c}$$

where $\phi(\varsigma)$, $\varsigma = x, y, z$ denotes the cubic spline Battle–Lemarié scaling function, and $h_n(t)$ is a rectangular pulse function. In the following sections, we analyze and discuss the update equations of APML face, edge, and corner regions, respectively.

MRTD Scheme in Face-APML Regions

The face-APMLs are defined at the interfaces of two ends of each Cartesian axis, and there are six of them in total as shown in Figure 6.1. Without losing generality, we can consider a wave absorption in the z-direction, and the associated APML regions are the face-APML$_z$. This implies $[S] = [S_z]$ and $[S_x] = [S_y] = [I]$, where $[I]$ is a unitary matrix. Hence, the Maxwell equations in the two face-APML$_Z$ regions can be expressed as

$$\begin{bmatrix} 0 & -\partial_z & \partial_y \\ \partial_z & 0 & -\partial_x \\ -\partial_y & \partial_x & 0 \end{bmatrix} \begin{bmatrix} H_x \\ H_y \\ H_z \end{bmatrix} = j\omega\varepsilon_0 \begin{bmatrix} s_z & 0 & 0 \\ 0 & s_z & 0 \\ 0 & 0 & s_z^{-1} \end{bmatrix} \begin{bmatrix} E_x \\ E_y \\ E_z \end{bmatrix} \tag{6.6a}$$

$$\begin{bmatrix} 0 & -\partial_z & \partial_y \\ \partial_z & 0 & -\partial_x \\ -\partial_y & \partial_x & 0 \end{bmatrix} \begin{bmatrix} E_x \\ E_y \\ E_z \end{bmatrix} = -j\omega\mu_0 \begin{bmatrix} s_z & 0 & 0 \\ 0 & s_z & 0 \\ 0 & 0 & s_z^{-1} \end{bmatrix} \begin{bmatrix} H_x \\ H_y \\ H_z \end{bmatrix} \tag{6.6b}$$

For the x- and y-components of the above equations, we can directly convert them into the time domain by using the frequency–time converter, $j\omega \Leftrightarrow \partial/\partial t$, and then using Galerkin's method to derive the MRTD update equations. For example, expanding the first equation of (6.6a) in the frequency domain yields

$$\frac{\partial H_z}{\partial y} - \frac{\partial H_y}{\partial z} = j\omega\varepsilon_0 s_z E_x = j\omega\varepsilon_0 \left(1 + \frac{\sigma_z}{j\omega\varepsilon_0}\right) E_x \tag{6.7a}$$

We easily obtain its corresponding time domain equation:

$$\frac{\partial H_z}{\partial y} - \frac{\partial H_y}{\partial z} = \varepsilon_0 \frac{\partial E_x}{\partial t} + \sigma_z E_x \tag{6.7b}$$

Substituting the MRTD field expansions in (6.4) into (6.7b), then sampling and testing (6.7b) with the function $\phi_{i+1/2}(x)\phi_j(y)\phi_k(z)h_{n+1/2}(t)$, and, finally, using the following orthogonal and integral relations

$$\int_{-\infty}^{+\infty} \phi_i(\varsigma)\phi_{i'}(\varsigma)\, d\varsigma = \delta_{i,i'}\Delta\varsigma, \qquad \varsigma = x, y, z \tag{6.8a}$$

$$\int_{-\infty}^{+\infty} h_{n'+1/2}(t) \frac{\partial}{\partial t} h_n(t)\, dt = \delta_{n',n+1} - \delta_{n',n} \tag{6.8b}$$

$$\int_{-\infty}^{+\infty} \phi_i(x) \frac{\partial}{\partial x} \phi_{i'+1/2}(x)\, dx = \sum_{\nu=-\infty}^{+\infty} a(\nu)\delta_{i+\nu,i'} \tag{6.8c}$$

we can derive the update equations for ${}_{\phi x}E$:

$$\begin{aligned} {}_{\phi x}E^{n+1}_{i+1/2,j,k} &= {}_{\phi x}E^{n}_{i+1/2,j,k} + \sum_{k'=-\infty}^{+\infty} (\sigma_z)_{k,k'} \left({}_{\phi x}E^{n+1}_{i+1/2,j,k'} + {}_{\phi,x}E^{n}_{i+1/2,j,k'} \right) \frac{\Delta t}{2\varepsilon_0} \\ &= \frac{1}{\varepsilon_0} \sum_{\nu} a(\nu) \left({}_{\phi z}H^{n+1/2}_{i+1/2,j+\nu+1/2,k} \frac{\Delta t}{\Delta y} - {}_{\phi y}H^{n+1/2}_{i+1/2,j,k+\nu+1/2} \frac{\Delta t}{\Delta z} \right) \end{aligned} \tag{6.9}$$

where $(\sigma_z)_{k,k'}$ is given as

$$(\sigma_z)_{k,k'} = \frac{1}{\Delta z}\int_{-\infty}^{+\infty} \phi_k(z)\sigma_z(z)\phi_{k'}(z)\,dz = \int_{\mathrm{APML}} \phi(z'-k)\sigma_z(z'\Delta z)\phi(z'-k')\,dz' \tag{6.10}$$

Obviously, it would be difficult to derive an update equation if we directly substitute $(\sigma_z)_{k,k'}$ defined in (6.10) into (6.9). Instead, a new APML parameter is defined, which is characterized only by the diagonal terms of (6.10); namely,

$$(\sigma_z)_{k,k'} = \delta_{k,k'}\int_{\mathrm{APML}} \phi^2(z'-k)\sigma_z(z'\Delta z)\,dz' = \delta_{k,k'}\sigma_z^k \equiv \delta_{k,k'}\sigma_{\alpha\max}\left|\frac{k-k_0}{N_p}\right|^m \tag{6.11}$$

where N_p is the cell number of the APML region, k_0 is the starting position of the APML region in the z-direction, and m is the order of polynomial variation. Note that this simplification does not change the defined physics problem at all, since it is performed purely in order to change the absorption material in the APML region. In the following, we use σ_z^k instead of $(\sigma_z)_{k,k}$ for simplicity and, similarly, σ_x^i for $(\sigma_x)_{i,i}$ and σ_y^j for $(\sigma_y)_{j,j}$. Using the above simplified equation, we can easily derive the update equation for (6.6) that reads

$$\begin{aligned}{}_{\phi x}E^{n+1}_{i+1/2,j,k} &= \left(\frac{1-\dfrac{\sigma_z^k\Delta t}{2\varepsilon_0}}{1+\dfrac{\sigma_z^k\Delta t}{2\varepsilon_0}}\right){}_{\phi x}E^{n}_{i+1/2,j,k} + \frac{1}{\varepsilon_0}\frac{1}{\left(1+\dfrac{\sigma_z^k\Delta t}{2\varepsilon_0}\right)} \\ &\quad\times\left[\sum_\nu a(\nu)\left({}_{\phi z}H^{n+1/2}_{i+1/2,j+\nu+1/2,k}\frac{\Delta t}{\Delta y} - {}_{\phi y}H^{n+1/2}_{i+1/2,j,k+\nu+1/2}\frac{\Delta t}{\Delta z}\right)\right]\end{aligned} \tag{6.12}$$

Next, consider the z-component of (6.6). Due to the reciprocal relation in the z-components of (6.6), we adopt a two-step approach by introducing an intermediate parameter [8]:

$$D_z = \frac{\varepsilon_0}{(1+\sigma_z/j\omega\varepsilon_0)}E_z \tag{6.13}$$

Again converting (6.6) and (6.13) into the time domain, we can obtain the following update equations:

$$\begin{aligned}{}_{\phi z}D^{n+1}_{i,j,k+1/2} &= {}_{\phi z}D^{n}_{i,j,k+1/2} + \sum_\nu a(\nu) \\ &\quad\times\left({}_{\phi y}H^{n+1/2}_{i+\nu+1/2,j,k+1/2}\frac{\Delta t}{\Delta x} - {}_{\phi x}H^{n+1/2}_{i,j+\nu+1/2,k+1/2}\frac{\Delta t}{\Delta y}\right)\end{aligned} \tag{6.14}$$

$$
{}_{\phi z}E^{n+1}_{i,j,k+1/2} = {}_{\phi z}E^{n}_{i,j,k+1/2} + \frac{1}{\varepsilon_0} \times \left[\left(1 + \frac{\sigma_z^k \Delta t}{2\varepsilon_0}\right) {}_{\phi z}D^{n+1}_{i,j,k+1/2} - \left(1 - \frac{\sigma_z^k \Delta t}{2\varepsilon_0}\right) {}_{\phi z}D^{n}_{i,j,k+1/2} \right] \tag{6.15}
$$

Through similar processes, we can derive the rest of the update equations for all face-APMLs (see Appendix C). Note that herein we have used an approach that requires eight components in the APML face region to construct the MRTD update algorithm.

MRTD Scheme in Edge-APML Regions

As shown in the Figure 6.1, the edge-APMLs in a rectangular computational volume are the intersections between any two face-APMLs. For simplicity, we consider that an electromagnetic wave propagates in a direction where electromagnetic fields will be absorbed along both the y- and z-axes. Its associated edge-APML region refers to edge-APML$_{yz}$, where $[S] = [S_y][S_z]$ and $[S_x] = [I]$. In this case, the Maxwell equations in the edge-APML$_{yz}$ region become

$$
\begin{bmatrix} 0 & -\partial_z & \partial_y \\ \partial_z & 0 & -\partial_x \\ -\partial_y & \partial_x & 0 \end{bmatrix} \begin{bmatrix} H_x \\ H_y \\ H_z \end{bmatrix} = j\omega\varepsilon \begin{bmatrix} s_y s_z & 0 & 0 \\ 0 & s_y^{-1} s_z & 0 \\ 0 & 0 & s_y s_z^{-1} \end{bmatrix} \begin{bmatrix} E_x \\ E_y \\ E_z \end{bmatrix} \tag{6.16a}
$$

$$
\begin{bmatrix} 0 & -\partial_z & \partial_y \\ \partial_z & 0 & -\partial_x \\ -\partial_y & \partial_x & 0 \end{bmatrix} \begin{bmatrix} E_x \\ E_y \\ E_z \end{bmatrix} = -j\omega\mu_0 \begin{bmatrix} s_y s_z & 0 & 0 \\ 0 & s_y^{-1} s_z & 0 \\ 0 & 0 & s_y s_z^{-1} \end{bmatrix} \begin{bmatrix} H_x \\ H_y \\ H_z \end{bmatrix} \tag{6.16b}
$$

Since there are only two categories of constitutive relations, we likewise only consider the x and z components of the above equations by using the two-step approach. Following the method presented in [8], we define

$$
D_x = \left(1 + \frac{\sigma_y}{j\omega\varepsilon_0}\right) \varepsilon_0 E_x \tag{6.17}
$$

$$
D_z = \left(\frac{1 + \dfrac{\sigma_y}{j\omega\varepsilon_0}}{1 + \dfrac{\sigma_z}{j\omega\varepsilon_0}} \right) \varepsilon_0 E_z \tag{6.18}
$$

and then combine (6.17) and (6.18) with (6.16) in order to obtain the MRTD update equations for the APML edge:

$$
{}_{\phi x}D^{n+1}_{i+1/2,j,k} = \left(\frac{1 - \dfrac{\sigma_z^k \Delta t}{2\varepsilon_0}}{1 + \dfrac{\sigma_z^k \Delta t}{2\varepsilon_0}} \right) {}_{\phi x}D^{n}_{i,j+1/2,k} + \frac{1}{\left(1 + \dfrac{\sigma_z^k \Delta t}{2\varepsilon_0}\right)} \sum_{\nu} a(\nu)
$$

$$
\times \left({}_{\phi z}H^{n+1/2}_{i+1/2,j+\nu+1/2,k} \frac{\Delta t}{\Delta y} - {}_{\phi y}H^{n+1/2}_{i+1/2,j,k+\nu+1/2} \frac{\Delta t}{\Delta z} \right) \tag{6.19}
$$

$$
{}_{\phi x}E^{n+1}_{i+1/2,j,k} = \left(\frac{1 - \dfrac{\sigma_y^j \Delta t}{2\varepsilon_0}}{1 + \dfrac{\sigma_y^j \Delta t}{2\varepsilon_0}} \right) {}_{\phi x}E^{n}_{i+1/2,j,k}
$$

$$
+ \frac{1}{\varepsilon_0} \frac{1}{\left(1 + \dfrac{\sigma_y^j \Delta t}{2\varepsilon_0}\right)} \left[{}_{\phi x}D^{n+1}_{i+1/2,j,k} - {}_{\phi x}D^{n}_{i+1/2,j,k} \right] \tag{6.20}
$$

$$
{}_{\phi z}D^{n+1}_{i,j,k+1/2} = {}_{\phi z}D^{n}_{i,j,k+1/2} + \sum_{\nu} a(\nu)
$$

$$
\times \left({}_{\phi y}H^{n+12}_{i+\nu+1/2,j,k+1/2} \frac{\Delta t}{\Delta x} - {}_{\phi x}H^{n+1/2}_{i,j+\nu+1/2,k+1/2} \frac{\Delta t}{\Delta y} \right) \tag{6.21}
$$

$$
{}_{\phi z}E^{n+1}_{i,j,k+1/2} = \left(\frac{1 - \dfrac{\sigma_y^j \Delta t}{2\varepsilon_0}}{1 + \dfrac{\sigma_y^j \Delta t}{2\varepsilon_0}} \right) {}_{\phi z}E^{n}_{i,j,k+1/2} + \frac{1}{\varepsilon_0} \frac{1}{\left(1 + \dfrac{\sigma_y^j \Delta t}{2\varepsilon_0}\right)}
$$

$$
\times \left[\left(1 + \frac{\sigma_z^k \Delta t}{2\varepsilon_0}\right) {}_{\phi z}D^{n+1}_{i,j,k+1/2} - \left(1 - \frac{\sigma_z^k \Delta t}{2\varepsilon_0}\right) {}_{\phi z}D^{n}_{i,j,k+1/2} \right] \tag{6.22}
$$

where σ_y^j is similar to the one defined in (6.11) by using a diagonal approximation.

Following the same procedure, we can obtain the remaining update equations for all electromagnetic fields in the edge-APML$_{\mathrm{yz}}$ (see Appendix C). Note that for all edge-APML regions we use the total twelve components to derive the MRTD update schemes.

MRTD Scheme in Corner-APML Regions

It is obvious that there are a total of eight different APML corners in a rectangular computational volume as shown in the Figure 6.1. We usually use corner-APML$_{\mathrm{xyz}}$ to indicate the APML regions where an electromagnetic wave must be absorbed in all three directions, such that $[S] = [S_x][S_y][S_z]$. The Maxwell equations in the frequency domain thus become

$$\begin{bmatrix} 0 & -\partial_z & \partial_y \\ \partial_z & 0 & -\partial_x \\ -\partial_y & \partial_x & 0 \end{bmatrix} \begin{bmatrix} H_x \\ H_y \\ H_z \end{bmatrix} = j\omega\varepsilon_0 \begin{bmatrix} s_y s_z s_x^{-1} & 0 & 0 \\ 0 & s_x s_z s_y^{-1} & 0 \\ 0 & 0 & s_x s_y s_z^{-1} \end{bmatrix} \begin{bmatrix} E_x \\ E_y \\ E_z \end{bmatrix} \tag{6.23a}$$

$$\begin{bmatrix} 0 & -\partial_z & \partial_y \\ \partial_z & 0 & -\partial_x \\ -\partial_y & \partial_x & 0 \end{bmatrix} \begin{bmatrix} E_x \\ E_y \\ E_z \end{bmatrix} = -j\omega\mu_0 \begin{bmatrix} s_y s_z s_x^{-1} & 0 & 0 \\ 0 & s_x s_z s_y^{-1} & 0 \\ 0 & 0 & s_x s_y s_z^{-1} \end{bmatrix} \begin{bmatrix} H_x \\ H_y \\ H_z \end{bmatrix} \tag{6.23b}$$

Since all the field components are similar, we only need to investigate one case. For instance, let us consider the case APML_{xyz} by deriving the y-components of the electric fields (6.23). Again, following the method presented in [8], we define the intermediate parameter

$$D_y = \left(\frac{1 + \dfrac{\sigma_x}{j\omega\varepsilon_0}}{1 + \dfrac{\sigma_y}{j\omega\varepsilon_0}} \right) \varepsilon_0 E_y \tag{6.24}$$

and combine (6.23) with (6.24), which leads to the update equations:

$$\begin{aligned} {}_{\phi y}D^{n+1}_{i,j+1/2,k} = {} & \left(\frac{1 - \dfrac{\sigma_z^k\,\Delta t}{2\varepsilon_0}}{1 + \dfrac{\sigma_z^k\,\Delta t}{2\varepsilon_0}} \right) {}_{\phi y}D^{n}_{i,j+1/2,k} + \frac{1}{\left(1 + \dfrac{\sigma_z^k\,\Delta t}{2\varepsilon_0}\right)} \sum_{v} a(v) \\ & \times \left({}_{\phi x}H^{n+1/2}_{i,j+1/2,k+v+1/2} \frac{\Delta t}{\Delta z} - {}_{\phi z}H^{n+1/2}_{i+v+1/2,j+1/2,k} \frac{\Delta t}{\Delta x} \right) \end{aligned} \tag{6.25}$$

$$\begin{aligned} {}_{\phi y}E^{n+1}_{i,j+1/2,k} = {} & \left(\frac{1 - \dfrac{\sigma_x^i\,\Delta t}{2\varepsilon_0}}{1 + \dfrac{\sigma_x^i\,\Delta t}{2\varepsilon_0}} \right) {}_{\phi y}E^{n}_{i,j+1/2,k} + \frac{1}{\varepsilon_0} \frac{1}{\left(1 + \dfrac{\sigma_x^i\,\Delta t}{2\varepsilon_0}\right)} \\ & \times \left[\left(1 + \frac{\sigma_y^j\,\Delta t}{2\varepsilon_0}\right) {}_{\phi y}D^{n+1}_{i,j+1/2,k} - \left(1 - \frac{\sigma_y^j\,\Delta t}{2\varepsilon_0}\right) {}_{\phi y}D^{n}_{i,j+1/2,k} \right] \end{aligned} \tag{6.26}$$

where σ_x^i is again similar to the one defined in (6.11). Note that we employ the same two-step approach and twelve components in all eight corner-APML regions (see Appendix C).

Excitation and APML Parameters

In order to extract the reflection coefficient from an APML wall, a pulse source backed by a perfect magnetic conductor (PMC) wall is excited to ensure that the electromagnetic wave propagates only along the forward direction. Due to the introduction of the PMC wall at one end of the structure, we have to use the appropriate images of fields to truncate this PMC boundary as shown in Figure 6.2.

To accurately evaluate the reflection coefficient of a transmission line from an APML wall, we adopt a two-run approach to generate incident and reflected fields. In the first run, we excite a one-way propagation field backed by a PMC wall by choosing the length of the microstrip line to be l_1. The field generated in this run is treated as the total field, $E^n_{y(\text{inc})}(z_i)$. Next, in order to consider the time delay of wave propagation and to remove the boundary effect, we retain the same structure and excitation but double the length of the microstrip line to l_2 ($l_2 \approx 2l_1$) and terminate the time updating before the field approaches the APML wall [11]. The field generated in the second run is treated as the incident field, $E^n_{y(\text{inc})}(z_i)$.

We use the Blackman–Harris window (BHW) function [12, 13] as the excitation function. This window function is defined as

$$W_{\text{BH}}(t_n) = a_1 + a_2 \cos\left(\frac{\pi(t_n - N_c)}{N_c}\right) + a_3 \cos\left(\frac{2\pi(t_n - N_c)}{N_c}\right) + a_4 \cos\left(\frac{3\pi(t_n - N_c)}{N_c}\right) \tag{6.27}$$

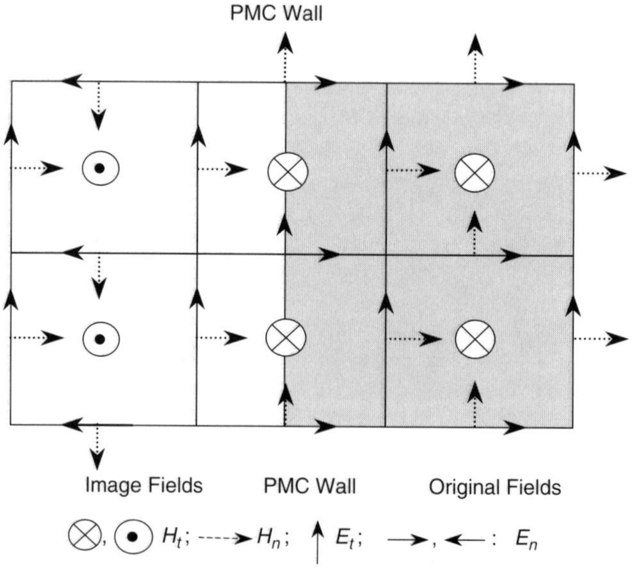

FIGURE 6.2 Image and original fields in the MRTD lattice with respect to a PMC wall.

where $N_c = N_t/2$, N_t is the width of the window function, and a_1, a_2, a_3, and a_4 are constants with the values

$$[\,a_1 \quad a_2 \quad a_3 \quad a_4\,] = [\,0.35875 \quad 0.48829 \quad 0.14128 \quad 0.01168\,] \tag{6.28}$$

A conventional way to determine the value of the constant $\sigma_{\alpha\,\max}$ in (6.11) is by designating the reflection coefficient at a normal incidence on an APML wall [8–11]. For a wave impinging on the APML at angle θ relative to the α-directed surface normal, this reflection factor can be computed using transmission line analysis, since the APML is highly lossy along the direction. Then

$$R(\theta) = \exp\left[-2\frac{\cos\theta}{\varepsilon_0\varepsilon_r V}\int_0^{N_P}\sigma_\alpha\,d\alpha\right] \tag{6.29}$$

where V is an electromagnetic wave propagating velocity in the medium space, N_p is APML thickness, and σ_α is the APML characteristic conductivity referring to propagation in α-direction.

Considering that the wave is normally incident on the APML wall, the reflection coefficient in the propagation z-direction then becomes

$$\begin{aligned} R(0) = R &= \exp\left[-\frac{2}{\varepsilon_0\varepsilon_r V}\int_0^{N_p}\sigma_z^k\,dk\right] \\ &= \exp\left[-\frac{2\sigma_{\alpha\,\max}N_p}{\varepsilon_0\varepsilon_r V(m+1)}\right] = \exp\left[-\frac{2\sigma_{\alpha\,\max}N_p}{\varepsilon_0\sqrt{\varepsilon_r}C(m+1)}\right] \end{aligned} \tag{6.30}$$

Usually, N_p, m, and $R(0)$ are predetermined, so (6.30) can be solved for $\sigma_{\alpha\,\max}$ as follows:

$$\sigma_{\alpha\,\max} = -\frac{\varepsilon_0 C(m+1)}{2N_p\sqrt{\varepsilon_{r\ \mathrm{eff}}}}\ln(R(0)) \quad \alpha = x, y, z \tag{6.31}$$

where C denotes the speed of electromagnetic wave propagation in free space.

6.3 NUMERICAL RESULTS

Empty and Partially Dielectric-Loaded Rectangular Waveguides

In this section, we investigate the absorption performance of the APML for an empty waveguide and for a partially dielectric-loaded rectangular waveguide (WR-90), whose dimensions are $a = 2.286$ cm by $b = 1.016$ cm, as shown in the Figure 6.3. For the dielectric-load waveguide, the dielectric constant of the polystyrene substrate is $\varepsilon_r = 2.56$ with a height of $h = b/3$. The spatial steps are chosen to be $\Delta x = 3.81$, $\Delta y = 3.487$, and $\Delta z = 3.0$ mm, and the time step is

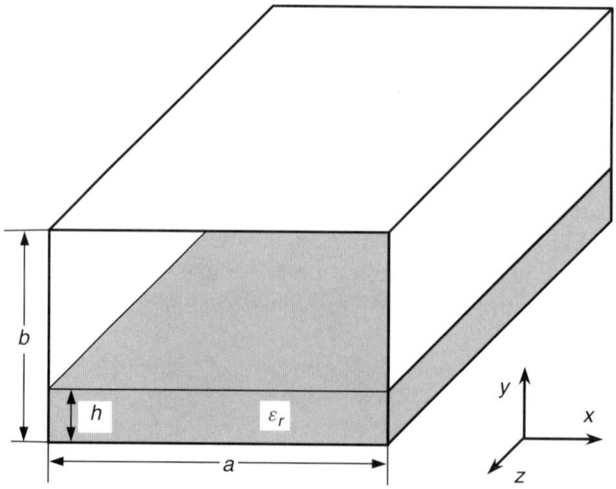

FIGURE 6.3 Geometry of an empty waveguide or a partially dielectric-load rectangular waveguide (WR-90, $a = 2.286$ cm, $b = 1.016$ cm, $h = b/3$, and $\varepsilon_r = 2.56$).

$\Delta t = 2.158$ ps for both cases. Total mesh dimensions are $6\ \Delta x \times 3\ \Delta y \times 100\ \Delta z$ and $6\ \Delta x \times 3\ \Delta y \times 300\ \Delta z$, respectively, for calculation of the total fields and the incident fields. The structures are excited as TE_z and TE_y modes for the empty and dielectric-loaded waveguides, respectively.

Figures 6.4 and 6.5 display the reflection coefficients R_{APML} in decibels (dB) as a function of N_{APML} and $R(0)$, or simply R, for the empty waveguide.

We observe a similar variation for the partially dielectric-loaded waveguide as seen in the Figures 6.6 and 6.7.

As the value of the N_{APML} increases from 4 to 14 layers, the magnitude of R_{APML} dramatically decreases. However, although the magnitude of R_{APML} is also very sensitive to the variation of R, it significantly increases as R decreases.

Open Microstrip Lines with Low and High Dielectric Substrates

To study the low and high dielectric substrate effects on the performance of the APML, we investigate an open microstrip line printed on substrates with $\varepsilon_r = 2.2$ and $\varepsilon_r = 13$, respectively, shown in Figure 6.8. For both cases, the computational volume is discretized with $\Delta x = \Delta y = 1.27$ and $\Delta z = 3.60$ mm, and the time step is chosen as $\Delta t = 0.1059$ ps. A four-layer APML is added to the top and two side walls and an image is placed on the bottom plate. For the case of the substrate with relative permittivity $\varepsilon_r = 2.2$, the computational mesh dimensions are given as $21\Delta x \times 12\Delta y \times 48\Delta z$ and $21\Delta x \times 12\Delta y \times 95\Delta z$ for calculation of the total fields and the incident fields, respectively; while for the case of the substrate with $\varepsilon_r = 13.0$, the mesh dimensions are $21\Delta x \times 12\Delta y \times 40\Delta z$ and $21\Delta x \times 12\Delta y \times 85\Delta z$, correspondingly. For both cases, the conductor strip is characterized by $w = 3\Delta x$, $h = 2\Delta y$, and infinitely thin thickness.

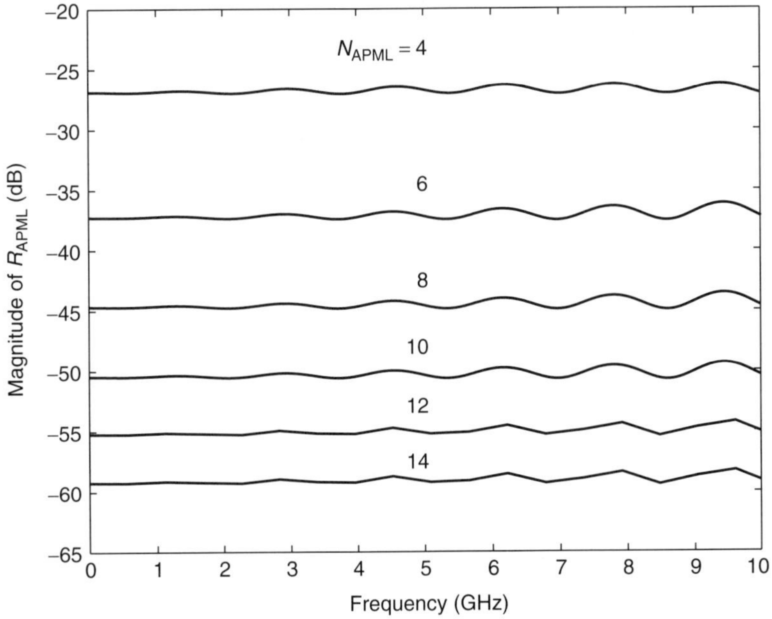

FIGURE 6.4 $|R_{APML}|$ with different values of N_{APML} ($R = 10^{-7}$, $m = 2$) for the empty waveguide.

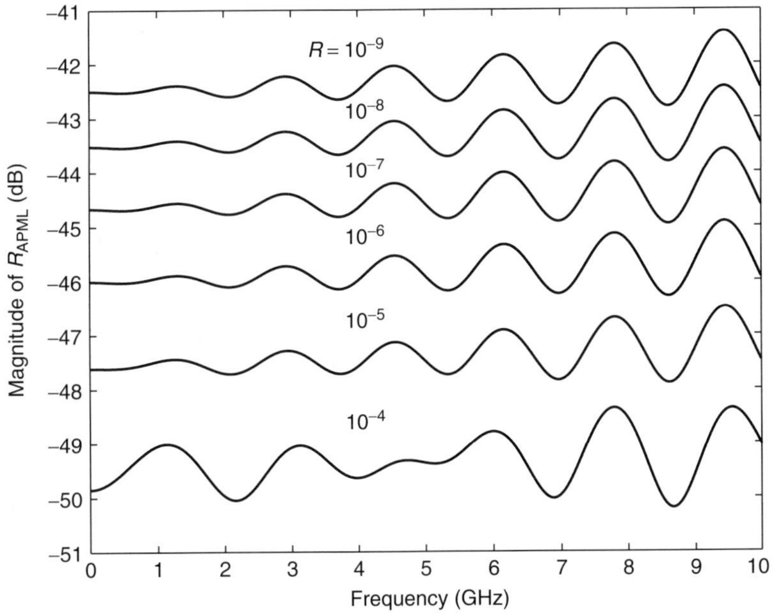

FIGURE 6.5 $|R_{APML}|$ with different values of R ($N_{APML} = 8$, $m = 2$) for the empty waveguide.

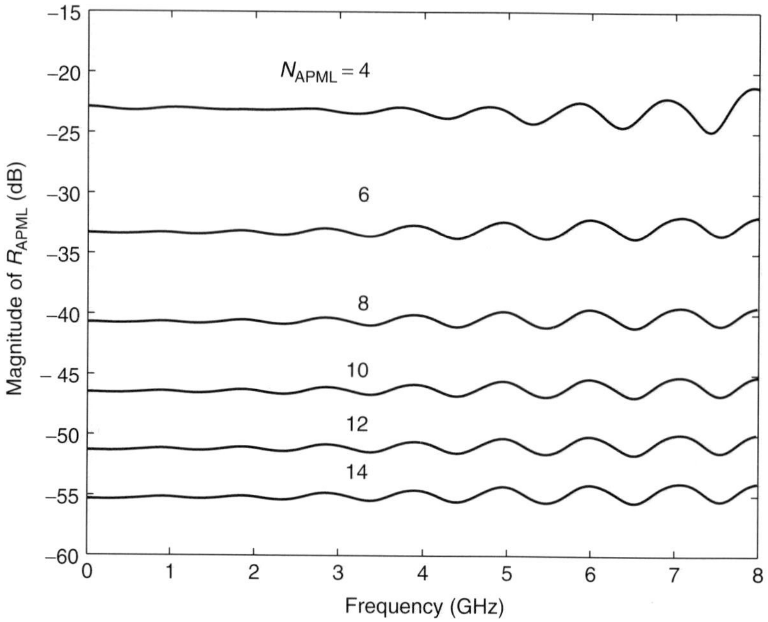

FIGURE 6.6 $|R_{APML}|$ with different values of N_{APML} ($R = 10^{-7}$, $m = 2$) for the dielectric-loaded waveguide.

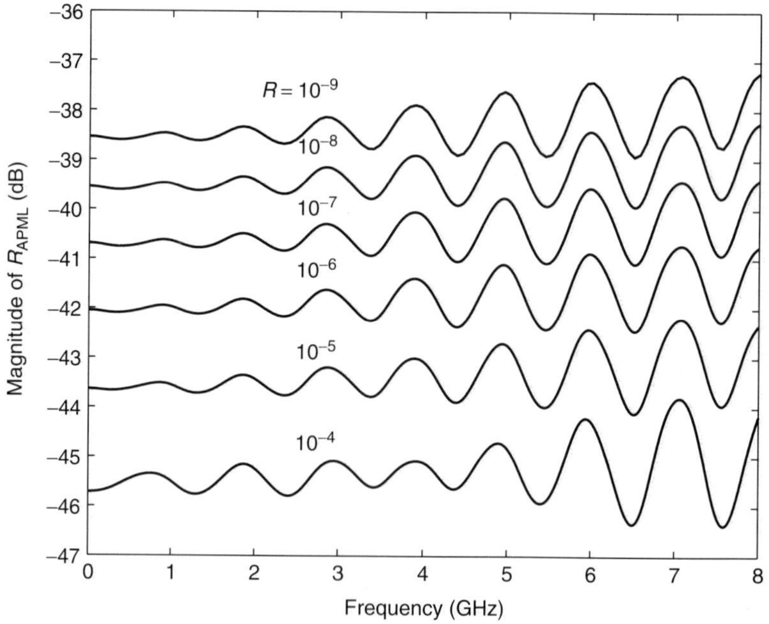

FIGURE 6.7 $|R_{APML}|$ with different values of R ($N_{APML} = 8$, $m = 2$) for the dielectric-loaded waveguide.

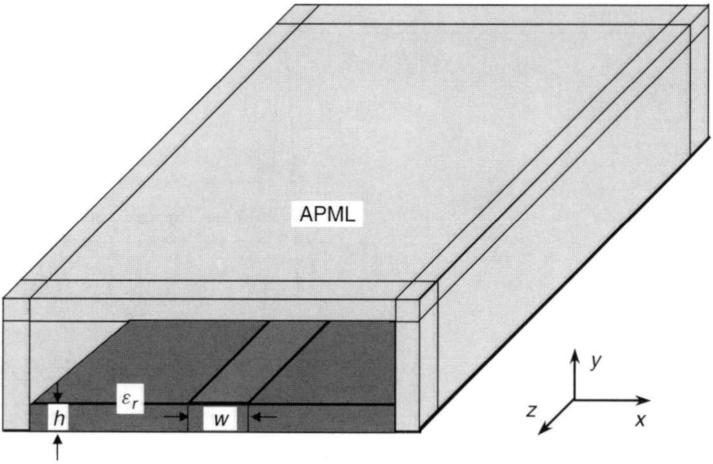

FIGURE 6.8 Geometry of a microstrip line with substrate ε_r.

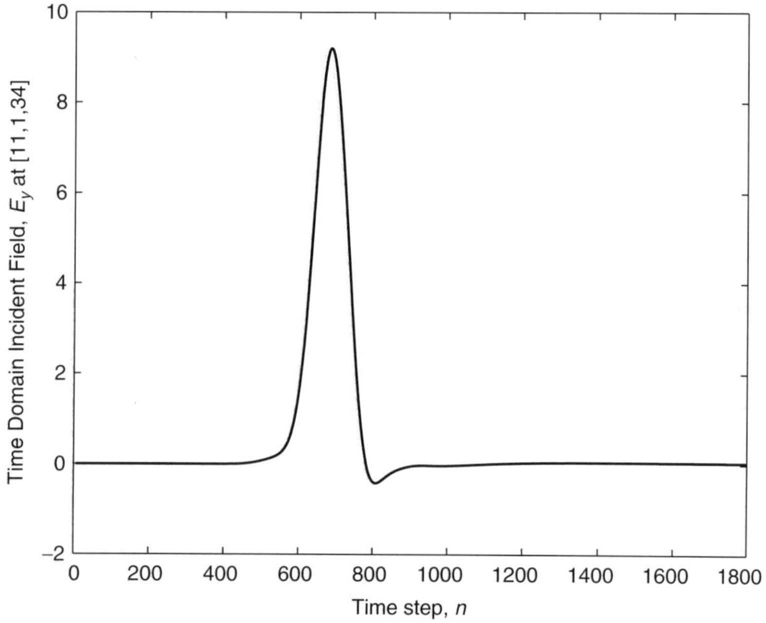

FIGURE 6.9 Incident field in the time domain ($R = 10^{-7}$, $m = 2$) for an open microstrip line with the substrate $\varepsilon_r = 2.2$.

Figures 6.9 and 6.10 show the time domain incident and total field signatures for different APML layers for the structure substrate with $\varepsilon_r = 2.2$. In Figure 6.11 we plot the extracted R_{APML} as a function of APML layers in the frequency range up to 60 GHz. Obviously, as N_{APML} increases, the magnitude of R_{APML}

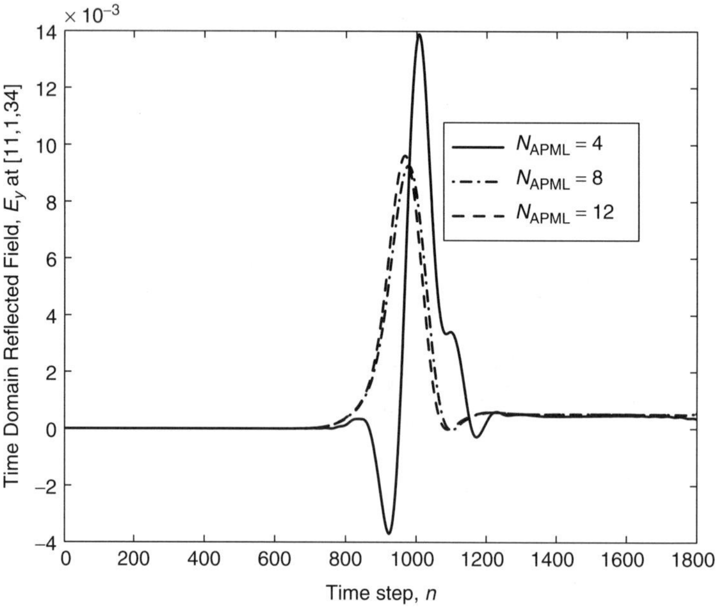

FIGURE 6.10 Reflected fields in the time domain ($R = 10^{-7}$, $m = 2$) for an open microstrip line with the substrate $\varepsilon_r = 2.2$.

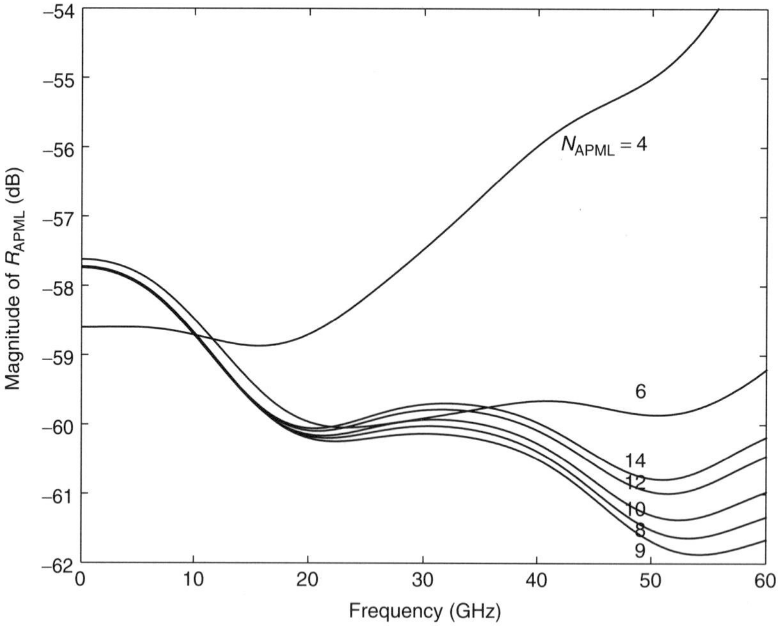

FIGURE 6.11 $|R_{\mathrm{APML}}|$ with different values of N_{APML} ($R = 10^{-7}$, $m = 2$) for an open microstrip line with the substrate $\varepsilon_r = 2.2$.

dramatically decreases in the high-frequency range. However, little significant change is seen in the low-frequency region. Note that there is an absorbing saturation state for the APML at $N_{\mathrm{APML}} = 9$ since the APML absorbing performance deteriorates when $N_{\mathrm{APML}} > 9$.

Next, as shown in Figure 6.12, R_{APML} is insensitive to the variation of R in spite of the fact that it varies greatly from 10^{-4} to 10^{-7}. Finally, as seen in Figure 6.13, we obtain the best APML performance when the order of the polynomial variation in the APML is $m = 2$.

Likewise, Figures 6.14 and 6.15 display the time domain incident and reflected fields for the substrate of microstrip line with $\varepsilon_r = 13$. As shown in Figure 6.16 and by comparison with the previous case, the APML thickness still largely determines its absorption performance, although this effect is weakened when the dielectric constant of the substrate is high.

Again, as shown in Figure 6.17, little significant change in R_{APML} is seen as R varies from 10^{-4} to 10^{-9}.

In order to demonstrate the effect of the dielectric constants on the APML absorption performance, we plot the magnitude of R_{APML} in Figure 6.18 for an open microstrip line with different substrate dielectric constants, namely, $\varepsilon_r = 1$, 2.2, 5, 10, and 13, respectively, by using the same geometrical structure, discretization, and APML. It is apparent that the APML with $\varepsilon_r = 1$ exhibits the best absorbing performance since it presents the lowest value of the magnitude of reflection coefficient.

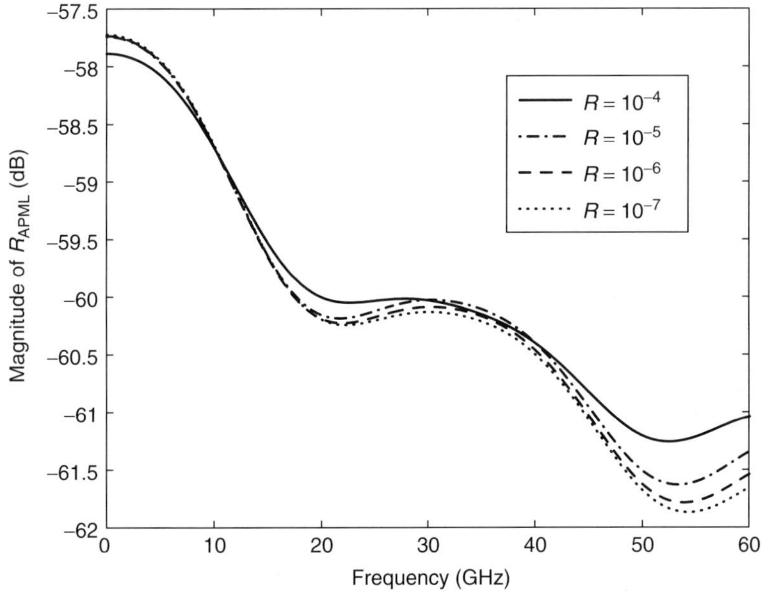

FIGURE 6.12 $|R_{\mathrm{APML}}|$ with different values of R ($N_{\mathrm{APML}} = 8$, $m = 2$) for an open microstrip line with the substrate $\varepsilon_r = 2.2$.

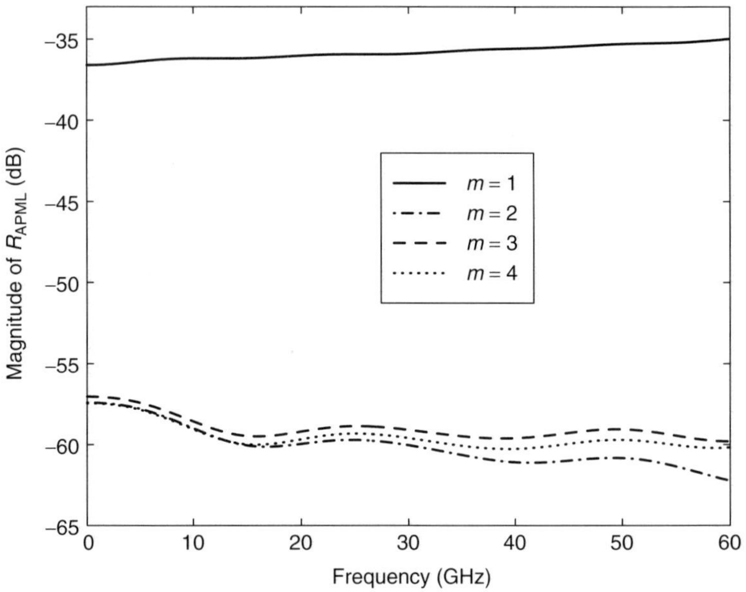

FIGURE 6.13 $|R_{APML}|$ with different values of m ($N_{APML} = 8$ and $R = 10^{-7}$) for an open microstrip line with the substrate $\varepsilon_r = 2.2$.

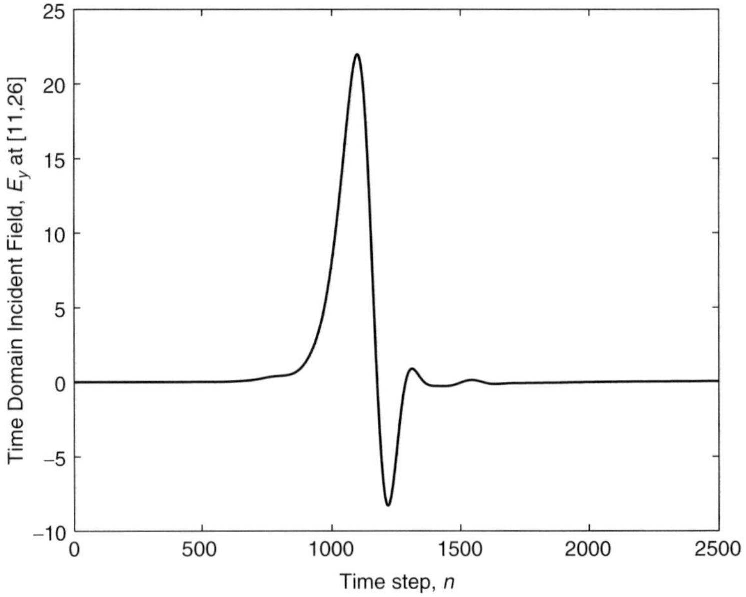

FIGURE 6.14 Incident field in the time domain ($R = 10^{-7}, m = 2$) for an open microstrip line with the substrate $\varepsilon_r = 13$.

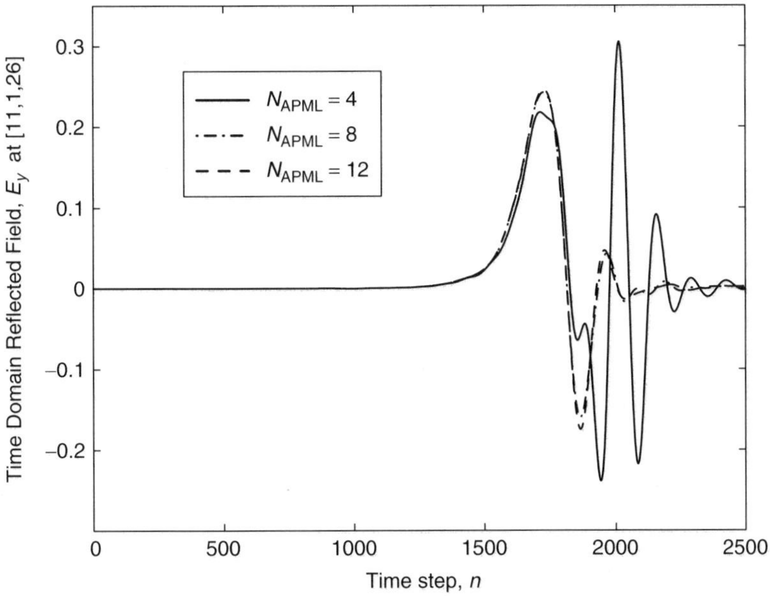

FIGURE 6.15 Reflected fields in the time domain ($R = 10^{-7}$, $m = 2$) for an open microstrip line with the substrate $\varepsilon_r = 13$.

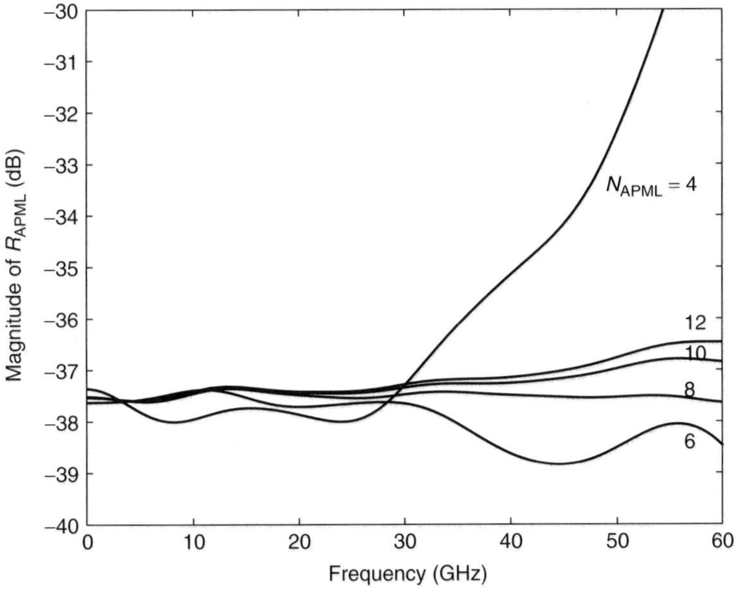

FIGURE 6.16 $|R_{APML}|$ with different values of N_{APML} ($R = 10^{-7}$, $m = 2$) for an open microstrip line with the substrate $\varepsilon_r = 13$.

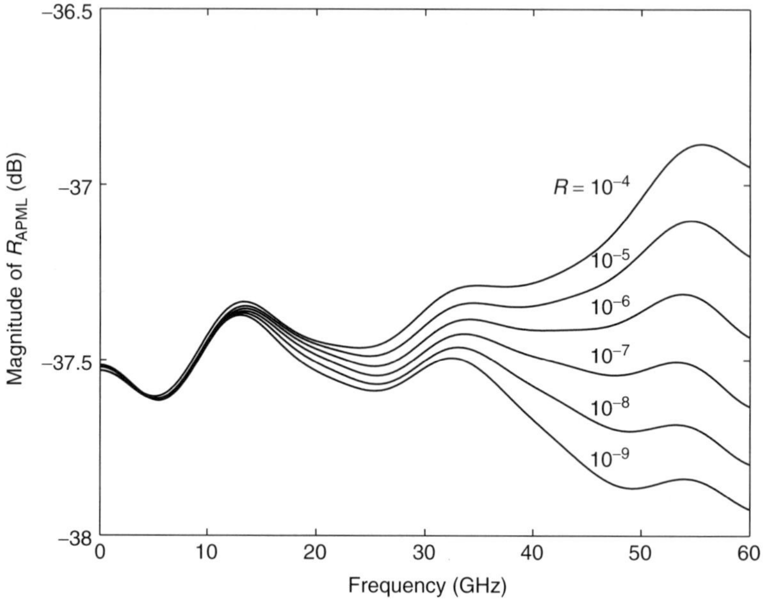

FIGURE 6.17 $|R_{APML}|$ with different values of R ($N_{APML} = 8$, $m = 2$) for an open microstrip line with the substrate $\varepsilon_r = 13$.

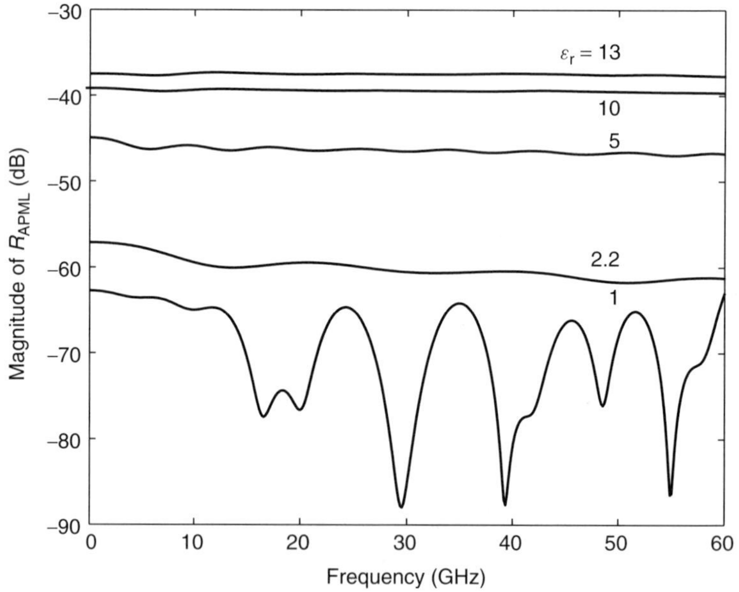

FIGURE 6.18 Magnitude of APML reflection coefficient for an open microstrip line with different substrate dielectric constants ($N_{APML} = 8$, $R = 10^{-7}$).

6.4 CONCLUSION

In this chapter, we successfully applied an APML in the implementation of the MRTD scheme for mesh truncation. To truncate boundaries for a grounded open structure and to form a forward propagation wave, we have combined this APML technique with applications of the images of PMC walls. The absorption performance of the APML as a function of N_{APML}, R, m, as well as the substrate dielectric constant ε_r has been examined extensively. It is found that R_{APML} is significantly affected by both N_{APML} and R and is optimized with $m = 2$. It is also found that the APML performs better in a structure with a low dielectric constant rather than a high one.

REFERENCES

[1] K. S. Yee, "Numerical solution of initial boundary value problems involving Maxwell's equations in isotropic media," *IEEE Trans. Antennas Propag.*, vol. AP-14, no. 3, pp. 302–307, May 1966.

[2] Q. H. Liu, "The PSTD algorithm: A time-domain method requiring only two cells per wavelength," *Microwave Opt. Technol. Lett.*, vol. 15, no. 3, pp. 158–165, June 1997.

[3] M. Krumpholz and L. P. B. Katehi, "MRTD: New time domain schemes based on multiresolution analysis," *IEEE Trans. Microwave Theory Tech.*, vol. 44, no. 4, pp. 555–571, Apr. 1996.

[4] Q. Cao and Y. Chen, "MRTD analysis of a transient electromagnetic pulse propagating through a dielectric layer," *Int. J. Electron.*, vol. 86, no. 4, pp. 459–474, 1999.

[5] Q. Cao, Y. Chen, and R. Mittra, "Multiple image technique (MIT) and anisotropic perfectly matched layer (APML) in implementation of MRTD scheme for boundary truncations of microwave structures," *IEEE Trans. Microwave Theory Tech.*, vol. 50, no. 6, pp. 1578–1589, June 2002.

[6] J. P. Berenger, "A perfectly matched layer for the absorption of electromagnetic waves," *J. Comput. Phys.*, vol. 114, pp. 185–200, Oct. 1994.

[7] Z. S. Sacks, D. M. Kingsland, R. Lee, and J. F. Lee, "A perfectly matched layer anisotropic absorber for use as an absorbing boundary condition," *IEEE Trans. Antennas Propag.*, vol. 43, no. 12, pp. 1460–1463, Dec. 1995.

[8] S. D. Gedney, "An anisotropic perfectly matched layered-absorbing medium for the truncation of FDTD lattice," *IEEE Trans. Antennas Propag.*, vol. 44, no. 12, pp. 1630–1939, Dec. 1996.

[9] L. P. B. Katehi, J. F. Harvey, and E. Tentzeris, "Time-domain analysis using multiresolution expansions," Chapter 3, in *Advances in Computational Electrodynamics: The Finite-Difference Time-Domain Method* (ed. A. Taflove), Artech House, Norwood, MA, 1998.

[10] Q. Cao and Y. Chen, "Application of a perfectly matched layer absorber for open boundary truncation in multiresolution time domain scheme," *IEEE Trans. Antennas Propag.*, vol. 51, no. 2, pp. 350–357, Feb. 2003.

[11] A. Taflove and S. C. Hagness, *Computational Electrodynamics: The Finite-Difference Time-Domain Method*, Artech House, Norwood, MA, 2000.

[12] Y. Chen, R. Mittra, and P. Harms, "Finite difference time domain algorithm for solving Maxwell's equations in rotationally symmetric geometries," *IEEE Trans. Microwave Theory Tech.*, vol. MTT-44, no. 6, pp. 832–839, June 1996.

[13] F. J. Harris, "On the use of windows for harmonic analysis with discrete Fourier transform," *Proc. IEEE*, vol. 66, pp. 51–83, Jan. 1978.

CHAPTER SEVEN

One-Dimensional MRTD Analysis

7.1 INTRODUCTION

As its name indicates, the multiresolution time domain (MRTD) method is based on what is known in the literature as multiresolution analysis (MRA) and has been applied successfully to a variety of electromagnetic problems [1–3]. It has been shown that the MRTD analysis can offer significant savings in computer memory as well as time. However, because of the complexity of implementation, applications of the MRTD scheme have been rather limited, especially when the MRTD is used with high-level multiresolution wavelets.

In this chapter, we discuss the MRTD approach based on the cubic spline Battle–Lemarié scaling and high-level multiresolution wavelet functions. We present a set of generalized orthogonality and integral relationships among the basis functions in order to derive the MRTD update equations. In particular, to increase computational accuracy with little addition of computational burden, we introduce the concept of the extended discontinuity subregion, where the electromagnetic fields change rapidly. We expand the fields in such a region by using both the scaling function (ϕ) and the wavelet function (ψ), while employing the scaling function expansions in the entire computational region. For didactic reasons we demonstrate the above technique by considering the problem of one-dimensional (1D) analysis of a transient wave pulse propagating through dielectric layers [3, 4].

7.2 MRTD FORMULATIONS

Maxwell Governing Equations

We consider a TEM_x plane wave propagating along the x-direction as shown in Figure 7.1 by using generalized formulations that include both the MRTD

Multiresolution Time Domain Scheme for Electromagnetic Engineering
By Yinchao Chen, Qunsheng Cao, and Raj Mittra
ISBN 0-471-27230-2

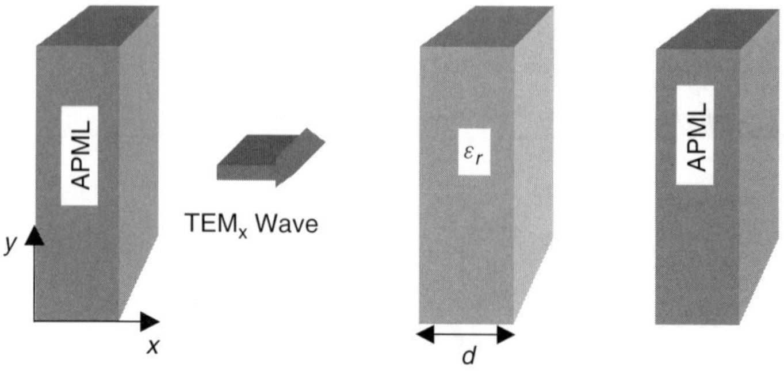

FIGURE 7.1 TEM$_x$ wave incident on a dielectric slab, where d is the width of the dielectric slab.

and anisotropic perfectly matched layer (APML) absorbing regions [5]. We can simplify Maxwell equations to obtain

$$\frac{\partial H_z}{\partial x} = -j\omega s_x D_y \tag{7.1}$$

$$D_y = \varepsilon E_y \tag{7.2}$$

$$\frac{\partial E_y}{\partial x} = -j\omega \mu_0 s_x H_z \tag{7.3}$$

with

$$s_x = \begin{cases} 1 + \dfrac{\sigma_x}{j\omega\varepsilon_0} & \text{in APML regions} \\ 1 & \text{non-APML region } (\sigma_x = 0) \end{cases} \tag{7.4}$$

where σ_x is the APML electric conductivity, and the permittivity ε is given as

$$\varepsilon = \varepsilon_0 \varepsilon_r(x) = \begin{cases} \varepsilon_0 \varepsilon_r & \text{in the dielectric regions} \\ \varepsilon_0 & \text{elsewhere} \end{cases} \tag{7.5}$$

SW-MRTD Field Expansions

We begin by expanding the electromagnetic field quantities in terms of the scaling and wavelet functions as follows:

$$E_y(x,t) = \sum_{i,n=-\infty}^{+\infty} \left[{}_{\phi y}E_i^n \phi_i(x) + \sum_{s=0}^{s_{\max}} \sum_{l=0}^{2^s-1} {}_{\psi y}{}_l^s E_{i+2^{-s}(l+1/2)}^n \psi_{s,i+2^{-s}(l+1/2)}(x) \right] h_n(t) \tag{7.6}$$

$$D_y(x,t) = \sum_{i,n=-\infty}^{+\infty} \left[{}_{\phi y}D_i^n \phi_i(x) + \sum_{s=0}^{s_{\max}} \sum_{l=0}^{2^s-1} {}_{\psi y}{}_l^s D_{i+2^{-s}(l+1/2)}^n \psi_{s,i+2^{-s}(l+1/2)}(x) \right] h_n(t) \tag{7.7}$$

$$
\begin{aligned}
H_z(x,t) = \sum_{i,n=-\infty}^{+\infty} \Bigg[{}_{\phi z}H_{i+1/2}^{n+1/2}\phi_{i+1/2}(x) \\
+ \sum_{s=0}^{s_{\max}} \sum_{l=0}^{2^s-1} {}_{\psi z l}^{\;s}H_{i-1/2+2^{-s}(l+1/2)}^{n+1/2} \psi_{s,i-1/2+2^{-s}(l+1/2)}(x) \Bigg] h_{n+1/2}(t)
\end{aligned}
\tag{7.8}
$$

where ${}_{\phi y}E_i^n$, ${}_{\psi y l}^{\;s}E_{i+2^{-s}(l+1/2)}^n$, ${}_{\phi z}H_{i+1/2}^{n+1/2}$, and ${}_{\psi z l}^{\;s}H_{i-1/2+2^{-s}(l+1/2)}^{n+1/2}$ are the field expansion coefficients in terms of the scaling and wavelet functions. Herein, we define the translated scaling and wavelet functions as

$$
\phi_i(x) = \phi\left(\frac{x}{\Delta x} - i\right) \tag{7.9a}
$$

$$
\psi_{i+1/2}(x) = \psi\left(\frac{x}{\Delta x} - i\right) \tag{7.9b}
$$

where the symmetrical points of the scaling and wavelet functions are $x = 0$ and $x = \frac{1}{2}$ for $i = 0$, respectively.

In general, we define an arbitrary resolution level s of the wavelet function as

$$
\psi_{s,i+2^{-s}(l+1/2)}(x) = 2^{s/2}\psi\left[2^s\left(\frac{x}{\Delta x} - i - \frac{l}{2^s}\right)\right] = 2^{s/2}\psi\left[2^s\left(\frac{x}{\Delta x} - i\right) - l\right] \tag{7.10}
$$

where there exist 2^s wavelets for a fixed value of $s_{\max}$ as given in (7.6) through (7.8). In other words, the symmetric point of the s level wavelet is located at $x = l + 1/2^{s+1}$, with $l = 0, 1, \ldots, 2^{s_{\max}} - 1$. For example if $s_{\max} = 1$, then $l = 0$, 1, and the symmetric points of the wavelets are located at $x = \frac{1}{4}$ and $x = \frac{3}{4}$. For $s_{\max} = 2$, $l = 0$, 1, 2, 3, the symmetrical point centers are located at $x = \frac{1}{8}$, $x = \frac{3}{8}$, $x = \frac{5}{8}$, and $x = \frac{7}{8}$. Figure 7.2 shows the positions of the E- and H-field expansion coefficients as given in (7.6) through (7.8) in correspondence with the scaling and wavelet functions in a 1D cell distribution pattern.

SW-MRTD Update Equations for TEM$_x$ Model

Substituting the field expansions (7.6)–(7.8) into Maxwell equations (7.1) and (7.3), and applying Galerkin's method [6], we can derive the field update equations

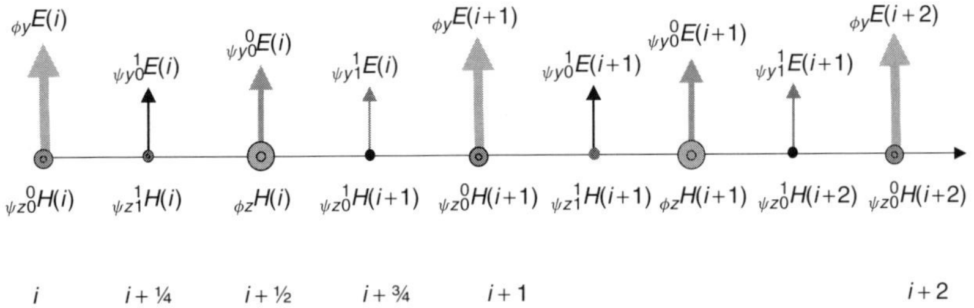

FIGURE 7.2 Position pattern of E- and H-field expansion coefficients in 1D cell distribution.

for time stepping in the context of the leapfrog scheme for field computation. For example, for a generalized medium we can rewrite (7.3) in the time domain as

$$\frac{\partial E_y}{\partial x} = -\mu_0 \frac{\partial H_z}{\partial t} - \frac{\mu_0 \sigma_x}{\varepsilon_0} H_z \tag{7.11}$$

Choosing the complex conjugant of the function $\phi_{i+1/2}(x)h_n(t)$ as the test function, sampling (7.11) in space and time, and using the appropriate orthogonal and integration relations (see Appendix B), we obtain the left-hand side (LHS) of (7.11) as

$$\begin{aligned}
&\int_{-\infty}^{+\infty}\int_{-\infty}^{+\infty} \phi_{i+1/2}(x)h_n(t)\left(\frac{\partial}{\partial x}\sum_{i',n'=-\infty}^{+\infty}\left[{}_{\phi y}E_{i'}^{n'}\phi_{i'}(x)\right.\right.\\
&\quad\left.\left.+\sum_{s=0}^{s_{\max}}\sum_{l=0}^{2^s-1}{}_{\psi y}{}^{s}_{l}E^{n'}_{i'+2^{-s}(l+1/2)}\psi_{s,i'+2^{-s}(l+1/2)}(x)\right]h_{n'}(t)\right)dx\,dt\\
&=\sum_{i'=-\infty}^{+\infty}\left[{}_{\phi y}E_{i'}^{n}\int_{+\infty}^{+\infty}\phi_{i+1/2}(x)\frac{\partial\phi_{i'}(x)}{\partial x}dx\right.\\
&\quad\left.+\sum_{s=0}^{s_{\max}}\sum_{l=0}^{2^s-1}{}_{\psi y}{}^{s}_{l}E^{n}_{i'+2^{-s}(l+1/2)}\int_{+\infty}^{+\infty}\phi_{i+1/2}(x)\frac{\partial\psi_{s,i'+2^{-s}(l+1/2)}(x)}{\partial x}dx\right]\Delta t\\
&=\sum_{\nu=-\infty}^{+\infty}{}_{\phi y}E_{i'}^{n}a(\nu)\delta_{i',i+\nu+1}\Delta t+\sum_{\nu=-\infty}^{+\infty}\sum_{s=0}^{s_{\max}}\sum_{l=0}^{2^S-1}{}_{\psi y}{}^{s}_{l}E^{n}_{i'+2^{-s}(l+1/2)}d_{s,l}(\nu)\delta_{i',i+\nu}\Delta t\\
&=\sum_{\nu=-\infty}^{+\infty}{}_{\phi y}E_{i+\nu+1}^{n}a(\nu)\ \Delta t+\sum_{\nu=-\infty}^{+\infty}\sum_{s=0}^{s_{\max}}\sum_{l=0}^{2^S-1}{}_{\psi y}{}^{s}_{l}E^{n}_{i+\nu+2^{-s}(1/2+l)}d_{s,l}(\nu)\ \Delta t
\end{aligned} \tag{7.12}$$

where we have used the following integral relations for the scaling and wavelet functions:

$$\int_{+\infty}^{+\infty}\phi_{i+1/2}(x)\frac{\partial\phi_{i'}(x)}{\partial x}dx=\sum_{\nu=-\infty}^{\nu=+\infty}a(\nu)\delta_{i',i+\nu+1} \tag{7.13a}$$

$$a(\nu)=\frac{1}{\pi}\int_{0}^{\infty}|\phi(x)|^2\omega\sin\left[\nu+\frac{1}{2}\right]d\omega \tag{7.13b}$$

$$\int_{+\infty}^{+\infty}\phi_{i+1/2}(x)\frac{\partial\psi_{s,i'+2^{-s}(l+1/2)}(x)}{\partial x}dx=\sum_{\nu=-\infty}^{+\infty}d_{s,l}(\nu)\delta_{i',i+\nu} \tag{7.14a}$$

$$d_{s,l}(\nu) = \frac{1}{\pi}\int_0^{\infty} \phi(\omega)|\psi_s(\omega)|\omega \sin\left[\omega\left(\nu - \frac{1}{2} + \frac{l}{2^s} + \frac{1}{2^{s+1}}\right)\right] d\omega \quad (7.14b)$$

The Kronecker symbol $\delta_{n',n}$ in (7.12) is defined as

$$\delta_{n',n} = \begin{cases} 1 & \text{for } n = n' \\ 0 & \text{for } n \neq n' \end{cases}$$

Similar operations on the right-hand side (RHS) of (7.11) lead to

$$\begin{aligned}
&-\mu_0 \int_{-\infty}^{+\infty}\int_{-\infty}^{+\infty} \phi_{i+1/2}(x)h_n(t) \sum_{i,n=-\infty}^{+\infty} \\
&\times \left[\begin{array}{l} {}_{\phi z}H_{i'+1/2}^{n'+1/2}\phi_{i'+1/2}(x) \\ + \displaystyle\sum_{s=0}^{s_{\max}}\sum_{l=0}^{2^s-1} {}_{\psi z l}^{\ s}H_{i'-1/2+2^{-s}(l+1/2)}^{n'+1/2}\psi_{s,i'-1/2+2^{-s}(l+1/2)}(x) \end{array}\right] \frac{\partial h_{n'+1/2}(t)}{\partial t}\, dx\, dt \\
&- \int_{-\infty}^{+\infty}\int_{-\infty}^{+\infty} \phi_{i+1/2}(x)h_n(t)\frac{\mu_0\sigma_x}{\varepsilon_0} \sum_{i,n=-\infty}^{+\infty} \\
&\times \left[\begin{array}{l} {}_{\phi z}H_{i'+1/2}^{n'+1/2}\phi_{i'+1/2}(x) \\ + \displaystyle\sum_{s=0}^{s_{\max}}\sum_{l=0}^{2^s-1} {}_{\psi z l}^{\ s}H_{i'-1/2+2^{-s}(l+1/2)}^{n'+1/2}\psi_{s,i'-1/2+2^{-s}(l+1/2)}(x) \end{array}\right] h_{n'+1/2}(t)\, dx\, dt \\
&= -\mu_0 \sum_{n=-\infty}^{+\infty} {}_{\phi z}H_{i+1/2}^{n'+1/2}\Delta x \left(\int_{-\infty}^{+\infty} h_n(t)\frac{\partial h_{n'+1/2}(t)}{\partial t}\, dt\right) \qquad (7.15) \\
&- \frac{\mu_0\Delta t}{\varepsilon_0}\sum_{i'=-\infty}^{+\infty} {}_{\phi z}H_{i'+1/2}^{n+1/2}\left(\int_{\text{APML}} \sigma_x\phi_{i+1/2}(x)\phi_{i'+1/2}(x)\, dx\right) \\
&- \mu_0 \sum_{n=-\infty}^{+\infty}\left({}_{\phi z}H_{i+1/2}^{n+1/2} - {}_{\phi z}H_{i+1/2}^{n-1/2}\right)\Delta x \\
&- \frac{\mu_0\Delta t}{\varepsilon_0}\sum_{i'=-\infty}^{+\infty}\left({}_{\phi z}H_{i'+1/2}^{n+1/2} + {}_{\phi z}H_{i'+1/2}^{n-1/2}\right)(\sigma_x)_{i,i'}
\end{aligned}$$

where, in addition to the relations summarized in (7.13) and (7.14), we have applied an integral relation for the rectangular pulse function:

$$\int_{-\infty}^{+\infty} h_n(t)\frac{\partial h_{n'+1/2}(t)}{\partial t}\, dt = (\delta_{n',n} - \delta_{n',n-1}) \qquad (7.16)$$

The equivalent conductivity now becomes

$$
\begin{aligned}
(\sigma_x)_{i,i'} &= \frac{1}{\Delta x} \int_{-\infty}^{+\infty} \phi_{i+1/2}(x)\sigma_x(x)\phi_{i'+1/2}(x)\,dx \\
&= \int_{\text{APML}} \phi(x' - i - \tfrac{1}{2})\sigma_x(x'\ \Delta x)\phi(x' - i' - \tfrac{1}{2})\,dx'
\end{aligned}
\tag{7.17}
$$

Next we define a new APML material, which is characterized only by the diagonal terms of (7.17), namely,

$$
(\sigma_x)_{i,i'} = \delta_{i,i'} \int_{\text{APML}} \phi^2(x' - i)\sigma_x(x'\ \Delta x)\,dx' = \delta_{i,i'}\sigma_x^i \equiv \delta_{i,i'}\sigma_{\alpha\,\max} \left| \frac{i - i_0}{d} \right|^m \tag{7.18}
$$

where d is the thickness of the APML region and m is the power of the polynomial with the value of 2. As a result, the update equation for ${}_{\phi z}H_{i+1/2}^{n+1/2}$ is derived as

$$
\begin{aligned}
{}_{\phi z}H_{i+1/2}^{n+1/2} &= \left(\frac{1 - \dfrac{\sigma_x^i\ \Delta t}{2\varepsilon_0}}{1 + \dfrac{\sigma_x^i\ \Delta t}{2\varepsilon_0}} \right) {}_{\phi z}H_{i+1/2}^{n-1/2} - \frac{1}{\mu_0 \left(1 + \dfrac{\sigma_x^i\ \Delta t}{2\varepsilon_0} \right)} \\
&\times \left(\sum_{\nu=-9}^{+8} a(\nu)_{\phi y}E_{i+\nu+1}^n + \sum_{s=0}^{s_{\max}} \sum_{l=0}^{2^s-1} \sum_{\nu=-\infty}^{+\infty} d_{s,l}(\nu)_{\psi y l}^{\ s}E_{i+\nu+2^{-s}(l+1/2)}^n \right) \frac{\Delta t}{\Delta x}
\end{aligned}
\tag{7.19}
$$

Similarly, we can obtain the update equations for the remaining components, which read

$$
\begin{aligned}
{}_{\psi z l}^{\ s}H_{i-1/2+2^{-s}(l+1/2)}^{n+1/2} &= \left(\frac{1 - \dfrac{\sigma_x^i\ \Delta t}{2\varepsilon_0}}{1 + \dfrac{\sigma_x^i\ \Delta t}{2\varepsilon_0}} \right) {}_{\psi z l}^{\ s}H_{i-1/2+2^{-s}(l+1/2)}^{n-1/2} - \frac{1}{\mu_0 \left(1 + \dfrac{\sigma_x^i\ \Delta t}{2\varepsilon_0} \right)} \\
&\times \left(\sum_{\nu=-\infty}^{\nu=+\infty} c_{s,l}(\nu)_{\phi y}E_{i+\nu}^n + \sum_{s'=0}^{s_{\max}} \sum_{l'}^{2^s-1} \sum_{\nu=-\infty}^{\nu=+\infty} b_{l,l'}^{s,s'}\ {}_{\psi y l'}^{\ s'}E_{i+\nu+2^{-s'}(l'+1/2)}^n \right) \frac{\Delta t}{\Delta x}
\end{aligned}
\tag{7.20}
$$

where we use the integral relations

$$
\int_{+\infty}^{+\infty} \psi_{s,i+2^{-s}(l+1/2)}(x) \frac{\partial \phi_{i'+1/2}(x)}{\partial x}\,dx = \sum_{\nu=-\infty}^{+\infty} c_{s,l}(\nu)\delta_{i',i+\nu} \tag{7.21a}
$$

$$c_{s,l}(\nu) = \frac{1}{\pi} \int_{0}^{\infty} \phi(\omega)|\psi_s(\omega)|\omega \sin\left[\omega\left(\nu + \frac{1}{2} - \frac{l}{2^s} - \frac{1}{2^{s+1}}\right)\right] d\omega \tag{7.21b}$$

$$\int_{+\infty}^{+\infty} \psi_{s,i-1/2+2^{-s}(l+1/2)}(x) \frac{\partial \psi_{s,i'+2^{-s'}(l'+1/2)}(x)}{\partial x} dx = \sum_{\nu=-\infty}^{+\infty} b_{l,l'}^{s,s'}(\nu)\delta_{i',i+\nu} \tag{7.22a}$$

$$b_{l,l'}^{s,s'}(\nu) = \frac{1}{\pi} \int_{0}^{\infty} |\psi_s(\omega)||\psi_{s'}(\omega)|\omega \sin \times \left[\omega\left(\nu + \frac{1}{2} + \frac{l'}{2^{s'}} - \frac{l}{2^s} + \frac{1}{2^{s'+1}} - \frac{1}{2^{s+1}}\right)\right] d\omega \tag{7.22b}$$

$$_{\phi y}D_i^{n+1} = \left(\frac{1 - \dfrac{\sigma_x^i \Delta t}{2\varepsilon_0}}{1 + \dfrac{\sigma_x^i \Delta t}{2\varepsilon_0}}\right) {_{\phi y}D_i^n} - \frac{1}{\left(1 + \dfrac{\sigma_x^i \Delta t}{2\varepsilon_0}\right)} \times \left(\sum_{\nu=-\infty}^{\nu=+\infty} a(\nu)_{\phi z}H_{i+\nu+1/2}^{n+1/2} + \sum_{s=0}^{s_{\max}} \sum_{l=0}^{2^s-1} \sum_{\nu=-\infty}^{\nu=+\infty} d_{s,l}(\nu)_{\psi z}{}_l^s H_{(i-1/2+\nu)+2^{-s}(l+1/2)}^{n+1/2}\right) \frac{\Delta t}{\Delta x} \tag{7.23}$$

$$_{\psi y}^{\ s}D_{i+2^{-s}(l+1/2)}^{n+1} = \left(\frac{1 - \dfrac{\sigma_x^i \Delta t}{2\varepsilon_0}}{1 + \dfrac{\sigma_x^i \Delta t}{2\varepsilon_0}}\right) {_{\psi y}^{\ s}D_{i+2^{-s}(l+1/2)}^n} - \frac{1}{\left(1 + \dfrac{\sigma_x^i \Delta t}{2\varepsilon_0}\right)} \left(\sum_{\nu=-\infty}^{\nu=+\infty} c_{s,l}(\nu)_{\phi z}H_{i+\nu+1/2}^{n+1/2} + \sum_{s'=0}^{s_{\max}} \sum_{l'=0}^{2^s-1} \sum_{\nu=-\infty}^{\nu=+\infty} b_{l,l'}^{s,s'}(\nu)_{\psi z}{}_{l'}^{s'} H_{(i+1/2+\nu)+2^{-s'}(l'+1/2)}^{n+1/2}\right) \frac{\Delta t}{\Delta x} \tag{7.24}$$

The above update equations are valid in all regions, which include both the APML as well as the normal MRTD regions. We can obtain the update equations in non-APML regions by simply setting $\sigma_x^i \equiv 0$ $(s_x(x) \equiv 1)$ in (7.19) and (7.20) and (7.22) and (7.23).

In practice, to avoid complexity of formulation, we only apply $s_{\max} \equiv 1$ in electromagnetic update equations. In order to solve the update equation for the remaining E- and H-fields, we apply the same discretization procedure to (7.2), obtaining

$$
{}_{\phi z}H_{i+1/2}^{n+1/2} = \left(\frac{1 - \dfrac{\sigma_x^i \Delta t}{2\varepsilon_0}}{1 + \dfrac{\sigma_x^i \Delta t}{2\varepsilon_0}} \right) {}_{\phi z}H_{i+1/2}^{n-1/2} - \frac{1}{\mu_0 \left(1 + \dfrac{\sigma_x^i \Delta t}{2\varepsilon_0} \right)}
\times \left[\begin{array}{l} \displaystyle\sum_{v=-\infty}^{+\infty} a(v)_{\phi y}E_{i+v+1}^{n} + \sum_{v=-\infty}^{+\infty} d_{0,0}(v)_{\psi y}{}_0^0E_{i+v+1/2}^{n} \\ \displaystyle + \sum_{v=-\infty}^{+\infty} d_{1,0}(v)_{\psi y}{}_0^1E_{i+v+1/4}^{n} + \sum_{v=-\infty}^{+\infty} d_{1,1}(v)_{\psi y}{}_1^1E_{i+v+3/4}^{n} \end{array} \right] \frac{\Delta t}{\Delta x} \tag{7.25}
$$

$$
{}_{\psi z}{}_0^0H_{i}^{n+1/2} = \left(\frac{1 - \dfrac{\sigma_x^i \Delta t}{2\varepsilon_0}}{1 + \dfrac{\sigma_x^i \Delta t}{2\varepsilon_0}} \right) {}_{\psi z}{}_0^0H_{i}^{n-1/2} - \frac{1}{\mu_0 \left(1 + \dfrac{\sigma_x^i \Delta t}{2\varepsilon_0} \right)}
\times \left[\begin{array}{l} \displaystyle\sum_{v=-\infty}^{+\infty} c_{0,0}(v)_{\phi y}E_{i+v}^{n} + \sum_{v=-\infty}^{+\infty} b_{00}^{00}(v)_{\psi y}{}_0^0E_{i+v+1/2}^{n} \\ \displaystyle + \sum_{v=-\infty}^{+\infty} b_{00}^{01}(v)_{\psi y}{}_0^1E_{i+v+1/4}^{n} + \sum_{v=-\infty}^{+\infty} b_{01}^{01}(v)_{\psi y}{}_1^1E_{i+v+3/4}^{n} \end{array} \right] \frac{\Delta t}{\Delta x} \tag{7.26}
$$

$$
{}_{\psi z}{}_0^1H_{i-1/4}^{n+1/2} = \left(\frac{1 - \dfrac{\sigma_x^i \Delta t}{2\varepsilon_0}}{1 + \dfrac{\sigma_x^i \Delta t}{2\varepsilon_0}} \right) {}_{\psi z}{}_0^1H_{i-1/4}^{n-1/2} - \frac{1}{\mu_0 \left(1 + \dfrac{\sigma_x^i \Delta t}{2\varepsilon_0} \right)}
\times \left[\begin{array}{l} \displaystyle\sum_{v=-\infty}^{+\infty} c_{1,0}(v)_{\phi y}E_{i+v}^{n} + \sum_{v=-\infty}^{+\infty} b_{0,0}^{1,0}(v)_{\psi y}{}_0^0E_{i+v+1/2}^{n} \\ \displaystyle + \sum_{v=-\infty}^{+\infty} b_{0,0}^{1,1}(v)_{\psi y}{}_0^1E_{i+v+1/4}^{n} + \sum_{v=-\infty}^{+\infty} b_{0,1}^{1,1}(v)_{\psi y}{}_1^1E_{i+v+3/4}^{n} \end{array} \right] \frac{\Delta t}{\Delta x} \tag{7.27}
$$

$$
{}_{\psi z}{}_1^1H_{i+1/4}^{n+1/2} = \left(\frac{1 - \dfrac{\sigma_x^i \Delta t}{2\varepsilon_0}}{1 + \dfrac{\sigma_x^i \Delta t}{2\varepsilon_0}} \right) {}_{\psi z}{}_1^1H_{i+1/4}^{n-1/2} - \frac{1}{\mu_0 \left(1 + \dfrac{\sigma_x^i \Delta t}{2\varepsilon_0} \right)}
\times \left[\begin{array}{l} \displaystyle\sum_{v=-\infty}^{+\infty} c_{1,1}(v)_{\phi y}E_{i+v}^{n} + \sum_{v=-\infty}^{+\infty} b_{1,0}^{1,0}(v)_{\psi y}^{\ 0}E_{i+v+1/2}^{n} \\ \displaystyle + \sum_{v=-\infty}^{+\infty} b_{1,0}^{1,1}(v)_{\psi y}^{\ 1}E_{i+v+1/4}^{n} + \sum_{v=-\infty}^{+\infty} b_{1,1}^{1,1}(v)_{\psi y}^{\ 1}E_{i+v+3/4}^{n} \end{array} \right] \frac{\Delta t}{\Delta x} \tag{7.28}
$$

$$
{}_{\phi y}D_i^{n+1} = \left(\frac{1 - \dfrac{\sigma_x^i \Delta t}{2\varepsilon_0}}{1 + \dfrac{\sigma_x^i \Delta t}{2\varepsilon_0}} \right) {}_{\phi y}D_i^n - \frac{1}{\left(1 + \dfrac{\sigma_x^i \Delta t}{2\varepsilon_0}\right)}
$$

$$
\times \left[\begin{array}{l} \displaystyle\sum_{\nu=-\infty}^{+\infty} a(\nu)_{\phi z} H_{i+1/2+\nu}^{n+1/2} + \sum_{\nu=-\infty}^{+\infty} d_{0,0}(\nu)_{\psi z}{}_0^0 H_{i+\nu}^{n+1/2} \\ \displaystyle + \sum_{\nu=-\infty}^{+\infty} d_{1,0}(\nu)_{\psi z}{}_0^1 H_{i+\nu+1/4}^{n+1/2} + \sum_{\nu=-\infty}^{+\infty} d_{1,1}(\nu)_{\psi z}{}_1^1 H_{i+\nu+3/4}^{n+1/2} \end{array} \right] \frac{\Delta t}{\Delta x} \quad (7.29)
$$

$$
{}_{\psi y}{}_0^0 D_{i+1/2}^{n+1} = \left(\frac{1 - \dfrac{\sigma_x^i \Delta t}{2\varepsilon_0}}{1 + \dfrac{\sigma_x^i \Delta t}{2\varepsilon_0}} \right) {}_{\psi y}{}_0^0 D_{i+1/2}^n - \frac{1}{\left(1 + \dfrac{\sigma_x^i \Delta t}{2\varepsilon_0}\right)}
$$

$$
\times \left[\begin{array}{l} \displaystyle\sum_{\nu=-\infty}^{+\infty} c_{0,0}(\nu)_{\phi z} H_{i+\nu+1/2}^{n+1/2} + \sum_{\nu=-\infty}^{+\infty} b_{0,0}^{0,0}(\nu)_{\psi z}{}_0^0 H_{i+\nu+1}^{n+1/2} \\ \displaystyle + \sum_{\nu=-\infty}^{+\infty} b_{0,0}^{0,1}(\nu)_{\psi z}{}_0^1 H_{i+\nu+3/4}^{n+1/2} + \sum_{\nu=-\infty}^{+\infty} b_{0,1}^{0,1}(\nu)_{\psi z}{}_1^1 H_{i+\nu+5/4}^{n+1/2} \end{array} \right] \frac{\Delta t}{\Delta x} \quad (7.30)
$$

$$
{}_{\psi y}{}_0^1 D_{i+1/4}^{n+1} = \left(\frac{1 - \dfrac{\sigma_x^i \Delta t}{2\varepsilon_0}}{1 + \dfrac{\sigma_x^i \Delta t}{2\varepsilon_0}} \right) {}_{\psi y}{}_0^1 D_{i+1/4}^n - \frac{1}{\left(1 + \dfrac{\sigma_x^i \Delta t}{2\varepsilon_0}\right)}
$$

$$
\times \left[\begin{array}{l} \displaystyle\sum_{\nu=-\infty}^{+\infty} c_{1,0}(\nu)_{\phi z} H_{i+\nu+1/2}^{n+1/2} + \sum_{\nu=-\infty}^{+\infty} b_{0,0}^{1,0}(\nu)_{\psi z}{}_0^0 H_{i+\nu+1}^{n+1/2} \\ \displaystyle + \sum_{\nu=-\infty}^{+\infty} b_{0,0}^{1,1}(\nu)_{\psi z}{}_0^1 H_{i+\nu+3/4}^{n+1/2} + \sum_{\nu=-\infty}^{+\infty} b_{0,1}^{1,1}(\nu)_{\psi z}{}_1^1 H_{i+\nu+5/4}^{n+1/2} \end{array} \right] \frac{\Delta t}{\Delta x} \quad (7.31)
$$

$$
{}_{\psi y}{}_1^1 D_{i+3/4}^{n+1} = \left(\frac{1 - \dfrac{\sigma_x^i \Delta t}{2\varepsilon_0}}{1 + \dfrac{\sigma_x^i \Delta t}{2\varepsilon_0}} \right) {}_{\psi y}{}_1^1 D_{i+3/4}^n - \frac{1}{\left(1 + \dfrac{\sigma_x^i \Delta t}{2\varepsilon_0}\right)}
$$

$$
\times \left[\begin{array}{l} \displaystyle\sum_{\nu=-\infty}^{+\infty} c_{1,1}(\nu)_{\phi z} H_{i+\nu+1/2}^{n+1/2} + \sum_{\nu=-\infty}^{+\infty} b_{1,0}^{1,0}(\nu)_{\psi z}{}_0^0 H_{i+\nu+1}^{n+1/2} \\ \displaystyle + \sum_{\nu=-\infty}^{+\infty} b_{1,0}^{1,1}(\nu)_{\psi z}{}_0^1 H_{i+\nu+3/4}^{n+1/2} + \sum_{\nu=-\infty}^{+\infty} b_{1,1}^{1,1}(\nu)_{\psi z}{}_1^1 H_{i+\nu+5/4}^{n+1/2} \end{array} \right] \frac{\Delta t}{\Delta x} \quad (7.32)
$$

The coefficients of $a(\nu)$, $c_{s,l}(\nu)$, $d_{s,l}(\nu)$, and $b_{l,l'}^{s,s'}(\nu)$ have to be evaluated accurately to ensure accuracy of the updates for each field. We have numerically calculated these coefficients using the Gaussian quadrature method [7]; for the reader's convenience, we summarize the values of these coefficients at the integer points in Tables 7.1–7.6.

TABLE 7.1 Coefficients $a(\nu)$, $b_0(\nu)$, and $c_0(\nu)$

ν	$a(\nu)$	$b_0(\nu)$	$c_0(\nu)$
0	1.29184622	2.47253880	0.0
1	−0.15607611	0.95622823	−0.04659726
2	0.05963908	0.16605866	0.05453940
3	−0.02930993	0.09392450	−0.03699956
4	0.01537167	0.00314122	0.02057443
5	−0.00818941	0.01349379	−0.01115290
6	0.00437913	−0.00285935	0.00597662
7	−0.00234395	0.00277920	−0.00320213
8	0.00125550	−0.00113156	0.00171317
9	−0.00067370	0.00070658	−0.00091592
10	0.00036364	−0.00036348	0.00048927
11	−0.00019931	−0.00016602	−0.00025660
12		−0.00030293	0.00012898

TABLE 7.2 Coefficients $c_{1,0}(\nu)$ ($s = 1$, $l = 0$) and $c_{1,1}(\nu)$ ($s = 1$, $l = 1$)

ν	$c_{1,0}(\nu)$	$c_{1,1}(\nu)$
−8	0.000077791	−0.000107917
−7	−0.000154260	0.000216187
−6	0.000313336	−0.000440914
−5	−0.000625694	0.000906723
−4	0.001348953	−0.001959836
−3	−0.002586165	0.004273722
−2	0.007532622	−0.011329640
−1	−0.020265614	0.024321565
0	0.030300911	−0.030300911
1	−0.024321565	0.020265614
2	0.011329640	−0.007532622
3	−0.004273722	0.002586165
4	0.001959836	−0.001348953
5	−0.000906723	0.00088486
6	0.000440914	−0.000313336
7	−0.000216187	0.000154260
8	0.000107917	−0.000077791

TABLE 7.3 Coefficients $d_{1,0}(\nu)$ ($s = 1$, $l = 0$) and $d_{1,1}(\nu)$ ($s = 1$, $l = 1$)

ν	$d_{1,0}(\nu)$	$d_{1,1}(\nu)$
−8	−0.000107917	0.000077791
−7	0.000216187	−0.000154260
−6	−0.000440914	0.000313336
−5	0.000906723	−0.000625694
−4	−0.001959836	0.001348953
−3	0.004273722	−0.002586165
−2	−0.011329640	0.007532622
−1	0.024321565	−0.020265614
0	−0.030300911	0.030300911
1	0.020265614	−0.024321565
2	−0.007532622	0.011329640
3	0.002586165	−0.004273722
4	−0.001348953	0.001959836
5	0.00088486	−0.000906723
6	−0.000313336	0.00062356
7	0.000154260	−0.000216187
8	−0.000077791	0.000107917

TABLE 7.4 Coefficients $b_{00}^{01}(\nu)$ ($s = 0$, $l = 0$, $s' = 1$, $l' = 0$) and $b_{01}^{01}(\nu)$ ($s = 0$, $l = 0$, $s' = 1$, $l' = 1$)

ν	$b_{00}^{01}(\nu)$	$b_{01}^{01}(\nu)$
−9	0.000068126	−0.000157979
−8	0.000143468	−0.000352128
−7	0.001101551	−0.001591902
−6	0.002106587	−0.004918699
−5	0.010876491	−0.018197365
−4	0.030774267	−0.062187652
−3	0.121900158	−0.216773548
−2	0.394611424	−0.739278550
−1	1.225829337	−1.609301081
0	1.609301081	−1.225829337
1	0.739278549	−0.394611424
2	0.216773549	−0.121900158
3	0.062187653	−0.030774267
4	0.018197366	−0.010876491
5	0.004918699	−0.002106587
6	0.001591902	−0.001101551
7	0.000352128	−0.000143468
8	0.000157979	−0.000068126

TABLE 7.5 Coefficients $b_{00}^{10}(\nu)$ ($s = 1$, $l = 0$, $s' = 0$, $l' = 0$), $b_{00}^{11}(\nu)$ ($s = 1$, $l = 0$, $s' = 1$, $l' = 0$), and $b_{01}^{11}(\nu)$ ($s = 1$, $l = 0$, $s' = 1$, $l' = 1$)

ν	$b_{00}^{10}(\nu)$	$b_{00}^{11}(\nu)$	$b_{01}^{11}(\nu)$
−9	−0.000157979	0.000062393	−0.000116529
−8	−0.000352128	0.000217810	−0.000406512
−7	−0.001591902	0.000760823	−0.001416527
−6	−0.004918699	0.002663166	−0.004916515
−5	−0.018197365	0.009390025	−0.016826645
−4	−0.062187652	0.033934368	−0.054686440
−3	−0.216773548	0.132668009	−0.142270550
−2	−0.739278550	0.641536536	0.056428700
−1	−1.609301081	5.039549683	0.000000000
0	−1.225829337	−5.039549683	−0.056428700
1	−0.394611424	−0.641536536	0.142270550
2	−0.121900158	−0.132668009	0.054686440
3	−0.030774267	−0.033934368	0.016826645
4	−0.010876491	−0.009390025	0.004916515
5	−0.002106587	−0.002663166	0.001416527
6	−0.001101551	−0.000760823	0.000406512
7	−0.000143468	−0.000217810	0.000116529
8	−0.000068126	−0.000062393	0.000003339

TABLE 7.6 Coefficients $b_{10}^{10}(\nu)$ ($s = 1$, $l = 1$, $s' = 0$, $l' = 0$), $b_{10}^{11}(\nu)$ ($s = 1$, $l = 1$, $s' = 1$, $l' = 0$), and $b_{11}^{11}(\nu)$ ($s = 1$, $l = 1$, $s' = 1$, $l' = 1$)

ν	$b_{10}^{10}(\nu)$	$b_{10}^{11}(\nu)$	$b_{11}^{11}(\nu)$
−9	0.000068126	−0.000003339	0.000062393
−8	0.000143468	−0.000406512	0.000217810
−7	0.001101551	−0.001416527	0.000760823
−6	0.002106587	−0.004916515	0.002663166
−5	0.010876491	−0.016826645	0.009390025
−4	0.030774267	−0.054686440	0.033934368
−3	0.121900158	−0.142270550	0.132668009
−2	0.394611424	0.056428700	0.641536536
−1	1.225829337	−0.000406512	5.039549683
0	1.609301081	0.000000000	−5.039549683
1	0.739278550	−0.056428700	−0.641536536
2	0.216773548	0.142270550	−0.132668009
3	0.062187652	0.054686440	−0.033934368
4	0.018197365	0.016826645	−0.009390025
5	0.004918699	0.004916515	−0.002663166
6	0.001591902	0.001416527	−0.000760823
7	0.000352128	0.000406512	−0.000217810
8	0.000157979	0.000116529	−0.000062393

In Table 7.1, the coefficients with values less than 10^{-5} are simply ignored. In the update equations, we use the coefficient symmetric relations, namely, $a(-\nu) = -a(\nu - 1)$, $b_0(-\nu) = -b_0(\nu - 1)$, and $c_0(-\nu) = -c_0(\nu)$.

In Table 7.5, the coefficients $b_{00}^{11}(\nu)$ and $b_{01}^{11}(\nu)$ satisfy the relations $b_{00}^{11}(-\nu) = -b_{00}^{11}(\nu)$ $(\nu > 0)$ and $b_{01}^{11}(-\nu) = -b_{01}^{11}(\nu - 2)$ $(\nu > -1)$, respectively.

In Table 7.6, the coefficients $b_{10}^{11}(\nu)$ and $b_{11}^{11}(\nu)$ satisfy the relations $b_{10}^{11}(-\nu) = -b_{10}^{11}(\nu)$ and $b_{11}^{11}(-\nu) = -b_{11}^{11}(\nu - 1)$ when $\nu > 0$, respectively.

Tables 7.1–7.5 illustrate that all the coefficients converge at different speeds. For example, $a(\nu)$ usually runs from -12 to 11 with absolute values greater than 10^{-4} for the Battle–Lemarié scaling and wavelet bases.

To further solve for the update equation for E-fields, we apply the same discretization procedure on (7.2) and obtain

$$\varepsilon_0 \sum_{i'=-\infty}^{+\infty} \left[\left(\varepsilon_r^{\phi,\phi}\right)_{i',i\phi y} E_{i'}^n + \sum_{s=0}^{s_{\max}} \sum_{l=0}^{2^s-1} \left({}_l^s\varepsilon_r^{\psi,\phi}\right)_{i',i\psi y}\, {}_l^s E_{i'+2^{-s}(l+1/2)}^n \right] = {}_{\phi y}D_i^n \quad (7.33)$$

$$\varepsilon_0 \sum_{i'=-\infty}^{+\infty} \left[\left({}_l^s\varepsilon_r^{\psi,\phi}\right)_{i,i'\phi y} E_{i'}^n + \sum_{s'=0}^{s_{\max}} \sum_{l'=0}^{2^{-s'}-1} \left({}_{l,l'}^{s,s'}\varepsilon_r^{\psi,\psi}\right)_{i,i'\psi y}\, {}_{l'}^{s'} E_{i'+2^{-s'}(l'+1/2)}^n \right]$$
$$= {}_{\psi y}{}_l^s D_{i+2^{-s}(l+1/2)}^n \quad (7.34)$$

The dielectric-related coefficients in the above equations are defined as follows:

$$\left(\varepsilon_r^{\phi,\phi}\right)_{i',i} = \frac{1}{\Delta x} \int_{-\infty}^{+\infty} \varepsilon_r(x)\phi_{i'}(x)\phi_i(x)\, dx$$
$$= \delta_{i,i'} + \sum_{\kappa=0}^{\kappa_{\max}} (\varepsilon_r(\kappa) - 1) \frac{1}{\Delta x} \int_{x_2(\kappa)}^{x_2(\kappa)} \phi_{i'}(x)\phi_i(x)\, dx$$
$$= \delta_{i,i'} + \sum_{\kappa=0}^{\kappa_{\max}} (\varepsilon_r(\kappa) - 1)\alpha_{i,i'}^{\phi\phi}(\kappa) \quad (7.35a)$$

$$\left({}_l^s\varepsilon_r^{\phi,\psi}\right)_{i,i'+2^{-s}(l+1/2)} = \frac{1}{\Delta x} \int_{-\infty}^{+\infty} \varepsilon_r(x)\phi_i(x)\psi_{s,i'+2^{-s}(l+1/2)}(x)\, dx$$
$$= \sum_{\kappa=0}^{\kappa_{\max}} (\varepsilon_r(\kappa) - 1) \frac{1}{\Delta x} \int_{x_2(\kappa)}^{x_2(\kappa)} \phi_i(x)\psi_{s,i'+2^{-s}(l+1/2)}(x)\, dx$$
$$= \sum_{\kappa=0}^{\kappa_{\max}} (\varepsilon_r(\kappa) - 1)\alpha_{i,i'+2^{-s}(l+1/2)}^{\phi\psi}(\kappa) \quad (7.35b)$$

$$\left({}^{s}_{l}\varepsilon_r^{\psi,\phi}\right)_{i+2^{-s}(l+1/2),i'} = \frac{1}{\Delta x}\int_{-\infty}^{+\infty} \varepsilon_y(x)\psi_{s,i+2^{-s}(l+1/2)}(x)\phi_{i'}(x)\,dx$$

$$= \sum_{\kappa=0}^{\kappa_{\max}} (\varepsilon_r(\kappa)-1)\frac{1}{\Delta x}\int_{x_2(\kappa)}^{x_2(\kappa)} \psi_{s,i+2^{-s}(l+1/2)}(x)\phi_{i'}(x)\,dx$$

$$= \sum_{\kappa=0}^{\kappa_{\max}} (\varepsilon_r(\kappa)-1)\alpha^{\psi\phi}_{i'+2^{-s}(l+1/2),i'}(\kappa) \tag{7.35c}$$

$$\left({}^{s,s'}_{l,l'}\varepsilon_r^{\psi,\psi}\right)_{i+2^{-s}(l+1/2),i'+2^{-s'}(l+1/2)}$$

$$= \frac{1}{\Delta x}\int_{-\infty}^{+\infty} \varepsilon_r(x)\psi_{s,i+2^{-s'}(l'+1/2)}(x)\psi_{s,i+2^{-s}(l+1/2)}(x)\,dx$$

$$= \delta_{i,i'}\delta_{s,s'}\delta_{l,l'} + \sum_{\kappa=0}^{\kappa_{\max}} (\varepsilon_r(\kappa)-1)\frac{1}{\Delta x}\int_{x_2(\kappa)}^{x_2(\kappa)} \psi_{i'+2^{-s'}(l'+1/2)}(x)\psi_{s,i+2^{-s}(l+1/2)}(x)\,dx$$

$$= \delta_{i,i'}\delta_{s,s'}\delta_{l,l'} + \sum_{\kappa=0}^{\kappa_{\max}} (\varepsilon_r(\kappa)-1)\alpha^{\psi\psi}_{i,+2^{-s'}(l+1/2),s,i+2^{-s}(l+1/2)}(\kappa) \tag{7.35d}$$

where $\kappa_{\max}$ is the maximum number of dielectric objects, and $x_1(\kappa)$ and $x_2(\kappa)$ are the start and end positions of the κth dielectric region.

For the sake of computational efficiency and simplicity, all field quantities are expanded in terms of scaling functions in the entire computational region; wavelets at levels $s = 0$ and 1 are used only for the extended discontinuity subregions (see Figure 7.3), where the fields change rapidly.

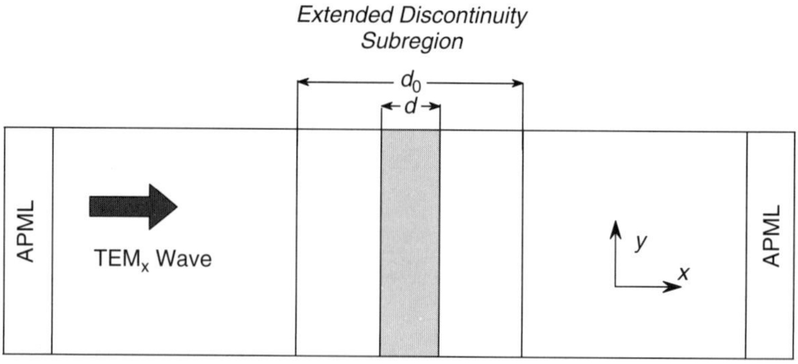

FIGURE 7.3 TEM$_x$ wave incident on a dielectric slab, where d is the width of the dielectric slab and d_0 is the extended dielectric region.

Consider, for example, the κth dielectric region $[x_1(\kappa), x_2(\kappa)] = [i_1\,\Delta x, i_2\,\Delta x]$ in a multilayer dielectric system. Taking into account the stencil effect of the scaling and wavelet functions [8], we can truncate the wavelet expansions in the extended discontinuity subregion, that is, $[(i_1 - M)\,\Delta x, (i_2 + M)\,\Delta x]$, where M is the stencil size or the number of expansion coefficients per side in the update equations, which is adjusted in accordance with the accuracy requirement. The simplest choice is $M = 0$; however, this value will not accurately represent scaling and wavelet expansions. It is usually unnecessary to set $M > 9$ for the cubic spline Battle–Lemarié MRTD because the scaling and wavelet functions are localized around the discontinuity subregions. The value of M is usually chosen to achieve the desired level of accuracy, and we set it equal to 9 in what follows. Thus, for $s_{\max} \equiv 1$, the update equations (7.33) and (7.34) read

$$
\begin{aligned}
{}_{\phi y}D_i^n = \varepsilon_0 \sum_{i'=-\infty}^{+\infty} \Big[&(\varepsilon_r^{\phi,\phi})_{i',i}\,{}_{\phi y}E_{i'}^n + ({}_0^0\varepsilon_r^{\psi,\phi})_{i,i'+1/2}\,{}_{\psi y 0}^{\;\;0}E_{i'+1/2}^n \\
&+ ({}_0^1\varepsilon_r^{\psi,\phi})_{i,i+1/4}\,{}_{\psi y 0}^{\;\;1}E_{i'+1/4}^n + ({}_1^1\varepsilon_r^{\psi,\phi})_{i,i+3/4}\,{}_{\psi y 1}^{\;\;1}E_{i'+3/4}^n \Big]
\end{aligned} \tag{7.36}
$$

$$
\begin{aligned}
{}_{\psi y 0}^{\;\;0}D_{i+1/2}^n = \varepsilon_0 \sum_{i'=-\infty}^{+\infty} [&({}_0^0\varepsilon_r^{\psi,\phi})_{i+1/2,i'}\,{}_{\phi y}E_{i'}^n + ({}_{0,0}^{0,0}\varepsilon_r^{\psi,\psi})_{i+1/2,i'+1/2}\,{}_{\psi y 0}^{\;\;0}E_{i'+1/2}^n \\
&+ ({}_{0,0}^{0,1}\varepsilon_r^{\psi,\psi})_{i+1/2,i'+1/4}\,{}_{\psi y 0}^{\;\;1}E_{i'+1/4}^n + ({}_{0,1}^{0,1}\varepsilon_r^{\psi,\psi})_{i+1/2,i'+3/4}\,{}_{\psi y 1}^{\;\;1}E_{i'+3/4}^n]
\end{aligned} \tag{7.37}
$$

$$
\begin{aligned}
{}_{\psi y 0}^{\;\;1}D_{i+1/4}^n = \varepsilon_0 \sum_{i'=-\infty}^{+\infty} [&({}_0^1\varepsilon_r^{\psi,\phi})_{i+1/4,i'}\,{}_{\phi y}E_{i'}^n + ({}_{0,0}^{1,0}\varepsilon_r^{\psi,\psi})_{i+1/4,i'+1/2}\,{}_{\psi y 0}^{\;\;0}E_{i'+1/2}^n \\
&+ ({}_{0,0}^{1,1}\varepsilon_r^{\psi,\psi})_{i+1/4,i'+1/4}\,{}_{\psi y 0}^{\;\;1}E_{i'+1/4}^n + ({}_{0,1}^{1,1}\varepsilon_r^{\psi,\psi})_{i+1/4,i'+3/4}\,{}_{\psi y 1}^{\;\;1}E_{i'+3/4}^n]
\end{aligned} \tag{7.38}
$$

$$
\begin{aligned}
{}_{\psi y 1}^{\;\;1}D_{i+3/4}^n = \varepsilon_0 \sum_{i'=-\infty}^{+\infty} [&({}_1^1\varepsilon_r^{\psi,\phi})_{i+3/4,i'}\,{}_{\phi y}E_{i'}^n + ({}_{1,0}^{1,0}\varepsilon_r^{\psi,\psi})_{i+3/4,i'+1/2}\,{}_{\psi y 0}^{\;\;0}E_{i'+1/2}^n \\
&+ ({}_{1,0}^{1,1}\varepsilon_r^{\psi,\psi})_{i+3/4,i'+1/4}\,{}_{\psi y 0}^{\;\;1}E_{i'+1/4}^n + ({}_{1,1}^{1,1}\varepsilon_r^{\psi,\psi})_{i+3/4,i'+3/4}\,{}_{\psi y 1}^{\;\;1}E_{i'+3/4}^n]
\end{aligned} \tag{7.39}
$$

with the dielectric-related coefficients defined as

$$
\begin{aligned}
(\varepsilon_r^{\phi,\phi})_{i',i} &= \frac{1}{\Delta x}\int_{-\infty}^{+\infty} \varepsilon_r(x)\phi_{i'}(x)\phi_i(x)\,dx = \delta_{i,i'} + \sum_{\kappa=0}^{\kappa_{\max}} (\varepsilon_r(\kappa) - 1) \\
&\times \frac{1}{\Delta x}\int_{x_2(\kappa)}^{x_2(\kappa)} \phi_{i'}(x)\phi_i(x)\,dx = \delta_{i,i'} + \sum_{\kappa=0}^{\kappa_{\max}} (\varepsilon_r(\kappa) - 1)\alpha_{i,i'}^{\phi\phi}(\kappa)
\end{aligned} \tag{7.40}
$$

$$
\begin{aligned}
({}_0^0\varepsilon_r^{\phi,\psi})_{i,i'+1/2} &= \frac{1}{\Delta x}\int_{-\infty}^{+\infty} \varepsilon_r(x)\phi_i(x)\psi_{0,i'+1/2}(x)\,dx \\
&= \sum_{\kappa=0}^{\kappa_{\max}}(\varepsilon_r(\kappa)-1)\frac{1}{\Delta x}\int_{x_2(\kappa)}^{x_2(\kappa)} \phi_i(x)\psi_{0,i'+1/2}(x)\,dx \qquad (7.41)\\
&= \sum_{\kappa=0}^{\kappa_{\max}}(\varepsilon_r(\kappa)-1)\alpha_{i,i'+1/2}^{\phi\psi}(\kappa)
\end{aligned}
$$

$$
\begin{aligned}
({}_0^1\varepsilon_r^{\phi,\psi})_{i,i'+1/4} &= \frac{1}{\Delta x}\int_{-\infty}^{+\infty} \varepsilon_r(x)\phi_i(x)\psi_{1,i'+1/4}(x)\,dx \\
&= \sum_{\kappa=0}^{\kappa_{\max}}(\varepsilon_r(\kappa)-1)\frac{1}{\Delta x}\int_{x_2(\kappa)}^{x_2(\kappa)} \phi_i(x)\psi_{1,i'+1/4}(x)\,dx \qquad (7.42)\\
&= \sum_{\kappa=0}^{\kappa_{\max}}(\varepsilon_r(\kappa)-1)\alpha_{i,i'+1/4}^{\phi\psi}(\kappa)
\end{aligned}
$$

$$
\begin{aligned}
({}_1^1\varepsilon_r^{\phi,\psi})_{i,i'+3/4} &= \frac{1}{\Delta x}\int_{-\infty}^{+\infty} \varepsilon_r(x)\phi_i(x)\psi_{1,i'+3/4}(x)\,dx \\
&= \sum_{\kappa=0}^{\kappa_{\max}}(\varepsilon_r(\kappa)-1)\frac{1}{\Delta x}\int_{x_2(\kappa)}^{x_2(\kappa)} \phi_i(x)\psi_{1,i'+3/4}(x)\,dx \qquad (7.43)\\
&= \sum_{\kappa=0}^{\kappa_{\max}}(\varepsilon_r(\kappa)-1)\alpha_{i,i'+3/4}^{\phi\psi}(\kappa)
\end{aligned}
$$

$$
\begin{aligned}
({}_0^0\varepsilon_r^{\psi,\phi})_{i+1/2,i'} &= \frac{1}{\Delta x}\int_{-\infty}^{+\infty} \varepsilon_y(x)\psi_{0,i+1/2}(x)\phi_{i'}(x)\,dx \\
&= \sum_{\kappa=0}^{\kappa_{\max}}(\varepsilon_r(\kappa)-1)\frac{1}{\Delta x}\int_{x_2(\kappa)}^{x_2(\kappa)} \psi_{0,i+1/2}(x)\phi_{i'}(x)\,dx \qquad (7.44)\\
&= \sum_{\kappa=0}^{\kappa_{\max}}(\varepsilon_r(\kappa)-1)\alpha_{i+1/2,i'}^{\psi\phi}(\kappa)
\end{aligned}
$$

$$\left({}_{0}^{1}\varepsilon_r^{\psi,\phi}\right)_{i+1/4,i'} = \frac{1}{\Delta x}\int_{-\infty}^{+\infty} \varepsilon_y(x)\psi_{1,i+1/4}(x)\phi_{i'}(x)\,dx$$

$$= \sum_{\kappa=0}^{\kappa_{\max}} (\varepsilon_r(\kappa) - 1)\frac{1}{\Delta x}\int_{x_2(\kappa)}^{x_2(\kappa)} \psi_{1,i+1/4}(x)\phi_{i'}(x)\,dx \tag{7.45}$$

$$= \sum_{\kappa=0}^{\kappa_{\max}} (\varepsilon_r(\kappa) - 1)\alpha_{i+1/4,i'}^{\psi\phi}(\kappa)$$

$$\left({}_{1}^{1}\varepsilon_r^{\psi,\phi}\right)_{i+3/4,i'} = \frac{1}{\Delta x}\int_{-\infty}^{+\infty} \varepsilon_y(x)\psi_{1,i+3/4}(x)\phi_{i'}(x)\,dx$$

$$= \sum_{\kappa=0}^{\kappa_{\max}} (\varepsilon_r(\kappa) - 1)\frac{1}{\Delta x}\int_{x_2(\kappa)}^{x_2(\kappa)} \psi_{1,i+3/4}(x)\phi_{i'}(x)\,dx \tag{7.46}$$

$$= \sum_{\kappa=0}^{\kappa_{\max}} (\varepsilon_r(\kappa) - 1)\alpha_{i'+3/4,i'}^{\psi\phi}(\kappa)$$

$$\left({}_{0,0}^{0,0}\varepsilon_r^{\psi,\psi}\right)_{i'+1/2,i+1/2} = \frac{1}{\Delta x}\int_{-\infty}^{+\infty} \varepsilon_r(x)\psi_{0,i'+1/2}(x)\psi_{0,i+1/2}(x)\,dx$$

$$= \delta_{i,i'} + \sum_{\kappa=0}^{\kappa_{\max}} (\varepsilon_r(\kappa) - 1)\frac{1}{\Delta x}\int_{x_2(\kappa)}^{x_2(\kappa)} \psi_{0,i'+1/2}(x)\psi_{0,i+1/2}(x)\,dx \tag{7.47}$$

$$= \delta_{i,i'} + \sum_{\kappa=0}^{\kappa_{\max}} (\varepsilon_r(\kappa) - 1)\alpha_{i'+1/2,i+1/2}^{\psi\psi}(\kappa)$$

$$\left({}_{0,0}^{0,1}\varepsilon_r^{\psi,\psi}\right)_{i'+1/2,i+1/4} = \frac{1}{\Delta x}\int_{-\infty}^{+\infty} \varepsilon_r(x)\psi_{0,i'+1/2}(x)\psi_{1,i+1/4}(x)\,dx$$

$$= \sum_{\kappa=0}^{\kappa_{\max}} (\varepsilon_r(\kappa) - 1)\frac{1}{\Delta x}\int_{x_2(\kappa)}^{x_2(\kappa)} \psi_{0,i'+1/2}(x)\psi_{1,i+1/4}(x)\,dx \tag{7.48}$$

$$= \sum_{\kappa=0}^{\kappa_{\max}} (\varepsilon_r(\kappa) - 1)\alpha_{i'+1/2,i+1/4}^{\psi\psi}(\kappa)$$

$$
\begin{aligned}
({}^{0,1}_{0,1}\varepsilon_r^{\psi,\psi})_{i'+1/2,i+3/4} &= \frac{1}{\Delta x}\int_{-\infty}^{+\infty} \varepsilon_r(x)\psi_{0,i'+1/2}(x)\psi_{1,i+3/4}(x)\,dx \\
&= \sum_{\kappa=0}^{\kappa_{\max}} (\varepsilon_r(\kappa)-1)\frac{1}{\Delta x}\int_{x_2(\kappa)}^{x_2(\kappa)} \psi_{0,i'+1/2}(x)\psi_{1,i+3/4}(x)\,dx \qquad (7.49)\\
&= \sum_{\kappa=0}^{\kappa_{\max}} (\varepsilon_r(\kappa)-1)\alpha^{\psi\psi}_{i'+1/2,i+3/4}(\kappa)
\end{aligned}
$$

$$
\begin{aligned}
({}^{1,0}_{0,0}\varepsilon_r^{\psi,\psi})_{i'+1/4,i+1/2} &= \frac{1}{\Delta x}\int_{-\infty}^{+\infty} \varepsilon_r(x)\psi_{1,i'+1/4}(x)\psi_{0,i+1/2}(x)\,dx \\
&= \sum_{\kappa=0}^{\kappa_{\max}} (\varepsilon_r(\kappa)-1)\frac{1}{\Delta x}\int_{x_2(\kappa)}^{x_2(\kappa)} \psi_{1,i'+1/4}(x)\psi_{0,i+1/2}(x)\,dx \qquad (7.50)\\
&= \sum_{\kappa=0}^{\kappa_{\max}} (\varepsilon_r(\kappa)-1)\alpha^{\psi\psi}_{i'+1/4,i+1/2}(\kappa) = ({}^{0,1}_{0,0}\varepsilon_r^{\psi,\psi})_{i'+1/2,i+1/4}
\end{aligned}
$$

$$
\begin{aligned}
({}^{1,1}_{0,0}\varepsilon_r^{\psi,\psi})_{i'+1/4,i+1/2} &= \frac{1}{\Delta x}\int_{-\infty}^{+\infty} \varepsilon_r(x)\psi_{1,i'+1/4}(x)\psi_{1,i+1/2}(x)\,dx \\
&= \delta_{i,i'} + \sum_{\kappa=0}^{\kappa_{\max}} (\varepsilon_r(\kappa)-1)\frac{1}{\Delta x}\int_{x_2(\kappa)}^{x_2(\kappa)} \psi_{1,i'+1/4}(x)\psi_{1,i+1/2}(x)\,dx \qquad (7.51)\\
&= \delta_{i,i'} + \sum_{\kappa=0}^{\kappa_{\max}} (\varepsilon_r(\kappa)-1)\alpha^{\psi\psi}_{i'+1/4,i+1/2}(\kappa)
\end{aligned}
$$

$$
\begin{aligned}
({}^{1,1}_{0,1}\varepsilon_r^{\psi,\psi})_{i'+1/4,i+3/4} &= \frac{1}{\Delta x}\int_{-\infty}^{+\infty} \varepsilon_r(x)\psi_{1,i'+1/4}(x)\psi_{1,i+3/4}(x)\,dx \\
&= \sum_{\kappa=0}^{\kappa_{\max}} (\varepsilon_r(\kappa)-1)\frac{1}{\Delta x}\int_{x_2(\kappa)}^{x_2(\kappa)} \psi_{1,i'+1/4}(x)\psi_{1,i+3/4}(x)\,dx \qquad (7.52)\\
&= \sum_{\kappa=0}^{\kappa_{\max}} (\varepsilon_r(\kappa)-1)\alpha^{\psi\psi}_{i'+1/4,i+3/4}(\kappa)
\end{aligned}
$$

$$\begin{aligned}
({}^{1,0}_{1,0}\varepsilon_r^{\psi,\psi})_{i'+3/4,i+1/2} &= \frac{1}{\Delta x}\int_{-\infty}^{+\infty} \varepsilon_r(x)\psi_{i'+3/4}(x)\psi_{0,i+1/2}(x)\,dx \\
&= \sum_{\kappa=0}^{\kappa_{\max}} (\varepsilon_r(\kappa)-1)\frac{1}{\Delta x}\int_{x_2(\kappa)}^{x_2(\kappa)} \psi_{1,i'+3/4}(x)\psi_{0,i+1/2}(x)\,dx \qquad (7.53)\\
&= \sum_{\kappa=0}^{\kappa_{\max}} (\varepsilon_r(\kappa)-1)\alpha^{\psi\psi}_{i'+3/4,i+1/2}(\kappa)
\end{aligned}$$

$$\begin{aligned}
({}^{1,1}_{1,0}\varepsilon_r^{\psi,\psi})_{i'+3/4,i+1/4} &= \frac{1}{\Delta x}\int_{-\infty}^{+\infty} \varepsilon_r(x)\psi_{1,i'+3/4}(x)\psi_{1,i+1/4}(x)\,dx \\
&= \sum_{\kappa=0}^{\kappa_{\max}} (\varepsilon_r(\kappa)-1)\frac{1}{\Delta x}\int_{x_2(\kappa)}^{x_2(\kappa)} \psi_{1,i'+3/4}(x)\psi_{1,i+1/4}(x)\,dx \qquad (7.54)\\
&= \sum_{\kappa=0}^{\kappa_{\max}} (\varepsilon_r(\kappa)-1)\alpha^{\psi\psi}_{i'+1/3,i+1/4}(\kappa) = ({}^{1,1}_{0,1}\varepsilon_r^{\psi,\psi})_{i'+1/4,i+3/4}
\end{aligned}$$

$$\begin{aligned}
({}^{1,1}_{1,1}\varepsilon_r^{\psi,\psi})_{i'+3/4,i+3/4} &= \frac{1}{\Delta x}\int_{-\infty}^{+\infty} \varepsilon_r(x)\psi_{1,i'+3/4}(x)\psi_{1,i+3/4}(x)\,dx \\
&= \delta_{i,i'} + \sum_{\kappa=0}^{\kappa_{\max}} (\varepsilon_r(\kappa)-1)\frac{1}{\Delta x}\int_{x_2(\kappa)}^{x_2(\kappa)} \psi_{1,i'+3/4}(x)\psi_{1,i+3/4}(x)\,dx \qquad (7.55)\\
&= \delta_{i,i'} + \sum_{\kappa=0}^{\kappa_{\max}} (\varepsilon_r(\kappa)-1)\alpha^{\psi\psi}_{i'+3/4,i+3/4}(\kappa)
\end{aligned}$$

The update equations (7.36) through (7.39) can be written in matrix form

$$\varepsilon_0 \sum_{i'=i_1-M}^{i_2+M} \begin{bmatrix}
[\varepsilon^{\phi,\phi}_{i,i'}] & [{}^0_0\varepsilon^{\phi,\psi}_{i,i'+1/2}] & [{}^1_0\varepsilon^{\phi,\psi}_{i,i'+1/4}] & [{}^1_1\varepsilon^{\phi,\psi}_{i,i'+3/4}] \\
[{}^0_0\varepsilon^{\psi,\phi}_{i+1/2,i'}] & [{}^{0,0}_{0,0}\varepsilon^{\psi,\psi}_{i+1/2,i'+1/2}] & [{}^{0,1}_{0,0}\varepsilon^{\psi,\psi}_{i+1/2,i'+1/4}] & [{}^{0,1}_{0,1}\varepsilon^{\psi,\psi}_{i+1/2,i'+3/4}] \\
[{}^1_0\varepsilon^{\psi,\phi}_{i+1/4,i'}] & [{}^{1,0}_{0,0}\varepsilon^{\psi,\psi}_{i+1/4,i'+1/2}] & [{}^{1,1}_{0,0}\varepsilon^{\psi,\psi}_{i+1/4,i'+1/4}] & [{}^{1,1}_{0,1}\varepsilon^{\psi,\psi}_{i+1/4,i'+3/4}] \\
[{}^1_1\varepsilon^{\psi,\phi}_{i+3/4,i'}] & [{}^{1,0}_{1,0}\varepsilon^{\psi,\psi}_{i+3/4,i'+1/2}] & [{}^{1,1}_{1,0}\varepsilon^{\psi,\psi}_{i+3/4,i'+1/4}] & [{}^{1,1}_{1,1}\varepsilon^{\psi,\psi}_{i+3/4,i'+3/4}]
\end{bmatrix}$$

$$\times \begin{bmatrix} [{}_{\phi y}E^n_{i'}] \\ [{}_{\psi y}{}^0_0E^n_{i'+1/2}] \\ [{}_{\psi y}{}^1_0E^n_{i'+1/4}] \\ [{}_{\psi y}{}^1_1E^n_{i'+3/4}] \end{bmatrix} = \begin{bmatrix} [{}_{\phi y}D^n_{i}] \\ [{}_{\psi y}{}^0_0D^n_{i+1/2}] \\ [{}_{\psi y}{}^1_0D^n_{i+1/4}] \\ [{}_{\psi y}{}^1_1D^n_{i+3/4}] \end{bmatrix} \qquad (7.56)$$

where the coefficient matrix $\left[\varepsilon_{i,i'}^{\phi,\phi}\right]$ may be written as

$$\left[\varepsilon_{i,i'}^{\phi,\phi}\right]=\begin{bmatrix} (\varepsilon_r^{\phi,\phi})_{i_1-M,i_1-M} & (\varepsilon_r^{\phi,\phi})_{i_1-M,i_1-M+1} & \cdots & (\varepsilon_r^{\phi,\phi})_{i_1-M,i_2+M-1} & (\varepsilon_r^{\phi,\phi})_{i_1-M,i_2+M} \\ (\varepsilon_r^{\phi,\phi})_{i_1-M+1,i_1-M} & (\varepsilon_r^{\phi,\phi})_{i_1-M+1,i_1-M+1} & \cdots & (\varepsilon_r^{\phi,\phi})_{i_1-M+1,i_2+M-1} & (\varepsilon_r^{\phi,\phi})_{i_1-M+1,i_2+M} \\ \vdots & \vdots & \vdots & \vdots & \vdots \\ (\varepsilon_r^{\phi,\phi})_{i_2+M-1,i_1-M} & (\varepsilon_r^{\phi,\phi})_{i_2+M-1,i_1-M+1} & \cdots & (\varepsilon_r^{\phi,\phi})_{i_2+M-1,i_2+M-1} & (\varepsilon_r^{\phi,\phi})_{i_2+M-1,i_2+M} \\ (\varepsilon_r^{\phi,\phi})_{i_2+M,i_1-M} & (\varepsilon_r^{\phi,\phi})_{i_2+M,i_1-M+1} & \cdots & (\varepsilon_r^{\phi,\phi})_{i_2+M,i_2+M-1} & (\varepsilon_r^{\phi,\phi})_{i_2+M,i_2+M} \end{bmatrix} \tag{7.57}$$

and $\left[{}_{\phi y}E_{i'}^{n}\right]$ is expressed as

$$\left[{}_{\phi y}E_{i'}^{n}\right]^T=\left[{}_{\phi y}E_{i_1-M}^{n} \quad {}_{\phi y}E_{i_1-M+1}^{n} \quad \cdots \quad {}_{\phi y}E_{i_2+M-1}^{n} \quad {}_{\phi y}E_{i_2+M}^{n}\right]^T \tag{7.58}$$

As seen from (7.56), the fields at discrete points are coupled, and we have to update the fields by solving a matrix equation with the dimensions $4(i_2-i_1+2M)\times 4(i_2-i_1+2M)$ at each time step. Note that all elements in the coefficient matrix (7.56) are constants for a specified system and that they can be precalculated and stored in a constant array. Thus, we can update E-fields in (7.56) simply by performing a matrix multiplication if the inverse matrix of the coefficient matrix in (7.56) is stored.

Through the inverse matrix operation, we can obtain the update equation of the E-field in the dielectric region $i\in[i_1\,\Delta x, i_2\,\Delta x]$ as

$$\begin{bmatrix} [{}_{\phi y}E_{i}^{n}] \\ [{}_{\psi y}{}^{0}_{0}E_{i+1/2}^{n}] \\ [{}_{\psi y}{}^{1}_{0}E_{i+1/4}^{n}] \\ [{}_{\psi y}{}^{1}_{1}E_{i+3/4}^{n}] \end{bmatrix}=\frac{1}{\varepsilon_0}\sum_{i'=i_1-M}^{i_2+M}$$

$$\times\begin{bmatrix} [\varepsilon_{i,i'}^{\phi,\phi}] & [{}^{0}_{0}\varepsilon_{i,i'+1/2}^{\phi,\psi}] & [{}^{1}_{0}\varepsilon_{i,i'+1/4}^{\phi,\psi}] & [{}^{1}_{1}\varepsilon_{i,i'+3/4}^{\phi,\psi}] \\ [{}^{0}_{0}\varepsilon_{i+1/2,i'}^{\psi,\phi}] & [{}^{0,0}_{0,0}\varepsilon_{i+1/2,i'+1/2}^{\psi,\psi}] & [{}^{0,1}_{0,0}\varepsilon_{i+1/2,i'+1/4}^{\psi,\psi}] & [{}^{0,1}_{0,1}\varepsilon_{i+1/2,i'+3/4}^{\psi,\psi}] \\ [{}^{1}_{0}\varepsilon_{i+1/4,i'}^{\psi,\phi}] & [{}^{1,0}_{0,0}\varepsilon_{i+1/4,i'+1/2}^{\psi,\psi}] & [{}^{1,1}_{0,0}\varepsilon_{i+1/4,i'+1/4}^{\psi,\psi}] & [{}^{1,1}_{0,1}\varepsilon_{i+1/4,i'+3/4}^{\psi,\psi}] \\ [{}^{1}_{1}\varepsilon_{i+3/4,i'}^{\psi,\phi}] & [{}^{1,0}_{1,0}\varepsilon_{i+3/4,i'+1/2}^{\psi,\psi}] & [{}^{1,1}_{1,0}\varepsilon_{i+3/4,i'+1/4}^{\psi,\psi}] & [{}^{1,1}_{1,1}\varepsilon_{i+3/4,i'+3/4}^{\psi,\psi}] \end{bmatrix}^{-1}\begin{bmatrix} [{}_{\phi y}D_{i'}^{n}] \\ [{}_{\psi y}{}^{0}_{0}D_{i'+1/2}^{n}] \\ [{}_{\psi y}{}^{1}_{0}D_{i'+1/4}^{n}] \\ [{}_{\psi y}{}^{1}_{1}D_{i'+3/4}^{n}] \end{bmatrix} \tag{7.59}$$

or

$$[E]=\frac{1}{\varepsilon_0}[\varepsilon]^{-1}[D] \tag{7.60}$$

where both the $[\varepsilon]$ and its inverse $[\varepsilon]^{-1}$ can be precomputed before the time updating and can be saved in the MRTDMaxwell solver.

Also, note that the diagonal approximation is the simplest means of evaluating (7.56), with the goal of generating leapfrog update equations similar to those in the FDTD algorithm. We employ the diagonal approximations by using $(\varepsilon_r^{\phi,\phi})_{i',i}\approx(\varepsilon_r^{\phi,\phi})_{i',i}\delta_{i,i'}$, $({}^{s}_{l}\varepsilon_r^{\phi,\psi})_{i,i'}\approx({}^{s}_{l}\varepsilon_r^{\phi,\psi})_{i,i'}\delta_{i,i'}$, $({}^{s}_{l}\varepsilon_r^{\psi,\phi})_{i,i'}\approx({}^{s}_{l}\varepsilon_r^{\psi,\phi})_{i,i'}\delta_{i,i'}$, and $({}^{s,s'}_{l,l'}\varepsilon_r^{\psi,\psi})_{i',i}\approx({}^{s,s'}_{l,l'}\varepsilon_r^{\psi,\psi})_{i',i}\delta_{i,i'}$. The update equations (7.33) and (7.34) can then be written as follows:

$$\left(\varepsilon_r^{\phi,\phi}\right)_{i,i}\,{}_{\phi y}E_i^n + \sum_{s=0}^{s_{\max}}\sum_{l=0}^{2^s-1}\left({}_l^s\varepsilon_r^{\psi,\phi}\right)_{i,i}\,{}_{\psi y}{}_l^s E^n_{i+2^{-s}(l+1/2)} = \frac{1}{\varepsilon_0}\,{}_{\phi y}D_i^n \tag{7.61}$$

$$\left({}_l^s\varepsilon_r^{\psi,\phi}\right)_{i,i}\,{}_{\phi y}E_i^n + \sum_{s'=0}^{s_{\max}}\sum_{l'=0}^{2^{-s'}-1}\left({}_{l,l'}^{s,s'}\varepsilon_r^{\psi,\psi}\right)_{i,i}\,{}_{\psi y}{}_{l'}^{s'} E^n_{i+2^{-s'}(l'+1/2)} = \frac{1}{\varepsilon_0}\,{}_{\psi y}{}_l^s D^n_{i+2^{-s}(l+1/2)} \tag{7.62}$$

Their corresponding matrix forms are given as

$$\varepsilon_0\begin{bmatrix} \varepsilon^{\phi,\phi}_{i,i} & {}^{0}_{0}\varepsilon^{\phi,\psi}_{i,i+1/2} & {}^{1}_{0}\varepsilon^{\phi,\psi}_{i,i+1/4} & {}^{1}_{1}\varepsilon^{\phi,\psi}_{i,i+3/4} \\ {}^{0}_{0}\varepsilon^{\psi,\phi}_{i+1/2,i} & {}^{0,0}_{0,0}\varepsilon^{\psi,\psi}_{i+1/2,i+1/2} & {}^{0,1}_{0,0}\varepsilon^{\psi,\psi}_{i+1/2,i+1/4} & {}^{0,1}_{0,1}\varepsilon^{\psi,\psi}_{i+1/2,i+3/4} \\ {}^{1}_{0}\varepsilon^{\psi,\phi}_{i+1/4,i} & {}^{1,0}_{0,0}\varepsilon^{\psi,\psi}_{i+1/4,i+1/2} & {}^{1,1}_{0,0}\varepsilon^{\psi,\psi}_{i+1/4,i+1/4} & {}^{1,1}_{0,1}\varepsilon^{\psi,\psi}_{i+1/4,i+3/4} \\ {}^{1}_{1}\varepsilon^{\psi,\phi}_{i+3/4,i} & {}^{1,0}_{1,0}\varepsilon^{\psi,\psi}_{i+3/4,i+1/2} & {}^{1,1}_{1,0}\varepsilon^{\psi,\psi}_{i+3/4,i+1/4} & {}^{1,1}_{1,1}\varepsilon^{\psi,\psi}_{i+3/4,i+3/4} \end{bmatrix}\begin{bmatrix} {}_{\phi y}E_i^n \\ {}_{\psi y}{}^{0}_{0}E^n_{i+1/2} \\ {}_{\psi y}{}^{1}_{0}E^n_{i+1/4} \\ {}_{\psi y}{}^{1}_{1}E^n_{i+3/4} \end{bmatrix} = \begin{bmatrix} {}_{\phi y}D_i^n \\ {}_{\psi y}{}^{0}_{0}D^n_{i+1/2} \\ {}_{\psi y}{}^{1}_{0}D^n_{i+1/4} \\ {}_{\psi y}{}^{1}_{1}D^n_{i+3/4} \end{bmatrix} \tag{7.63}$$

$$\begin{bmatrix} {}_{\phi y}E_i^n \\ {}_{\psi y}{}^{0}_{0}E^n_{i+1/2} \\ {}_{\psi y}{}^{1}_{0}E^n_{i+1/4} \\ {}_{\psi y}{}^{1}_{1}E^n_{i+3/4} \end{bmatrix} = \frac{1}{\varepsilon_0}\begin{bmatrix} \varepsilon^{\phi,\phi}_{i,i} & {}^{0}_{0}\varepsilon^{\phi,\psi}_{i,i+1/2} & {}^{1}_{0}\varepsilon^{\phi,\psi}_{i,i+1/4} & {}^{1}_{1}\varepsilon^{\phi,\psi}_{i,i+3/4} \\ {}^{0}_{0}\varepsilon^{\psi,\phi}_{i+1/2,i} & {}^{0,0}_{0,0}\varepsilon^{\psi,\psi}_{i+1/2,i+1/2} & {}^{0,1}_{0,0}\varepsilon^{\psi,\psi}_{i+1/2,i+1/4} & {}^{0,1}_{0,1}\varepsilon^{\psi,\psi}_{i+1/2,i+3/4} \\ {}^{1}_{0}\varepsilon^{\psi,\phi}_{i+1/4,i} & {}^{1,0}_{0,0}\varepsilon^{\psi,\psi}_{i+1/4,i+1/2} & {}^{1,1}_{0,0}\varepsilon^{\psi,\psi}_{i+1/4,i+1/4} & {}^{1,1}_{0,1}\varepsilon^{\psi,\psi}_{i+1/4,i+3/4} \\ {}^{1}_{1}\varepsilon^{\psi,\phi}_{i+3/4,i} & {}^{1,0}_{1,0}\varepsilon^{\psi,\psi}_{i+3/4,i+1/2} & {}^{1,1}_{1,0}\varepsilon^{\psi,\psi}_{i+3/4,i+1/4} & {}^{1,1}_{1,1}\varepsilon^{\psi,\psi}_{i+3/4,i+3/4} \end{bmatrix}^{-1}\begin{bmatrix} {}_{\phi y}D_i^n \\ {}_{\psi y}{}^{0}_{0}D^n_{i+1/2} \\ {}_{\psi y}{}^{1}_{0}D^n_{i+1/4} \\ {}_{\psi y}{}^{1}_{1}D^n_{i+3/4} \end{bmatrix} \tag{7.64}$$

At this stage, we should point out that the quantities in the update equations are not physical fields (expansion coefficients only) and that the real fields should be calculated by adding all expansion terms. For example, the total field $E_y(x_0, t_0)$ at an arbitrary point x_0 at time t_0 with $|t_0/\Delta t - k| < \frac{1}{2}$ is given as

$$\begin{aligned} E_y(x_0, t_0) &= \iint_\infty E_y(x,t)\delta(x-x_0)\delta(t-t_0)\,dx\,dt = \iint_\infty \sum_{i,n=-\infty}^{+\infty} \\ &\times\left[{}_{\phi y}E_i^n\phi_i(x) + \sum_{s=0}^{s_{\max}}\sum_{l=0}^{2^s-1}{}_{\psi y}{}_l^s E^n_{i+2^{-s}(1/2+l)}\psi_{s,i+2^{-s}(1/2+l)}(x)\right]h^n(t)\delta_{x,x_0}\delta_{t,t_0}dx\,dt \\ &= \sum_{i=-\infty}^{+\infty}\left[{}_{\phi y}E_i^n\phi_i(x_0) + \sum_{s=0}^{s_{\max}}\sum_{l=0}^{2^s-1}{}_{\psi y}{}_l^s E^n_{i+2^{-s}(1/2+l)}\psi_{s,i+2^{-s}(1/2+l)}(x_0)\right] \\ &= \sum_{i=-\infty}^{+\infty}\left[{}_{\phi y}E_i^n\phi\left(\frac{x_0}{\Delta x}-i\right) + \sum_{s=0}^{s_{\max}}\sum_{l=0}^{2^s-1}{}_{\psi y}{}_l^s E^n_{i+2^{-s}(1/2+l)}2^{s/2}\psi\left[2^s\left(\frac{x_0}{\Delta x}-i\right)-l\right]\right] \\ &= \sum_{i=-\infty}^{+\infty}\left[{}_{\phi y}E_i^n\phi(i_0-i) + \sum_{s=0}^{s_{\max}}\sum_{l=0}^{2^s-1}{}_{\psi y}{}_l^s E^n_{i+2^{-s}(1/2+l)}2^{s/2}\psi[2^s(i_0-i)-l]\right] \\ &= \sum_{i=-\infty}^{+\infty}{}_{\phi y}E^n_{i_0-i}\phi(i) + \sum_{i=-\infty}^{+\infty}\sum_{s=0}^{s_{\max}}\sum_{l=0}^{2^s-1}{}_{\psi y}{}_l^s E^n_{i_0-i+2^{-s}(1/2+l)}2^{s/2}\psi(2^s i-l) \end{aligned} \tag{7.65}$$

For the case of $s_{\max} = 1$,

$$
\begin{aligned}
E_y(x_0, t_0) &= \sum_{i=-\infty}^{+\infty} {}_{\phi y}E^n_{i_0-i}\phi(i) + \sum_{i=-\infty}^{+\infty} {}_{\psi y0}^{\ 0}E^n_{i_0-i+1/2}\psi(i) \\
&\quad + \sum_{i=-\infty}^{+\infty} {}_{\psi y0}^{\ 1}E^n_{i_0-i+1/4}2^{1/2}\psi(2i) + \sum_{i=-\infty}^{+\infty} {}_{\psi y1}^{\ 1}E^n_{i_0-i+3/4}2^{1/2}\psi(2i-1) \\
&= \sum_{i=-\infty}^{+\infty} {}_{\phi y}E^n_{i_0-i}c_s(i) + \sum_{i=-\infty}^{+\infty} {}_{\psi y0}^{\ 0}E^n_{i_0-i+1/2}c_{w00}(i) \\
&\quad + \sum_{i=-\infty}^{+\infty} {}_{\psi y0}^{\ 1}E^n_{i_0-i+1/4}c_{w10}(i) + \sum_{i=-\infty}^{+\infty} {}_{\psi y1}^{\ 1}E^n_{i_0-i+3/4}c_{w11}(i) \\
&= E_{y\_s} + E_{y\_w00} + E_{y\_w10} + E_{y\_w11}
\end{aligned} \tag{7.66}
$$

where E_{y_s}, E_{y_w00}, and E_{y_w10} and E_{y_w11} are the fields corresponding to scaling, zero-level, and one-level wavelet functions. Tables 7.7a and 7.7b give

TABLE 7.7a Coefficients $c_s(i)$ and $c_{w00}(i)$ ($s = 0$), where $c_s(-i) = c_s(i)$ and $c_{w00}(-i) = c_{w00}(i+1)$

i	$c_s(i)$	$c_{w00}(i)$
0	1.08903098	−0.07644715
1	−0.06114826	−0.07644715
2	0.04505332	0.10634707
3	−0.02231131	−0.07275711
4	0.01185276	0.02259060
5	−0.00614264	−0.01178303
6	0.00291179	
7	−0.00151689	
8	0.00079093	
9	−0.00038209	
10	0.00026870	

TABLE 7.7b Coefficients $c_{w10}(i)$ ($s = 1$, $l = 0$) and $c_{w11}(i)$ ($s = 1$, $l = 1$)

i	$c_{w10}(i)$	$c_{w11}(i)$
−2	−0.01530218	
−1	−0.09842077	0.03145128
0	−1.08112586	0.15039746
1	0.16240631	−1.00651109
2	0.03086846	−0.09303439
3		−0.01442576

the values for the coefficients $c_s(i)$, $c_{w00}(i)$, $c_{w10}(i)$, and $c_{w11}(i)$ of the scaling and wavelet functions.

S-MRTD Update Equations for TEM$_x$ Model

Let us consider a simple case, where we use only the scaling functions in the field expansions. From (7.19) and (7.23), we then obtain the S-MRTD update equations:

$$
\begin{aligned}
{}_{\phi y}E_i^{n+1} &= \left(\frac{1 - \dfrac{\sigma_x \, \Delta t}{2\varepsilon_0}}{1 + \dfrac{\sigma_x \, \Delta t}{2\varepsilon_0}} \right) {}_{\phi y}E_i^n \\
&\quad - \frac{1}{\varepsilon_0 \left(1 + \dfrac{\sigma_x \, \Delta t}{2\varepsilon_0}\right)} \left(\sum_{i'=i_1-M}^{i_2+M} \sum_{\nu=-\infty}^{+\infty} a(\nu) \left(\left[\varepsilon_r^{\phi\phi}(\kappa) \right]^{-1} \right)_{i,i'} {}_{\phi z}H_{i'+\nu+1/2}^{n+1/2} \right) \frac{\Delta t}{\Delta x}
\end{aligned}
\tag{7.67}
$$

$$
\begin{aligned}
{}_{\phi z}H_{i+1/2}^{n+1/2} &= \left(\frac{1 - \dfrac{\sigma_x \, \Delta t}{2\varepsilon_0}}{1 + \dfrac{\sigma_x \, \Delta t}{2\varepsilon_0}} \right) {}_{\phi z}H_{i+1/2}^{n-1/2} - \frac{1}{\mu_0 \left(1 + \dfrac{\sigma_x \, \Delta t}{2\varepsilon_0}\right)} \\
&\quad \times \left(\sum_{\nu=-\infty}^{+\infty} a(\nu) {}_{\phi y}E_{i+\nu+1}^n \right) \frac{\Delta t}{\Delta x}
\end{aligned}
\tag{7.68}
$$

where the matrix $[\varepsilon_r^{\phi\phi}(\kappa)]^{-1}$ is the inverse of $\lfloor \varepsilon_r^{\phi\phi}(\kappa) \rfloor$ and its dimensions are $(i_2 - i_1 + 2M) \times (i_2 - i_1 + 2M)$. We note that this version of the S-MRTD scheme is much simpler, and it is more convenient for many practical applications.

The constant $\sigma_{\max}$ is determined by the designated reflection coefficient for a wave normally incident on the APML wall [1, 5]:

$$
\sigma_{\max} = -\frac{\varepsilon_0 c(m+1)}{2d} \ln(R(0)) \tag{7.69}
$$

where c is the speed of electromagnetic wave propagation and $R(0)$ is the designated reflection coefficient with a normal incidence. Here we assume $R(0) = 1.0 \times 10^{-8}$.

Before closing this section, we point out that an alternative choice (compared to (7.18)) for the conductivity σ_x is

$$
\sigma_x = c_0 \sigma_{x\max} \left(\frac{x}{d}\right)^m = c_0 \frac{\varepsilon_0}{\Delta t} \left(\frac{x}{d}\right)^m \qquad (c_0 = 0.8 \sim 1.6) \tag{7.70}
$$

which assures good absorption of electromagnetic waves in the x-direction.

7.3 APPLICATION RESULTS

As shown in Figure 7.3, a plane TEM$_x$ transient pulse, generated by an exciting E_y field, is incident on a dielectric slab characterized with relative permittivity

TABLE 7.8 Discretization in the MRTD and FDTD Methods

Method	Δx (mm)	Δt (ps)	N_x (cells)
FDTD	1.5	2.0	312
S-MRTD	6.0	2.0	74
SW-MRTD ($s = 0$)	6.0	2.0	74
SW-MRTD ($s = 1$)	6.0	2.0	74

$\varepsilon_y = 4$ and width $d = 6 \times 10^{-2}$m. The entire computational domain contains 74 cells with the cell size of $\Delta x = 6.0$ mm. Assuming that the maximum frequency of the incident wave $f_{\max} = 5$ GHz, the corresponding minimum wavelength is $\lambda_{\min} = 3 \times 10^{-2}$ m inside the medium. We adopt an eight-layered APML to truncate boundaries. Table 7.8 lists the discretization for the MRTD and FDTD. The ratio of cell size of the FDTD to that of the MRTD is 1:4.

In order to see the field distribution of electric fields, we plot field distributions at time step $n = 350$ as shown in Figure 7.4 for the fields constructed with the scaling functions and wavelets. It can be seen that the contribution from wavelets is at a much lower level, and this is the reason why we expand wavelets only in an extended discontinuity region. In Figure 7.5, we only enlarge the wavelet effect on the electric field in the time domain. The enlarged part shows that the

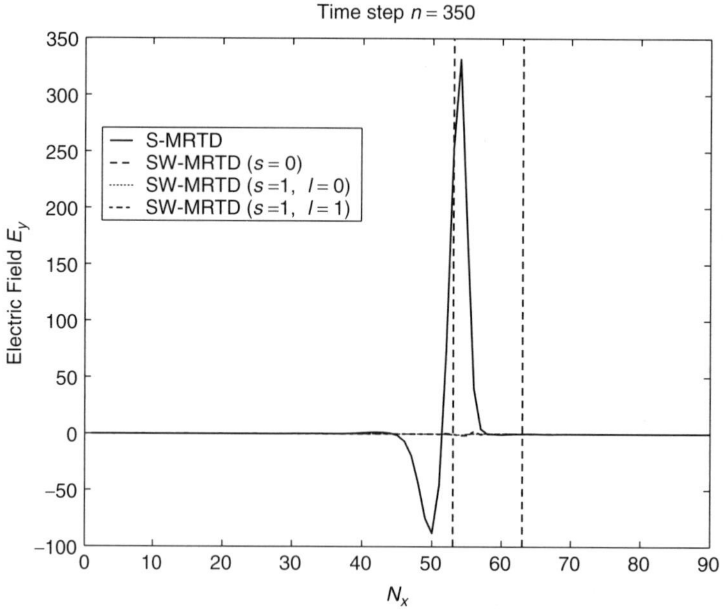

FIGURE 7.4 Time domain electric field for a single-layer dielectric slab system at the time step $n = 350$ ($d = 60$ mm, $\Delta x = 6$ mm, $\varepsilon_r = 4.0$).

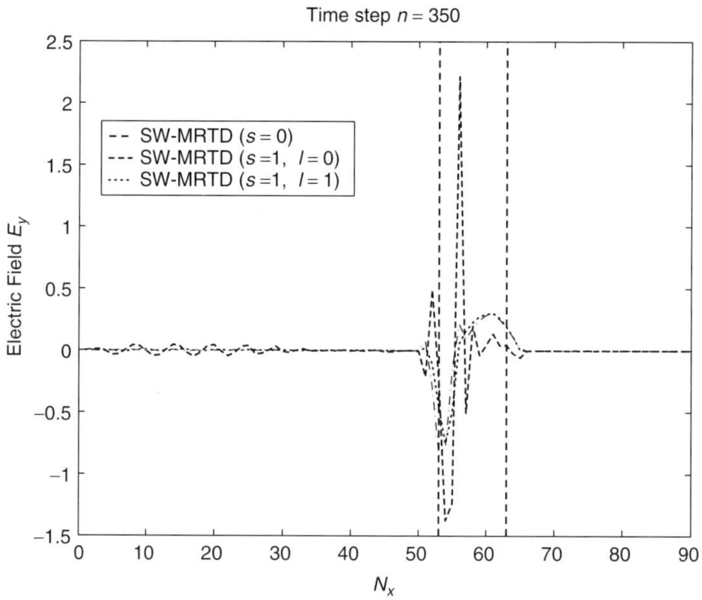

FIGURE 7.5 Enlarged time domain electric field at the time step $n = 350$.

wavelet function used in the MRTD scheme display the much detail information of the electromagnetic wave propagating.

In Figure 7.6, we show the reflection coefficients for a single-layer dielectric system calculated using the FDTD, S-MRTD, and SW-MRTD ($s = 0, 1$) methods. To emphasize the differences among these methods, we have enlarged part of the figure in Figure 7.7. Observe that there is excellent agreement between all methods; however, the SW-MRTD with $s = 1$ shows the best accuracy with relative error less than 0.02% calculated at the peak with respect to the analytic solution as shown in Figure 7.7.

In fact, the scaling function component in the MRTD scheme based on the Battle–Lemarié wavelet family is dominant relative to wavelet components. Figures 7.8 and 7.9 display the E-field distributions in the time domain. As shown in these figures, the magnitudes contributed by the wavelet components are much smaller compared to those derived from the scaling components.

Next, we investigate a two-layer dielectric slab system where the two slabs are separated by a distance of $d = 1.2 \times 10^{-1}$m, and each of them is characterized by $\varepsilon_{r1} = \varepsilon_{r2} = 4.0$ and $w_1 = w_2 = 6 \times 10^{-2}$m, as shown in Figure 7.10.

As shown in Figure 7.11, the reflection coefficients for the two-layer dielectric slab system derived from various methods show great consistency compared to the analytical results. Again, the SW-MRTD method is more efficient than the FDTD method.

Another practical example, as illustrated in Figure 7.12, displays the reflection coefficient distribution for a ten-dielectric-layered, periodic stratified system. The results are obtained by using both the S-MRTD and FDTD methods. The width

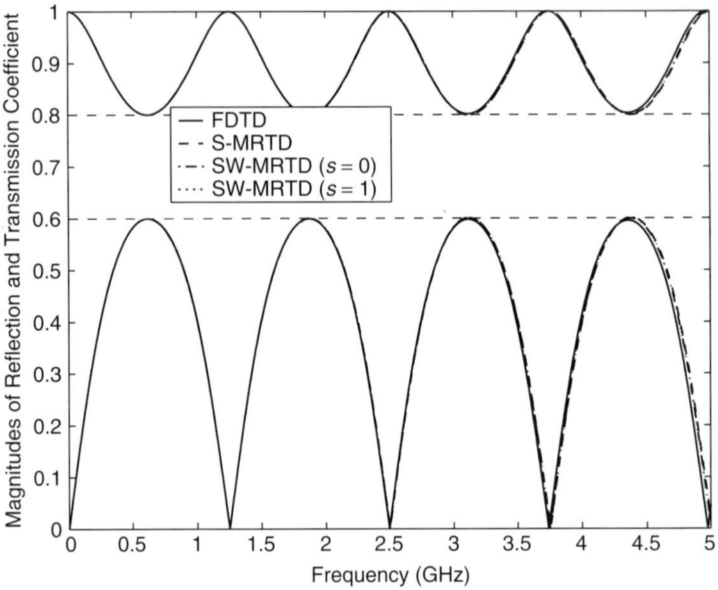

FIGURE 7.6 Magnitude of reflection and transmission coefficients R is for a single-layer dielectric slab system versus frequency for full frequency bandwidth of interest.

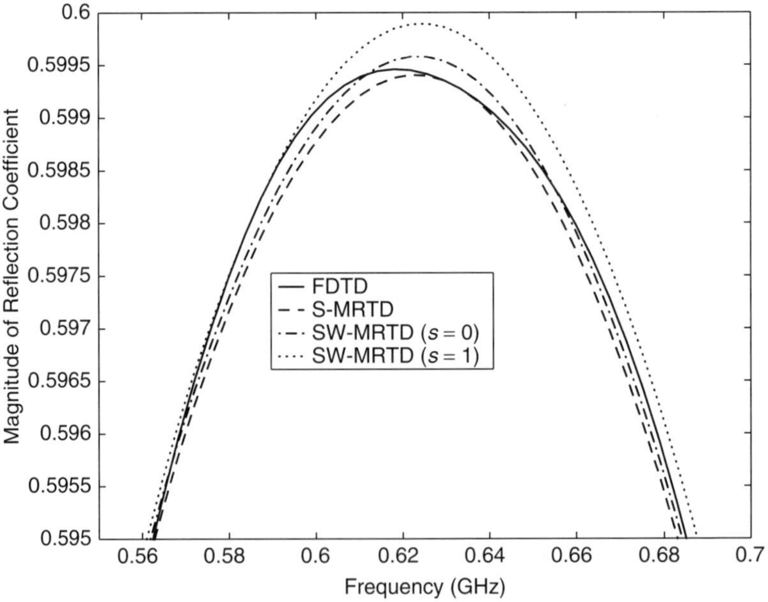

FIGURE 7.7 Magnitude of reflection coefficient R for a single-layer dielectric slab system versus frequency for enlarged frequency response.

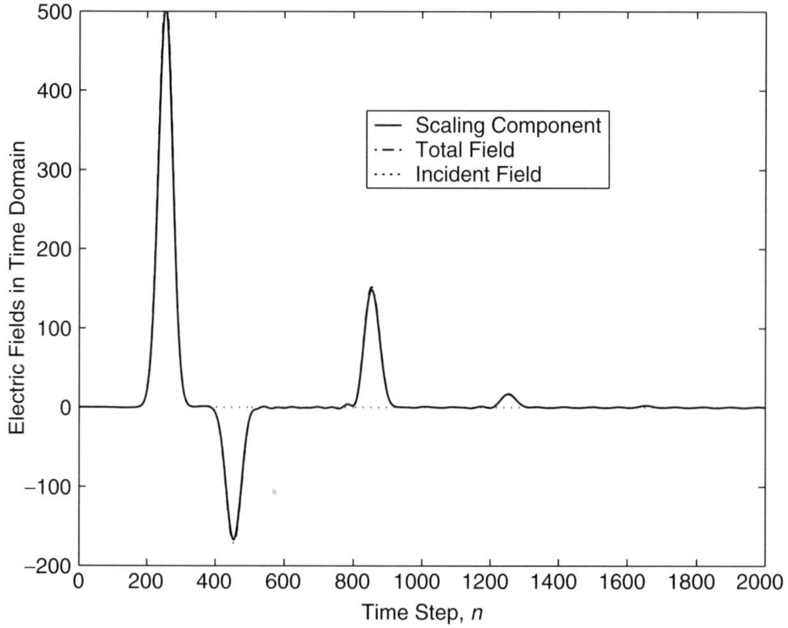

FIGURE 7.8 Time domain incident field E_y^{inc}, scaling component ${}_sE_y$, and total field E_y.

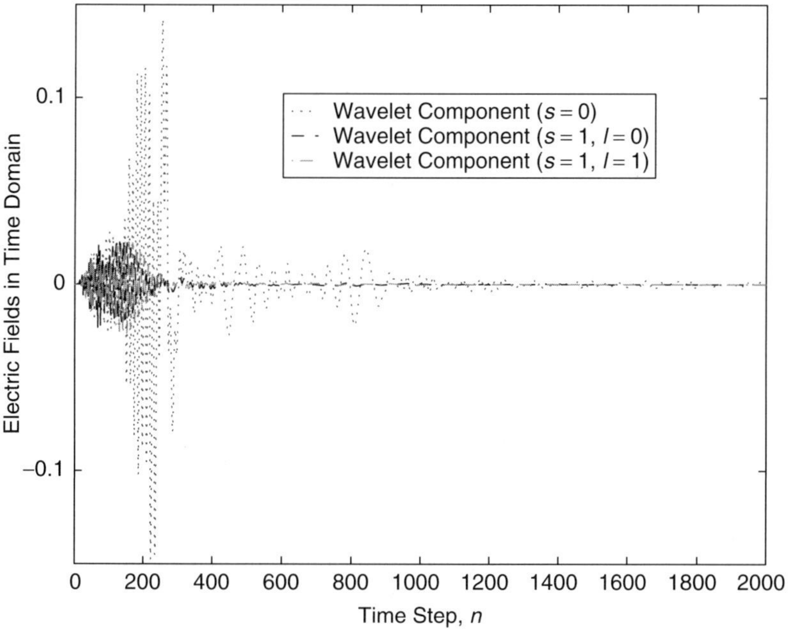

FIGURE 7.9 Time domain wavelet components ${}_wE_y$.

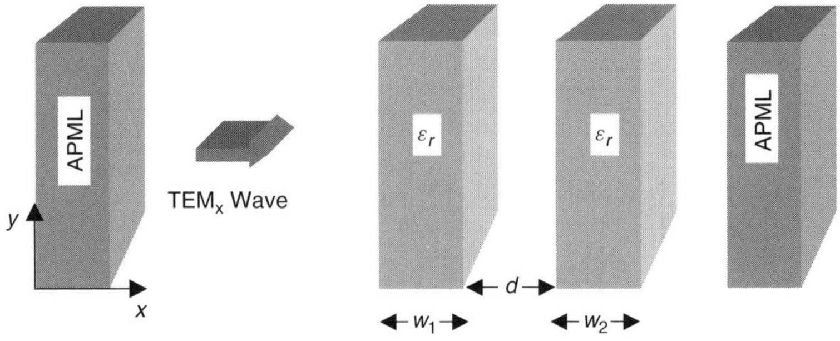

FIGURE 7.10 TEM$_x$ wave incident on two-slab dielectric system.

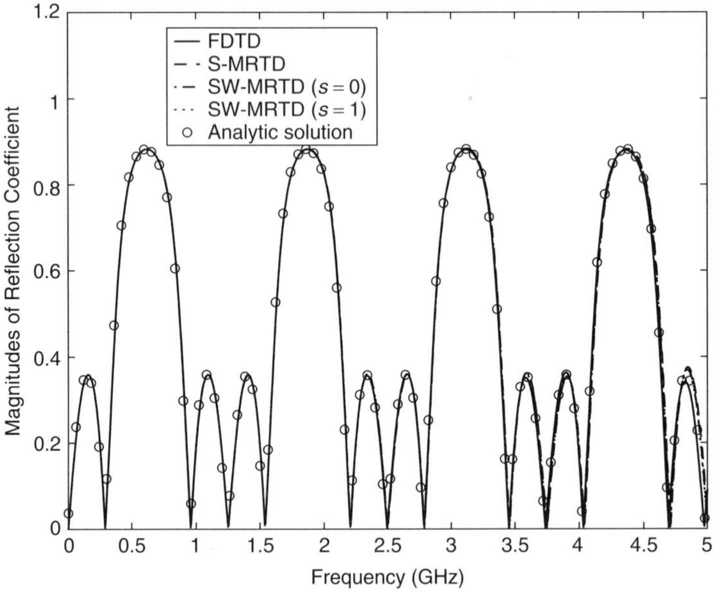

FIGURE 7.11 Magnitude of the reflection coefficient R for a two-slab dielectric system versus frequency.

of the slab is $w_1 = 6 \times 10^{-3}$ m and the separation between any two slabs is $w_2 = 6 \times 10^{-3}$ m; the computed values are compared with the analytic solution. The MRTD presents the best accuracy for such a complicated structure.

Finally, in order to compare different wavelet bases, we compare the MRTD results derived from the cubic spline Battle–Lemarié basis with those from the Haar wavelet basis. We expand electromagnetic fields as a summation of a Haar scaling function with zero-level wavelets:

$$E_y(x,t) = \sum_{i,n=-\infty}^{+\infty} \left[{}_{\phi y}E_i^n \phi_i(x) + {}_{\psi y}E_i^n \psi_i(x) \right] h^n(t) \tag{7.71a}$$

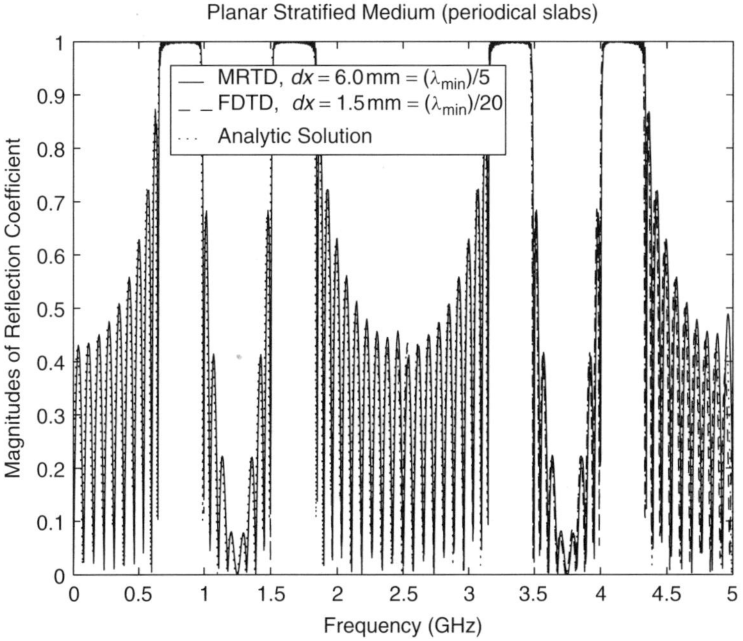

FIGURE 7.12 Magnitude of the reflection coefficients for a ten-dielectric-layered, periodic stratified system.

$$H_z(x,t) = \sum_{i,n=-\infty}^{+\infty} \left[{}_{\phi z}H^{n+1/2}_{i+1/4}\,\phi_{i+1/4}(x) + {}_{\psi z}H^{n+1/2}_{i+1/4}\,\psi_{i+1/4}(x) \right] h^{n+1/2}(t) \quad (7.71b)$$

where we represent fields in terms of the rectangular pulse function $h(t)$ in time, and the rectangular scaling function $\phi(x)$ and rectangular wavelet function $\psi(x)$ in space, respectively. As shown in Figure 7.13, the E-fields are expanded in integer grid points, and the H-fields are located at one-fourth the cell size from integer grid points.

Using Galerkin's method, we obtain the update equations:

$$\begin{aligned} {}_{\phi y}E_i^{n+1} &= {}_{\phi y}E_i^{n} \\ &- \frac{1}{\varepsilon}\left[\left({}_{\phi z}H^{n+1/2}_{i+1/2} - {}_{\phi z}H^{n+1/2}_{i-1/2}\right) + \left({}_{\psi z 1}^{\;0}H^{n+1/2}_{i+1/4} - {}_{\psi z\, 0}^{\;\;0}H^{n+1/2}_{i-1/4}\right)\right]\frac{\Delta t}{\Delta x} \end{aligned} \quad (7.72)$$

$$\begin{aligned} {}_{\psi y 0}^{\;0}E_i^{n+1} &= {}_{\psi y}E_i^{n} \\ &- \frac{1}{\varepsilon}\left[\left({}_{\phi z}H^{n+1/2}_{i+1/4} - {}_{\phi z}H^{n+1/2}_{i-1/4}\right) + \left(3\,{}_{\psi z}H^{n+1/2}_{i+1/4} + {}_{\psi z}H^{n+1/2}_{i-1/4}\right)\right]\frac{\Delta t}{\Delta x} \end{aligned} \quad (7.73)$$

$$\begin{aligned} {}_{\phi z}H^{n+1/2}_{i+1/4} &= {}_{\phi z}E^{n-1/2}_{i+1/4} \\ &- \frac{1}{\mu}\left[\left({}_{\phi y}E_i^{n} - {}_{\phi y}E_i^{n}\right) + \left({}_{\psi y}E_i^{n} - {}_{\psi y}E_i^{n}\right)\right]\frac{\Delta t}{\Delta x} \end{aligned} \quad (7.74)$$

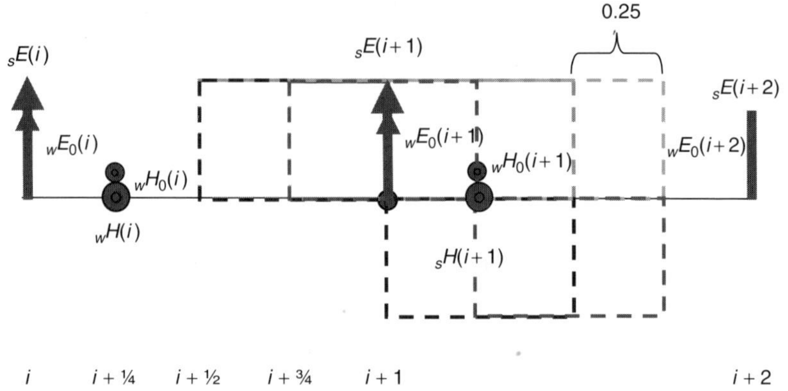

FIGURE 7.13 Positions of E- and H-fields in 1D Haar-based MRTD system.

$$
{}_{\psi z}H_{i+1/4}^{n+1/2} = {}_{\psi z}H_{i+1/4}^{n-1/2} + \frac{1}{\mu}\left[\left(-{}_{\phi y}E_{i+1}^{n} + {}_{\phi y}E_{i}^{n}\right) + \left(-{}_{\psi y}E_{i+1}^{n} - 3{}_{\psi y}E_{i}^{n}\right)\right]\frac{\Delta t}{\Delta x} \tag{7.75}
$$

where we have used the integral relations as follows:

$$
\int_{-\infty}^{+\infty} \phi_i(x)\frac{\partial \phi_{i'+1/4}(x)}{\partial x}\,dx = \delta_{i',i} - \delta_{i',i-1} \tag{7.76}
$$

$$
\int_{-\infty}^{+\infty} \phi_i(x)\frac{\partial \psi_{i'+1/4}(x)}{\partial x}\,dx = \delta_{i',i+1} - \delta_{i',i} \tag{7.77}
$$

$$
\int_{-\infty}^{+\infty} \psi_i(x)\frac{\partial \phi_{i'+1/4}(x)}{\partial x}\,dx = \delta_{i',i} - \delta_{i',i+1} \tag{7.78}
$$

$$
\int_{-\infty}^{+\infty} \psi_i(x)\frac{\partial \psi_{i'+1/4}(x)}{\partial x}\,dx = \delta_{i',i+1} + 3\delta_{i',i}, \tag{7.79}
$$

Figure 7.14 shows the results derived from the FDTD, Haar based MRTD, Battle–Lemarié based MRTD, and the analytical solution. The discretization sizes for the FDTD and Harr based MRTD methods are 1.5×10^{-3} m; the discretization size of the Battle–Lemarié based MRTD is 6×10^{-3} m; and the time interval is 2.5×10^{-12} s for all cases. The dielectric layer is characterized by $\varepsilon_r = 4.0$ and the thickness $d = 6 \times 10^{-2}$ m.

The results in Figure 7.14 indicate that the Battle–Lemarié based MRTD is more accurate than other numerical methods even when using a much sparser grid, although the update equations of the Haar based MRTD are much simpler.

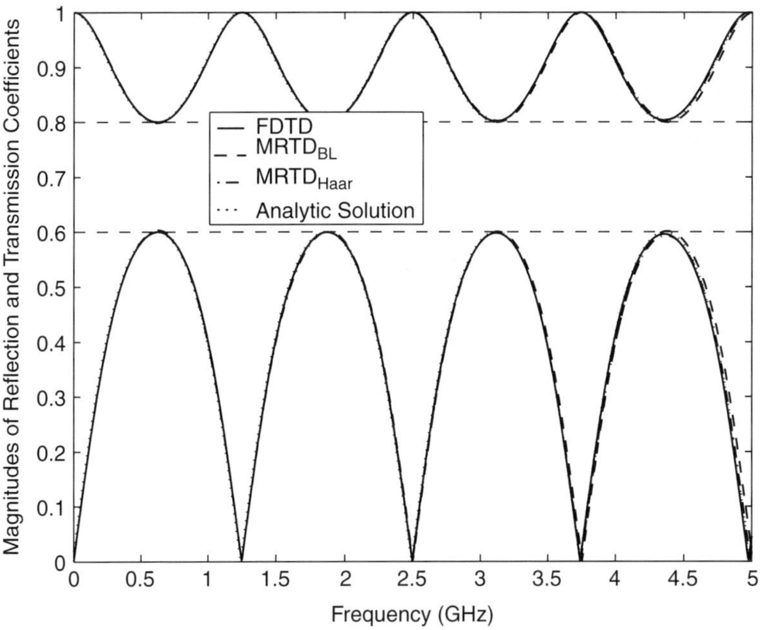

FIGURE 7.14 Magnitude of the reflection and transmission coefficients for a one-slab dielectric system versus frequency.

7.4 CONCLUSION

In this chapter, we have explored the MRTD based on the cubic spline Battle–Lemarié scaling and wavelet functions and obtained generalized update equations for one-dimensional, multilayer wave propagation problems. The computed results have been compared with those derived from the FDTD and the analytical solutions. The results are in excellent agreement with the analytical solutions. Although the wavelet expansions add accuracy to numerical computations, it could greatly increase complexity and computational load in MRTD applications. The numerical results obtained by the Battle–Lemarié based MRTD scheme are more accurate than those derived from the Haar based MRTD scheme under the same discretization.

REFERENCES

[1] M. Krumpholz and L. P. B. Katehi, "MRTD: New time-domain schemes based on multiresolution analysis," *IEEE Trans. Microwave Theory Tech.*, vol. 44. no. 4, pp. 555–571, Apr. 1996.

[2] L. P. B. Katehi, J. F. Harvey, and E. Tentzeris, "Time-Domain Analysis Using Multiresolution Expansions," Chapter 3, in *Advances in Computational Electrodynamics: The Finite-Difference Time-Domain Method* (ed. A. Taflove), Artech House, Norwood, MA 1998.

[3] Q. Cao and Y. Chen, "MRTD analysis of a transient electromagnetic pulse propagation through a dielectric layer," *Int. J. Electron.*, vol. 86, no. 4, pp. 459–474, 1999.

[4] Q. Cao and Y. Chen, "Fundamental research on multiresolution time domain scheme and its application," *Electromagnetics,* vol. 21, no. 6, pp. 485–496, Aug.–Sept. 2001.

[5] S. D. Gendeny, "An anisotropic perfectly matched layer-absorbing medium for the truncation of FDTD lattices," *IEEE Trans. Antennas Propag.*, vol. 44, pp. 1630–1639, 1996.

[6] R. F. Harrington, *Field Computation by Moment Methods*, 2nd ed., Krieger, Melbourne, FL, 1982.

[7] R. L. Burden and J. D. Faires, *Numerical Analysis*, 6th ed., Brooks/Cole Publishing, Belmont, CA, 1997.

[7] C. K. Chui, *An Introduction to Wavelets*, Academic Press, San Diego, 1992.

[8] R. J. Luebbers, K. S. Kunz, and K. A. Chamberlin, "An interactive demonstration of electromagnetic wave propagation using time-domain finite differences," *IEEE Trans. Educ.*, vol. 33. pp. 60–68, Feb. 1990.

CHAPTER EIGHT

Two-Dimensional MRTD Analysis

8.1 INTRODUCTION

The multiresolution time domain scheme (MRTD) [1], based on the expansions of unknown fields in terms of scaling and wavelet functions, is employed to improve computational efficiency and accuracy relative to the conventional finite difference time domain (FDTD) method. The efficiency of the MRTD method in analyzing a variety of microwave structures has been demonstrated in many recent publications [1–5].

In this chapter, a version of the two-dimensional MRTD (2D-MRTD) scheme, in conjunction with an unsplit anisotropic perfectly matched layer (APML) [6], will be presented for the analysis of printed uniform transmission lines and TE_z and TM_z wave propagation along a parallel waveguide. The MRTD algorithm described in this chapter retains the philosophy of the leapfrog algorithm with the representation form similar to that used in the FDTD. In addition, we introduce the formulation of the current 2D-MRTD algorithm, in conjunction with an APML for truncating open regions. Also, we present a simple and effective way to determine the APML parameters. Finally, we apply the imaging concept that can be used for the analysis of microwave structures with inhomogeneous materials, including different dielectrics and conductors with infinite conductivity. Such imaging is needed to truncate the boundaries in the MRTD formulation [4, 7].

8.2 MRTD ANALYSIS FOR PRINTED TRANSMISSION LINES

Field Construction

In order to analyze uniform transmission lines, we adopt the algorithm of the MRTD that is similar to those applied in the two-dimensional FDTD (2D-FDTD)

Multiresolution Time Domain Scheme for Electromagnetic Engineering
By Yinchao Chen, Qunsheng Cao, and Raj Mittra
ISBN 0-471-27230-2

methods [8, 9]. Assuming a lossless transmission line sitting along the z-direction (i.e., the direction of propagation), the phase factor $e^{-j\beta z}$ is introduced into the Maxwell equations so that a general three-dimensional MRTD (3D-MRTD) lattice can be transformed into a 2D planar cell by the suppression of the $e^{-j\beta z}$ factor. Figure 8.1 depicts a 2D unit cell of the MRTD scheme.

Following the procedure described in two earlier publications [8] and [9], we begin with the MRTD analysis by deriving a set of governing equations for the problem at hand, namely, a lossless guided-wave structure, shown in Figure 8.2. To ensure the MRTD update equations contain only real quantities, we expand

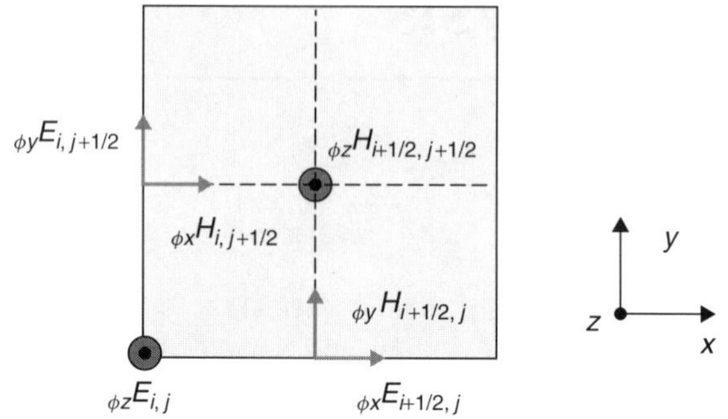

FIGURE 8.1 Compressed 2D-MRTD lattice on the x-y plane.

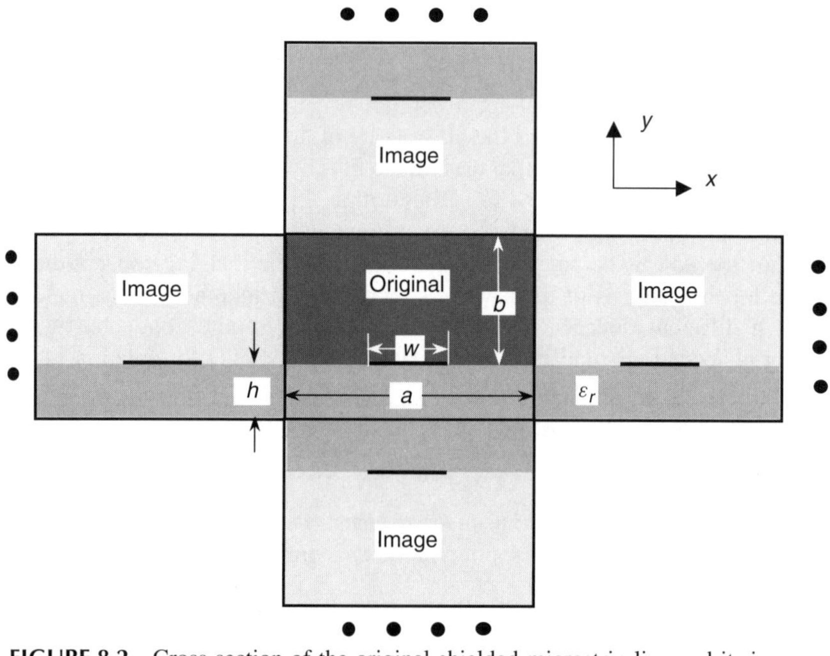

FIGURE 8.2 Cross section of the original shielded microstrip line and its images.

the electromagnetic field quantities as follows:

$$\begin{aligned}&[E_x(x,y,z,t), E_y(x,y,z,t), E_z(x,y,z,t)]^t\\&\quad = [jE_x(x,y,t), jE_y(x,y,t), E_z(x,y,t)]^t e^{-j\beta z}\end{aligned} \tag{8.1a}$$

$$\begin{aligned}&[H_x(x,y,z,t), H_y(x,y,z,t), H_z(x,y,z,t)]^t\\&\quad = [H_x(x,y,t), H_y(x,y,t), jH_z(x,y,t)]^t e^{-j\beta z}\end{aligned} \tag{8.1b}$$

where β is the propagation constant for the structure.

Using the generalized differential matrix operators (GDMOs) [10], we can obtain the following governing equations from the Maxwell curl equations:

$$\begin{aligned}&\begin{bmatrix} 0 & j\beta & \partial_y \\ -j\beta & 0 & -\partial_x \\ -\partial_y & \partial_x & 0 \end{bmatrix}\begin{bmatrix} H_x \\ H_y \\ jH_z \end{bmatrix}\\&= \left(\varepsilon_0 \frac{\partial}{\partial t}\begin{bmatrix} \varepsilon_{xx} & 0 & 0 \\ 0 & \varepsilon_{yy} & 0 \\ 0 & 0 & \varepsilon_{zz} \end{bmatrix} + \begin{bmatrix} \sigma_x^e & 0 & 0 \\ 0 & \sigma_y^e & 0 \\ 0 & 0 & \sigma_z^e \end{bmatrix}\right)\begin{bmatrix} jE_x \\ jE_y \\ E_z \end{bmatrix}\end{aligned} \tag{8.2a}$$

$$\begin{aligned}&\begin{bmatrix} 0 & j\beta & \partial_y \\ -j\beta & 0 & -\partial_x \\ -\partial_y & \partial_x & 0 \end{bmatrix}\begin{bmatrix} jE_x \\ jE_y \\ E_z \end{bmatrix}\\&= \left(-\mu_0 \frac{\partial}{\partial t}\begin{bmatrix} \mu_{xx} & 0 & 0 \\ 0 & \mu_{yy} & 0 \\ 0 & 0 & \mu_{zz} \end{bmatrix} + \begin{bmatrix} \sigma_x^m & 0 & 0 \\ 0 & \sigma_y^m & 0 \\ 0 & 0 & \sigma_z^m \end{bmatrix}\right)\begin{bmatrix} H_x \\ H_y \\ jH_z \end{bmatrix}\end{aligned} \tag{8.2b}$$

Note that, for the sake of generality, the electric and magnetic conductivities, relative permittivity, and relative permeability are designated in the form of a diagonal tensor. In practical applications, σ_α^m is set to zero. Note also that a printed planar transmission line is usually inhomogeneous in the vertical direction but homogeneous in the horizontal direction, which leads to $\varepsilon_{\alpha\alpha} = \varepsilon_{\alpha\alpha}(y)$ $(\alpha = x, y, z)$.

MRTD Algorithm for Printed Transmission Lines

In the next stage, we expand all the field quantities by using the scaling functions in space and the pulse functions in time. The field expansions read as follows:

$$E_x(\vec{r},t) = \sum_{n,i,j=-\infty}^{+\infty} {}_{\phi x}E^n_{i+1/2,j}\phi_{i+1/2}(x)\phi_j(y)h_n(t) \tag{8.3a}$$

$$E_y(\vec{r},t) = \sum_{n,i,j=-\infty}^{+\infty} {}_{\phi y}E^n_{i,j+1/2}\phi_i(x)\phi_{j+1/2}(y)h_n(t) \tag{8.3b}$$

$$E_z(\vec{r}, t) = \sum_{n,i,j=-\infty}^{+\infty} {}_{\phi z}E^{n}_{i,j}\phi_i(x)\phi_j(y)h_n(t) \tag{8.3c}$$

$$H_x(\vec{r}, t) = \sum_{n,i,j=-\infty}^{+\infty} {}_{\phi x}H^{n+1/2}_{i,j+1/2}\phi_i(x)\phi_{j+1/2}(y)h_{n+1/2}(t) \tag{8.4a}$$

$$H_y(\vec{r}, t) = \sum_{n,i,j=-\infty}^{+\infty} {}_{\phi y}H^{n+1/2}_{i+1/2,j}\phi_{i+1/2}(x)\phi_j(y)h_{n+1/2}(t) \tag{8.4b}$$

$$H_z(\vec{r}, t) = \sum_{n,i,j=-\infty}^{+\infty} {}_{\phi z}H^{n+1/2}_{i+1/2,j+1/2}\phi_{i+1/2}(x)\phi_{j+1/2}(y)h_{n+1/2}(t) \tag{8.4c}$$

where $\phi(x)$ denotes the scaling function, which in general can be the cubic spline Battle–Lemarié, Daubechies, biorthogonal Cohen–Daubechies–Feauveau (CDF), or biorthogonal scaling functions, and $h(t)$ is a rectangular pulse function. Substitution of the field expansions into the Maxwell equations and application of Galerkin's method lead to a set of field update equations. For example, the x-components of the update equations are given by

$${}_{\phi x}H^{n+1/2}_{i,j+1/2} = {}_{\phi x}H^{n-1/2}_{i,j+1/2} + \frac{\Delta t}{\mu_0\mu_x}\left[\beta_{\phi y}E^{n}_{i,j+1/2} - \sum_{\nu} a(\nu)_{\phi z}E^{n}_{i,j+\nu+1}\frac{1}{\Delta y}\right] \tag{8.5}$$

$$\begin{aligned}&\sum_{i',j'=-\infty}^{+\infty} (\sigma^e_x)_{ii',jj'}\frac{\Delta t}{2}\left({}_{\phi x}E^{n+1}_{i'+1/2,j'} + {}_{\phi x}E^{n}_{i'+1/2,j'}\right)\\ &+ \varepsilon_0 \sum_{j'=-\infty}^{+\infty} (\varepsilon_{xx})_{j,j'}\left({}_{\phi x}E^{n+1}_{i+1/2,j'} - {}_{\phi x}E^{n}_{i+1/2,j'}\right)\\ &= \Delta t\left[\beta_{\phi y}H^{n+1/2}_{i+1/2,j} + \sum_{\nu} a(\nu)_{\phi z}H^{n+1/2}_{i+1/2,j+\nu+1}\frac{1}{\Delta y}\right]\end{aligned} \tag{8.6}$$

where the coefficients $a(\nu)$ are expansion coefficients that can be obtained by the orthogonal and integral relations of the scaling functions [1, 4].

Assuming the conducting strip is characterized in the original structure with width $(i_2 - i_1)\ \Delta x$ and thickness τ, its conductivity can be expressed as

$$\sigma^e_\alpha = \begin{cases} \sigma, & i_1\,\Delta x \le x \le i_2\,\Delta x,\ \left(j_0\,\Delta y - \frac{\tau}{2}\right) < y < \left(j_0\,\Delta y + \frac{\tau}{2}\right) \\ 0, & \text{otherwise} \end{cases} \quad (\alpha = x, y) \tag{8.7}$$

and the relative permittivity is distributed in the original structure by

$$\varepsilon_{\alpha\alpha} = \varepsilon_{\alpha\alpha}(y) = \begin{cases} \varepsilon_r, & 0 \le y \le h \\ 1, & \text{otherwise} \end{cases} \quad (\alpha = x, y) \tag{8.8}$$

By considering image contributions in the whole space, including original and image spaces, the $(\varepsilon_{xx})_{j,j'}$ and $(\sigma_x^e)_{ii',jj'}$ are given as

$$(\varepsilon_{xx})_{j,j'} = \int_{-\infty}^{+\infty} \phi_j(y)\varepsilon_{xx}(y)\phi_{j'}(y)\frac{dy}{\Delta y} = \delta_{j,j'} + (\varepsilon_r - 1)\int_{-h}^{+h} \phi_j(y)\phi_{j'}(y)\frac{dy}{\Delta y}$$

$$= \delta_{j,j'} + (\varepsilon_r - 1)\alpha_{j,j'}^{\phi\phi}(-h, h) \tag{8.9}$$

with the definition of

$$\alpha_{j,j'}^{\phi\phi}(-h, h) = \int_{-h}^{+h} \phi_j(y)\phi_{j'}(y)\frac{dy}{\Delta y} \tag{8.10}$$

$$(\sigma_x^e)_{ii',jj'} = \int_{-\infty}^{+\infty}\int_{-\infty}^{+\infty} \phi_{i+1/2}(x)\phi_j(y)\sigma_x^e\phi_{i'+1/2}(x)\phi_{j'}(y)\frac{dx\,dy}{\Delta x\ \Delta y}$$

$$= \sigma^e \int_{i_1\Delta x}^{i_2\Delta x} \phi_{i+1/2}(x)\phi_{i'+1/2}(x)dx \int_{j_0\Delta y-\tau/2}^{j_0\Delta y+\tau/2} \tag{8.11}$$

$$\times\ \phi_j(y)\phi_{j'}(y)\,dy(\delta_{j,j_0} + \delta_{j,-j_0})$$

$$= \sigma^e\alpha_{i+1/2,i'+1/2}^{\phi\phi}(x_1, x_2)\alpha_{j,j'}^{\phi\phi}(-\tau/2, \tau/2)(\delta_{j,j_0} + \delta_{j,-j_0})$$

In the above integral relation, we use the diagonal matrix approximation, that is,

$$\sigma^e\alpha_{i+1/2,i'+1/2}^{\phi\phi}(x_1, x_2)\alpha_{j,j'}^{\phi\phi}(-\tau/2, \tau/2)(\delta_{j,j_0} + \delta_{j,-j_0})$$

$$\approx \sigma^e\alpha_{i+1/2,i'+1/2}^{\phi\phi}(x_1, x_2)\alpha_{j,j'}^{\phi\phi}(-\tau/2, \tau/2)(\delta_{j,j_0} + \delta_{j,-j_0})\delta_{i,i'}\delta_{j,j'} \tag{8.12}$$

$$= \sigma_e(\delta_{j,j_0} + \delta_{j,-j_0})\delta_{i,i'}\delta_{j,j'}$$

and the relative permittivity is

$$\sigma_e = \sigma^e\alpha_{i+1/2,i'+1/2}^{\phi\phi}(x_1, x_2)\alpha_{j,j'}^{\phi\phi}(-\tau/2, \tau/2) \tag{8.13}$$

We can further rewrite the time domain equation (8.6) as

$$\left({}_{\phi x}E_{i+1/2,j}^{n+1} + {}_{\phi x}E_{i+1/2,j}^{n}\right)\frac{\sigma_e(\delta_{j,j_0} + \delta_{j,-j_0} + \delta_{j,2j_0} + \cdots)\Delta t}{2}$$

$$+\ \varepsilon_0 \sum_{j'=-\infty}^{+\infty} (\varepsilon_{xx})_{j,j'}\left({}_{\phi x}E_{i+1/2,j'}^{n+1} - {}_{\phi x}E_{i+1/2,j'}^{n}\right)$$

$$= \Delta t\left[\beta_{\phi y}H_{i+1/2,j}^{n+1/2} + \sum_{\nu} a(\nu)_{\phi z}H_{i+1/2,j+\nu+1}^{n+1/2}\frac{1}{\Delta y}\right] \tag{8.14}$$

The summation index ν in (8.5), (8.6), and (8.14) is applicable not only to regions in the interior of the original structure but also to the image regions. Two types of boundary are frequently used for truncating the grids—the absorbing boundary condition for open structures and the perfectly electric conductor (PEC) for shielded ones. Usually, the PEC boundary is handled by using the image principle, as described in Chapter 5 as well as in [1] and [11]. Figure 8.2 shows the original structures, namely, the shielded microstrip lines and its images. By using the image technique, we can express all the image field quantities in terms of the fields in the original region $[0,N_y]$ as follows:

$$\left({}_{\phi x}E^{n+1}_{i+1/2,j} + {}_{\phi x}E^{n}_{i+1/2,j}\right)\frac{\sigma_e\delta_{j,j_0}\Delta t}{2} + \varepsilon_0\sum_{j'=0}^{Ny}(\varepsilon^*_{xx})_{j,j'}\left({}_{\phi x}E^{n+1}_{i+1/2,j'} - {}_{\phi x}E^{n}_{i+1/2,j'}\right)$$

$$= \Delta t\left[\beta_{\phi y}H^{n+1/2}_{i+1/2,j} + \sum_{\nu} a(\nu)_{\phi z}H^{n+1/2}_{i+1/2,j+\nu+1}\frac{1}{\Delta y}\right] \tag{8.15}$$

with

$$\sum_{j'=0}^{Ny}(\varepsilon^*_{xx})_{j,j'} = \sum_{j'=0}^{Ny}(\varepsilon_{xx})_{j,j'} - \sum_{j'=-1}^{-Ny}(\varepsilon_{xx})_{j,j'} - \sum_{j'=Ny+1}^{2Ny}(\varepsilon_{xx})_{j,j'}$$

$$- \cdots + (-1)^m\sum_{j'=-(m-1)Ny-1}^{-j-M_0}(\varepsilon_{xx})_{j,j'} + (-1)^m\sum_{j'=m\cdot Ny+1}^{j+M_0}(\varepsilon_{xx})_{j,j'} \tag{8.16}$$

where M_0 is the number of sampled points around index j, the first term in (8.16) is associated with the original structure, and the remaining terms represent the m images. By taking the inverse of (8.15), we can obtain the update equation for the x-component of the E field:

$${}_{\phi x}E^{n+1}_{i+1/2,j'} = \sum_{j'=0}^{Ny}([A])_{j,j'}\,{}_{\phi x}E^{n}_{i+1/2,j'}$$

$$+ \frac{1}{\varepsilon_0}\sum_{j'=0}^{Ny}([B])_{j,j'}\left[\beta_{\phi y}H^{n+1/2}_{i+1/2,j'}\Delta t + \sum_{\nu} a(\nu)_{\phi z}H^{n+1/2}_{i+1/2,j'+\nu+1}\frac{\Delta t}{\Delta y}\right] \tag{8.17}$$

where

$$[A] = \left[[\varepsilon^*_{xx}]_{j,j'} + [I]\frac{\sigma_e\delta_{j,j_0}\Delta t}{2\varepsilon_0}\right]^{-1}\cdot\left[[\varepsilon^*_{xx}]_{j,j'} - [I]\frac{\sigma_e\delta_{j,j_0}\Delta t}{2\varepsilon_0}\right] \tag{8.18a}$$

$$[B] = \left[[\varepsilon^*_{xx}]_{j,j'} + [I]\frac{\sigma_e\delta_{j,j_0}\Delta t}{2\varepsilon_0}\right]^{-1} \tag{8.18b}$$

For a shielded structure, the multiple image technique (MIT) method is used to construct the $\lfloor \varepsilon_{xx}^{*} \rfloor$ matrix [5]. Since the matrices $[A]$ and $[B]$ depend only on the basis functions and the material properties of the structure, they can be computed in advance of the time update computation and be saved in the MRTD Maxwell solver for the update equations.

APML Algorithm for Open Structures

In this section, we address the problem of mesh truncation in the context of the 2D-MRTD scheme for an open microstrip line as shown in Figure 8.3.

Since the APML is applied only in the x- and y-directions, the APML parameter matrix is $[S] = [s_x][s_y]$. The Maxwell curl equations in this APML medium in the frequency domain can be written as [6]

$$\begin{bmatrix} j\dfrac{\partial H_z}{\partial y} + j\beta H_y \\ -j\beta H_x - j\dfrac{\partial H_z}{\partial x} \\ \dfrac{\partial H_y}{\partial x} - \dfrac{\partial H_x}{\partial y} \end{bmatrix} = j\omega\varepsilon_0 \begin{bmatrix} \varepsilon_{xx} s_y s_x^{-1} & 0 & 0 \\ 0 & \varepsilon_{yy} s_x s_y^{-1} & 0 \\ 0 & 0 & \varepsilon_{zz} s_x s_y \end{bmatrix} \begin{bmatrix} jE_x \\ jE_y \\ E_z \end{bmatrix} \tag{8.19a}$$

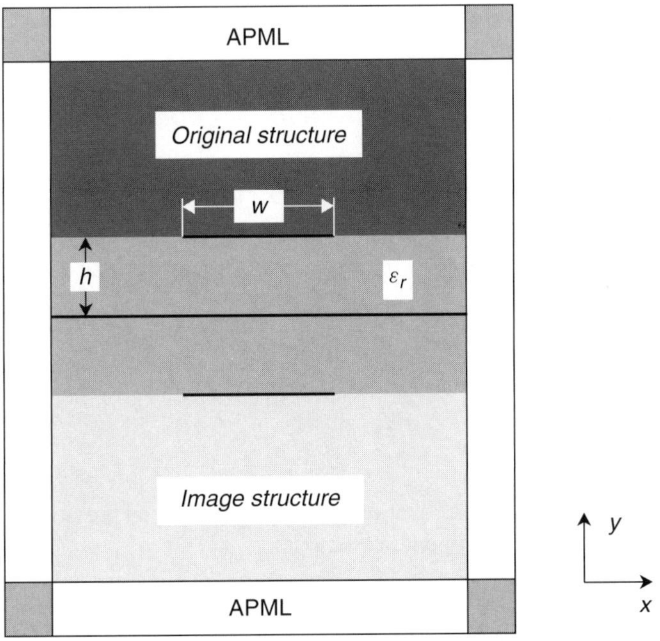

FIGURE 8.3 Cross section of the original printed transmission lines of the open microstrip line and its image.

$$\begin{bmatrix} \dfrac{\partial E_z}{\partial y} + \beta E_y \\ \beta E_x - \dfrac{\partial E_z}{\partial x} \\ j\left(\dfrac{\partial E_y}{\partial x} - \dfrac{\partial E_x}{\partial y}\right) \end{bmatrix} = -j\omega\mu_0 \begin{bmatrix} \mu_{xx}s_y s_x^{-1} & 0 & 0 \\ 0 & \mu_{yy}s_x s_y^{-1} & 0 \\ 0 & 0 & \mu_{zz}s_x s_y \end{bmatrix} \begin{bmatrix} H_x \\ H_y \\ jH_z \end{bmatrix} \tag{8.19b}$$

with

$$s_\alpha = \begin{cases} 1 + \dfrac{\sigma_\alpha}{j\omega\varepsilon_0} & \text{for absorption in } \alpha \text{ direction} \\ 1 & \text{elsewhere} \end{cases} \quad (\alpha = x, y) \tag{8.20}$$

Depending on the combination of the values of s_x and s_y, the Maxwell equations mentioned above are valid in all of the APML regions, which include the two side edges, the top wall, and the two corners.

For generality, we derive the MRTD update equations by considering the case of a corner region. Here we use a two-step update approach to obtain time stepping fields inside the APML regions. First, we define

$$D_x = \varepsilon_0\varepsilon_{xx}\left(1 + \frac{\sigma_x}{j\omega\varepsilon_0}\right)^{-1} E_x \tag{8.21}$$

Then we immediately simplify the x-components of the governing equations for the E- and D-fields given as

$$\frac{\partial H_z}{\partial y} + \beta H_y = j\omega D_x + \frac{\sigma_y}{\varepsilon_0} D_x \tag{8.22a}$$

$$\varepsilon_0\varepsilon_{xx}\frac{\partial E_x}{\partial t} = j\omega D_x + \frac{\sigma_x}{\varepsilon_0} D_x \tag{8.22b}$$

Similarly, we also obtain the x-components of the governing equations for the H- and B-fields:

$$\frac{\partial E_z}{\partial y} + \beta E_y = -j\omega B_x - \frac{\sigma_y}{\varepsilon_0} B_x \tag{8.23a}$$

$$\mu_0\mu_{xx}\frac{\partial H_x}{\partial t} = j\omega B_x + \frac{\sigma_x}{\varepsilon_0} B_x \tag{8.23b}$$

Following the MRTD discretization procedure, described previously, we derive the desired E- and D-field update equations:

$$_{\phi x}D^{n+1}_{i+1/2,j} = \frac{1 - \dfrac{\sigma_y\,\Delta t}{2\varepsilon_0}}{1 + \dfrac{\sigma_y\,\Delta t}{2\varepsilon_0}}\, {}_{\phi x}D^{n}_{i+1/2,j} + \frac{1}{1 + \dfrac{\sigma_y\,\Delta t}{2\varepsilon_0}}$$

$$\times \left(\beta_{\phi y} H^{n+1/2}_{i+1/2,j}\ \Delta t + \sum_m a(m)_{\phi z} H^{n+1/2}_{i+1/2,j+m+1} \frac{\Delta t}{\Delta y} \right) \tag{8.24}$$

$$_{\phi x}E^{n+1}_{i+1/2,j} = {}_{\phi x}E^{n}_{i+1/2,j} + \frac{1}{\varepsilon_0} \sum_{j'} ([\varepsilon_x]^{-1})_{j,j'}$$
$$\times \left[\left(1 + \frac{\sigma_x\ \Delta t}{2\varepsilon_0}\right) {}_{\phi x}D^{n+1}_{i+1/2,j'} - \left(1 - \frac{\sigma_x\ \Delta t}{2\varepsilon_0}\right) {}_{\phi x}D^{n}_{i+1/2,j'} \right] \tag{8.25}$$

By assuming the relative permeability $\mu_{\alpha\alpha} \equiv 1$ in APML regions, we obtain the desired H- and B-field update equations:

$$_{\phi x}B^{n+1/2}_{i,j+1/2} = \frac{1 - \dfrac{\sigma_y\ \Delta t}{2\varepsilon_0}}{1 + \dfrac{\sigma_y\ \Delta t}{2\varepsilon_0}} {}_{\phi x}B^{n-1/2}_{i,j+1/2}$$
$$- \frac{1}{1 + \dfrac{\sigma_y\ \Delta t}{2\varepsilon_0}} \left(\beta_{\phi y} E^{n}_{i,j+1/2}\ \Delta t + \sum_m a(m)_{\phi z} E^{n}_{i,j+m+1} \frac{\Delta t}{\Delta y} \right) \tag{8.26}$$

$$_{\phi x}H^{n+1/2}_{i,j+1/2} = {}_{\phi x}H^{n-1/2}_{i,j+1/2} + \frac{1}{\mu_0} \left[\left(1 + \frac{\sigma_x\ \Delta t}{2\varepsilon_0}\right) {}_{\phi x}B^{n+1/2}_{i,j+1/2} \right.$$
$$\left. - \left(1 - \frac{\sigma_x\ \Delta t}{2\varepsilon_0}\right) {}_{\phi x}B^{n-1/2}_{i,j+1/2} \right] \tag{8.27}$$

Similarly, we can employ the same procedure to obtain the remaining E- and H-field update equations as follows:

$$_{\phi y}D^{n+1}_{i,j+1/2} = \frac{1 - \dfrac{\sigma_x\ \Delta t}{2\varepsilon_0}}{1 + \dfrac{\sigma_x\ \Delta t}{2\varepsilon_0}} {}_{\phi y}D^{n}_{i,j+1/2} + \frac{1}{1 + \dfrac{\sigma_x\ \Delta t}{2\varepsilon_0}}$$
$$\times \left(\beta_{\phi x} H^{n+1/2}_{i,j+1/2}\ \Delta t - \sum_m a(m)_{\phi z} H^{n+1/2}_{i+m+1/2,j} \frac{\Delta t}{\Delta x} \right) \tag{8.28}$$

$$_{\phi y}E^{n+1}_{i,j+1/2} = {}_{\phi y}E^{n}_{i,j+1/2} + \frac{1}{\varepsilon_0}$$
$$\times \left[\left(1 + \frac{\sigma_y\ \Delta t}{2\varepsilon_0}\right) {}_{\phi y}D^{n+1}_{i,j+1/2} - \left(1 - \frac{\sigma_y\ \Delta t}{2\varepsilon_0}\right) {}_{\phi y}D^{n}_{i,j+1/2} \right] \tag{8.29}$$

$$_{\phi z}D^{n+1}_{i,j} = \frac{1 - \dfrac{\sigma_x\ \Delta t}{2\varepsilon_0}}{1 + \dfrac{\sigma_x\ \Delta t}{2\varepsilon_0}} {}_{\phi z}D^{n}_{i,j} + \frac{1}{1 + \dfrac{\sigma_x\ \Delta t}{2\varepsilon_0}}$$

$$\times \left(\sum_m a(m)_{\phi y} H^{n+1/2}_{i+m+1/2,j} \frac{\Delta t}{\Delta x} - \sum_m a(m)_{\phi x} H^{n+1/2}_{i,j+m+1/2} \frac{\Delta t}{\Delta y} \right) \tag{8.30}$$

$${}_{\phi z}E^{n+1}_{i,j} = \frac{1 - \dfrac{\sigma_y \ \Delta t}{2\varepsilon_0}}{1 + \dfrac{\sigma_y \ \Delta t}{2\varepsilon_0}} {}_{\phi z}E^{n}_{i,j} + \frac{1}{\varepsilon_0} \frac{1}{1 + \dfrac{\sigma_y \ \Delta t}{2\varepsilon_0}} \left[{}_{\phi z}D^{n+1}_{i,j} - {}_{\phi z}D^{n}_{i,j} \right] \tag{8.31}$$

$${}_{\phi y}B^{n+1/2}_{i+1/2,j} = \frac{1 - \dfrac{\sigma_x \ \Delta t}{2\varepsilon_0}}{1 + \dfrac{\sigma_x \ \Delta t}{2\varepsilon_0}} {}_{\phi y}B^{n-1/2}_{i+1/2,j}$$

$$+ \frac{1}{1 + \dfrac{\sigma_x \ \Delta t}{2\varepsilon_0}} \left(\beta_{\phi x} E^{n+1}_{i+1/2,j} \ \Delta t - \sum_m a(m)_{\phi z} E^{n+1/2}_{i+m1,j} \frac{\Delta t}{\Delta x} \right) \tag{8.32}$$

$${}_{\phi y}H^{n+1/2}_{i+1/2,j} = {}_{\phi y}H^{n-1/2}_{i+1/2,j} + \frac{1}{\mu_0}$$

$$\times \left[\left(1 + \frac{\sigma_y \ \Delta t}{2\varepsilon_0} \right) {}_{\phi y}B^{n+1/2}_{i+1/2,j+1/2} - \left(1 - \frac{\sigma_y \ \Delta t}{2\varepsilon_0} \right) {}_{\phi y}B^{n+1/2}_{i+1/2,j+1/2} \right] \tag{8.33}$$

$${}_{\phi z}B^{n+1/2}_{i+1/2,j+1/2} = \frac{1 - \dfrac{\sigma_x \ \Delta t}{2\varepsilon_0}}{1 + \dfrac{\sigma_x \ \Delta t}{2\varepsilon_0}} {}_{\phi z}B^{n-1/2}_{i+1/2,j+1/2} + \frac{1}{1 + \dfrac{\sigma_x \ \Delta t}{2\varepsilon_0}}$$

$$\times \left(\sum_m a(m)_{\phi y} E^{n+1}_{i,j+m+1/2} \frac{\Delta t}{\Delta x} - \sum_m a(m)_{\phi x} E^{n+1}_{i+1/2,j+m+1} \frac{\Delta t}{\Delta y} \right) \tag{8.34}$$

$${}_{\phi z}H^{n+1/2}_{i+1/2,j+1/2} = \frac{1 - \dfrac{\sigma_y \ \Delta t}{2\varepsilon_0}}{1 + \dfrac{\sigma_y \ \Delta t}{2\varepsilon_0}} {}_{\phi z}H^{n-1/2}_{i+1/2,j+1/2}$$

$$+ \frac{1}{\mu_0} \frac{1}{1 + \dfrac{\sigma_y \ \Delta t}{2\varepsilon_0}} \left[{}_{\phi z}B^{n+1/2}_{i+1/2,j+1/2} - {}_{\phi z}B^{n+1/2}_{i+1/2,j+1/2} \right] \tag{8.35}$$

In applying the MRTD method, we may choose the time step Δt to satisfy the stability and limit condition discussed earlier in all simulations. Starting from the convergence condition of the above D-field update equations, we have proved empirically that the following APML parameter yields good absorption [6]:

$$\sigma_\alpha = \sigma_{\max} \left| \frac{\alpha - \alpha_0}{N_p \ \Delta\alpha} \right|^p \qquad (\alpha = x, y) \tag{8.36a}$$

or

$$\sigma_\alpha \approx \gamma \frac{\varepsilon_0}{\Delta t} \left| \frac{i - i_0}{N_p} \right|^p \tag{8.36b}$$

where α_0 is the starting position in the α-direction, and N_p is the number of cells within the APML region. In principle, this relation is consistent with the standard form derived from a plane wave, normally incident on an APML region. We can minimize the reflection error with good absorption when $p = 2$ and $\gamma = 0.8 \sim 1.6$ are chosen.

Computation of the Propagation Characteristics

Typically, for printed transmission lines, the two parameters of interest are the effective dielectric constant, ε_{eff}, and the characteristic impedance, Z_0. The former can be evaluated from

$$\varepsilon_{\text{eff}}(f) = \left(\frac{c\beta}{2\pi f} \right)^2 \tag{8.37}$$

where c is the wave propagation velocity in free space and β is the propagation constant of the waveguide structure.

If the structure supports a TEM or a quasi-TEM wave, its characteristic impedance Z_0 may be defined as

$$Z_0(f) = \frac{V(f)}{I(f)} = \frac{FFT\left[-\int_{C_v} \vec{E} \cdot d\vec{l} \right]}{FFT\left[\oint_{C_i} \vec{H} \cdot d\vec{l} \right]} = \frac{FFT\left[-\int_{C_v} E_y \, dy \right]}{FFT\left[\oint_{C_i} \vec{H} \cdot d\vec{l} \right]} \tag{8.38}$$

where the integral contours C_v and C_i are integral paths as displayed in Figure 8.4 for shielded or open microstrip lines, and $V(f)$ and $I(f)$ are the frequency

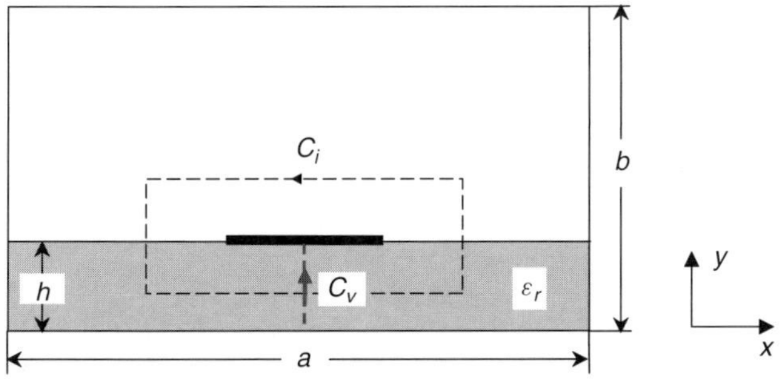

FIGURE 8.4 Cross section of a shielded microstrip line with integral contours.

domain voltage and current, respectively. Defining the time domain voltage $V(t)$ as an integral from the PEC ground to the strip line, at an arbitrary space point (x_0, y) and at time t_0 with $\left(n - \frac{1}{2}\right) \Delta t < t_0 < \left(n + \frac{1}{2}\right) \Delta t$, we can calculate the voltage as follows:

$$
\begin{aligned}
V &= -\int_0^h E_y(x_0, y, t_0)\, dy = -\sum_{l=0}^{N_s} \int_{\Delta y}^{(l+1)\Delta y} E_y(x_0, y, t_0)\, dy \\
&= -\sum_{l=0}^{N_s} \int_{\Delta y}^{(l+1)\Delta y} \left[\iint E_y(x, y, t)\delta(x - x_0)\delta(t - t_0)\, dx\, dt \right] dy \\
&= -\sum_{l=0}^{N_s} \int_{\Delta y}^{(l+1)\Delta y} \left[\sum_{n,i,j=-\infty}^{+\infty} {}_{\phi y}E^n_{i,j+1/2} \right. \\
&\quad \times \left. \iint_{\infty} \phi_i(x)\phi_{j+1/2}(y)h_n(t)\delta(x - x_0)\delta(t - t_0)\, dx\, dt \right] dy \qquad (8.39) \\
&= -\sum_{l=0}^{N_s} \sum_{i,j=-\infty}^{+\infty} {}_{\phi y}E^n_{i,j+1/2}\phi_i(x_0) \int_{\Delta y}^{(l+1)\Delta y} \phi_{j+1/2}(y)\, dy \\
&= -\sum_{l=0}^{N_s} \left[\sum_{i,j-l=-\infty}^{\infty} {}_{\phi y}E^n_{i,j-l}\phi_i(x_0)b(j)\, \Delta y \right]
\end{aligned}
$$

with

$$
b(j) = \int_{-1/2}^{1/2} \phi(y - j)\, dy \qquad (8.40)
$$

where N_s is the number of cells from the bottom ground to the PEC strip. The indices i and j of the summation span the original microstrip line and its image regions. Finally, the integral coefficients $b(j)$, calculated by (8.40), form a localized distribution and are even symmetric; that is, $b(j) = b(-j)$. The integral coefficients are listed in Table 8.1.

We can also apply the same procedure to solve for the current density distribution on the PEC strip. However, since the integral contour C_i consists of four segments that surround the PEC strip, we have to carry out the integrals in a stepwise manner as in the following:

$$
I = \oint_c \vec{H} \cdot dl = \left[\int_{(i_1-1/2)\Delta x}^{(i_1+1/2)\Delta x} H_x(x, y_1, t_0)\, dx + \cdots + \int_{(i_2-1/2)\Delta x}^{(i_2+1/2)\Delta x} H_x(x, y_1, t_0)\, dx \right]
$$

TABLE 8.1 Integral Coefficient $b(j)$

j	$b(j)$
0	0.9143952
1	0.0385998
2	0.0095740
3	−0.00865792
4	0.00506293
5	−0.00270341
6	0.0014053
7	−0.0007203
8	0.0000695

$$- \left[\int_{(i_1-1/2)\Delta x}^{(i_1+1/2)\Delta x} H_x(x, y_2, t_0)\, dx + \cdots + \int_{(i_2-1/2)\Delta x}^{(i_2+1/2)\Delta x} H_x(x, y_2, t_0)\, dx \right] \tag{8.41}$$
$$+ \left[- \int_{(j_0-1/2)\Delta y}^{(j_0+1/2)\Delta y} H_y(x_1, y, t_0)\, dy + \int_{(j_0-1/2)\Delta y}^{(j_0+1/2)\Delta y} H_y(x_2, y, t_0)\, dy \right]$$

where the first integration can be derived as

$$\int_{(i_1-1/2)\Delta x}^{(i_1+1/2)\Delta x} H_x(x, y_1, t_0)\, dx$$
$$= \int_{(i_1-1/2)\Delta x}^{(i_1+1/2)\Delta x} \left[\iint H_x(x, y, t)\delta(y - y_1)\delta(t - t_0)\, dy\, dt \right] dx$$
$$= \int_{(i-1/2)\Delta x}^{(i_1+1/2)\Delta x} \sum_{i,j=-\infty}^{+\infty} {}_{\phi x}H^{n+1/2}_{i,j_1-j}\phi(j)\phi_i(x)\, dx \tag{8.42}$$
$$= \sum_{i,j=-\infty}^{+\infty} {}_{\phi x}H^{n+1/2}_{i_1+i,j_1-j}\phi(j)b(i)$$

Finally, we have the time domain current as follows:

$$I = \sum_{l=i_1}^{i_2} \left[\sum_{i,j=-\infty}^{+\infty} \left({}_{\phi x}H^{n+1/2}_{l+i,j_1-j} - {}_{\phi x}H^{n+1/2}_{l+i,j_2-j} \right) \phi(j)b(i) \right]$$
$$+ \sum_{i,j=-\infty}^{+\infty} \left[\left({}_{\phi y}H^{n+1/2}_{i_2-i,j_0+j} - {}_{\phi y}H^{n+1/2}_{i_1-i,j_0+j} \right) \phi(i)b(j) \right] \tag{8.43}$$

The characteristic impedance can be obtained from (8.38) after the frequency domain voltage and current have been derived.

Window Modulation and Truncation

Conventionally, the MRTD time-stepping iterations are truncated by a direct termination. Indeed, such truncation may be viewed as adding a rectangular window (RGW) function onto the original time signals. The window function of W_{RG} may be defined as [11]

$$W_{\mathrm{RG}}(t_n) = H(t_n) - H(t_n - N_t) = \begin{cases} 1, & t_n \in [0, N_t] \\ 0, & \text{otherwise} \end{cases} \tag{8.44}$$

where $H(x)$ is the Heaviside function, which is equal to one when x is greater than zero, and zero otherwise.

The total system response in the time domain is the product of the original signal and the window function:

$$[E_i^W(t_n), H_i^W(t_n)] = [E_i(t_n), H_i(t_n)] \cdot W_{\mathrm{RG}}(t_n) \quad (i = x, y, or\ z) \tag{8.45}$$

where $[E_i(t_n), H_i(t_n)]$ denotes the original time domain signatures of the electric and magnetic fields derived by using the MRTD algorithm, and $[E_i^W(t_n), H_i^W(t_n)]$ represents the windowed versions of the electromagnetic fields.

Consequently, the corresponding frequency response is a convolution, in the frequency domain, of the original signal and the RGW function [12]:

$$\begin{aligned} [E_i^W(\omega), H_i^W(\omega)] &= [E_i(\omega), H_i(\omega)] * W_{\mathrm{RG}}(\omega) \\ &= \int_0^\infty [E_i(\tau), H_i(\tau)] \cdot W_{\mathrm{RG}}(\omega - \tau)\, d\tau \end{aligned} \tag{8.46}$$

Note that only when $N_t \to \infty$ does the expression become

$$W_{\mathrm{RG}}(\omega) \to \delta(\omega) \tag{8.47a}$$

$$\lfloor E_i^W(\omega), H_i^W(\omega) \rfloor \to [E_i(\omega), H_i(\omega)] \tag{8.47b}$$

where ω is the radian frequency. Normally, the truncation with the RGW does not pose a problem for a time response that decays sufficiently rapidly in time. However, such decay is relatively slow for a resonant structure, and early truncation can lead to significant errors in the results due to the high level of RGW sidelobes in the frequency domain response, which can be identified as the Gibbs phenomenon.

To demonstrate the Gibbs phenomenon arising from the window effects, a RGW frequency response is investigated with a time step of 0.1 ps, a frequency band of 100 GHz, and total of 2048 time steps. Figure 8.5 shows both its time

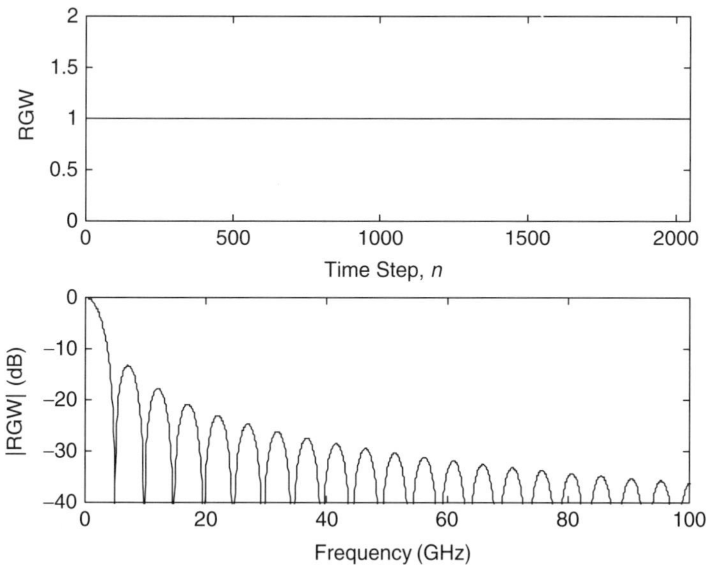

FIGURE 8.5 Normalized RGW function in the time and frequency domains.

domain distribution and the corresponding frequency response, in terms of decibels (dB). It should be evident that the RGW sidelobes can distort the frequency domain results unless the number of time iterations, N_t, is chosen to be sufficiently large.

To circumvent the difficulties with the RGW truncation, some concepts of signal processing technique are proposed by utilizing a low sidelobe function to modulate and truncate the entire computed FDTD signatures. A function that is well suited for this purpose is the Blackman–Harris window (BHW) function [13, 14]. This window function is defined as

$$W_{\rm BH}(t_n) = a_1 + a_2 \cos\left(\frac{\pi(t_n - N_c)}{N_c}\right) + a_3 \cos\left(\frac{2\pi(t_n - N_c)}{N_c}\right) + a_4 \cos\left(\frac{3\pi(t_n - N_c)}{N_c}\right) \quad (8.48)$$

where $N_c = N_t/2$, and a_1, a_2, a_3, and a_4 are constants with the values

$$[a_1 \quad a_2 \quad a_3 \quad a_4] = [0.35875 \quad 0.48829 \quad 0.14128 \quad 0.01168] \quad (8.49)$$

Figure 8.6 shows the time domain distribution and the corresponding frequency response of a BHW function based on the parameters used for generating the RGW function as shown in Figure 8.5. The sidelobe level of this function is extremely low—less than −92 dB. Such a low sidelobe level and its corresponding smooth main beam guarantee that the ripples will introduce little corruption in the convoluted signals.

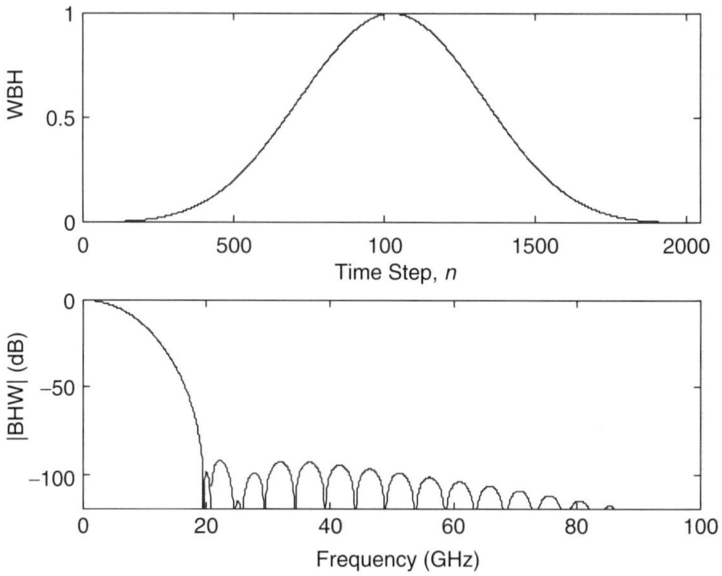

FIGURE 8.6 Normalized BHW function in the time and frequency domains.

We find the following: (1) The BHW is far superior to the RGW for eliminating both the initial transients due to excitation and window truncation effects. (2) The BHW has little effect on the location of the peaks, that is, resonant frequencies. In contrast, the RGW could introduce a slight shift in the location of the resonant frequency because of its high sidelobe levels. (3) The fields $[E_i^W(\omega), H_i^W(\omega)]$, modulated and truncated by using the BHW, approximate the original signals $[E_i(\omega), H_i(\omega)]$ better than those treated by using the RGW, and this, in turn, helps reduce the iteration time in the time domain by as much as a factor of 10 [11].

8.3 APPLICATION RESULTS FOR PRINTED TRANSMISSION LINES

In this section we analyze various shielded and open printed transmission lines. In the first example, we study a shielded microstrip line as shown in Figure 8.2, whose trace is assumed to be very thin and a perfect electric conductor. Figure 8.7 shows that the propagation characteristics derived from an application of the present 2D-MRTD scheme agree quite well with those derived from the spectral domain approach (SDA) [15].

Next, we study the propagation characteristics of a shielded microstrip line with $w = h = 1.5$, $a = 6.5$, and $b = 3.5$ mm, whose substrate is characterized by $\varepsilon_{xx} = \varepsilon_{zz} = 9.4$ and $\varepsilon_{yy} = 11.6$, as shown in Figure 8.2. We observe, from Figures 8.8 and 8.9, that the MRTD-computed results are in good agreement with those published in the literature, derived by using the FDTD method [9]. For

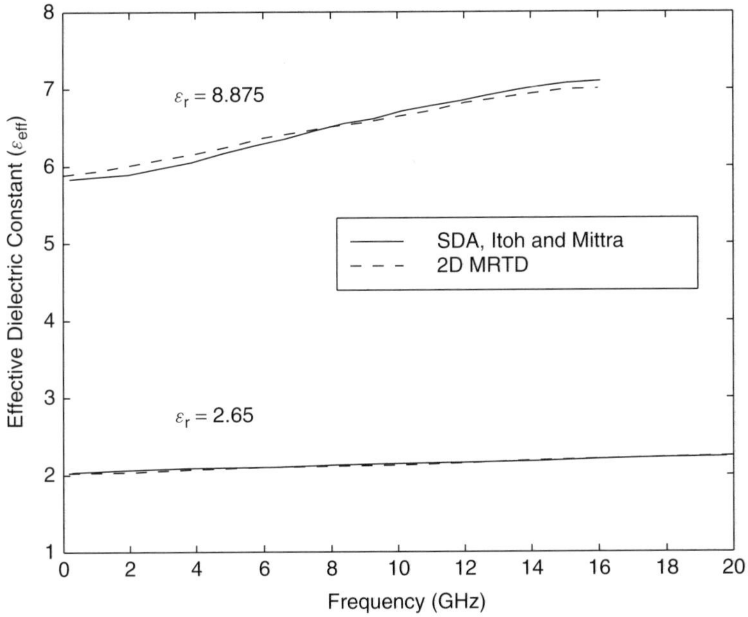

FIGURE 8.7 Effective dielectric constant ε_{eff} versus frequency for a shielded microstrip line with $w = h = 1.27$ and $a = b = 12.7$ mm.

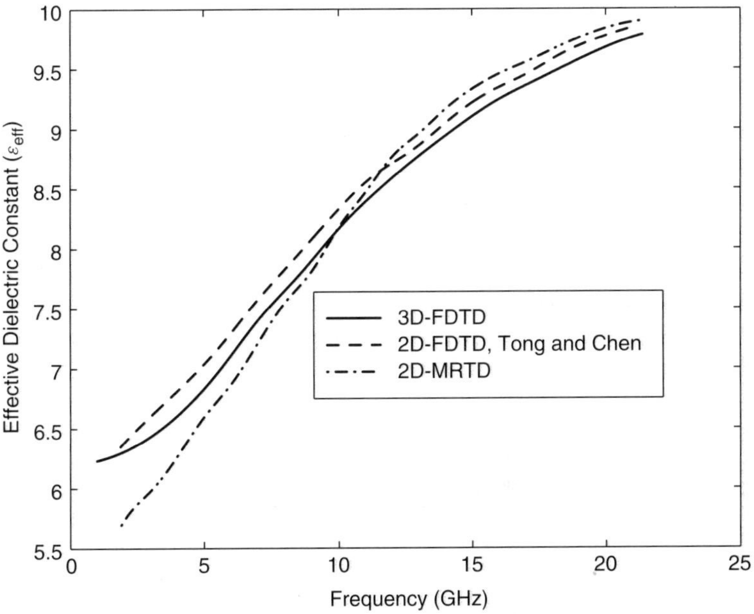

FIGURE 8.8 Frequency dependence of propagation characteristics of a shielded microstrip line for ε_{eff}.

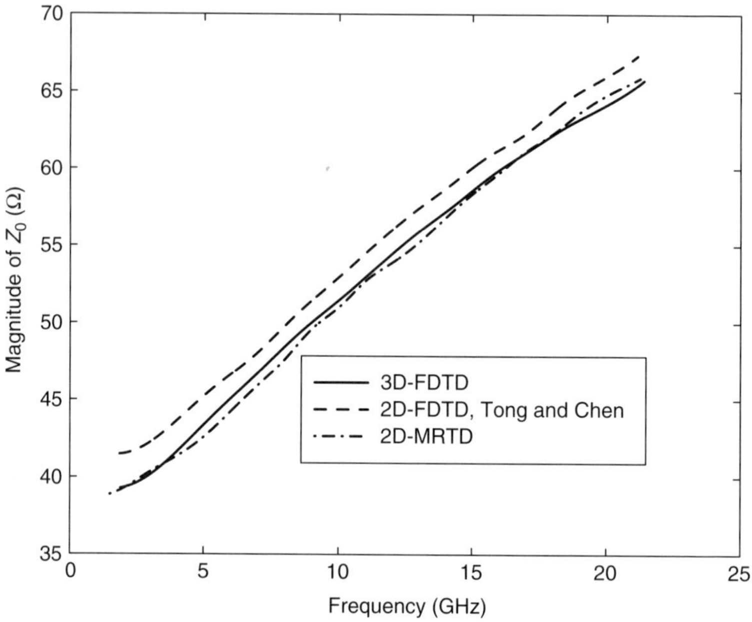

FIGURE 8.9 Frequency dependence of propagation characteristics of a shielded microstrip line for magnitude of Z_0 in Ω.

the structure defined above, the distance from the air–dielectric interface of the structure to the top of the PEC wall is set with seven cells. The corresponding discretization parameters employed in the FDTD and the MRTD methods are summarized, and a comparison of the total number of cells needed in the FDTD and MRTD techniques is presented in Table 8.2. The relative advantage of the MRTD is evident from this table.

Finally, in the last example of this section, we investigate an open microstrip line, whose geometry is specified in Figure 8.3. Once again, the computed results as shown in Figures 8.10–8.12 are observed to be in good agreement with those derived from the FDTD [9, 16, 17], SDA [18], and empirical approaches [19, 20]. The corresponding discretization parameters employed in the FDTD and MRTD methods are summarized in Table 8.3. Note that when derived from the

TABLE 8.2 Discretization Dimensions of a Shielded Microstrip Line

Method	Δx (mm)	Δy (mm)	N_x	N_y
2D-MRTD	0.5	0.5	13	7
2D-FDTD	0.125	0.125	52	28
3D-FDTD	0.125	0.125	52	28

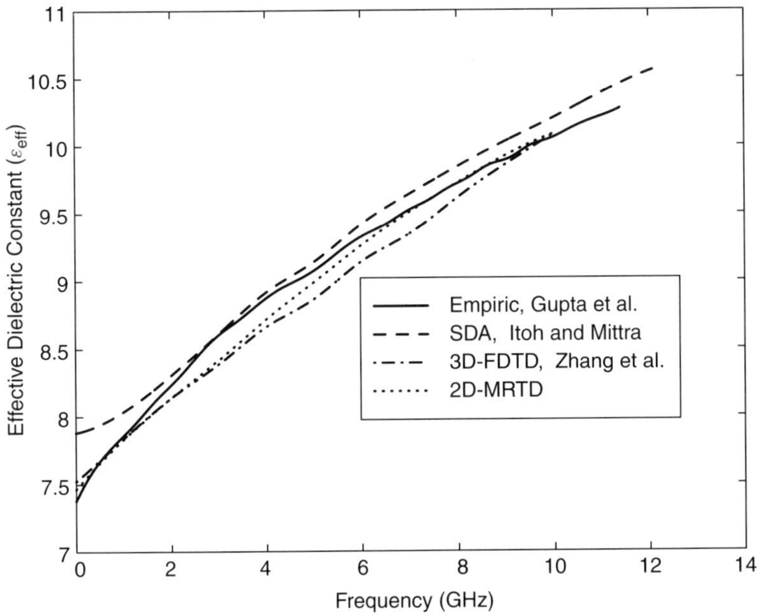

FIGURE 8.10 Effective dielectric constant ε_{eff} versus frequency of an open microstrip line with $w/h = 0.96$, $h = 3.17$ mm, and $\varepsilon_r = 11.7$.

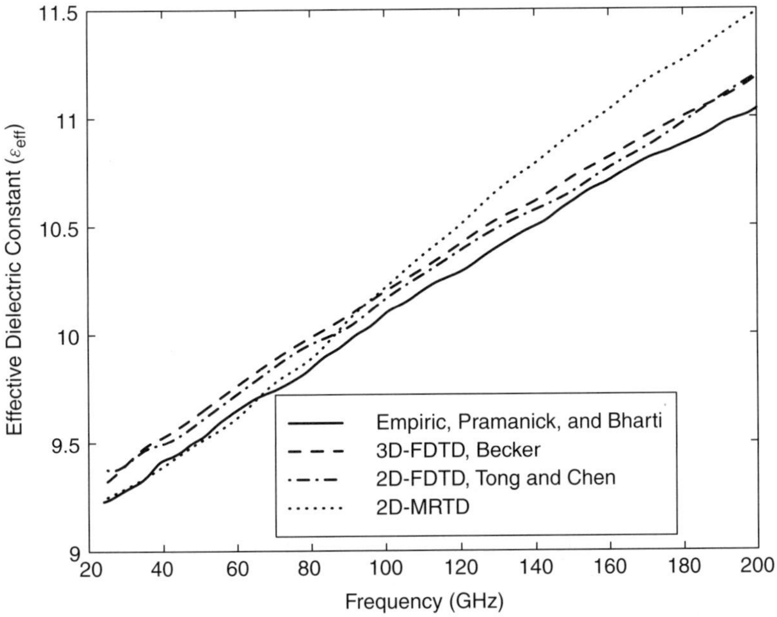

FIGURE 8.11 Frequency dependence of propagation characteristics of an open microstrip line for ε_{eff} with $w/h = 1.5$, $h = 0.10$ mm, and $\varepsilon_r = 13$.

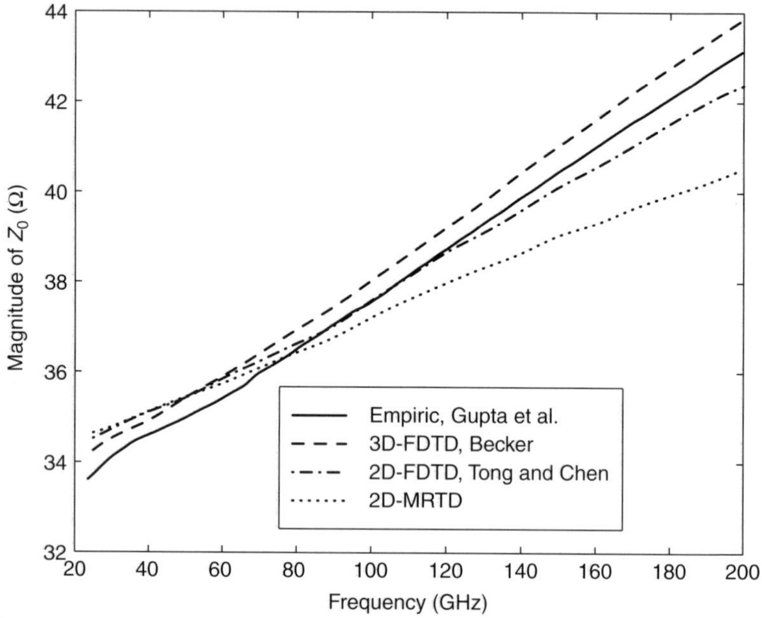

FIGURE 8.12 Frequency dependence of propagation characteristics of an open microstrip line for magnitude of Z_0 in Ω with $w/h = 1.5$, $h = 0.10$ mm, and $\varepsilon_r = 13.0$.

TABLE 8.3 Discretization Dimensions of an Open Microstrip Line ($\varepsilon_r = 13.0$)

Method	Δx (mm)	Δy (mm)	N_x	N_y
2D-MRTD	0.05	0.03333	21	12
2D-FDTD	0.0125	0.0125	110	30
3D-FDTD	0.0125	0.0125	55	30

2D-MRTD method, the same shielded and open microstrip lines require only about 6.25% and 7.64% of the 2D-FDTD scheme, respectively. To ensure computational accuracy, we have to use at least two cells along the width of the PEC strip, even for a very narrow strip.

8.4 MRTD ANALYSIS FOR PARALLEL WAVEGUIDE STRUCTURES

In this section, we analyze a waveguide system that uniformly extends to infinity along the z-direction as shown in Figure 8.13, while remaining discontinuous along the x-direction. Assuming all fields are uniformly distributed along the z-direction, we can use a two-dimensional model to analyze this waveguide structure.

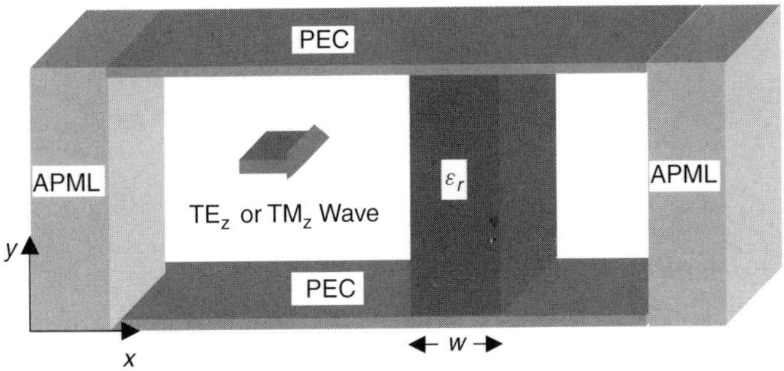

FIGURE 8.13 Transient wave propagation systems for a TE$_z$ or TM$_z$ wave.

Governing Equations for TE$_z$ and TM$_z$ Waves

Let us consider a wave normally incident on a parallel-plate waveguide boundary dielectric layer as shown in Figure 8.13. In general, the Maxwell curl equations in the frequency domain can be expressed as

$$\begin{bmatrix} 0 & -\partial_z & \partial_y \\ \partial_z & 0 & -\partial_x \\ -\partial_y & \partial_x & 0 \end{bmatrix} \begin{bmatrix} H_x \\ H_y \\ H_z \end{bmatrix} = j\omega\varepsilon_0 \begin{bmatrix} \varepsilon_x & 0 & 0 \\ 0 & \varepsilon_y & 0 \\ 0 & 0 & \varepsilon_z \end{bmatrix} \begin{bmatrix} {s_x}^{-1} & 0 & 0 \\ 0 & s_x & 0 \\ 0 & 0 & s_x \end{bmatrix} \begin{bmatrix} E_x \\ E_y \\ E_z \end{bmatrix}$$
$$= j\omega\varepsilon_0 \begin{bmatrix} \varepsilon_x {s_x}^{-1} & 0 & 0 \\ 0 & \varepsilon_y s_x & 0 \\ 0 & 0 & \varepsilon_z s_x \end{bmatrix} \begin{bmatrix} E_x \\ E_y \\ E_z \end{bmatrix} \tag{8.50a}$$

$$\begin{bmatrix} 0 & -\partial_z & \partial_y \\ \partial_z & 0 & -\partial_x \\ -\partial_y & \partial_x & 0 \end{bmatrix} \begin{bmatrix} E_x \\ E_y \\ E_z \end{bmatrix} = -j\omega\mu_0 \begin{bmatrix} \mu_x & 0 & 0 \\ 0 & \mu_y & 0 \\ 0 & 0 & \mu_z \end{bmatrix} \begin{bmatrix} {s_x}^{-1} & 0 & 0 \\ 0 & s_x & 0 \\ 0 & 0 & s_x \end{bmatrix} \begin{bmatrix} H_x \\ H_y \\ H_z \end{bmatrix}$$
$$= -j\omega\mu_0 \begin{bmatrix} \mu_x {s_x}^{-1} & 0 & 0 \\ 0 & \mu_y s_x & 0 \\ 0 & 0 & \mu_z s_x \end{bmatrix} \begin{bmatrix} H_x \\ H_y \\ H_z \end{bmatrix} \tag{8.50b}$$

where s_α is the same as defined in (8.20).

Given the case of a TE$_z$ wave normally incident, the existing components of the electromagnetic fields are E_x, E_y, H_z, and D_x (in the APML regions only), and the corresponding governing equations in the time domain can be simplified as

$$\frac{\partial E_x}{\partial y} - \frac{\partial E_y}{\partial x} = \frac{\mu_0}{\varepsilon_0}\sigma_x H_z + \mu_0 \frac{\partial H_z}{\partial t} \tag{8.51a}$$

$$\frac{\partial D_x}{\partial t} + \frac{\sigma_x}{\varepsilon_0} D_x = \varepsilon_0 \varepsilon_x \frac{\partial E_x}{\partial t} \tag{8.51b}$$

$$\frac{\partial H_z}{\partial y} = \frac{\partial D_x}{\partial t} \tag{8.51c}$$

$$\frac{\partial H_z}{\partial x} = -\sigma_x \varepsilon_y E_y - \varepsilon_0 \varepsilon_y \frac{\partial E_y}{\partial t} \tag{8.51d}$$

Similarly, in the case of a TM_z wave normally incident on a parallel-plate boundary dielectric layer, the fields E_z, H_x, H_y, and B_x exist in the electromagnetic fields and the governing equations for the TM_z mode can be expressed as

$$\frac{\partial H_y}{\partial x} - \frac{\partial H_x}{\partial y} = \sigma_x E_z + \varepsilon_0 \varepsilon_z \frac{\partial E_z}{\partial t} \tag{8.52a}$$

$$\frac{\partial B_x}{\partial t} + \frac{\sigma_x}{\varepsilon_0} B_x = \mu_0 \frac{\partial H_x}{\partial t} \tag{8.52b}$$

$$\frac{\partial E_z}{\partial y} = -\frac{\partial B_x}{\partial t} \tag{8.52c}$$

$$\frac{\partial E_z}{\partial x} = \frac{\mu_0}{\varepsilon_0} \sigma_x H_y + \mu_0 \frac{\partial H_y}{\partial t} \tag{8.52d}$$

In principle, all of the above governing equations are valid for all regions that can be differentiated by setting values of Δx, Δy, Δz, and σ_x.

MRTD Algorithm for TE_z Wave

Let us consider the case of a TE_z polarized wave normally incident on a parallel-plate bounded dielectric layer as shown in Figure 8.13. As the first step, we expand all the field quantities in terms of scaling functions in space and rectangular pulse functions in time as follows:

$$D_x(\vec{r}, t) = \sum_{n,i,j=-\infty}^{+\infty} {}_{\phi x} D^n_{i+1/2,j} \phi_{i+1/2}(x) \phi_j(y) h_n(t) \tag{8.53a}$$

$$E_x(\vec{r}, t) = \sum_{n,i,j=-\infty}^{+\infty} {}_{\phi x} E^n_{i+1/2,j} \phi_{i+1/2}(x) \phi_j(y) h_n(t) \tag{8.53b}$$

$$E_y(\vec{r}, t) = \sum_{n,i,j=-\infty}^{+\infty} {}_{\phi y} E^n_{i,j+1/2} \phi_i(x) \phi_{j+1/2}(y) h_n(t) \tag{8.53c}$$

$$H_z(\vec{r}, t) = \sum_{n,i,j=-\infty}^{+\infty} {}_{\phi z} H^{n+1/2}_{i+1/2,j+1/2} \phi_{i+1/2}(x) \phi_{j+1/2}(y) h_{n+1/2}(t) \tag{8.53d}$$

Applying Galerkin's method, we immediately obtain the MRTD update equations in a dielectric medium:

$$
{}_{\phi x}E^{n+1}_{i+1/2,j} = {}_{\phi x}E^{n}_{i+1/2,j} + \frac{1}{\varepsilon_0}\sum_{i'=0}^{N_R-N_L-1}([\varepsilon_x]^{-1})_{i,i'} \times \left(\sum_{m=-M_y}^{M_y} a(m)\,{}_{\phi z}H^{n+1/2}_{i'+1/2,j+m+1/2}\frac{\Delta t}{\Delta y}\right) \tag{8.54a}
$$

$$
{}_{\phi y}E^{n+1}_{i,j+1/2} = {}_{\phi y}E^{n}_{i,j+1/2} + \frac{1}{\varepsilon_0}\sum_{i'=0}^{N_R-N_L}([\varepsilon_y]^{-1})_{i,i'} \times \left(\sum_{m=-M_x}^{M_x} a(m)\,{}_{\phi z}H^{n+1/2}_{i'+m+1/2,j+1/2}\frac{\Delta t}{\Delta x}\right) \tag{8.54b}
$$

$$
{}_{\phi z}H^{n+1/2}_{i+1/2,j+1/2} = {}_{\phi z}H^{n-1/2}_{i+1/2,j+1/2} + \frac{1}{\mu_0}\sum_{m=-M_y}^{M_y} a(m)\left({}_{\phi x}E^{n}_{i+1/2,j+m+1}\frac{\Delta t}{\Delta y}\right) - \frac{1}{\mu_0}\sum_{m=-M_y}^{M_y} a(m)\left({}_{\phi y}E^{n}_{i+m+1,j+1/2}\frac{\Delta t}{\Delta x}\right) \tag{8.54c}
$$

where M_x and M_y, with a value of 9, represent the number of cells used in the summation, which is determined by the convergence of the sampled scaling function; N_L and N_R represent the number of cells counted from left and right interfaces of the dielectric layer, which indicate the coupling regions between the cells around the interfaces. Even though the above summation in the y-direction accounts for the number of cells in all original and image regions as shown in Figure 8.14, the symmetrical properties of the fields lead to a degeneration of the summation that only needs to include the field quantities inside the real structure.

The relative permittivity matrices in the MRTD transform domain $[\varepsilon_{x,y}]$ are determined by

$$
[\varepsilon_x]_{i,i'} = \frac{1}{\Delta x}\int_{-\infty}^{+\infty} \varepsilon_x(x)\phi_{i+1/2}(x)\phi_{i'+1/2}(x)\,dx \tag{8.55a}
$$

$$
[\varepsilon_y]_{i,i'} = \frac{1}{\Delta x}\int_{-\infty}^{+\infty} \varepsilon_y(x)\phi_{i}(x)\phi_{i'}(x)\,dx \tag{8.55b}
$$

Since the matrices $[\varepsilon_x]$ and $[\varepsilon_y]$ are functions of the scaling basis and the configuration materials, they can be computed in advance and saved in the MRTD Maxwell solver for update calling.

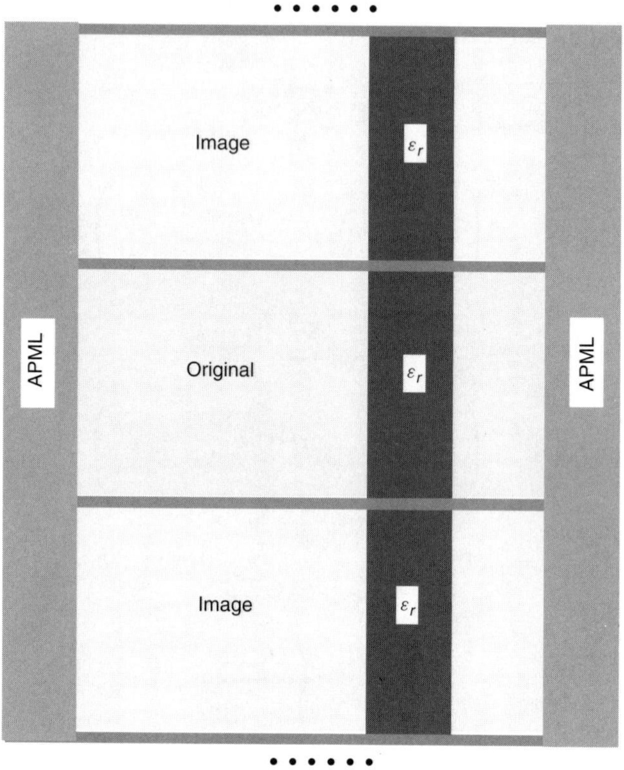

FIGURE 8.14 PEC parallel-plate bounded structure and its images.

We can derive the update equations in the APML regions as follows:

$$
{}_{\phi x}D^{n+1}_{i+1/2,j} = {}_{\phi x}D^{n}_{i+1/2,j} + \sum_{m=-M_y}^{M_y} a(m)\,{}_{\phi z}H^{n+1/2}_{i+1/2,j+m+1/2}\frac{\Delta t}{\Delta y} \tag{8.56a}
$$

$$
\begin{aligned}
{}_{\phi x}E^{n+1}_{i+1/2,j} = {}_{\phi x}E^{n}_{i+1/2,j} + \frac{1}{\varepsilon_0}\Bigg[\left(1 + \frac{\sigma_x\ \Delta t}{2\varepsilon_0}\right){}_{\phi x}D^{n+1}_{i+1/2,j} \\
- \left(1 - \frac{\sigma_x\ \Delta t}{2\varepsilon_0}\right){}_{\phi x}D^{n}_{i+1/2,j}\Bigg]
\end{aligned} \tag{8.56b}
$$

$$
\begin{aligned}
{}_{\phi y}E^{n+1}_{i,j+1/2} = \frac{1 - \dfrac{\sigma_x\ \Delta t}{2\varepsilon_0}}{1 + \dfrac{\sigma_x\ \Delta t}{2\varepsilon_0}}\,{}_{\phi y}E^{n}_{i,j+1/2} - \frac{1}{\varepsilon_0}\frac{1}{1 + \dfrac{\sigma_x\ \Delta t}{2\varepsilon_0}} \\
\times \sum_{m=-M_x}^{M_x} a(m)\,{}_{\phi z}H^{n+1/2}_{i+m+1/2,j+1/2}\frac{\Delta t}{\Delta x}
\end{aligned} \tag{8.56c}
$$

$$
{}_{\phi z}H^{n+1/2}_{i+1/2,j+1/2} = \frac{1-\dfrac{\sigma_x\,\Delta t}{2\varepsilon_0}}{1+\dfrac{\sigma_x\,\Delta t}{2\varepsilon_0}}\,{}_{\phi z}H^{n-1/2}_{i+1/2,j+1/2} + \frac{1}{\mu_0}\frac{1}{1+\dfrac{\sigma_x\,\Delta t}{2\varepsilon_0}} \times \left\{ \sum_{m=-M_y}^{M_y} a(m)\left({}_{\phi x}E^{n}_{i+1/2,j+m+1}\frac{\Delta t}{\Delta y}\right) - \sum_{m=-M_x}^{M_x} a(m)\left({}_{\phi y}E^{n}_{i+m+1,j+1/2}\frac{\Delta t}{\Delta x}\right)\right\} \tag{8.56d}
$$

To construct the total fields at an arbitrary node (i_0, j_0), we must include field contributions from the node itself, as well as those from its neighboring nodes within a small region. For example, at $x_0 = i_0\ \Delta x$, $y_0 = \left(j+\frac{1}{2}\right)\Delta y$, and $t = t_0\left(|t_0/\Delta t - n| < \frac{1}{2}\right)$, we can obtain the total field, E_y, by sampling their expansions with the space and time impulse functions

$$
\begin{aligned}
E_y(i_0, j_0, t_0) &= \iiint\limits_{\infty} E_y(x, y, t)\delta(x-x_0)\delta(y-y_0)\delta(t-t_0)\,dx\,dy\,dt \\
&= \sum_{i,j=-\infty}^{\infty} {}_{\phi y}E^{n}_{i,j+1/2}\phi_i(x_0)\phi_{j+1/2}(y_0) \\
&= \sum_{i,j=-\infty}^{\infty} {}_{\phi y}E^{n}_{i_0-i,j-j_0}\phi(i)\phi(j) \\
&\approx \sum_{i,j=-M_x,M_y}^{M_x,M_y} {}_{\phi y}E^{n}_{i_0-i,j-j_0}\phi(i)\phi(j)
\end{aligned} \tag{8.57}
$$

Actually, we only need to take a few terms in the summation to compute the total fields where the values of $\phi(m)$ are listed in Table 4.5 of Chapter 4 for various values of the integer variable.

MRTD Algorithm for TM_z Wave

For the case of the TM_z polarized wave, in the same manner, we expand all the field quantities as

$$
D_x(\vec{r}, t) = \sum_{n,i,j=-\infty}^{+\infty} {}_{\phi x}D^{n}_{i+1/2,j}\phi_{i+1/2}(x)\phi_j(y)h_n(t) \tag{8.58a}
$$

$$
E_x(\vec{r}, t) = \sum_{n,i,j=-\infty}^{+\infty} {}_{\phi x}E^{n}_{i+1/2,j}\phi_{i+1/2}(x)\phi_j(y)h_n(t) \tag{8.58b}
$$

$$E_y(\vec{r}, t) = \sum_{n,i,j=-\infty}^{+\infty} {}_{\phi y}E^n_{i,j+1/2}\phi_i(x)\phi_{j+1/2}(y)h_n(t) \tag{8.58c}$$

$$H_z(\vec{r}, t) = \sum_{n,i,j=-\infty}^{+\infty} {}_{\phi z}H^{n+1/2}_{i+1/2,j+1/2}\phi_{i+1/2}(x)\phi_{j+1/2}(y)h_{n+1/2}(t) \tag{8.58d}$$

Applying Galerkin's method, we can also obtain the MRTD update equations:

$$\begin{aligned} {}_{\phi x}E^{n+1}_{i+1/2,j} &= {}_{\phi x}E^n_{i+1/2,j} + \frac{1}{\varepsilon_0}\sum_{i'=0}^{N_R-N_L-1}([\varepsilon_x]^{-1})_{i,i'} \\ &\quad\times\left(\sum_{\nu=-M_y}^{M_y} a(\nu)\,{}_{\phi z}H^{n+1/2}_{i'+1/2,j+\nu+1/2}\frac{\Delta t}{\Delta y}\right) \end{aligned} \tag{8.59a}$$

$$\begin{aligned} {}_{\phi y}E^{n+1}_{i,j+1/2} &= {}_{\phi y}E^n_{i,j+1/2} + \frac{1}{\varepsilon_0}\sum_{i'=0}^{N_R-N_L}([\varepsilon_y]^{-1})_{i,i'} \\ &\quad\times\left(\sum_{\nu=-M_x}^{M_x} a(\nu)\,{}_{\phi z}H^{n+1/2}_{i'+\nu+1/2,j+1/2}\frac{\Delta t}{\Delta x}\right) \end{aligned} \tag{8.59b}$$

$$\begin{aligned} {}_{\phi z}H^{n+1/2}_{i+1/2,j+1/2} &= {}_{\phi z}H^{n-1/2}_{i+1/2,j+1/2} + \frac{1}{\mu_0}\sum_{\nu=-M_y}^{M_y} a(\nu)\left({}_{\phi x}E^n_{i+1/2,j+\nu+1}\frac{\Delta t}{\Delta y}\right) \\ &\quad - \frac{1}{\mu_0}\sum_{\nu=-M_y}^{M_y} a(\nu)\left({}_{\phi y}E^n_{i+\nu+1,j+1/2}\frac{\Delta t}{\Delta x}\right) \end{aligned} \tag{8.59c}$$

This time, the relative permittivity matrix in the MRTD transform domain, $[\varepsilon_{x,y}]$, is expressed as

$$[\varepsilon_x]_{i,i'} = \frac{1}{\Delta x}\int_{-\infty}^{+\infty} \varepsilon_x(x)\phi_{i+1/2}(x)\phi_{i'+1/2}(x)\,dx \tag{8.60a}$$

$$[\varepsilon_y]_{i,i'} = \frac{1}{\Delta x}\int_{-\infty}^{+\infty} \varepsilon_y(x)\phi_i(x)\phi_{i'}(x)\,dx \tag{8.60b}$$

8.5 APPLICATION RESULTS FOR PARALLEL-WAVEGUIDE STRUCTURES

In this section, we first look at a propagation system that is bounded by a pair of PEC parallel plates. Along this structure a TE_z wave, generated by uniformly exciting E_y at an input port, is incident on a dielectric layer with a cross section of 9×7.5 cm^2 and $\varepsilon_r = 4$ as shown in Figure 8.13. The entire computational domain is discretized into 50×5 cells with $\Delta x = \Delta y = 1.5$ cm, and the time step, Δt, is chosen to be 1.81×10^{-11} s.

As shown in Figure 8.15, 8.16, and 8.17, although the propagation system is bounded by a pair of PEC parallel plates, its time domain field signatures, E_y and E_y^{inc}, still retain a fast convergent rate; the magnitude of the reflection coefficient for the dielectric layer agrees well with the analytical solution; and the APML layers perform satisfactorily with the magnitude of the reflection coefficient R_{APML} as low as -47 dB for $N_p = 4$, 5, 6, and 7, respectively.

Next, considering a structure having two dielectric slabs in the parallel-plate bounded system, each of two slabs is characterized by a cross section of 3 cm $\times$ 4 cm and dielectric constants of $\varepsilon_{r1} = \varepsilon_{r2} = 4$ in layers 1 and 2. The distance between the two slabs is 2 cm. The entire computational area for the cross section of the structure is chosen as 39 cm $\times$ 4 cm. Two open sides along the direction of wave propagation are placed with an eight-layer APML absorbing layer.

Figure 8.18 shows fairly good agreement among the magnitudes of the reflection coefficients computed using a 2D-MRTD scheme, a 2D-FDTD method, and an analytical technique. As seen in Table 8.4, the MRTD requires only a fraction of the computational space used in the FDTD technique and presents more accurate results.

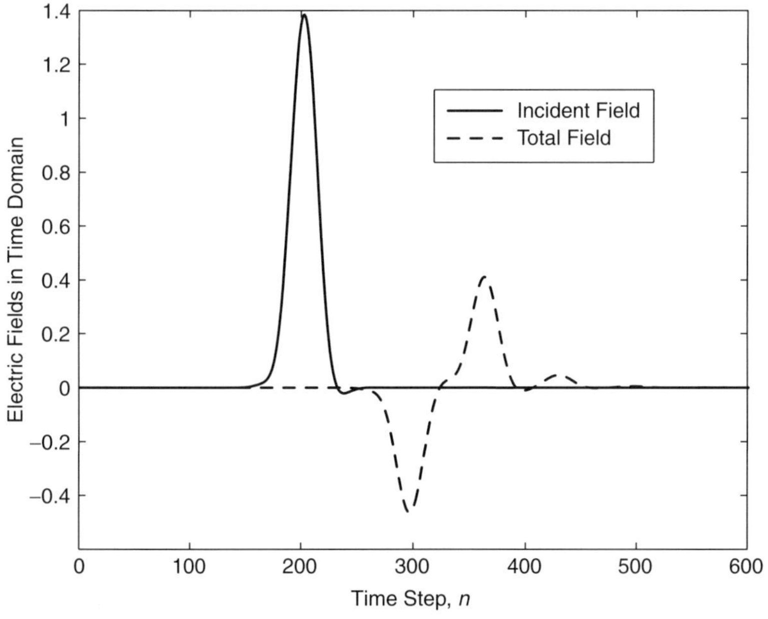

FIGURE 8.15 Time domain incident field E_y^{inc} and total field E_y for the case of the TE_z wave.

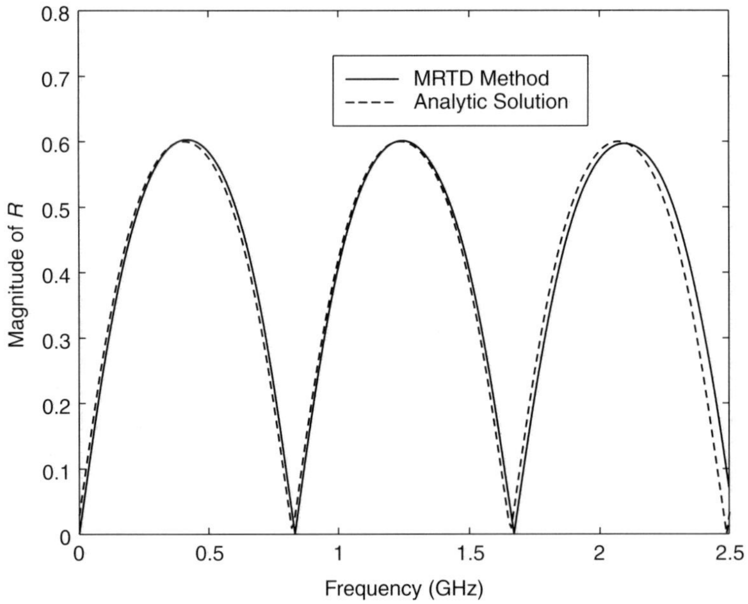

FIGURE 8.16 Magnitude of the reflection coefficient R for a dielectric layer versus frequency for the case of the TE_z wave.

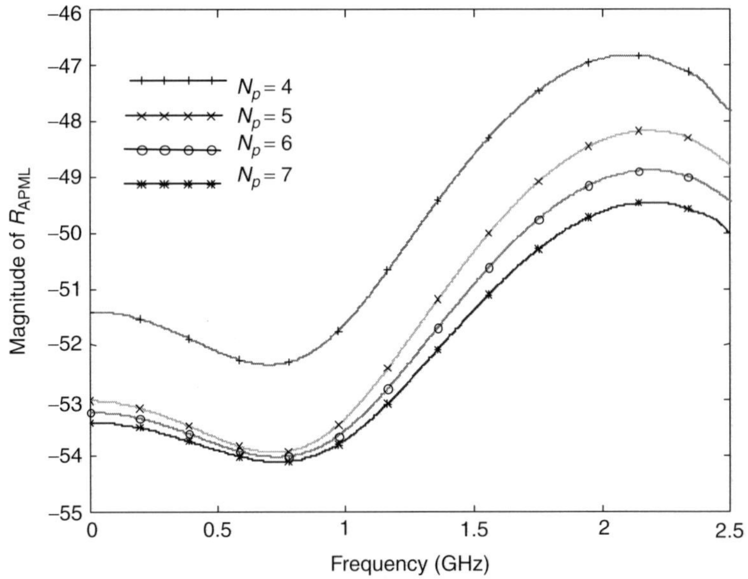

FIGURE 8.17 Magnitude of the reflection coefficient R_{APML} for the right APML wall versus frequency for the case of the TE_z wave.

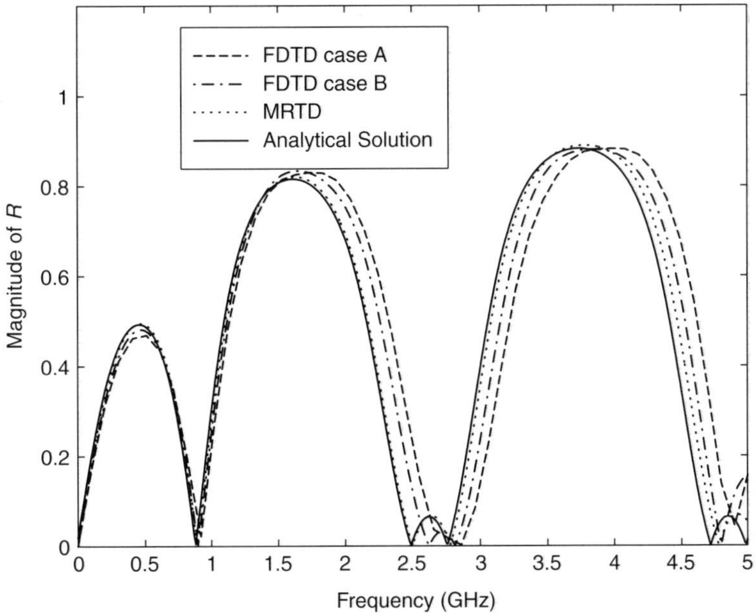

FIGURE 8.18 Magnitude of the reflection coefficient R for a two-slab dielectric system versus frequency ($d = 9$ cm, $\varepsilon_r = 4.0$).

TABLE 8.4 Discretization for Parallel-Plate Bounded Dielectric Layer System

Technique	$\Delta x = \Delta y$ (cm)	Δt (ps)	$N_x \times N_y$
FDTD Case A	0.2 ($\lambda_{min}/15$)	4.48	211×20
FDTD Case B	0.12 ($\lambda_{min}/25$)	2.69	341×33
MRTD	1 ($\lambda_{min}/3$)	14.3	55×4

(N_x, N_y: Cell numbers in the x and y directions)

Unlike the previous case, the system for the TM_z wave is uniformly excited with the E_z component at an input port position, and the cross section of the dielectric layer is 12×10 cm^2. We still retain $\varepsilon_r = 4$ for the dielectric constant, and 50×5 cells for the computational domain, but $\Delta x = \Delta y = 2$ cm and $\Delta t = 2.41 \times 10^{-11}$ s. Once again, the computed reflection coefficient agrees excellently with the analytical solution as shown in Figure 8.19.

8.6 CONCLUSIONS

In this chapter, we explored the 2D-MRTD method based on the cubic spline Battle–Lemarié scaling function expansions. We derived 2D-MRTD analysis formulations for both open and shielded transmission lines and analyzed the TE_z

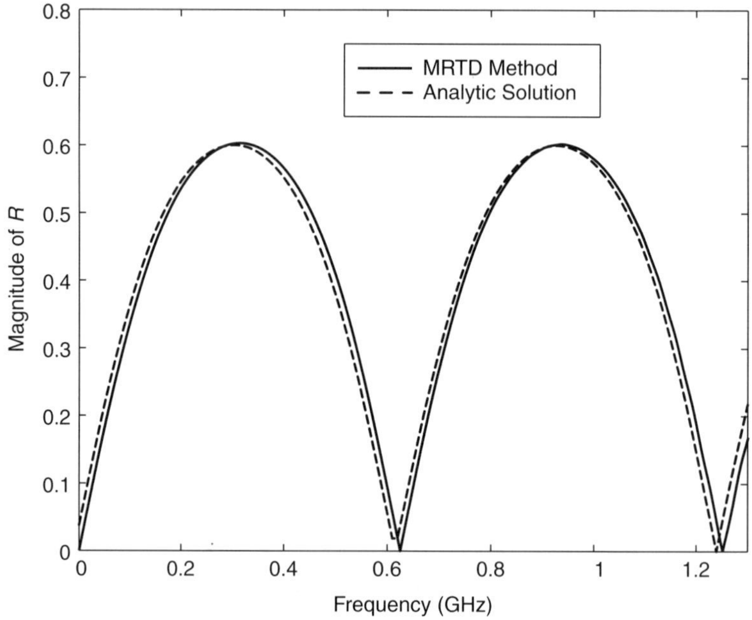

FIGURE 8.19 Magnitude of the reflection coefficient R for a dielectric slab versus frequency for the case of the TM_z wave.

and TM_z wave propagation problems associated with a parallel-plate waveguide structure system. A comparison of the computed 2D-MRTD results with both those derived from the FDTD and the analytical methods shows that they are in excellent agreement and the MRTD scheme can significantly reduce the computer memory requirement.

REFERENCES

[1] M. Krumpholz and L. P. B. Katehi, "MRTD: New time domain schemes based on multiresolution analysis," *IEEE Trans. Microwave Theory Tech.*, vol. 44, pp. 555–571, Apr. 1996.

[2] E. M. Tentzeris, M. Krumpholz, and L. P. B. Katehi, "Application of MRTD to printed transmission line," *IEEE MTT-S Dig.*, pp. 573–575, 1996.

[3] E. M. Tentzeris, R. L. Roberson, J. F. Harvey, and L. P. B. Katehi, "Stability and dispersion analysis of Battle–Lemarie-based MRTD schemes," *IEEE Trans. Microwave Theory Tech.*, vol. 47, pp. 1004–1013, July 1999.

[4] Q. Cao and Y. Chen, "MRTD analysis of a transient electromagnetic pulse propagating through a dielectric layer," *Int. J. Electron.*, vol. 86, pp. 459–474, Apr. 1999.

[5] R. L. Roberson, E. M. Tentzeris, M. Krumpholz, and L. P. B. Katehi, "Modeling of dielectric cavity structures using multiresolution time domain analysis," *Int. J.*

Numerical Modeling: Electronic Networks, Devices Fields, vol. 11, pp. 55–68, Apr. 1998.

[6] S. D. Gedney, "An anisotropic perfectly matched layered-absorbing medium for the truncation of FDTD lattices," *IEEE Trans. Antennas Propag.*, vol. 44, pp. 1630–1939, 1996.

[7] Q. Cao, Y. Chen, and R. Mittra, "Multiple image technique (MIT) and anisotropic perfectly matched layer (APML) in implementation of MRTD scheme for boundary truncations of microwave structures," *IEEE Trans. Microwave Theory Tech.*, vol. 50, no. 6, pp. 1578–1589, June 2002.

[8] S. Xiao and R. Vahldieck, "An efficient 2-D FDTD algorithm using real variables," *IEEE Microwave Guided Wave Lett.*, vol. 3, pp. 127–129, May 1993.

[9] M. Tong and Y. Chen, "Analysis of propagation characteristics and field images for printed transmission lines printed on anisotropic substrates using a 2D-FDTD method," *IEEE Trans. Microwave Theory Tech.*, vol. MTT-46, pp. 1507–1510, Oct. 1998.

[10] Y. Chen, K. Sun, B. Beker, and R. Mittra, "Unified matrix presentation of Maxwell's and wave equations using generalized differential matrix operators," *IEEE Trans. Educ.*, vol. 41, pp. 61–69, Feb. 1998.

[11] Y. Chen, M. Tong, and R. Mittra, "Efficient and accurate finite difference time domain analysis of resonant structures using the Blackman–Harris window function," *Microwave Opt. Technol. Lett.*, vol. 15, no. 6, pp. 389–392, Aug. 1997.

[12] A. V. Opppenheim and R. W. Schafer, *Discrete-Time Signal Processing*, Prentice-Hall, New York, 1989.

[13] Y. Chen, R. Mittra, and P. Harms, "Finite difference time domain algorithm for solving Maxwell's equations in rotationally symmetric geometries," *IEEE Trans. Microwave Theory Tech.*, vol. MTT-44, no. 6, pp. 832–839, June 1996.

[14] F. J. Harris, "On the use of windows for harmonic analysis with discrete Fourier transform," *Proc. IEEE*, vol. 66, pp. 51–83, Jan. 1978.

[15] T. Itoh and R. Mittra, "A technique for computing dispersion characteristics of shielded microstrip line," *IEEE Trans. Microwave Theory Tech.*, vol. MTT-22, pp. 896–898, Oct. 1974.

[16] W. D. Becker, *The Application of Time-Domain Electromagnetic Field Solvers to Computer Package Analysis and Design*, Ph.D. Dissertation, Department of Electrical and Computer Engineering, University of Illinois at Urbana-Champaign, 1993.

[17] X. Zhang, J. Fang, K. K. Mei, and Y. Liu, "Calculations of the dispersive characteristics of microstrips by the time-domain finite difference method," *IEEE Trans. Microwave Theory Tech.*, vol. 36, pp. 263–267, Feb. 1988.

[18] T. Itoh and R. Mittra, "Spectral-domain approach for calculating the dispersion characteristics of microstrip lines," *IEEE Trans. Microwave Theory Tech.*, vol. MTT-21, pp. 496–499, July 1973.

[19] P. Pramanick and P. Bharti, "An accurate description of dispersion in microstrip," *Microwave J.*, pp. 89–96, Dec. 1981.

[20] K. C. Gupta, R. Garg, and R. Chadha, *Computer-Aided Design of Microwave Circuits*, Artech House, Norwood, MA, 1981.

CHAPTER NINE

Three-Dimensional MRTD Analysis

9.1 INTRODUCTION

In this chapter, the multiresolution time domain (MRTD) scheme, in conjunction with the unsplit anisotropic perfectly matched layers (APMLs) [1, 2] and the multiple image technique (MIT) [3] for open and PEC-shielded boundary truncations, is applied to three-dimensional (3D) structures. Applications of the APML technique and MIT are illustrated by analyzing an open microstrip line and a double-layer dielectric-loaded cavity, respectively. In order to show how we obtain the 3D-MRTD update equations involving different materials and boundaries, we derive the MRTD update equations involving both dielectrics and conductors, in particular, for a planar microstrip-type structure. We also develop a systematic formulation technique for constructing the constitutive relations and update equations in the MRTD transformed domain by utilizing only the field quantities defined in the real structures [3, 4].

9.2 METHOD OF ANALYSIS

Maxwell Governing Equations

When a computational domain involves dielectrics and conductors simultaneously, it becomes much more difficult to derive update equations using the MRTD scheme rather than its counterpart in the FDTD algorithm. This is because the nonlocalized features of scaling and wavelet functions, such as the Battle–Lemarié wavelet basis function, lead to coupling between cells. Prior to deriving the MRTD update equations, we first represent the Maxwell curl equations in

Multiresolution Time Domain Scheme for Electromagnetic Engineering
By Yinchao Chen, Qunsheng Cao, and Raj Mittra
ISBN 0-471-27230-2

matrix form by using the generalized differential matrix operators (GDMOs) [5], which yields

$$\begin{bmatrix} 0 & -\partial_z & \partial_y \\ \partial_z & 0 & -\partial_x \\ -\partial_y & \partial_x & 0 \end{bmatrix} \begin{bmatrix} H_x \\ H_y \\ H_z \end{bmatrix} = \begin{bmatrix} \sigma_{xx} & 0 & 0 \\ 0 & \sigma_{yy} & 0 \\ 0 & 0 & \sigma_{zz} \end{bmatrix} \begin{bmatrix} E_x \\ E_y \\ E_z \end{bmatrix} + \varepsilon_0 \begin{bmatrix} \varepsilon_{xx} & 0 & 0 \\ 0 & \varepsilon_{yy} & 0 \\ 0 & 0 & \varepsilon_{zz} \end{bmatrix} \frac{\partial}{\partial t} \begin{bmatrix} E_x \\ E_y \\ E_z \end{bmatrix} \tag{9.1}$$

$$\begin{bmatrix} 0 & -\partial_z & \partial_y \\ \partial_z & 0 & -\partial_x \\ -\partial_y & \partial_x & 0 \end{bmatrix} \begin{bmatrix} E_x \\ E_y \\ E_z \end{bmatrix} = -\mu_0 \begin{bmatrix} \mu_{xx} & 0 & 0 \\ 0 & \mu_{yy} & 0 \\ 0 & 0 & \mu_{zz} \end{bmatrix} \frac{\partial}{\partial t} \begin{bmatrix} H_x \\ H_y \\ H_z \end{bmatrix} \tag{9.2}$$

Note that, for the sake of generality, we represent the relative permittivity, relative permeability, and conductivity as diagonal tensors. Note also that the computational domain for a planar printed circuit is usually inhomogeneous in the vertical direction but homogeneous in the horizontal direction.

MRTD Scheme in a Dielectric–Conductor Hybrid Region

Next, we expand all the field quantities in terms of the scaling functions in space and pulse functions in time, which read

$$E_x(\vec{r}, t) = \sum_{i,j,k,n=-\infty}^{\infty} {}_{\phi x}E^n_{i+1/2,j,k}\phi_{i+1/2}(x)\phi_j(y)\phi_k(z)h_n(t) \tag{9.3a}$$

$$E_y(\vec{r}, t) = \sum_{i,j,k,n=-\infty}^{\infty} {}_{\phi y}E^n_{i,j+1/2,k}\phi_i(x)\phi_{j+1/2}(y)\phi_k(z)h_n(t) \tag{9.3b}$$

$$E_z(\vec{r}, t) = \sum_{i,j,k,n=-\infty}^{\infty} {}_{\phi z}E^n_{i,j,k+1/2}\phi_i(x)\phi_j(y)\phi_{k+1/2}(z)h_n(t) \tag{9.3c}$$

$$H_x(\vec{r}, t) = \sum_{i,j,k,n=-\infty}^{\infty} {}_{\phi x}H^n_{i,j+1/2,k+1/2}\phi_i(x)\phi_{j+1/2}(y)\phi_{k+1/2}(z)h_{n+1/2}(t) \tag{9.4a}$$

$$H_y(\vec{r}, t) = \sum_{i,j,k,n=-\infty}^{\infty} {}_{\phi y}H^n_{i+1/2,j,k+1/2}\phi_{i+1/2}(x)\phi_j(y)\phi_{k+1/2}(z)h_{n+1/2}(t) \tag{9.4b}$$

$$H_z(\vec{r}, t) = \sum_{i,j,k,n=-\infty}^{\infty} {}_{\phi y}H^n_{i+1/2,j+1/2,k}\phi_{i+1/2}(x)\phi_{j+1/2}(y)\phi_k(z)h_{n+1/2}(t) \tag{9.4c}$$

where $\phi(x)$ denotes the cubic spline Battle–Lemarié scaling functions, which could be another type of scaling function, and $h_n(t)$ is a rectangular pulse function. Substitution of the above field expansions into the Maxwell equations, followed by the application of Galerkin's method, leads us to the following discretized equations, for example, the x-component of E- and H-fields:

$$\sum_{i',j',k'=-\infty}^{+\infty} (\varepsilon_{xx})_{ii',jj',kk'} \left({}_{\phi x}E^{n+1}_{i'+1/2,j',k'} - {}_{\phi x}E^{n}_{i'+1/2,j',k'} \right)$$

$$+ \sum_{i',j',k'=-\infty}^{+\infty} (\sigma_{xx})_{ii',jj',kk'} \left({}_{\phi x}E^{n+1}_{i'+1/2,j',k'} + {}_{\phi x}E^{n}_{i'+1/2,j',k'} \right) \frac{\Delta t}{2\varepsilon_0}$$

$$= \frac{1}{\varepsilon_0} \sum_{\nu} a(\nu) \left({}_{\phi z}H^{n+1/2}_{i+1/2,j+\nu+1/2,k} \frac{\Delta t}{\Delta y} - {}_{\phi y}H^{n+1/2}_{i+1/2,j,k+\nu+1/2} \frac{\Delta t}{\Delta z} \right) \tag{9.5}$$

$${}_{\phi x}H^{n+1/2}_{i,j+1/2,k+1/2} = {}_{\phi x}H^{n-1/2}_{i,j+1/2,k+1/2}$$

$$+ \frac{1}{\mu_0} \sum_{\nu} a(\nu) \left[{}_{\phi y}E^{n}_{i,j+1/2,k+\nu+1} \frac{\Delta t}{\Delta z} - {}_{\phi z}E^{n}_{i,j+\nu+1,k+1/2} \frac{\Delta t}{\Delta y} \right] \tag{9.6}$$

where the coefficients $(\varepsilon_{xx})_{ii',jj'kk'}$ and $(\sigma_{xx})_{ii',jj'kk'}$ are given as follows:

$$(\varepsilon_{xx})_{ii',jj',kk'} = \iiint_{\infty} \phi_{i+1/2}(x)\phi_j(y)\phi_k(z)\varepsilon_{xx}(\vec{r})\phi_{i'+1/2}(x)\phi_{j'}(y)\phi_{k'}(z) \frac{dx\,dy\,dz}{\Delta x\ \Delta y\ \Delta z} \tag{9.7a}$$

$$(\sigma_{xx})_{ii',jj',kk'} = \iiint_{\infty} \phi_{i+1/2}(x)\phi_j(y)\phi_k(z)\sigma_{xx}(\vec{r})\phi_{i'+1/2}(x)\phi_{j'}(y)\phi_{k'}(z) \frac{dx\,dy\,dz}{\Delta x\ \Delta y\ \Delta z} \tag{9.7b}$$

The coefficient $a(\nu)$ appearing in (9.5) and (9.6) is defined as an integration of a scaling basis function and its derivative as given in [6] and [7]. As seen in (9.7), we have to evaluate the coefficients of $(\varepsilon_{xx})_{ii',jj'kk'}$ and $(\sigma_{xx})_{ii',jj'kk'}$ in order to derive the MRTD update equations.

Now we consider a planar printed structure as shown in Figure 9.1, whose substrate relative permittivity and conducting strip patch conductivity can be characterized, respectively, as

$$\varepsilon_{\alpha\alpha}(\vec{r}) = \varepsilon_{\alpha\alpha}(y) = \begin{cases} \varepsilon_r, & -\infty < x, z < \infty, 0 \le y \le j_0\ \Delta y \\ 0, & \text{otherwise} \end{cases} \qquad (\alpha = x, y, z) \tag{9.8}$$

$$\sigma_{\alpha\alpha}(\vec{r}) = \begin{cases} \sigma, & i_1\ \Delta x \le x \le i_2\ \Delta x,\ j_0\ \Delta y - \dfrac{\tau}{2} \le y \le j_0\ \Delta y + \dfrac{\tau}{2},\ k_1\ \Delta z \le z \le k_2\ \Delta z \\ 0, & \text{otherwise} \end{cases} \tag{9.9}$$

where τ is the thickness of the conducting patch printed on the substrate and $\alpha = x, y, z$.

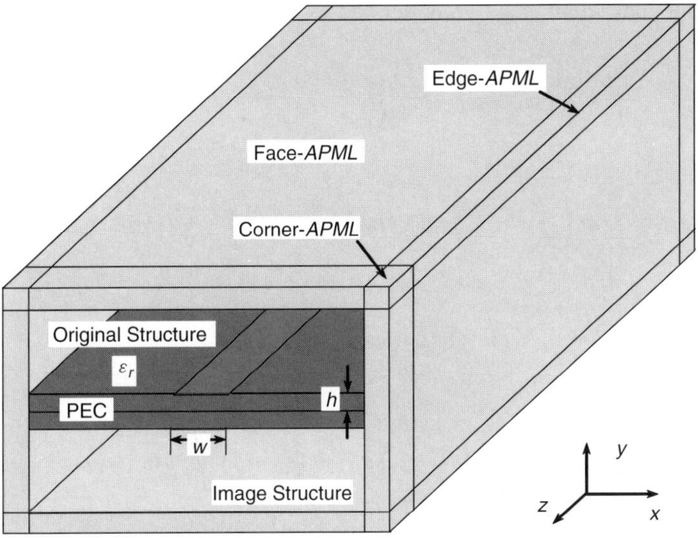

FIGURE 9.1 Geometry and image for an open grounded microstrip line enclosed by face, edge, and corner APML regions.

By applying the multiple image technique (MIT) to truncate the PEC boundary [3, 5], we can further rewrite the time domain equation (9.5) as

$$\sum_{j'=-\infty}^{+\infty} (\varepsilon_{xx})_{j,j'} \left({}_{\phi x}E^{n+1}_{i+1/2,j',k} - {}_{\phi x}E^{n}_{i+1/2,j',k} \right)$$

$$+ \sum_{i',k'=-\infty}^{+\infty} \sigma^{e}_{ii',kk'}(\delta_{j,j_0} + \delta_{j,-j_0}) \left({}_{\phi x}E^{n+1}_{i'+1/2,j,k'} + {}_{\phi x}E^{n}_{i'+1/2,j,k'} \right) \frac{\Delta t}{2\varepsilon_0}$$

$$= \frac{1}{\varepsilon_0} \sum_{\nu} a(\nu) \left({}_{\phi z}H^{n+1/2}_{i+1/2,j+\nu+1/2,k} \frac{\Delta t}{\Delta y} - {}_{\phi y}H^{n+1/2}_{i+1/2,j,k+\nu+1/2} \frac{\Delta t}{\Delta z} \right) \quad (9.10)$$

The coefficients $(\varepsilon_{xx})_{j,j'}$ and $\sigma^{e}_{ii',kk'}$ are explicitly described below. For a planar printed circuit structure, the integration domain of the coefficient $(\varepsilon_{xx})_{j,j'}$ defined in (9.7a) includes the entire real and image regions, which leads to

$$(\varepsilon_{xx})_{ii',jj',kk'}$$

$$= \left(\int_{-\infty}^{+\infty} \phi_{i+1/2}(x)\phi_{i'+1/2}(x) \frac{dx}{\Delta x} \right) \left(\int_{-\infty}^{+\infty} \phi_j(y)\varepsilon_{xx}(y)\phi_{j'}(y) \frac{dy}{\Delta y} \right)$$

$$\times \left(\int_{-\infty}^{+\infty} \phi_k(z)\phi_{k'}(z) \frac{dz}{\Delta z} \right) \quad (9.11)$$

$$= \delta_{i,i'}(\varepsilon_{xx})_{j,j'}\delta_{k,k'}$$

where the coefficient $(\varepsilon_{xx})_{j,j'}$ is defined as

$$
\begin{aligned}
(\varepsilon_{xx})_{j,j'} &= \int_{-\infty}^{+\infty} \phi_j(y)\varepsilon_{xx}(y)\phi_{j'}(y)\frac{dy}{\Delta y} \\
&= \int_{-\infty}^{+\infty} \phi\left(\frac{y}{\Delta y} - j\right)\varepsilon_{xx}(y)\phi\left(\frac{y}{\Delta y} - j'\right)\frac{dy}{\Delta y} \\
&= \int_{-\infty}^{+\infty} \phi(y' - j)\varepsilon_{xx}(y\Delta y)\phi(y' - j')\,dy' \qquad (9.12) \\
&= \int_{-\infty}^{+\infty} \phi(y' - j)\phi(y' - j')\,dy' + (\varepsilon_r - 1)\int_{-j_0}^{j_0} \phi(y' - j)\phi(y' - j')\,dy' \\
&= \delta_{j,j'} + (\varepsilon_r - 1)\alpha_{j,j'}^{\phi\phi}
\end{aligned}
$$

$$
\alpha_{j,j'}^{\phi\phi} = \frac{1}{\Delta y}\int_{-j_0\Delta y}^{j_0\Delta y} \phi_j(y)\phi_{j'}(y)\,dy \qquad (9.12b)
$$

where $\delta_{\alpha,\alpha'}$ $(\alpha, \alpha' = x, y, z)$ is the Kronecker symbol $\delta_{\alpha,\alpha'} = 0$, if $\alpha \neq \alpha'$; and $\delta_{\alpha,\alpha'} = 1$, if $\alpha = \alpha'$. Note that the values of $(\varepsilon_{xx})_{j,j'}$ form a rendered sparse matrix due to the exponential decay of the Battle–Lemarié function.

In the evaluation of the coefficient $(\sigma_{xx})_{ii',jj',kk'}$, the integration domain defined in (9.7b) comprises all conductors in both the real and image regions, which leads to

$$
\begin{aligned}
&(\sigma_{xx})_{ii',jj',kk'} \\
&= \sigma\left(\int_{i_1\Delta x}^{i_2\Delta x} \phi_{i+1/2}(x)\phi_{i'+1/2}(x)\frac{dx}{\Delta x}\right)\left(\int_{j_0\Delta y-\tau/2}^{j_0\Delta y+\tau/2} \phi_j(y)\phi_{j'}(y)\frac{dy}{\Delta y}\right) \\
&\quad \times (\delta_{j,j_0} + \delta_{j,-j_0})\left(\int_{k_1\Delta z}^{k_2\Delta z} \phi_k(z)\phi_{k'}(z)\frac{dz}{\Delta z}\right) \qquad (9.13) \\
&\approx \sigma\alpha_{i+1/2,i+1/2}^{\phi\phi}\delta_{j,j'}(\delta_{j,j_0} + \delta_{j,-j_0})\alpha_{k,k'}^{\phi\phi}
\end{aligned}
$$

with

$$
\alpha_{i+1/2,i+1/2}^{\phi\phi} = \frac{1}{\Delta x}\int_{i_1\Delta x}^{i_2\Delta x} \phi_{i+1/2}(x)\phi_{i'+1/2}(x)\,dx \qquad (9.14a)
$$

$$\alpha_{k,k'}^{\phi\phi} = \frac{1}{\Delta z} \int_{k_1 \Delta z}^{k_2 \Delta z} \phi_k(z)\phi_{k'}(z)\, dz \tag{9.14b}$$

If we approximate $\alpha_{i+1/2,i+1/2}^{\phi\phi}$ and $\alpha_{k,k'}^{\phi\phi}$ with only diagonal terms, in other words, with both of them existing only within the defined conductor region, we have

$$(\sigma_{xx})_{ii',jj',kk'} \approx \sigma^e \delta_{i,i'}\delta_{j,j'}(\delta_{j,j_0} + \delta_{j,-j_0})\delta_{k,k'} \tag{9.15}$$

where σ^e is the original conductivity multiplied by the constants derived from the diagonal approximations of $\alpha_{i+1/2,i+1/2}^{\phi\phi}$ and $\alpha_{k,k'}^{\phi\phi}$. In particular, note that σ^e is proportional to the conductivity σ and is defined in the same region as σ.

In principle, all image field quantities could be expressed in terms of the fields in the original structure in either a symmetric or an asymmetric manner as described in Figure 9.2 [3–6].

For an open grounded planar structure such as that shown in Figure 9.1 whose grids in the original region along the y-direction are defined in the region of $[0, N_y]$, the first term of (9.10) can then be degenerated and simplified as

$$\begin{aligned}
&\sum_{j'=-N_y}^{Ny} (\varepsilon_{xx})_{j,j'} \left({}_{\phi x}E_{i+1/2,j',k}^{n+1} - {}_{\phi x}E_{i+1/2,j',k}^{n} \right) \\
&= \left(\sum_{j'=0}^{Ny} (\varepsilon_{xx})_{j,j'} - \sum_{j'=-Ny}^{-1} (\varepsilon_{xx})_{j,j'} \right) \left({}_{\phi x}E_{i+1/2,j',k}^{n+1} - {}_{\phi x}E_{i+1/2,j',k}^{n} \right) \qquad (9.16) \\
&= \sum_{j'=0}^{Ny} (\varepsilon_{xx}^*)_{j,j'} \left({}_{\phi x}E_{i+1/2,j',k}^{n+1} - {}_{\phi x}E_{i+1/2,j',k}^{n} \right)
\end{aligned}$$

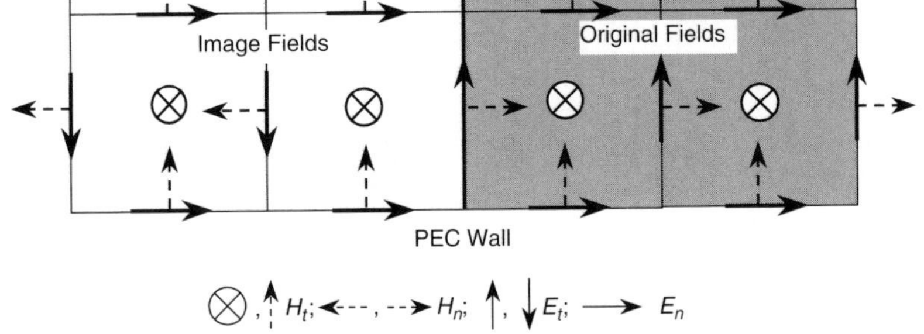

FIGURE 9.2 Image and original fields in the MRTD lattice for a PEC wall.

Hence, the time domain update equation (9.10) can be further derived as

$$
\begin{aligned}
&\sum_{j'=0}^{Ny}(\varepsilon_{xx}^{*})_{j,j'}\left({}_{\phi x}E_{i+1/2,j',k}^{n+1} - {}_{\phi x}E_{i+1/2,j',k}^{n}\right)\\
&\quad+\left({}_{\phi x}E_{i+1/2,j,k}^{n+1} + {}_{\phi x}E_{i+1/2,j,k}^{n}\right)\frac{\sigma^{e}\Delta t}{2\varepsilon_0}\delta_{j,j_0}\\
&\quad=\sum_{j'=0}^{Ny}\left((\varepsilon_{xx}^{*})_{j,j'}+\delta_{j,j_0}\delta_{j,j'}\frac{\sigma^{e}\Delta t}{2\varepsilon_0}\right){}_{\phi x}E_{i+1/2,j',k}^{n+1}\\
&\quad\quad-\sum_{j'=0}^{Ny}\left((\varepsilon_{xx}^{*})_{j,j'}-\delta_{j,j_0}\delta_{j,j'}\frac{\sigma^{e}\Delta t}{2\varepsilon_0}\right){}_{\phi x}E_{i+1/2,j',k}^{n}\\
&\quad=\frac{1}{\varepsilon_0}\sum_{\nu}a(\nu)\left({}_{\phi z}H_{i+1/2,j+\nu+1/2,k}^{n+1/2}\frac{\Delta t}{\Delta y} - {}_{\phi y}H_{i+1/2,j,k+\nu+1/2}^{n+1/2}\frac{\Delta t}{\Delta z}\right)
\end{aligned}
\tag{9.17}
$$

where only the fields in the original structure are used.

By taking an inverse of the matrix equation (9.17), we can obtain the update equation for the x-component of E-fields as follows:

$$
\begin{aligned}
{}_{\phi x}E_{i+1/2,j,k}^{n+1} &= \sum_{j'=0}^{Ny}[A_x]_{j,j'}\,{}_{\phi x}E_{i+1/2,j',k}^{n}\\
&\quad+\frac{1}{\varepsilon_0}\sum_{j'=0}^{Ny}[B_x]_{j,j'}\left[\sum_{\nu}a(\nu)\left({}_{\phi z}H_{i+1/2,j+\nu+1/2,k}^{n+1/2}\frac{\Delta t}{\Delta y}\right.\right.\\
&\quad\left.\left.-{}_{\phi y}H_{i+1/2,j,k+\nu+1/2}^{n+1/2}\frac{\Delta t}{\Delta z}\right)\right]
\end{aligned}
\tag{9.18}
$$

where $[A_x]_{j,j'}$ and $[B_x]_{j,j'}$ denote the elements of matrices $[A_x]$ and $[B_x]$ that correspond to the interaction of j and j' cells:

$$
[A_x] = \left[(\varepsilon_{xx}^{*})_{j,j'}+\delta_{j,j_0}\delta_{j,j'}\frac{\sigma^{e}\Delta t}{2\varepsilon_0}\right]^{-1}\cdot\left[(\varepsilon_{xx}^{*})_{j,j'}-\delta_{j,j_0}\delta_{j,j'}\frac{\sigma^{e}\Delta t}{2\varepsilon_0}\right] \tag{9.19a}
$$

$$
[B_x] = \left[(\varepsilon_{xx}^{*})_{j,j'}+\delta_{j,j_0}\delta_{j,j'}\frac{\sigma^{e}\Delta t}{2\varepsilon_0}\right]^{-1} \tag{9.19b}
$$

Due to the Stencil effect [8], the coefficient matrix $\lfloor(\varepsilon_{xx})_{j,j'}\rfloor$ forms a rendered sparse matrix. Since $(\varepsilon_{xx})_{j,j'}$ is proportional to $\xi_{j,j'}$ for all off-diagonal terms, typically it is sufficient to truncate it for all terms with $|j-j'|\geq M$ and $M=9$.

Similarly, we can derive all the update equations for the other components shown as follows:

$$
\begin{aligned}
&\sum_{j'=0}^{Ny-1}(\varepsilon^*_{yy})_{j,j'}\left({}_{\phi y}E^{n+1}_{i,j'+1/2,k}-{}_{\phi y}E^{n}_{i,j'+1/2,k}\right)\\
&\quad+\left({}_{\phi y}E^{n+1}_{i,j+1/2,k}+{}_{\phi y}E^{n}_{i,j+1/2,k}\right)\frac{\sigma^e\Delta t}{2\varepsilon_0}\delta_{j,j_0}\\
&\quad=\sum_{j'=0}^{Ny-1}\left((\varepsilon^*_{yy})_{j,j'}+\delta_{j,j_0}\delta_{j,j'}\frac{\sigma^e\Delta t}{2\varepsilon_0}\right){}_{\phi y}E^{n+1}_{i,j'+1/2,k}\\
&\quad-\sum_{j'=0}^{Ny-1}\left((\varepsilon^*_{yy})_{j,j'}-\delta_{j,j_0}\delta_{j,j'}\frac{\sigma^e\Delta t}{2\varepsilon_0}\right){}_{\phi y}E^{n}_{i,j'+1/2,k}\\
&\quad=\frac{1}{\varepsilon_0}\sum_{\nu}a(\nu)\left({}_{\phi x}H^{n+1/2}_{i,j+1/2,k+\nu+1/2}\frac{\Delta t}{\Delta z}-{}_{\phi z}H^{n+1/2}_{i+\nu+1/2,j+1/2,k}\frac{\Delta t}{\Delta x}\right)
\end{aligned}
\tag{9.20}
$$

$$
\begin{aligned}
{}_{\phi y}E^{n+1}_{i,j+1/2,k}&=\sum_{j'=0}^{Ny-1}[A_y]_{j,j'}\,{}_{\phi y}E^{n}_{i,j'+1/2,k}\\
&\quad+\frac{1}{\varepsilon_0}\sum_{j'=0}^{Ny-1}[B_y]_{j,j'}\left[\sum_{\nu}a(\nu)\left({}_{\phi x}H^{n+1/2}_{i,j+1/2,k+\nu+1/2}\frac{\Delta t}{\Delta z}\right.\right.\\
&\quad\left.\left.-{}_{\phi z}H^{n+1/2}_{i+\nu+1/2,j+1/2,k}\frac{\Delta t}{\Delta x}\right)\right]
\end{aligned}
\tag{9.21}
$$

where $[A_y]_{j,j'}$ and $[B_y]_{j,j'}$ are the elements of matrices $[A_y]$ and $[B_y]$ that correspond to the interaction of j and j' cells as follows:

$$
[A_y]=\left[(\varepsilon^*_{yy})_{j,j'}+\delta_{j,j_0}\delta_{j,j'}\frac{\sigma^e\Delta t}{2\varepsilon_0}\right]^{-1}\cdot\left[(\varepsilon^*_{yy})_{j,j'}-\delta_{j,j_0}\delta_{j,j'}\frac{\sigma^e\Delta t}{2\varepsilon_0}\right] \tag{9.22a}
$$

$$
[B_y]=\left[(\varepsilon^*_{yy})_{j,j'}+\delta_{j,j_0}\delta_{j,j'}\frac{\sigma^e\Delta t}{2\varepsilon_0}\right]^{-1} \tag{9.22b}
$$

$$
\begin{aligned}
&\sum_{j'=0}^{Ny}(\varepsilon^*_{zz})_{j,j'}({}_{\phi z}E^{n+1}_{i,j',k+1/2}-{}_{\phi z}E^{n}_{i,j',k+1/2})\\
&\quad+\left({}_{\phi z}E^{n+1}_{i,j,k+1/2}+{}_{\phi z}E^{n}_{i,j,k+1/2}\right)\frac{\sigma^e\Delta t}{2\varepsilon_0}\delta_{j,j_0}\\
&\quad=\sum_{j'=0}^{Ny}\left((\varepsilon^*_{zz})_{j,j'}+\delta_{j,j_0}\delta_{j,j'}\frac{\sigma^e\Delta t}{2\varepsilon_0}\right){}_{\phi z}E^{n+1}_{i,j',k+1/2}
\end{aligned}
$$

$$
\begin{aligned}
&-\sum_{j'=0}^{Ny}\left((\varepsilon_{zz}^{*})_{j,j'}-\delta_{j,j_0}\delta_{j,j'}\frac{\sigma^{e}\Delta t}{2\varepsilon_0}\right)\,{}_{\phi z}E^{n}_{i,j',k+1/2}\\
&=\frac{1}{\varepsilon_0}\sum_{\nu}a(\nu)\left({}_{\phi y}H^{n+1/2}_{i+\nu+1/2,j,k+1/2}\frac{\Delta t}{\Delta x}-{}_{\phi x}H^{n+1/2}_{i,j+\nu+1/2,k+1/2}\frac{\Delta t}{\Delta y}\right) \quad (9.23)
\end{aligned}
$$

$$
\begin{aligned}
{}_{\phi z}E^{n+1}_{i,j,k+1/2}&=\sum_{j'=0}^{Ny}[A_z]_{j,j'}\,{}_{\phi z}E^{n}_{i,j',k+1/2}\\
&\quad+\frac{1}{\varepsilon_0}\sum_{j'=0}^{Ny}[B_z]_{j,j'}\left[\sum_{\nu}a(\nu)\left({}_{\phi y}H^{n+1/2}_{i+\nu+1/2,j,k+1/2}\frac{\Delta t}{\Delta x}\right.\right.\\
&\quad\left.\left.-{}_{\phi x}H^{n+1/2}_{i,j+\nu+1/2,k+1/2}\frac{\Delta t}{\Delta y}\right)\right] \quad (9.24)
\end{aligned}
$$

where $[A_z]_{j,j'}$ and $[B_z]_{j,j'}$ are the elements of matrixes $[A_z]$ and $[B_z]$ that correspond to the interaction of j and j' cells as follows:

$$
[A_z]=\left[(\varepsilon_{zz}^{*})_{j,j'}+\delta_{j,j_0}\delta_{j,j'}\frac{\sigma^{e}\Delta t}{2\varepsilon_0}\right]^{-1}\cdot\left[(\varepsilon_{zz}^{*})_{j,j'}-\delta_{j,j_0}\delta_{j,j'}\frac{\sigma^{e}\Delta t}{2\varepsilon_0}\right] \quad (9.25a)
$$

$$
[B_z]=\left[(\varepsilon_{zz}^{*})_{j,j'}+\delta_{j,j_0}\delta_{j,j'}\frac{\sigma^{e}\Delta t}{2\varepsilon_0}\right]^{-1} \quad (9.25b)
$$

9.3 APPLICATION RESULTS

Double-Layer Dielectric-Filled Cavity

We apply the MRTD scheme in conjunction with the MIT [3] to analyze a two-layer dielectric-filled cavity. The dimensions of the cavity are set to $1 \times 2 \times 1.5$ m^3, in the x-, y-, and z-directions, as shown in Figure 9.3. The computational domain is discretized into $2 \times 6 \times 3$ cells. We investigate the following two cases: (i) $\varepsilon_{r1} = 64$, $\varepsilon_{r2} = 8$ and (ii) $\varepsilon_{r1} = 8$, $\varepsilon_{r2} = 64$, with $N_{y1} = N_{y2} = 1.5$ cells. The structure dimensions are $V = 1 \times 2 \times 1.5$ m^3. We also analyze the same problem using the traditional FDTD mesh with a discretization of $\Delta x = \Delta y = 2.5$ cm and $\Delta z = 3$ cm, that is, $40 \times 40 \times 50$ cells. As seen from Figures 9.4 and 9.5, and from Table 9.1, the two sets of computed results agree very well with each other; the percentage differences are 0.038% and 0.343%, for case (i) and case (ii), respectively. However, the MRTD scheme only uses 0.045% of the computational resources relative to the conventional FDTD algorithm.

Microstrip Lines

We also analyze the propagation characteristic for an open microstrip line shown in Figure 9.1, where the PEC strip is assumed to be infinitesimally thin. The

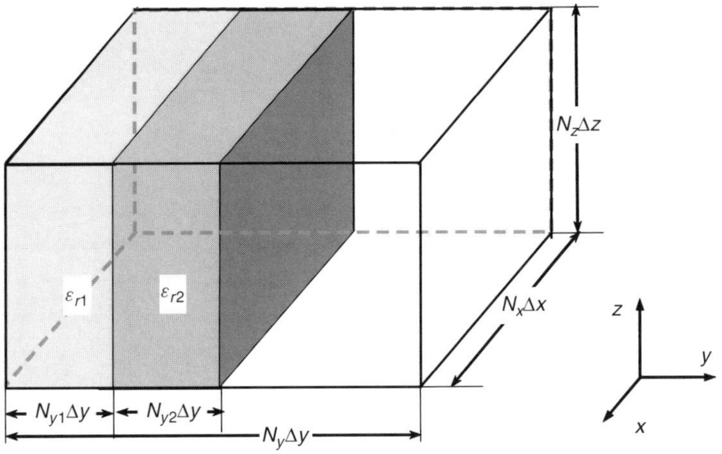

FIGURE 9.3 Geometry of a two-layer, dielectric-loaded cavity.

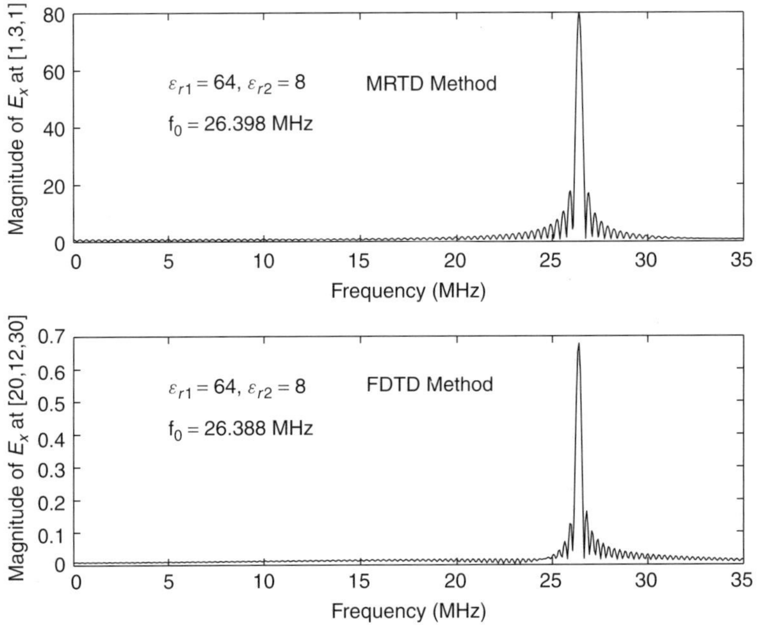

FIGURE 9.4 Frequency spectrum of E_x for a two-layer, dielectric-loaded cavity.

total computational volume is discretized into a grid with $36 \times 18 \times 68$ cells, which is only about 4.87% of that used in the FDTD method [9]. The substrate is characterized by $\varepsilon_r = 13$, and the PEC strip is specified by $w = 4\ \Delta x = 0.15$ mm and $h = 4\ \Delta y = 0.1$ mm as given in Table 9.2. Eight layers and four layers of APMLs are used to truncate the computational domain along the two wave propagation directions, and two side and top walls, respectively.

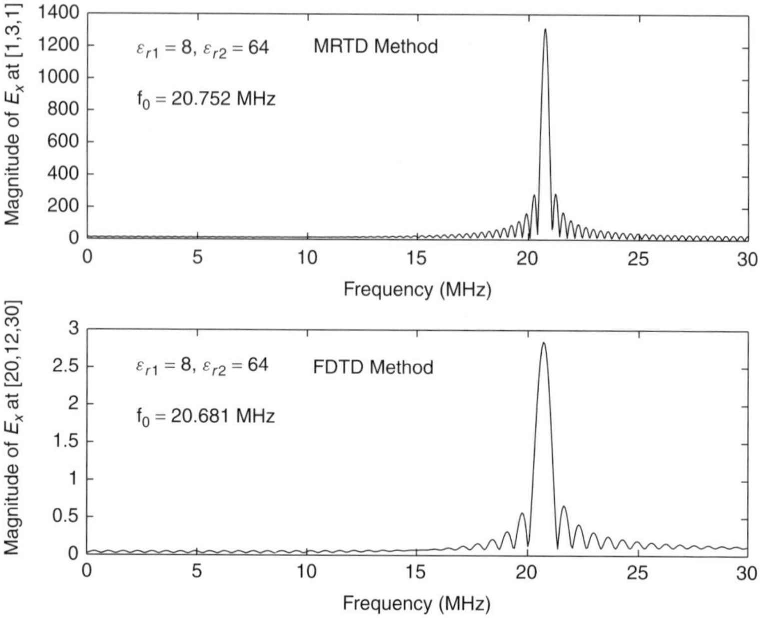

FIGURE 9.5 Frequency spectrum of E_x for a two-layer, dielectric-loaded cavity.

TABLE 9.1 Resonant Frequencies (MHz) for a Two-Layer, Dielectric-Filled Cavity ($N_{y1} = N_{y2} = 1.5$; $V = 1 \times 2 \times 1.5$ m^3; $\Delta x = \Delta z = \frac{1}{2}$, $\Delta y = \frac{1}{3}$ m)

Case	FDTD (Cell) (40 × 40 × 50)	MRTD (Cell) (2 × 6 × 3)	Percentage Difference
(i)	26.388	26.398	0.038%
(ii)	20.681	20.752	0.343%

TABLE 9.2 Spatial and Time Discretization for a Microstrip Line

Discretization	MRTD	FDTD, Schamberger et al.
Δx (mm)	0.0375	0.0125
Δy (mm)	0.0250	0.0125
Δz (mm)	0.0300	0.0125
Δt (ps)	0.0208	0.0280
$N_x \times N_y \times N_z$	36 × 18 × 68	110 × 40 × 160
VDR	4.87%	
CPU time (seconds)	7182.42	49881.59
TDR	14.40%	

(VDR: volume difference ratio; TDR: time difference ratio)

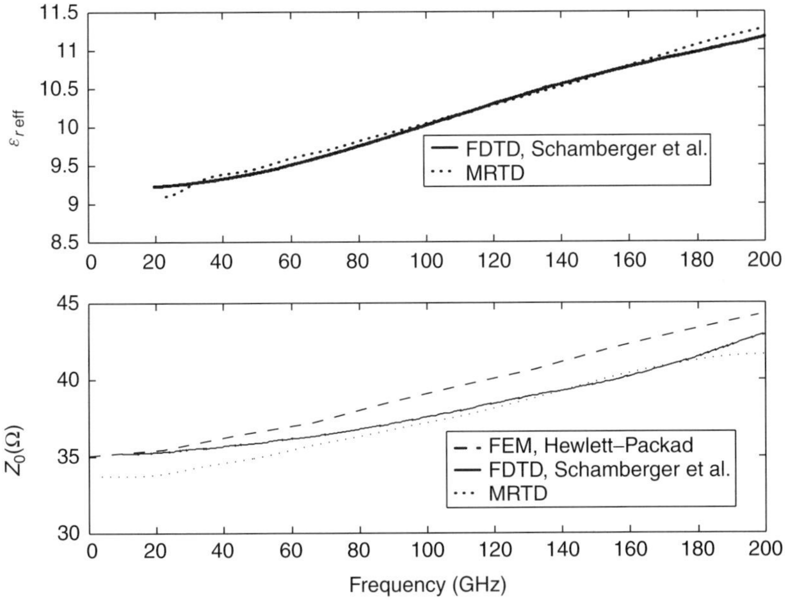

FIGURE 9.6 Characteristics ($\varepsilon_{\mathrm{eff}}$ and Z_0) as a function of frequency for a microstrip line ($\varepsilon_r = 13$, $w = 0.15$ mm, and $h = 0.1$ mm).

As shown in Figure 9.6, we observe good agreement between results for both the effective dielectric constant and characteristic impedance derived from the MRTD and FDTD [9], respectively, while the prior requires only a small fraction of the total discretization space used in the latter case. We have also compared the characteristic impedance derived from the MRTD with that obtained using Hewlett–Packard's FEM solver [10] and have observed fairly good agreement for a very broad frequency bandwidth as shown in Figure 9.6. In addition, for a total time step of $N_t = 2000$, the CPU time of the MRTD is only 7182.42 seconds, while that of the FDTD with the same total time step requires 49,881.59 seconds. Thus, the processing time required by the FDTD method is almost seven times longer than that required by the MRTD. All computations in this research are performed with a 333 MHz, 128M RAM Pentium II-MAX PC Compatible.

A further validation study is performed for an open microstrip line characterized by $w = 4\ \Delta x = 3.04 \times 10^{-3}$ m, $h = 3\ \Delta y = 3.171 \times 10^{-3}$ m, and the substrate $\varepsilon_r = 11.7$, where the total computational volume is discretized into a grid with $24 \times 14 \times 48$ cells. Again, as shown in Figure 9.7, we find good agreement for the results of the effective dielectric constants derived from the MRTD method and MoM [11], respectively.

To investigate grid discretization effects, we retain the geometry of the microstrip line unchanged with $w = 0.15$ mm, $h = 0.1$ mm, and the substrate $\varepsilon_r = 13$, and study the variation of the $\varepsilon_{r\ \mathrm{eff}}$ and Z_0 in terms of different cell sizes Δx and Δy as shown in Figures 9.8 and 9.9, respectively. A sufficiently fine

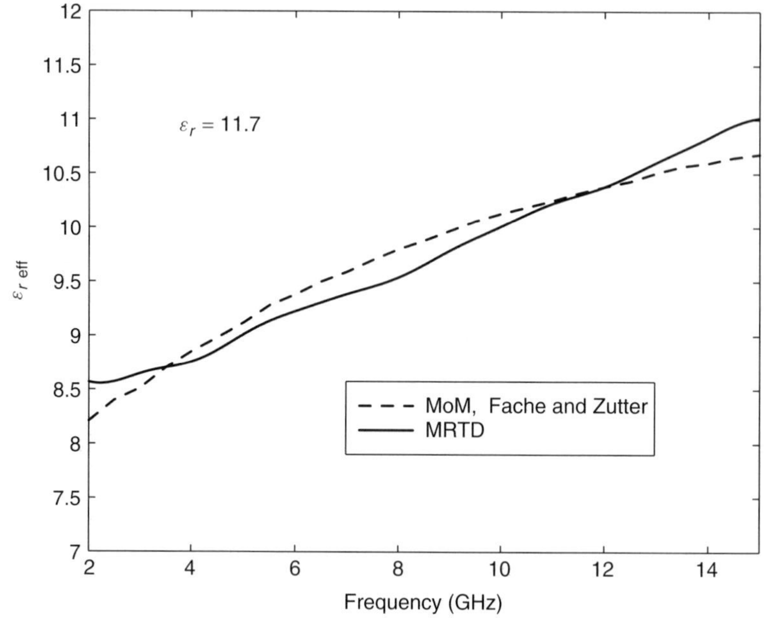

FIGURE 9.7 Relative dielectric constant $\varepsilon_{\mathrm{eff}}$ as a function of frequency for a microstrip line ($\varepsilon_r = 11.7$, $w = 3.04$ mm, and $h = 3.171$ mm).

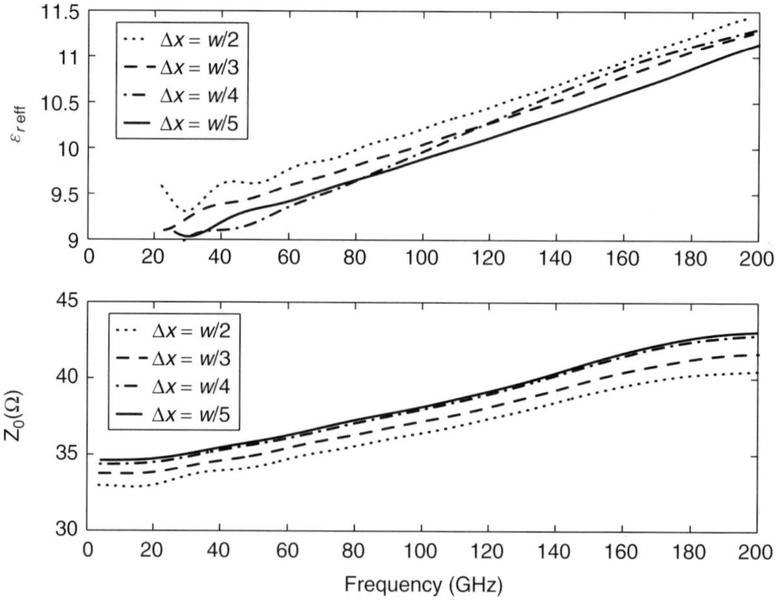

FIGURE 9.8 Characteristics ($\varepsilon_{\mathrm{eff}}$ and Z_0) with different Δx values for a microstrip line ($\varepsilon_r = 13$, $w = 0.15$, and $h = 0.1$ mm).

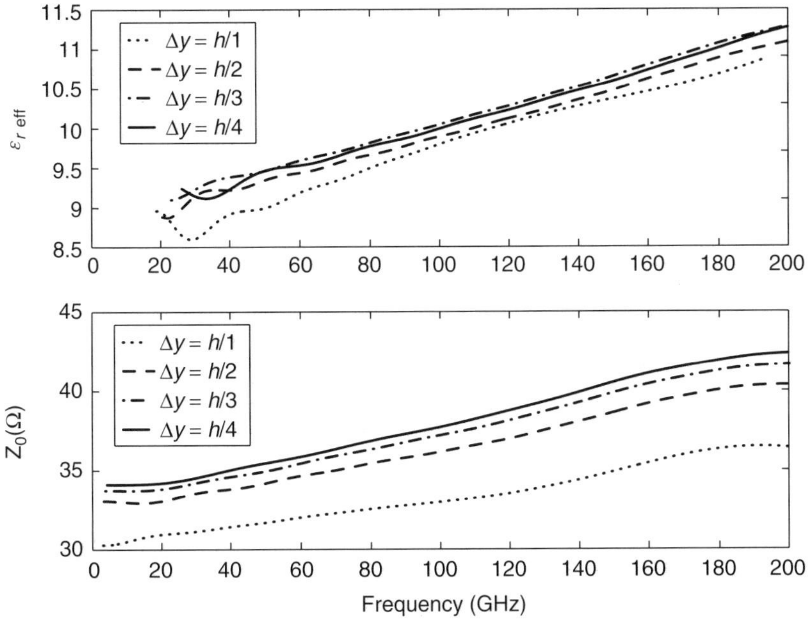

FIGURE 9.9 Characteristics (ε_{eff} and Z_0) with different Δy values for a microstrip line ($\varepsilon_r = 13$, $w = 0.15$ mm, and $h = 0.1$ mm).

discretization must be used to obtain good accuracy and the minimum number of cells for accurately modeling the PEC strip is two cells in both directions, although a finer grid leads to a better accuracy.

9.4 CONCLUSIONS

In this chapter, we developed the MRTD algorithm for a hybrid dielectric–conductor, which can be used as a general treatment method for practical structures. We analyzed a two-layer dielectric-filled cavity and an open microstrip line by using a scaling function based MRTD scheme in conjunction with the APML method and MIT. We found that the developed scaling function based MRTD is highly efficient and accurate for the analysis of dielectric cavities and microstrip lines. In particular, the MRTD scheme requires only a fraction of the computational space and CPU time that are used in the FDTD technique at the price of complexity in mathematical modeling.

REFERENCES

[1] Z. S. Sacks, D. M. Kingsland, R. Lee, and J. F. Lee, "A perfectly matched layer anisotropic absorber for use as an absorbing boundary condition," *IEEE Trans. Antennas Propag.*, vol. 43, no. 12, pp. 1460–1463, Dec. 1995.

[2] S. D. Gedney, "An anisotropic perfectly matched layered-absorbing medium for the truncation of FDTD lattice," *IEEE Trans. Antennas Propag.*, vol. 44, no. 12, pp. 1630–1939, Dec. 1996.

[3] Q. Cao, Y. Chen, and R. Mittra, "Multiple image technique (MIT) and anisotropic perfectly matched layer (APML) in implementation of MRTD scheme for boundary truncations of microwave structures," *IEEE Trans. Microwave Theory Tech.*, vol. 50, no. 6, pp. 1578–1589, June 2002.

[4] Q. Cao and Y. Chen, "Application of a perfectly matched layer absorber for open boundary truncation in multiresolution time domain scheme," *IEEE Trans. Antennas Propag.*, vol. 51, no. 2, pp. 350–357, Feb. 2003.

[5] Y. Chen, K. Sun, B. Beker, and R. Mittra, "Unified matrix presentation of Maxwell's and wave equations using generalized differential matrix operator," *IEEE Trans. Educ.*, vol. 41, no. 1, pp. 61–69, Feb. 1998.

[6] M. Krumpholz and L. P. B. Katehi, "MRTD: New time-domain schemes based on multiresolution analysis," *IEEE Trans. Microwave Theory Tech.*, vol. 44, no. 4, Apr. 1996.

[7] Q. Cao and Y. Chen, "MRTD analysis of a transient electromagnetic pulse propagating through a dielectric layer," *Int. J. Electron.*, vol. 86, no. 4, pp. 459–474, 1999.

[8] L. P. B. Katehi, J. F. Harvey, and E. Tentzeris, "Time-domain analysis using multiresolution expansions," Chapter 3, in *Advances in Computational Electrodynamics: The Finite-Difference Time-Domain Method* (ed. A. Taflove), Artech House, Norwood, MA, 1998.

[9] M. A. Schamberger, S. Kosanovich, and R. Mittra, "Parameter extraction and correction for transmission lines and discontinuities using the finite-difference time-domain method," *IEEE Trans. Microwave Theory Tech.*, vol. 44, no. 6, pp. 919–925, June 1996.

[10] Ansoft, *High-Frequency Structure Simulator* (HFSS).

[11] N. Fache and D. D. Zutter, "Rigorous full-wave space-domain solution for dispersive microstrip lines," *IEEE Trans. Microwave Theory Tech.*, vol. 36, no. 4, pp. 731–737, 1988.

CHAPTER TEN

MRTD Analysis for MMICs

10.1 INTRODUCTION

The multiresolution time domain (MRTD) scheme, introduced by Krumpholz and Katehi [1], is a time domain Maxwell solver based on the expansions of electromagnetic fields in terms of scaling and wavelet functions. The technique has been shown to have two major potential impacts: (1) a unified field expansion structure, including the finite difference time domain (FDTD) method and its high-order expansions, and (2) significant reduction of the requirements on computational space and CPU time [2]. The MRTD scheme has been applied successfully to the analysis of fundamental guided-wave structures and devices, such as cavities and parallel-plate bounded dielectric layers [1–3]. However, two major challenges hinder the implementation of the MRTD scheme for practical applications: (1) it is impossible for the MRTD to truncate a boundary by specifying a tangential field for both open and shielded structures, due to coupling between cells resulting from high-order expansions of fields; and (2) the orthogonal relations between scaling and wavelet basis functions are corrupted in an inhomogeneous region. This yields great complexity in deriving field update equations for a structure with different materials.

In this chapter, we first quickly review the concepts of microwave networks, and then apply the 3D-MRTD scheme for the analysis of various planar printed monolithic millimeter-wave integrated circuits (MMICs), including microstrip patch antennas, as well as microstrip low-pass and band-pass filters. In particular, we compare the scattering parameter results derived using the MRTD approach with those obtained from the FDTD. It is observed that good agreement among these methods is achieved; however, the MRTD technique requires only around 15% of the computational space and about 25% or less of the computational CPU time [4] used by the FDTD methods [5].

Multiresolution Time Domain Scheme for Electromagnetic Engineering
By Yinchao Chen, Qunsheng Cao, and Raj Mittra
ISBN 0-471-27230-2

10.2 MICROWAVE NETWORKS

Prior to discussing extracting time domain MMIC parameters, we quickly review the fundamental concepts used in transmission line theory and microwave networks, which are available in most microwave engineering textbooks [6, 7].

Voltages and Currents for a TEM Mode Structure

In the microwave frequency range the measurement of voltage or current for a transmission line is difficult, unless the structure supports the transverse electromagnetic (TEM) mode. Assuming the z-axis is the wave propagation direction, a TEM wave defines the z-components of electromagnetic fields as zero: $E_z = H_z = 0$. In general, a transmission line with two or more conductor traces supports TEM waves, whose electric and magnetic field lines can be distributed as shown in Figure 10.1.

In the cross-section plane, the voltage V of the +conductor relative to the −conductor can be defined as

$$V = \int_{+}^{-} \vec{E} \cdot d\vec{l} \tag{10.1}$$

where the integration path begins on the +conductor and ends on the −conductor. Note that the voltage defined in (10.1) does not depend on the integration path.

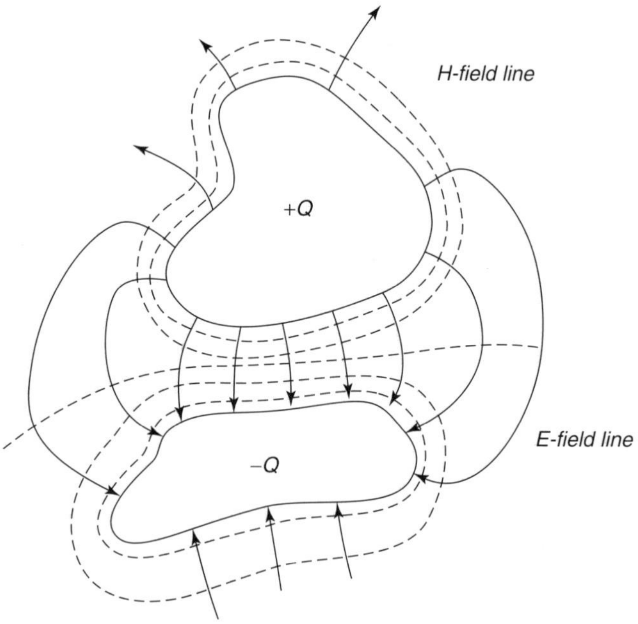

FIGURE 10.1 Electric and magnetic field distributions for a parallel two-conductor line.

The total current flowing on the +conductor can be determined from application of Ampere's law as

$$I = \oint_{C^+} \vec{H} \cdot d\vec{l} \tag{10.2}$$

where the integration contour is any closed path enclosing the +conductor. Correspondingly, the characteristic impedance Z_0 can then be defined for traveling waves as

$$Z_0 = \frac{V}{I} \tag{10.3}$$

Another very important parameter for a transmission line is the effective dielectric constant. For a transmission line whose cross section is inhomogeneous, such as a microstrip line, we usually use the effective dielectric constant ε_{eff} to describe the characteristics of the transmission line. These types of structures do not support a pure TEM wave but usually support a hybrid TM-TE wave; however, in many practical applications they do support a quasi-TEM mode when the dielectric substrate is electrically very thin. In other words, the fields are essentially the same as those in the case of the static electric fields. Therefore, good approximations for the phase velocity and propagation constant can be obtained from static or quasistatic solutions. The phase velocity and propagation constant can be expressed as

$$v_p = \frac{c}{\sqrt{\varepsilon_{\text{eff}}}} \tag{10.4}$$

$$\beta = \beta_0\sqrt{\varepsilon_{\text{eff}}} \text{ or } \varepsilon_{\text{eff}} = \left(\frac{\beta}{\beta_0}\right)^2 \tag{10.5}$$

where ε_{eff} is the effective dielectric constant for the transmission line of interest. Since some of the field lines are in the dielectric region and some are in air, the effective dielectric constant satisfies the relation

$$1 < \varepsilon_{\text{eff}} < \varepsilon_r \tag{10.6}$$

Impedance and Admittance Matrices

Once voltages and currents have been defined at various points in a microwave network, we can use the impedance or admittance matrices based on circuit theory to relate the terminal or port quantities to each other, which leads to a matrix representation of the microwave network. We frequently employ impedance, admittance, and scattering matrices to describe an arbitrary microwave network.

Now we consider an arbitrary N-port microwave network (as depicted in Figure 10.2) whose ports may be any type of transmission line, or equivalent transmission line of a single propagating waveguide mode. At a specific point on the nth port, that is, a terminal plane, we define equivalent voltages and

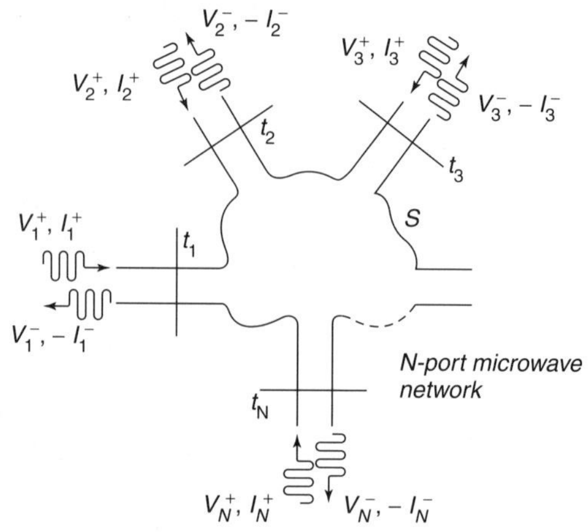

FIGURE 10.2 An N-port microwave network.

currents for the incident waves ($V_n{}^+$, I_n^+) and reflected waves (V_n^-, I_n^-). At the nth terminal plane, the total voltage and current are given by

$$V_n = V_n^+ + V_n^- \tag{10.7a}$$

$$I_n = I_n^+ + I_n^- \tag{10.7b}$$

as defined at the reference points with $z_i = 0$ ($i = 1, 2, \ldots, N$).

The impedance matrix $[Z]$ of the microwave network that relates these voltages and currents is defined as

$$\begin{bmatrix} V_1 \\ V_2 \\ \vdots \\ V_N \end{bmatrix} = \begin{bmatrix} Z_{11} & Z_{12} & \cdots & Z_{1N} \\ Z_{21} & & & \vdots \\ \vdots & & & \vdots \\ Z_{N1} & \cdots & \cdots & Z_{N1} \end{bmatrix} \begin{bmatrix} I_1 \\ I_2 \\ \vdots \\ I_N \end{bmatrix} \tag{10.8a}$$

or

$$[V] = [Z][I] \tag{10.8b}$$

Similarly, we define the admittance matrix as

$$\begin{bmatrix} I_1 \\ I_2 \\ \vdots \\ I_N \end{bmatrix} = \begin{bmatrix} Y_{11} & Y_{12} & \cdots & Y_{1N} \\ Y_{21} & & & \vdots \\ \vdots & & & \vdots \\ Y_{N1} & \cdots & \cdots & Y_{N1} \end{bmatrix} \begin{bmatrix} V_1 \\ V_2 \\ \vdots \\ V_N \end{bmatrix} \tag{10.9a}$$

or

$$[I] = [Y][V] \tag{10.9b}$$

The $[Z]$ and $[Y]$ matrices are the inverse of each other:

$$[Y] = [Z]^{-1} \tag{10.10}$$

Note that both $[Z]$ and $[Y]$ relate to the total port voltages and currents. From (10.8a), we obtain

$$Z_{ij} = \left.\frac{V_i}{I_j}\right|_{I_k=0 \text{ for } k\neq j} \tag{10.11}$$

where Z_{ij} can be found by driving port j with the current I_j, open-circuiting all other ports, and measuring the open-circuit voltage at port i. As a result, Z_{ij} is the input impedance seen looking into port i and is the transfer impedance between ports i and j, when all other ports are open circuits.

Similarly, we have

$$Y_{ij} = \left.\frac{I_i}{V_j}\right|_{V_k=0 \text{ for } k\neq j} \tag{10.12}$$

which indicates that Y_{ij} can be found by setting the driving port j with the voltage V_j, short-circuiting all other ports (so $V_k = 0$ for k not equal to j), and measuring the short-circuit current at port i.

Scattering Matrix

In practice, we are usually interested in finding reflection and transmission coefficients for a microwave network. Such quantities can be solved most conveniently by using scattering parameters to describe the microwave network. Like the impedance or admittance matrix for an N-port network, the scattering matrix provides a complete description of the network. An important difference in the scattering matrix—in comparison to impedance and admittance matrices—is that it relates incident and reflected voltages at all ports while the impedance and admittance matrices are defined based on the total voltage and current waves.

Consider the N-port network where V_n^+ is the amplitude of the voltage wave incident on port n, and V_n^- is the amplitude of the voltage wave reflected from port n. If all ports are featured with the same characteristic impedance, the scattering matrix, $[S]$, is defined in relation to these incident and reflected voltage waves as

$$\begin{bmatrix} V_1^- \\ V_2^- \\ \vdots \\ V_n^- \end{bmatrix} = \begin{bmatrix} S_{11} & S_{12} & \cdots & S_{1n} \\ S_{21} & & & \vdots \\ \vdots & & & \vdots \\ S_{n1} & \cdots & \cdots & S_{nn} \end{bmatrix} \begin{bmatrix} V_1^+ \\ V_2^+ \\ \vdots \\ V_n^+ \end{bmatrix} \tag{10.13a}$$

or

$$[V^-] = [S][V^+] \tag{10.13b}$$

where a specific element of the matrix can be defined as

$$S_{ij} = \left. \frac{V_i^-}{V_j^+} \right|_{V_k^+=0 \text{ for } k \neq j} \tag{10.14}$$

As seen in (10.14), S_{ij} can be solved by driving port j with an incident wave of voltage, V_j^+, and by measuring the reflected wave amplitude, V_i^-, at port i. The incident waves on all ports except the jth port are set to zero; therefore, all ports should be terminated in matched loads to avoid reflection. Obviously, S_{ii} is the reflection coefficient seen looking into port i when all other ports are terminated in matched loads, and S_{ij} is the transmission coefficient from port j to port i when all other ports are terminated in matched loads.

In general, the characteristic impedances of a multiport network may be different. In this case, the elements of the S parameters can be written as

$$S_{ij} = \left. \frac{V_i^- \sqrt{Z_{0j}}}{V_j^+ \sqrt{Z_{oi}}} \right|_{V_k^+=0 \text{ for } k \neq j} \tag{10.15}$$

10.3 EXTRACTION OF MMIC CHARACTERISTICS

Time Domain Voltage and Current

Most MMIC characteristics can be extracted from the time or frequency domain voltage and current. Since the time domain updated quantities in the MRTD do not represent fields but field expansion coefficients only, we must first construct the total fields from these coefficients. For instance, the time domain voltage for a microstrip line can be solved, as shown in Figure 10.3, by taking an integration of E_y from the ground plane to the strip along an integral path L with the specified values of x_0 and z_0. The time domain voltage is then given as

$$\begin{aligned} V &= -\int_0^h E_y(x_0, y, z_0, t_0)\, dy \\ &= -\sum_{l=0}^{N_s} \int_{l\Delta y}^{(l+1)\Delta y} \left[\iiint E_y(x, y, z, t)\delta(x - x_0)\delta(z - z_0)\delta(t - t_0)\, dx\, dz\, dt \right] dy \\ &= -\sum_{l=0}^{N_s} \left[\int_{l\Delta y}^{(l+1)\Delta y} \sum_{i,j,k=-\infty}^{+\infty} {}_{\phi y}E^n_{i_0-i,j+1/2,k_0-k}\phi(i)\phi(k)\phi_{j+1/2}(y)\, dy \right] \\ &= -\sum_{l=0}^{N_s} \left[\sum_{i,j,k=-\infty}^{+\infty} {}_{\phi y}E^n_{i_0-i,j+l+1/2,k}\phi(i)\phi(k)b(j)\ \Delta y \right] \end{aligned} \tag{10.16}$$

with

$$b(j) = \int_{-1/2}^{1/2} \phi(y - j)\,dy \tag{10.17}$$

where N_s is the number of cells counted from the ground plane to the PEC strip and $h = N_s\ \Delta y$. Note that the summation indexes i, j, and k must span over all real and image regions. The integral coefficient $b(j)$ forms a localized distribution with an even symmetry $b(j) = b(-j)$ as listed in Table 8.1 (Chapter 8). We can apply a similar procedure to solve for all required field quantities.

Similarly, we can solve for the time domain current by integrating the magnetic field intensity surrounding the strip along an enclosed contour C as shown in Figure 10.3, or explicitly:

$$\begin{aligned} I &= \oint_C \vec{H} \cdot d\vec{l} \\ &= \left[\int_{(i_1-1/2)\Delta x}^{(i_1+1/2)\Delta x} H_x(x, y_1, z_0, t_0)\,dx + \cdots + \int_{(i_2-1/2)\Delta x}^{(i_2+1/2)\Delta x} H_x(x, y_1, z_0, t_0)\,dx \right] \\ &\quad - \left[\int_{(i_1-1/2)\Delta x}^{(i_1+1/2)\Delta x} H_x(x, y_2, z_0, t_0)\,dx + \cdots + \int_{(i_2-1/2)\Delta x}^{(i_2+1/2)\Delta x} H_x(x, y_2, z_0, t_0)\,dx \right] \\ &\quad + \left[- \int_{(j_0-1/2)\Delta y}^{(j_0+1/2)\Delta y} H_y(x_1, y, z_0, t_0)\,dy + \int_{(j_0-1/2)\Delta y}^{(j_0+1/2)\Delta y} H_y(x_2, y, z_0, t_0)\,dy \right] \end{aligned} \tag{10.18}$$

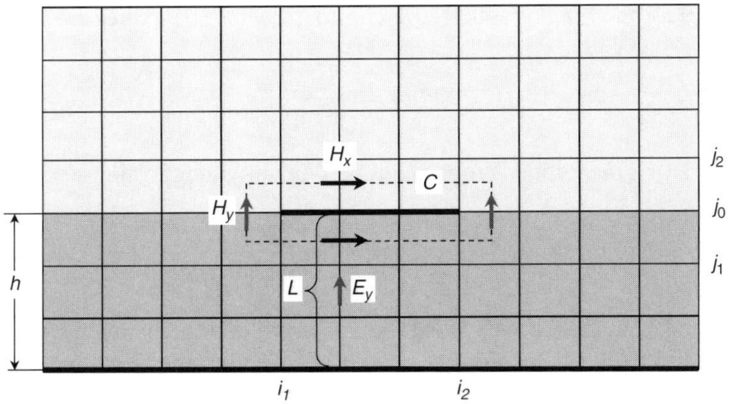

FIGURE 10.3 Integral paths, C and L, of the time domain voltage and current for a microstrip line.

where the first integration can be derived as

$$
\begin{aligned}
&\int_{(i_1-1/2)\Delta x}^{(i_1+1/2)\Delta x} H_x(x, y_1, z_0, t_0)\, dx \\
&= \int_{(i_1-1/2)\Delta x}^{(i_1+1/2)\Delta x} \left[\iiint H_x(x, y, z, t)\delta(y - y_1)\delta(z - z_0)\delta(t - t_0)\, dy\, dz\, dt \right] dx \\
&= \int_{(i-1/2)\Delta x}^{(i_1+1/2)\Delta x} \sum_{i,j,k=-\infty}^{+\infty} {}_{\phi x}H^{n+1/2}_{i,j_1-j,k_0-k}\phi(j)\phi(k)\phi_i(x)\, dx \qquad (10.19) \\
&= \sum_{i,j,k=-\infty}^{+\infty} {}_{\phi x}H^{n+1/2}_{i_1+i,j_1-j,k_0-k}\phi(j)\phi(k)b(i)
\end{aligned}
$$

As a result, the time domain current is given as

$$
\begin{aligned}
I = &\sum_{l=i_1}^{i_2} \left[\sum_{i,j,k=-\infty}^{+\infty} ({}_{\phi x}H^{n+1/2}_{l+i,j_1-j,k_0-k} - {}_{\phi x}H^{n+1/2}_{l+i,j_2-j,k_0-k})\phi(j)\phi(k)b(i) \right] \\
&+ \sum_{i,j,k=-\infty}^{+\infty} \left[({}_{\phi y}H^{n+1/2}_{i_2-i,j_0+j,k_0-k} - {}_{\phi y}H^{n+1/2}_{i_1-i,j_0+j,k_0-k})\phi(i)\phi(k)b(j) \right] \qquad (10.20)
\end{aligned}
$$

Propagation Characteristics for a Transmission Line

As mentioned earlier, the two most important characteristics for a transmission line are the effective dielectric constant ε_{eff} and characteristic impedance Z_0. As seen in (10.5), the effective dielectric constant ε_{eff} can be expressed as

$$
\varepsilon_{\text{eff}}(\omega) = \frac{\beta^2(\omega)}{\omega^2 \varepsilon_0 \mu_0} \qquad (10.21)
$$

where β can be determined from two time domain voltage values:

$$
\beta(\omega) = \frac{1}{(z_j - z_i)} \text{ angle} \left(\frac{\text{FFT } [V(z_i, t)]}{\text{FFT } [V(z_j, t)]} \right) \qquad (10.22)
$$

where z_i and z_j are two points along the transmission line separated by a distance $(z_j - z_i)$, and FFT denotes the fast Fourier transform to the time domain voltage signals, $V(z_i, t)$ and $V(z_j, t)$. It is not always necessary to extract the propagation characteristics using a postprocessing algorithm [5] when the APMLs are used for mesh truncation.

Subsequently, the characteristic impedance Z_0 is then calculated by

$$Z_0 = \frac{\text{FFT}\ \{V(z_i, t)\}}{\text{FFT}\ \{I(z_i, t)\}} \tag{10.23}$$

Both the time domain voltage and current can be solved at a specified position, z_i, as mentioned earlier in this chapter.

Scattering Parameters for a Multiport Millimeter-Wave Circuit

On the basis of the defined voltage signals at each port, we can solve the scattering parameters for a multiport millimeter-wave integrated circuit. The scattering matrix coefficients in the frequency domain for a multiport system are defined as

$$[V^-] = [S][V^+] \tag{10.24}$$

where $\lfloor V^- \rfloor$ and $\lfloor V^+ \rfloor$ are the reflected and incident voltage vectors, respectively, and $[S]$ denotes the scattering matrix [6]. An arbitrary element on the scattering matrix, S_{ij}, can be solved by

$$S_{ij} = \frac{\text{FFT}\ \{V_i^-(t)\}\sqrt{Z_{oj}}}{\text{FFT}\ \{V_j^+(t)\}\sqrt{Z_{oi}}} \tag{10.25}$$

where $V_i^-(t)$ and $V_j^+(t)$ are the reflected and incident voltage signals defined at the ports i and j, respectively, and $\sqrt{Z_{oi}}$ and $\sqrt{Z_{oj}}$ are their corresponding characteristic impedance values at these two ports.

10.4 APPLICATION RESULTS

In the following, we analyze three representative MMICs and compare the MRTD computed results with those derived by using the FDTD, in terms of computational efficiency and accuracy. Note that all computations presented in this section are performed with a 333 MHz, 128 M RAM Pentium II-MAX PC Compatible.

A Rectangular Microstrip Patch Antenna

The first practical structure to study is a rectangular microstrip patch antenna, whose geometry and dimensions are shown in Figure 10.4. As seen in Table 10.1, the computational domain is discretized into $35 \times 10 \times 40$ cells and the total time step is set to 5000 in order to ensure a satisfactory frequency resolution. The resulting area of rectangular patch is $15\ \Delta x \times 16\ \Delta z$. As seen in Figures 10.4–10.6, there is fairly good agreement between results derived from the MRTD and FDTD approaches for the magnitudes of the reflection coefficient, $S_{11}(f)$, and the input impedance $Z_{in}(f)$; however, the computational space and

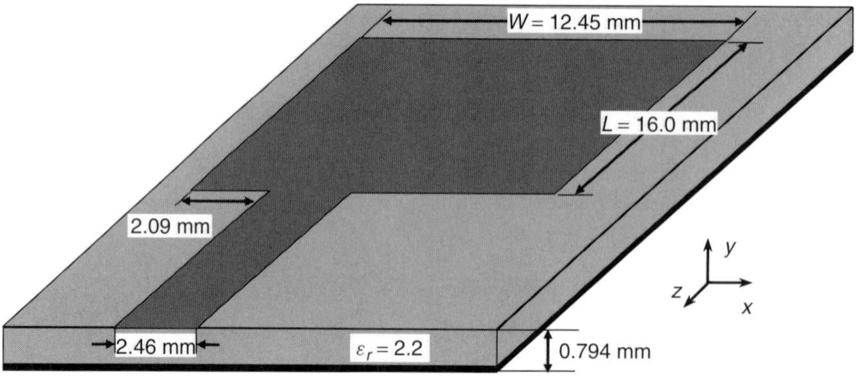

FIGURE 10.4 Geometry and dimensions for a microstrip patch antenna.

TABLE 10.1 Spatial and Time Discretization for a Microstrip Patch Antenna

Discretization	MRTD	FDTD (Sheen et al.)
Δx (mm)	0.830	0.389
Δy (mm)	0.397	0.265
Δz (mm)	1.000	0.400
Δt (ps)	0.4119	0.4410
$N_x \times N_y \times N_z$	35 × 10 × 40	60 × 16 × 100
VDR	14.58%	
CPU time (seconds)	4255.93	19254.33
TDR	22.10%	

(VDR: volume difference ratio; TDR: time difference ratio)

CPU time of the MRTD are only about 14.58% and 22.1%, respectively, of that used in the FDTD technique [7].

A Microstrip Low-Pass Filter

As shown in Figure 10.7, the next MMIC structure to be analyzed is a microstrip low-pass filter, whose computational volume is discretized into 45 × 10 × 45 cells as given in Table 10.2. The central strip segment is specified by the area 25 $\Delta x \times 3\ \Delta z$, and the time update is terminated at $N_t = 4000$. Once again, there is good consistency within a wide frequency range between the results of the scattering parameters derived from both the MRTD and FDTD approaches [8], as seen in Figure 10.8. The computational volume difference ratio (VDR) and CPU time difference ratio (TDR) between the MRTD scheme and FDTD method are 15.82% and 17.96%, respectively.

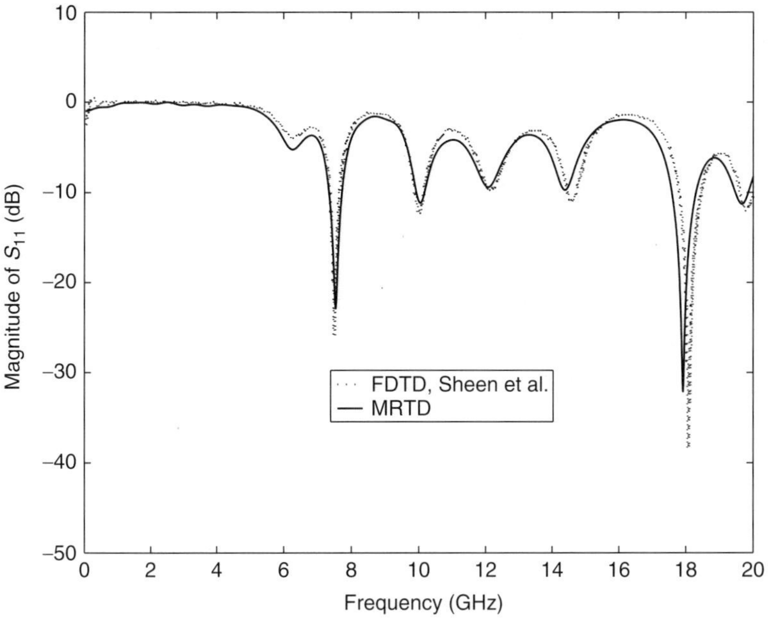

FIGURE 10.5 Magnitude of the reflection coefficient S_{11} for a microstrip patch antenna.

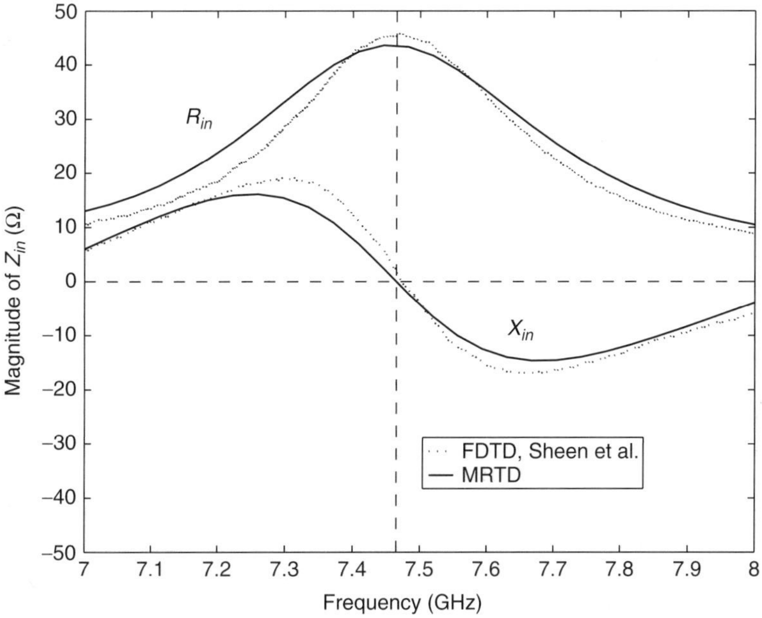

FIGURE 10.6 Magnitude of the input impedance Z_{in} for a microstrip patch antenna.

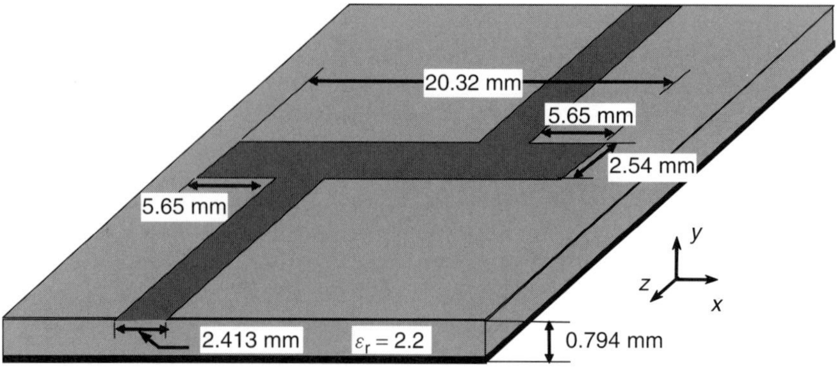

FIGURE 10.7 Geometry and dimensions for a microstrip low-pass filter.

TABLE 10.2 Spatial and Time Discretization for a Microstrip Low-Pass Filter

Discretization	MRTD	FDTD (Sheen et al.)
Δx (mm)	0.8128	0.4064
Δy (mm)	0.3970	0.2650
Δz (mm)	0.8467	0.4233
Δt (ps)	0.3996	0.4410
$N_x \times N_y \times N_z$	$45 \times 10 \times 45$	$80 \times 16 \times 100$
VDR	15.82%	
CPU time (seconds)	3553.77	19782.42*c*
TDR	17.96%	

(VDR: volume difference ratio; TDR: time difference ratio)

A Microstrip Band-Pass Filter

Finally, we investigate a microstrip band-pass filter, whose geometry, as shown in Figure 10.9, is discretized into $35 \times 12 \times 72$ cells, which is about 15.75% of that used in the FDTD [5] as given in Table 10.3. The resulting area for each strip is $3\ \Delta x \times 30\ \Delta z$. Due to the resonant features of the structure, a total time step of 30,000 is used to ensure that resonant signals stabilize. The corresponding MRTD CPU time is 59,976.41 seconds, while that of the FDTD CPU time with the same total time step is 238,768.52 seconds.

Figure 10.10 displays the magnitudes of the S-parameters for the band-pass filter computed by the MRTDand the FDTD methods [9]; these magnitudes are observed to be a good match for the entire shape of the plot, as well as the band-pass frequencies.

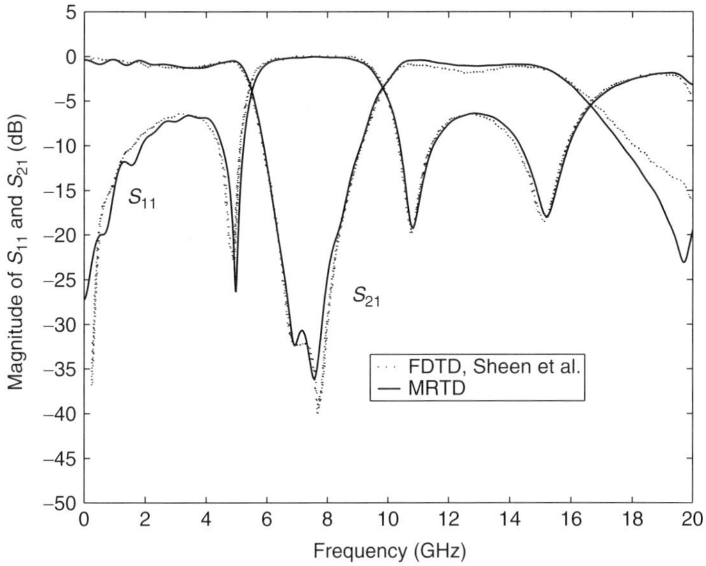

FIGURE 10.8 Magnitudes of the S-parameters S_{11} and S_{21} for a microstrip low-pass filter.

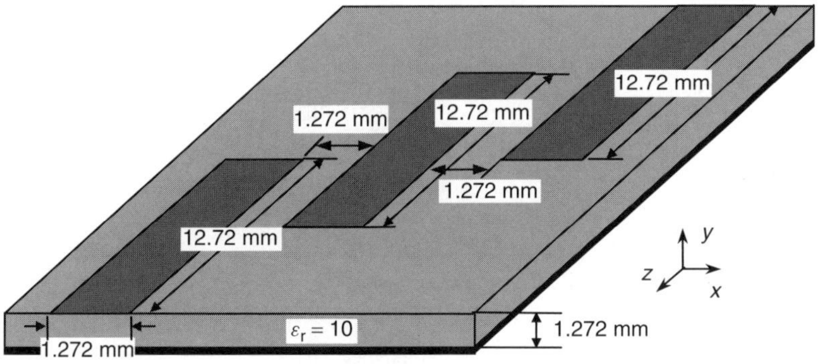

FIGURE 10.9 Geometry and dimensions for a microstrip band-pass filter.

10.5 CONCLUSIONS

In this chapter, we analyzed various millimeter-wave integrated circuits by using a scaling function based MRTD scheme in conjunction with an APML for termination of open boundaries. It is found through extensive investigations on various representative millimeter-wave integrated circuits that the developed scaling function based MRTD is highly efficient for the analysis of MMICs and requires only

TABLE 10.3 Spatial and Time Discretization for a Microstrip Band-Pass Filter

Discretization	MRTD	FDTD (Shibata et al.)
Δx (mm)	0.424	0.212
Δy (mm)	0.424	0.212
Δz (mm)	0.424	0.212
Δt (ps)	0.2991	0.1760
$N_x \times N_y \times N_z$	$35 \times 12 \times 72$	$80 \times 20 \times 120$
VDR		15.75%
CPU time (seconds)	59976.41	238768.52
TDR		25.12%

(VDR: volume difference ratio; TDR: time difference ratio)

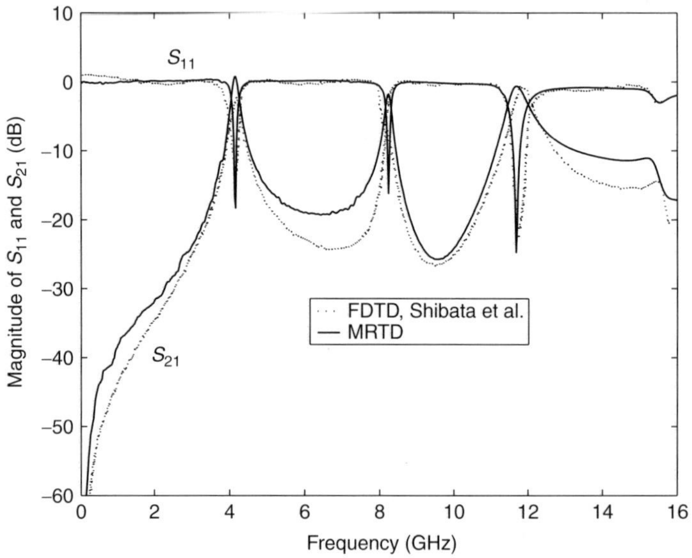

FIGURE 10.10 Magnitudes of the S-parameters, S_{11} and S_{21} for a microstrip band-pass filter.

a fraction of the computational space and CPU time compared to that used by the FDTD techniques—at the price of complexity in mathematical modeling.

REFERENCES

[1] M. Krumpholz and L. P. B. Katehi, "MRTD: New time domain schemes based on multiresolution analysis," *IEEE Trans. Microwave Theory Tech.,* vol. 44, no. 4, pp. 555–571, Apr. 1996.

[2] L. P. B. Katehi, J. F. Harvey, and E. Tentzeris, "Time-Domain Analysis Using Multiresolution Expansions," Chapter 3, in *Advances in Computational Electrodynamics: The Finite-Difference Time-Domain Method* (ed. A. Taflove), Artech House, Norwood, MA, 1998.

[3] Q. Cao and Y. Chen, "MRTD analysis of a transient electromagnetic pulse propagating through a dielectric layer," *Int. J. Electron.*, vol. 86, pp. 459–474, Apr. 1999.

[4] Q. Cao and Y. Chen, "Scaling function based multiresolution time domain analysis for planar printed millimeter-wave integrated circuits," *IEE Proc. Microwaves Antennas Propag.*, vol. 148 no. 3, pp. 179–187, June 2001.

[5] M. A. Schamberger, S. Kosanovich, and R. Mittra, "Parameter extraction and correction for transmission lines and discontinuities using the finite-difference time-domain method," *IEEE Trans. Microwave Theory Tech.*, vol. 44, no. 6, pp. 919–925, June 1996.

[6] D. M. Pozar, *Microwave Engineering*, Addison-Wesley, Boston, 1993.

[7] R. E. Collin, *Foundation for Microwave Engineering*, McGraw-Hill, New York, 1966.

[8] D. M. Sheen, S. M. Ali, M. D. Abouzahra, and J. A. Kong, "Application of the three-dimensional finite-difference time—domain method to the analysis of planar microwave circuits," *IEEE Trans. Microwave Theory Tech.*, vol. 38, no. 7, pp. 849–857, July 1990.

[9] T. Shibata, T. Hayashi, and T. Kimura, "Analysis of microstrip circuits using three-dimensional full-wave electromagnetic field analysis in the time domain," *IEEE Trans. Microwave Theory Tech.*, vol. 36, no. 6, pp. 1064–1070, 1988.

CHAPTER ELEVEN

MRTD Scattering Analysis: 2D Cases

11.1 INTRODUCTION

The finite difference time domain (FDTD) method, which has been extensively used to investigate electromagnetic scattering [1, 2], usually requires large amounts of computer resources and computational time. To improve the efficiency of the FDTD technique, the multiresolution time domain (MRTD) scheme, which is a higher-order expansion technique, has recently been developed for the analysis of practical electromagnetic scattering problems [3].

In this chapter, we begin by reviewing the fundamental concepts of electromagnetic wave scattering and then apply the MRTD scheme for the scattering analysis of two-dimensional targets, for example, perfectly electric conductors (PECs) and dielectric cylinders, with transverse magnetic (TM) and transverse electric (TE) polarized waves. To overcome the difficulties that arise because of the nonlocalized characteristics of the scaling and wavelet basis functions that are employed in the MRTD scheme, we adopt the scattered field formulation [4] in which the incident fields are present only either inside or on the surface of the scattering target. More importantly, we derive the generalized field update equations involving both the dielectric and PEC scattering targets and develop a two-dimensional MRTD (2D-MRTD) near-to-far-zone field transformation to compute the radar scattering cross section (RCS) or the scattering width (SW) [5]. We calculate the scattering width based on different wavelet bases and discuss the effects of these bases on computing accuracy of the scattering width. We find that the MRTD results derived by using the cubic spline Battle–Lemarié scaling and wavelet basis functions yield the best accuracy.

Multiresolution Time Domain Scheme for Electromagnetic Engineering
By Yinchao Chen, Qunsheng Cao, and Raj Mittra
ISBN 0-471-27230-2

11.2 SCATTERING FUNDAMENTALS

Incident, Scattered, and Total Fields

We begin by examining the phenomenon of wave propagation in the presence of a scatterer with an arbitrary geometry. In the presence of the scatterer, the total electromagnetic fields can be decomposed into two components—the incident fields ($\vec{E}^{\text{inc}}$, $\vec{H}^{\text{inc}}$) and scattered fields ($\vec{E}^{\text{scat}}$, $\vec{H}^{\text{scat}}$). The total ($\vec{E}^{\text{tot}}$, $\vec{H}^{\text{tot}}$) field is constructed by using a superposition of the scattered and incident (direct) fields as follows:

$$\vec{E}^{\text{tot}} = \vec{E}^{\text{inc}} + \vec{E}^{\text{scat}} \tag{11.1a}$$

$$\vec{H}^{\text{tot}} = \vec{H}^{\text{inc}} + \vec{H}^{\text{scat}} \tag{11.1b}$$

It is evident that the incident (direct) fields are also equal to the total fields produced by the sources in the absence of the scatterer.

Radar Cross Section (RCS)

An important parameter of interest in electromagnetic scattering studies by a target is usually referred to as the echo area or the radar cross section (RCS). The RCS is defined as the area intercepting the amount of power that, when scattered isotropically, would produce at the receiver a density that is equal to the density scattered by the actual target.

In particular, for a two-dimensional target the scattering parameter is referred to as the scattering width (SW) or, alternatively, as the radar cross section per unit length. Precisely, the scattering width for a two-dimensional target is defined as

$$\text{RCS}_{2\text{D}} = \left\{ \begin{array}{l} \lim_{\rho\to\infty} \left[2\pi\rho \dfrac{S^{\text{scat}}}{S^{\text{inc}}} \right] \\[2ex] \lim_{\rho\to\infty} \left[2\pi\rho \dfrac{|\vec{E}^{\text{scat}}|^2}{|\vec{E}^{\text{inc}}|^2} \right] \\[2ex] \lim_{\rho\to\infty} \left[2\pi\rho \dfrac{|\vec{H}^{\text{scat}}|^2}{|\vec{H}^{\text{inc}}|^2} \right] \end{array} \right\} \tag{11.2}$$

In general, the radar cross section for a three-dimensional target can be expressed as

$$\sigma_{3\text{D}} = \left\{ \begin{array}{l} \lim_{r\to\infty} \left[4\pi r^2 \dfrac{S^{\text{scat}}}{S^{\text{inc}}} \right] \\[2ex] \lim_{r\to\infty} \left[4\pi r^2 \dfrac{|\vec{E}^{\text{scat}}|^2}{|\vec{E}^{\text{inc}}|^2} \right] \\[2ex] \lim_{r\to\infty} \left[4\pi r^2 \dfrac{|\vec{H}^{\text{scat}}|^2}{|\vec{H}^{\text{inc}}|^2} \right] \end{array} \right\} \tag{11.3}$$

where ρ or r is the distance from the target to an observation point; S^{scat} and S^{inc} indicate the scattered and incident power densities, respectively; $\vec{E}^{\text{scat}}$ and $\vec{E}^{\text{inc}}$ represent the scattered and incident electric fields; and $\vec{H}^{\text{scat}}$ and $\vec{H}^{\text{inc}}$ are the scattered and incident magnetic fields.

Note that the unit of the two-dimensional SW is length (meters in the MKS system), whereas the three-dimensional RCS is area (meters squared in the MKS system). In practice, we usually adopt a common reference—one meter for the two-dimensional SW and one square meter for the three-dimensional RCS. Therefore, the most common designation is dB/m (or dBm) for the SW, and dB/m^2 (or dBsm) for the RCS.

Monostatic and Bistatic Scattering

In scattering analysis, we often use the terms *monostatic* and *bistatic* scattering to indicate scattering features of a target. When the transmitter and receiver are collocated, the RCS is usually referred to as monostatic or backscattered, while it is called bistatic when the two are at different locations.

Pure Scattered Electromagnetic Field Formulation

Both incident and scattered fields must satisfy the Maxwell equations independently if the materials in the working space are linear. For simplicity, we assume that the scattering takes place in free space, though the generalization to the case of a uniform medium is relatively straightforward. When the incident wave travels through free space and impinges on a target or scatterer, the total field propagates in free space, in the region outside the scatterer, and in the media of the scatterer when propagating inside.

Inside the scatterer, the total fields satisfy the Maxwell curl equations:

$$\nabla \times \vec{E}^{\text{tot}} = -\mu \frac{\partial \vec{H}^{\text{tot}}}{\partial t} - \sigma^* \vec{H}^{\text{tot}} \tag{11.4a}$$

$$\nabla \times \vec{H}^{\text{tot}} = \varepsilon \frac{\partial \vec{E}^{\text{tot}}}{\partial t} + \sigma \vec{E}^{\text{tot}} \tag{11.4b}$$

where ε, μ, σ, and σ^* denote permittivity, permeability, electric conductivity, and magnetic conductivity, respectively, at an arbitrary point of the space. If the outer side of the scatterer is free space, then the above equations simplify as follows:

$$\nabla \times \vec{E}^{\text{tot}} = -\mu_0 \frac{\partial \vec{H}^{\text{tot}}}{\partial t} \tag{11.5a}$$

$$\nabla \times \vec{H}^{\text{tot}} = \varepsilon_0 \frac{\partial \vec{E}^{\text{tot}}}{\partial t} \tag{11.5b}$$

The incident fields satisfy the Maxwell equations in free space, namely,

$$\nabla \times \vec{E}^{\text{inc}} = -\mu_0 \frac{\partial \vec{H}^{\text{inc}}}{\partial t} \tag{11.6a}$$

$$\nabla \times \vec{H}^{\text{inc}} = \varepsilon_0 \frac{\partial \vec{E}^{\text{inc}}}{\partial t} \tag{11.6b}$$

By superimposing the incident and scattered fields, the total fields expressed in (11.1) can be rewritten as

$$\nabla \times (\vec{E}^{\text{inc}} + \vec{E}^{\text{scat}}) = -\mu \frac{\partial (\vec{H}^{\text{inc}} + \vec{H}^{\text{scat}})}{\partial t} - \sigma^* (\vec{H}^{\text{inc}} + \vec{H}^{\text{scat}}) \tag{11.7a}$$

$$\nabla \times (\vec{H}^{\text{inc}} + \vec{H}^{\text{scat}}) = \varepsilon \frac{\partial (\vec{E}^{\text{inc}} + \vec{E}^{\text{scat}})}{\partial t} + \sigma (\vec{E}^{\text{inc}} + \vec{E}^{\text{scat}}) \tag{11.7b}$$

Subtracting the incident fields defined in (11.6) from (11.7), we obtain the following equations in the interior of the scatterer:

$$\nabla \times \vec{E}^{\text{scat}} = -\mu \frac{\partial \vec{H}^{\text{scat}}}{\partial t} - (\mu - \mu_0) \frac{\partial \vec{H}^{\text{inc}}}{\partial t} - \sigma^* (\vec{H}^{\text{inc}} + \vec{H}^{\text{scat}}) \tag{11.8a}$$

$$\nabla \times \vec{H}^{\text{scat}} = \varepsilon \frac{\partial \vec{E}^{\text{scat}}}{\partial t} + (\varepsilon - \varepsilon_0) \frac{\partial \vec{E}^{\text{inc}}}{\partial t} + \sigma (\vec{E}^{\text{inc}} + \vec{E}^{\text{scat}}) \tag{11.8b}$$

The corresponding equations for the scattered fields outside the scatterer become

$$\nabla \times \vec{E}^{\text{scat}} = -\mu \frac{\partial \vec{H}^{\text{scat}}}{\partial t} \tag{11.9a}$$

$$\nabla \times \vec{H}^{\text{scat}} = \varepsilon \frac{\partial \vec{E}^{\text{scat}}}{\partial t} \tag{11.9b}$$

Obviously, the incident fields only exist inside the target if it is dielectric or on the surface if it is a PEC scatterer. In the scattering analysis, we employ (11.8) and (11.9) to introduce the incident fields, and use them to derive the far-zone scattering fields. Note that this is quite different from the conventional procedure employed in the FDTD algorithm.

11.3 GOVERNING EQUATIONS OF MRTD

Here we adopt the pure scattered field formulation, which we discussed earlier to describe the time domain scattering electromagnetic fields by dividing the computational domain into the inside and outside of target regions [4], as shown in Figure 11.1. In this section, we focus primarily on two-dimensional scattering problems for both the TM_z and TE_z polarized waves incident on a scattering target where the medium parameters are (σ, ε).

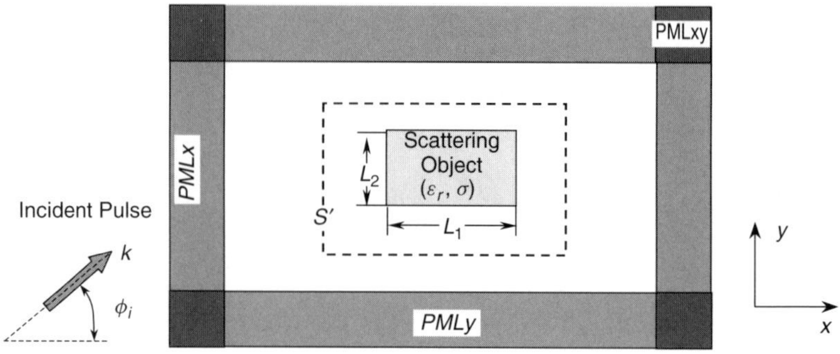

FIGURE 11.1 Cross section of a two-dimensional scattering geometry for a PEC or dielectric cylinder.

Both the scattered and incident fields exist inside the target region, and the governing equations are therefore given by

$$\varepsilon \frac{\partial \vec{E}^{\text{scat}}}{\partial t} = -\sigma \vec{E}^{\text{scat}} - \sigma \vec{E}^{\text{inc}} - (\varepsilon - \varepsilon_0) \frac{\partial \vec{E}^{\text{inc}}}{\partial t} + \nabla \times \vec{H}^{\text{scat}} \tag{11.10a}$$

$$\mu \frac{\partial \vec{H}^{\text{scat}}}{\partial t} = -\sigma^* \vec{H}^{\text{scat}} - \sigma^* \vec{H}^{\text{inc}} - (\mu - \mu_0) \frac{\partial \vec{H}^{\text{inc}}}{\partial t} - \nabla \times \vec{E}^{\text{scat}} \tag{11.10b}$$

where $(\vec{E}^{\text{inc}}, \vec{H}^{\text{inc}})$ and $(\vec{E}^{\text{scat}}, \vec{H}^{\text{scat}})$ are the incident and scattered fields, respectively; ε, μ, σ, and σ^* denote the permittivity, permeability, electric conductivity, and magnetic conductivity of the target, respectively. For the sake of simplicity, we assume that $\sigma^* = 0$ and $\mu = \mu_0$ in the discussion that follows. For the TM_z polarized wave, (11.10) can be simplified to

$$\varepsilon \frac{\partial E_z^{\text{scat}}}{\partial t} = -\sigma E_z^{\text{scat}} - \sigma E_z^{\text{inc}} - j\omega(\varepsilon - \varepsilon_0) E_z^{\text{inc}} + \left[\frac{\partial H_y^{\text{scat}}}{\partial x} - \frac{\partial H_x^{\text{scat}}}{\partial y} \right] \tag{11.11a}$$

$$\mu_0 \frac{\partial H_x^{\text{scat}}}{\partial t} = -\frac{\partial E_z^{\text{scat}}}{\partial y} \tag{11.11b}$$

$$\mu_0 \frac{\partial H_y^{\text{scat}}}{\partial t} = \frac{\partial E_z^{\text{scat}}}{\partial x} \tag{11.11c}$$

The spatial relationship among the updated expansion coefficients is shown in Figure 11.2 for both TM_z and TE_z waves.

Similarly, for the TE_z polarized wave, we can simplify (11.10) to

$$\mu_0 \frac{\partial H_z^{\text{scat}}}{\partial t} = -\left(\frac{\partial E_y^{\text{scat}}}{\partial x} - \frac{\partial E_x^{\text{scat}}}{\partial y} \right) \tag{11.12a}$$

$$\varepsilon \frac{\partial E_x^{\text{scat}}}{\partial t} = -\sigma E_x^{\text{scat}} - \sigma E_x^{\text{inc}} - (\varepsilon - \varepsilon_0) \frac{\partial E_x^{\text{inc}}}{\partial t} + \frac{\partial H_z^{\text{scat}}}{\partial y} \tag{11.12b}$$

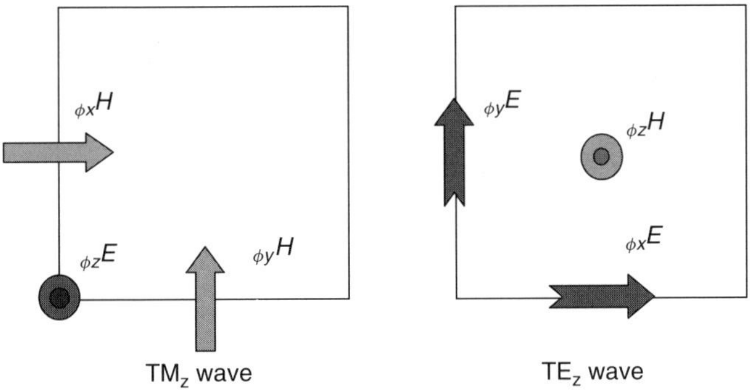

FIGURE 11.2 Fundamental topologies of TM_z and TE_z waves.

$$\varepsilon\frac{\partial E_y^{\rm scat}}{\partial t} = -\sigma E_y^{\rm scat} - \sigma E_y^{\rm inc} - (\varepsilon - \varepsilon_0)\frac{\partial E_y^{\rm inc}}{\partial t} + \frac{\partial H_z^{\rm scat}}{\partial x} \tag{11.12c}$$

In regions exterior to the scattering target, we introduce an anisotropic perfectly matched layer (APML) [6], and, by using the generalized differential matrix operators (GDMOs) [7] (given in Appendix A), we can express the governing field equations in a matrix form as

$$\begin{bmatrix} 0 & -\partial_z & \partial_y \\ \partial_z & 0 & -\partial_x \\ -\partial_y & \partial_x & 0 \end{bmatrix}\begin{bmatrix} H_x^{\rm scat} \\ H_y^{\rm scat} \\ 0 \end{bmatrix} = j\omega\varepsilon_0[S]\begin{bmatrix} 0 \\ 0 \\ E_z^{\rm scat} \end{bmatrix} \tag{11.13a}$$

$$\begin{bmatrix} 0 & -\partial_z & \partial_y \\ \partial_z & 0 & -\partial_x \\ -\partial_y & \partial_x & 0 \end{bmatrix}\begin{bmatrix} 0 \\ 0 \\ E_z^{\rm scat} \end{bmatrix} = -j\omega\mu_0[S]\begin{bmatrix} H_x^{\rm scat} \\ H_y^{\rm scat} \\ 0 \end{bmatrix} \tag{11.13b}$$

for a TM_z polarized wave, and

$$\begin{bmatrix} 0 & -\partial_z & \partial_y \\ \partial_z & 0 & -\partial_x \\ -\partial_y & \partial_x & 0 \end{bmatrix}\begin{bmatrix} 0 \\ 0 \\ H_z^{\rm scat} \end{bmatrix} = j\omega\varepsilon_0[S]\begin{bmatrix} E_x^{\rm scat} \\ E_y^{\rm scat} \\ 0 \end{bmatrix} \tag{11.14a}$$

$$\begin{bmatrix} 0 & -\partial_z & \partial_y \\ \partial_z & 0 & -\partial_x \\ -\partial_y & \partial_x & 0 \end{bmatrix}\begin{bmatrix} E_x^{\rm scat} \\ E_y^{\rm scat} \\ 0 \end{bmatrix} = -j\omega\mu_0[S]\begin{bmatrix} 0 \\ 0 \\ H_z^{\rm scat} \end{bmatrix} \tag{11.14b}$$

for a TE_z polarized wave. The parameter matrix $[S]$ in the above equations is defined as

$$[S] = \begin{bmatrix} S_{xx} & 0 & 0 \\ 0 & S_{yy} & 0 \\ 0 & 0 & S_{zz} \end{bmatrix} = \begin{bmatrix} s_x^{-1}s_y & 0 & 0 \\ 0 & s_x s_y^{-1} & 0 \\ 0 & 0 & s_x s_y \end{bmatrix} \tag{11.15a}$$

with

$$s_\alpha = \begin{cases} 1 + \dfrac{\sigma_\alpha^e}{j\omega\varepsilon_0}, & \text{in APML regions} \\ 1, & \text{non-APML region } (\sigma_\alpha^e = 0) \end{cases} \qquad (\alpha = x, y) \tag{11.15b}$$

where σ_α^e is spatially varying conductivity along the absorption axis. To achieve a smoothly varying evanescent behavior in the APML regions, we designate σ_α^e as

$$\sigma_\alpha^e = \sigma_{\alpha\,\max}^e \left| \frac{\alpha - \alpha_0}{d} \right|^2, \quad \alpha = x, y \tag{11.16}$$

where d is the thickness of the APML region. The constant $\sigma_{\alpha\,\max}^e$ is determined by a designated reflection coefficient $R(0)$ for the case of normal incidence on the APML, where

$$\begin{aligned} R(0) &= \exp\left[-\frac{2}{\varepsilon_0 v} \int_0^d \sigma_\alpha^e(\alpha)\, d\alpha \right] = \exp\left[-\frac{2\sigma_{\alpha\,\max}^e d}{3\varepsilon_0 v} \right] \\ &= \exp\left[-\frac{2\sigma_{\alpha\,\max}^e d\sqrt{\varepsilon_{r\ \text{eff}}}}{3\varepsilon_0 c} \right] \end{aligned} \tag{11.17}$$

We can solve for $\sigma_{\alpha\,\max}^e$ from the above equation to read

$$\sigma_{\alpha\,\max}^e = -\frac{3\varepsilon_0 c}{2d\sqrt{\varepsilon_{r\ \text{eff}}}} \ln(R(0)), \quad \alpha = x, y \tag{11.18}$$

where v and c denote the velocities of the electromagnetic wave propagation in the media and free space, respectively, and $\varepsilon_{r\ \text{eff}}$ is the effective relative permittivity.

11.4 MRTD SCATTERING ALGORITHM FOR TM$_Z$ WAVE

MRTD Update Equations Inside the Target

The first step to implementing the scaling function based MRTD (S-MRTD) is to expand all of the scattering field quantities in terms of the scaling functions in space and pulse functions in time as follows:

$$E_z^{\text{scat}}(\vec{r}, t) = \sum_{i,j,n=-\infty}^{\infty} {}_{\phi z}^{\text{scat}} E_{i,j}^n \tilde{\phi}_i(x) \tilde{\phi}_j(y) h_n(t) \tag{11.19a}$$

$$H_x^{\text{scat}}(\vec{r}, t) = \sum_{i,j,n=-\infty}^{\infty} {}_{\phi x}^{\text{scat}} H_{i,j+1/2}^{n+1/2} \tilde{\phi}_i(x) \tilde{\phi}_{j+1/2}(y) h_{n+1/2}(t) \tag{11.19b}$$

$$H_y^{\text{scat}}(\vec{r}, t) = \sum_{i,j,n=-\infty}^{\infty} {}_{\phi y}^{\text{scat}} H_{i+1/2,j}^{n+1/2} \tilde{\phi}_{i+1/2}(x) \tilde{\phi}_j(y) h_{n+1/2}(t) \tag{11.19c}$$

where $\tilde{\phi}_i(\xi)$ ($\xi = x, y$) denotes the dual scaling function shifted by i normalized units. The incident fields are expanded in a similar way as the scattering fields are expanded. The $h_n(t)$ and $\phi_i(\xi)$ or $\tilde{\phi}_i(\xi)$ are generated from the mother scaling functions by dilation and translation as follows:

$$h_n(t) = h\left(\frac{t}{\Delta t} - n + \frac{1}{2}\right) \tag{11.20}$$

and

$$\phi_i(\xi) = \phi\left(\frac{\xi}{\Delta\xi} - i\right) \quad \text{or} \quad \tilde{\phi}_i(\xi) = \tilde{\phi}\left(\frac{\xi}{\Delta\xi} - i\right) \tag{11.21}$$

If the scaling function $\phi_i(\xi) = \tilde{\phi}_i(\xi)$, then the expansion is called an orthogonal expansion; otherwise, it is referred to as a biorthogonal expansion. In this work, we considered a variety of scaling functions represented by $\phi(\xi)$ and their translated functions represented by $\phi_i(\xi)$ (or translated dual functions $\tilde{\phi}_i(\xi)$). In principle, we can use any of the scaling functions, such as the cubic spline Battle–Lemarié scaling function, the compact support Daubechies scaling function D_4 ($p = 2$) [8], the compact support spline biorthogonal (Coiflet wavelet) scaling function ($p = 2$, $\tilde{p} = 2$) [9], the biorthogonal Coiflet scaling function ($p = 4$, $\tilde{p} = 4$) [9], or the Deslauriers–Dubuc interpolating functions [10]. In particular, the Deslauriers–Dubuc scaling function has its dual scaling function, which can be chosen as the Dirac delta function: $\tilde{\phi}(\xi) = \delta(\xi)$ [11, 12]. The Deslauriers–Dubuc interpolating scaling functions are constructed with the autocorrelation function of the Daubechies' orthogonal scaling functions with $N = 2$ as shown in Figure 11.3.

Substitution of the above MRTD field expansions into the time domain Maxwell equations, followed by an application of the standard Galerkin's method, yields the following update equations for the two H-fields:

$${}^{\text{scat}}_{\phi x}H^{n+1/2}_{i,j+1/2} = {}^{\text{scat}}_{\phi x}H^{n-1/2}_{i,j+1/2} - \frac{1}{\mu_0}\sum_{\nu} a(\nu)^{\text{scat}}_{\phi z}E^{n}_{i,j+\nu+1/2}\frac{\Delta t}{\Delta y} \tag{11.22a}$$

$${}^{\text{scat}}_{\phi y}H^{n+1/2}_{i+1/2,j} = {}^{\text{scat}}_{\phi y}H^{n-1/2}_{i+1/2,j} - \frac{1}{\mu_0}\sum_{\nu} a(\nu)^{\text{scat}}_{\phi z}E^{n}_{i+\nu+1/2,j}\frac{\Delta t}{\Delta x} \tag{11.22b}$$

The expansion coefficients $\alpha(\nu)$ connecting the scaling functions and their derivatives can be obtained via a numerical integration in the Fourier domain [3]. Table 11.1 lists the coefficients $\alpha(\nu)$ of the Daubechies D_4 and biorthogonal Coiflet scaling functions derived in this manner.

Derivations of the update equations for the E-fields are much more involved, since the orthogonal relationship of the basis functions cannot be applied directly in inhomogeneous space. Starting from the time domain equation for a lossless dielectric target ($\sigma = 0$),

$$\varepsilon\frac{\partial E^{\text{scat}}_z}{\partial t} = -(\varepsilon - \varepsilon_0)\frac{\partial E^{\text{inc}}_z}{\partial t} + \left[\frac{\partial H^{\text{scat}}_y}{\partial x} - \frac{\partial H^{\text{scat}}_x}{\partial y}\right] \tag{11.23}$$

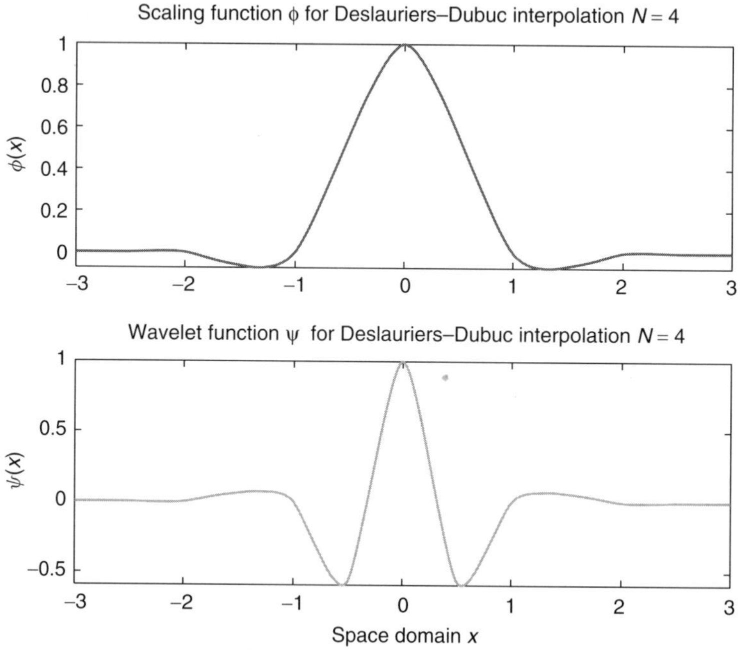

FIGURE 11.3 Deslauriers–Dubuc interpolating scaling and wavelet functions.

TABLE 11.1 Connection Coefficients $\alpha(\nu)$ with $\alpha(-\nu) = -\alpha(\nu - 1)$

ν	Daubechies	Coiflet ($p = 2$, $\tilde{p} = 2$)	Coiflet ($p = 4$, $\tilde{p} = 4$)
0	1.22953239	1.23464519	1.31103170
1	−0.09358996	−0.09715386	−0.15600966
2	0.01025133	0.01162914	0.04199606
3	0.00003558	−0.00019002	−0.00865439
4			0.00083094
5			0.00000989

we can derive the following update equation for the E-field:

$$
{}^{\text{scat}}_{\phi z}E^{n+1}_{i,j} = {}^{\text{scat}}_{\phi z}E^{n}_{i,j} + \frac{1}{\left[1 + \sum\limits_{\kappa=1}^{N} \left(\varepsilon_r^{\kappa} - 1\right) \alpha^{\kappa}_{i,i} \beta^{\kappa}_{j,j}\right]}
\times \sum_{i'=i-\Lambda}^{i'=i+\Lambda} \sum_{j'=j-\Lambda}^{j'=j+\Lambda} \sum_{\kappa=1}^{N} \left(\varepsilon_r^{\kappa} - 1\right) \alpha^{\kappa}_{i,i'} \beta^{\kappa}_{j,j'} \left({}^{\text{inc}}_{\phi z}E^{n+1}_{i',j'} - {}^{\text{inc}}_{\phi z}E^{n}_{i',j'}\right)
$$

$$+ \frac{1}{\varepsilon_0 \left[1 + \sum_{\kappa=1}^{N} (\varepsilon_r^\kappa - 1)\alpha_{i,i}^\kappa \beta_{j,j}^\kappa \right]} \sum_{\nu} a(\nu) \times \left({}_{\phi y}^{\text{scat}} H_{i+\nu+1/2,j}^{n+1/2} \frac{\Delta t}{\Delta x} - {}_{\phi x}^{\text{scat}} H_{i,j+\nu+1/2}^{n+1/2} \frac{\Delta t}{\Delta y} \right) \tag{11.24}$$

where N is the number of dielectric media in the computational domain, and Λ is a constant that is usually designated within the range of 6 to 9 for the Battle–Lemarié wavelet basis and 6 for the Daubechies D_4 and the biorthogonal Coiflet bases, depending on the localization property of the scaling functions. Equation (11.24) indicates that to update a space field we have to use the field information within an extended region that covers the area $[(i_2^k + \Lambda) - (i_1^k - \Lambda)] \times [(j_2^k + \Lambda) - (j_1^k - \Lambda)]$, as shown in Figure 11.4, for the contribution from the κth scattering target.

For simplicity, we have used the diagonal approximation in the evaluation of the update coefficients; that is, we have replaced $\alpha_{i,i'}^\kappa \beta_{j,j'}^\kappa$ by $\alpha_{i,i'}^\kappa \beta_{j,j'}^\kappa \delta_{i,i'} \delta_{j,j'}$. The approximation is justified because of the compact support of the basis functions. In general, the coefficients $\alpha_{i,i'}^\kappa$ and $\beta_{j,j'}^\kappa$ are defined as

$$\alpha_{i,i'}^\kappa = \frac{1}{\Delta x} \int_{x_1^\kappa}^{x_2^\kappa} \phi_{i'}(x)\tilde{\phi}_i(x)\, dx \tag{11.25a}$$

$$\beta_{j,j'}^\kappa = \frac{1}{\Delta y} \int_{y_1^\kappa}^{y_2^\kappa} \phi_{j'}(y)\tilde{\phi}_j(y)\, dy \tag{11.25b}$$

where (x_1^κ, x_2^κ) and (y_1^κ, y_2^κ) are the lower and upper limits of the κth dielectric medium along the x- and y-directions, respectively. Figures 11.5–11.8 show the

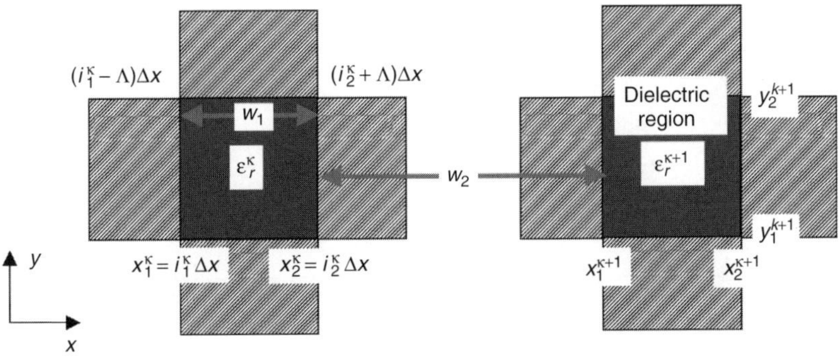

FIGURE 11.4 Geometry of a multiple-target system, where the κth and $(\kappa + 1)$th targets are displayed.

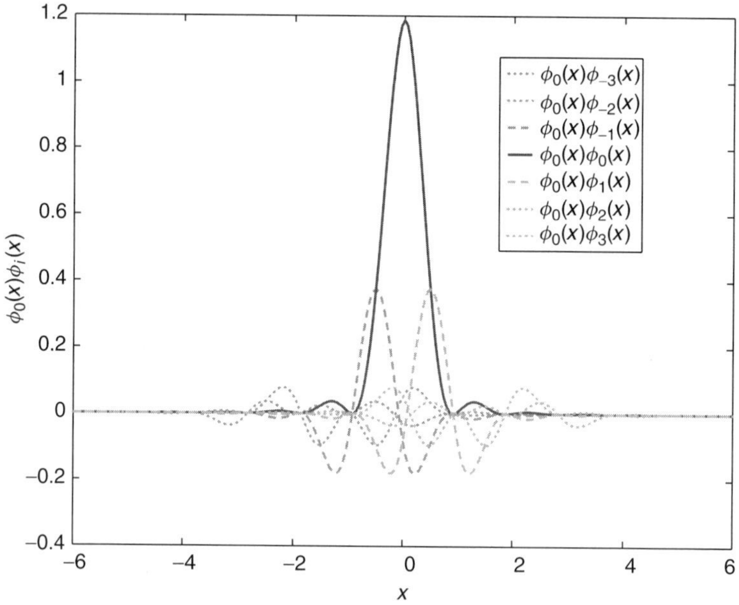

FIGURE 11.5 Multiplication of the scaling and its shifted functions of the cubic spline Battle–Lemarié basis.

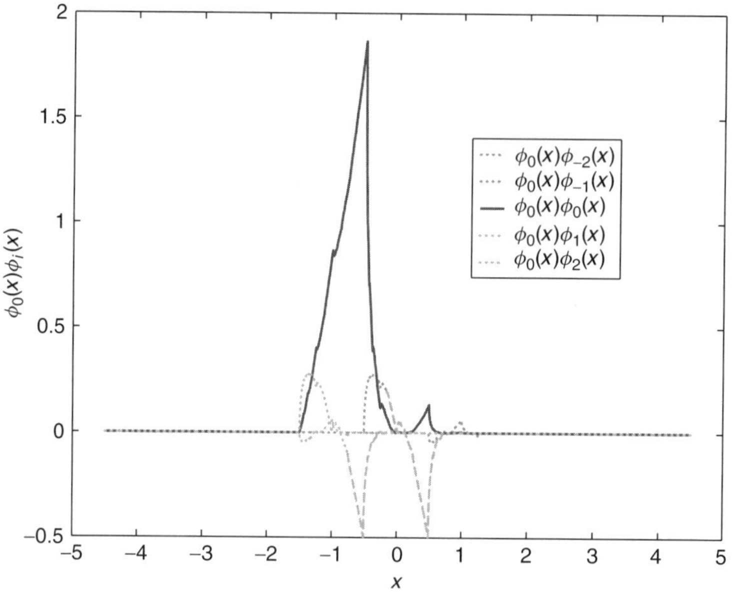

FIGURE 11.6 Multiplication of the scaling and its shifted functions of the Daubechies D_4 basis.

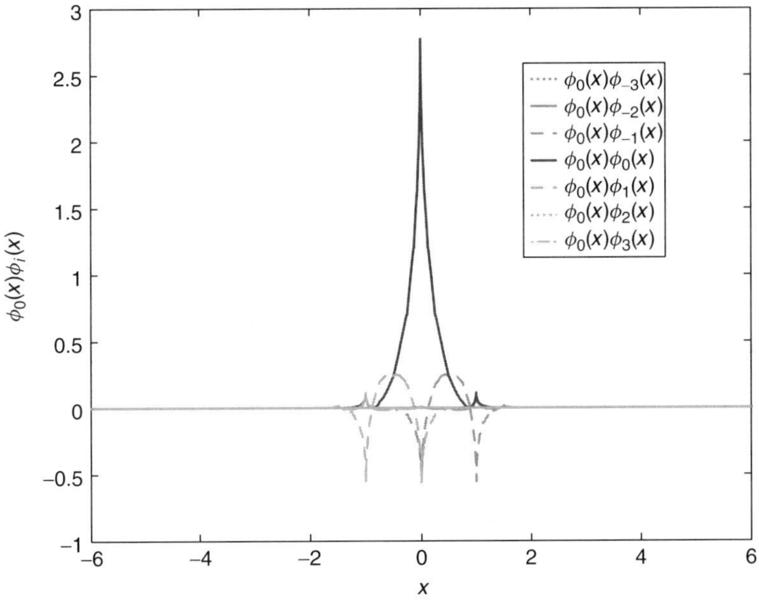

FIGURE 11.7 Multiplication of the scaling and its shifted functions of the biorthogonal Coiflet basis ($p = 2$, $\widetilde{p} = 2$).

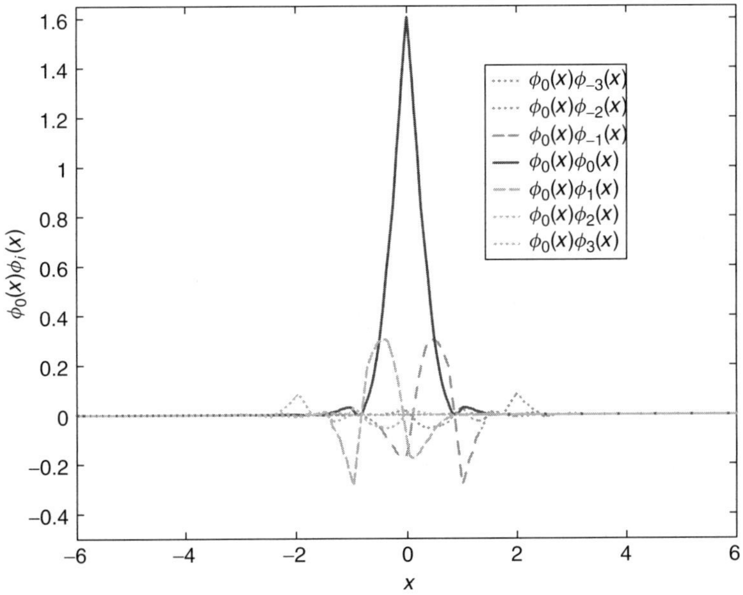

FIGURE 11.8 Multiplication of the scaling and its shifted functions of the biorthogonal Coiflet basis ($p = 4$, $\widetilde{p} = 4$).

distributions of the scaling function $\phi_i(x)$, the dual scaling function $\tilde{\phi}_{i'}(x)$, and the products of $\phi_i(x)\tilde{\phi}_{i'}(x)$ for the different wavelet bases, respectively.

Similarly, for a lossy dielectric target ($\sigma \neq 0$), from the governing equation (11.11), we can use the same procedure to develop the update equations for the lossless dielectric case and derive the following equations for the E-field:

$$
\begin{aligned}
{}^{\text{scat}}_{\phi z}E^{n+1}_{i,j} &= \frac{C_-}{C_+}{}^{\text{scat}}_{\phi z}E^{n}_{i,j} - \frac{C_0}{C_+}\sum_{i'=i-\Lambda}^{i'=i+\Lambda}\sum_{j'=j-\Lambda}^{j'=j+\Lambda}\sum_{\kappa=1}^{N}\alpha^{\kappa}_{i,i'}\beta^{\kappa}_{j,j'}\left({}^{\text{inc}}_{\phi z}E^{n+1}_{i',j'} + {}^{\text{inc}}_{\phi z}E^{n}_{i',j'}\right) \\
&\quad - \frac{1}{C_+}\sum_{i'=i-\Lambda}^{i'=i+\Lambda}\sum_{j'=j-\Lambda}^{j'=j+\Lambda}\sum_{\kappa=1}^{N}(\varepsilon^{\kappa}_r - 1)\alpha^{\kappa}_{i,i'}\beta^{\kappa}_{j,j'}\left({}^{\text{inc}}_{\phi z}E^{n+1}_{i,j'} - {}^{\text{inc}}_{\phi z}E^{n}_{i',j'}\right) \\
&\quad + \frac{1}{C_+}\frac{1}{\varepsilon_0}\sum_{\nu}a(\nu)\left({}^{\text{scat}}_{\phi y}H^{n+1/2}_{i+\nu+1/2,j}\frac{\Delta t}{\Delta y} - {}^{\text{scat}}_{\phi x}H^{n+1/2}_{i,j+\nu+\frac{1}{2}}\frac{\Delta t}{\Delta x}\right)
\end{aligned}
\tag{11.26}
$$

where

$$
C_- = \left[1 + \sum_{\kappa=1}^{N}\left((\varepsilon^{\kappa}_r - 1) - \frac{\sigma\,\Delta t}{2\varepsilon_r}\right)\alpha^{\kappa}_{i,i}\beta^{\kappa}_{j,j}\right] \tag{11.27a}
$$

$$
C_+ = \left[1 + \sum_{\kappa=1}^{N}\left((\varepsilon^{\kappa}_r - 1) + \frac{\sigma\,\Delta t}{2\varepsilon_r}\right)\alpha^{\kappa}_{i,i}\beta^{\kappa}_{j,j}\right] \tag{11.27b}
$$

$$
C_0 = \frac{\sigma\,\Delta t}{2\varepsilon_r} \tag{11.27c}
$$

and

$$
\alpha^{\kappa}_{i,i'} = \frac{1}{\Delta x}\int_{x_1^{\kappa}}^{x_2^{\kappa}}\phi_{i'}(x)\tilde{\phi}_i(x)\,dx \tag{11.28a}
$$

$$
\beta^{\kappa}_{j,j'} = \frac{1}{\Delta y}\int_{y_1^{\kappa}}^{y_2^{\kappa}}\phi_{j'}(y)\tilde{\phi}_j(y)\,dy \tag{11.28b}
$$

where the same diagonal approximation relations have been used.

MRTD Update Equations Outside the Target

In regions outside the target, the computational domain consists of both the free space and APML regions. Here, we use a set of unified governing equations to include these regions:

$$
\varepsilon_0 s_x s_y \frac{\partial E_z^{\text{scat}}}{\partial t} = \left[\frac{\partial H_y^{\text{scat}}}{\partial x} - \frac{\partial H_x^{\text{scat}}}{\partial y}\right] \tag{11.29a}
$$

$$\mu_0 s_x^{-1} s_y \frac{\partial H_x^{\text{scat}}}{\partial} = -\frac{\partial E_z^{\text{scat}}}{\partial y} \tag{11.29b}$$

$$\mu_0 s_x s_y^{-1} \frac{\partial H_y^{\text{scat}}}{\partial} = \frac{\partial E_z^{\text{scat}}}{\partial x} \tag{11.29c}$$

In common with the FDTD approach used in [6], we again use a two-step approach to derive the MRTD update equations in regions outside the target:

$$\begin{aligned} {}^{\text{scat}}_{\phi z}D^{n+1}_{i,j} &= \frac{1-\dfrac{\sigma_x^e \Delta t}{2\varepsilon_0}}{1+\dfrac{\sigma_x^e \Delta t}{2\varepsilon_0}} {}^{\text{scat}}_{\phi z}D^{n}_{i,j} + \frac{1}{1+\dfrac{\sigma_x^e \Delta t}{2\varepsilon_0}} \sum_\nu a(\nu) \\ &\times \left({}^{\text{scat}}_{\phi y}H^{n+1/2}_{i+\nu+1/2,j} \frac{\Delta t}{\Delta x} - {}^{\text{scat}}_{\phi x}H^{n+1/2}_{i,j+\nu+1/2} \frac{\Delta t}{\Delta y} \right) \end{aligned} \tag{11.30a}$$

$${}^{\text{scat}}_{\phi z}E^{n+1}_{i,j} = \frac{1-\dfrac{\sigma_y^e \Delta t}{2\varepsilon_0}}{1+\dfrac{\sigma_y^e \Delta t}{2\varepsilon_0}} {}^{\text{scat}}_{\phi z}E^{n}_{i,j} + \frac{1}{\varepsilon_0} \frac{1}{1+\dfrac{\sigma_y^e \Delta t}{2\varepsilon_0}} \left[{}^{\text{scat}}_{\phi z}D^{n+1}_{i,j} - {}^{\text{scat}}_{\phi z}D^{n}_{i,j} \right] \tag{11.30b}$$

$${}^{\text{scat}}_{\phi x}B^{n+1/2}_{i,j+1/2} = {}^{\text{scat}}_{\phi x}B^{n-1/2}_{i,j+1/2} - \sum_\nu a(\nu) {}^{\text{scat}}_{\phi z}E^{n}_{i,j+\nu+1/2} \frac{\Delta t}{\Delta y} \tag{11.31a}$$

$$\begin{aligned} {}^{\text{scat}}_{\phi x}H^{n+1/2}_{i,j+1/2} &= \frac{1-\dfrac{\sigma_x^e \Delta t}{2\varepsilon_0}}{1+\dfrac{\sigma_x^e \Delta t}{2\varepsilon_0}} {}^{\text{scat}}_{\phi z}H^{n-1/2}_{i,j+1/2} + \frac{1}{\mu_0} \frac{1}{1+\dfrac{\sigma_x^e \Delta t}{2\varepsilon_0}} \\ &\times \left[\left(1+\frac{\sigma_y^e \Delta t}{2\varepsilon_0}\right) {}^{\text{scat}}_{\phi z}B^{n+1/2}_{i,j+1/2} - \left(1-\frac{\sigma_y^e \Delta t}{2\varepsilon_0}\right) {}^{\text{scat}}_{\phi z}B^{n-1/2}_{i,j+1/2} \right] \end{aligned} \tag{11.31b}$$

$${}^{\text{scat}}_{\phi y}B^{n+1/2}_{i+1/2,j} = {}^{\text{scat}}_{\phi y}B^{n-1/2}_{i+1/2,j} - \sum_\nu a(\nu) {}^{\text{scat}}_{\phi z}E^{n}_{i+\nu+1/2,j} \frac{\Delta t}{\Delta x} \tag{11.32a}$$

$$\begin{aligned} {}^{\text{scat}}_{\phi y}H^{n+1/2}_{i+1/2,j} &= \frac{1-\dfrac{\sigma_x^e \Delta t}{2\varepsilon_0}}{1+\dfrac{\sigma_x^e \Delta t}{2\varepsilon_0}} {}^{\text{scat}}_{\phi y}H^{n-1/2}_{i+1/2,j} + \frac{1}{\mu_0} \frac{1}{1+\dfrac{\sigma_x^e \Delta t}{2\varepsilon_0}} \\ &\times \left[\left(1+\frac{\sigma_y^e \Delta t}{2\varepsilon_0}\right) {}^{\text{scat}}_{\phi y}B^{n+1/2}_{i+1/2,j} - \left(1-\frac{\sigma_x^e \Delta t}{2\varepsilon_0}\right) {}^{\text{scat}}_{\phi y}B^{n-1/2}_{i+1/2,j} \right] \end{aligned} \tag{11.32b}$$

where, as shown in Figure 11.1, there are four face-APML and four corner-APML regions. In the non-APML regions, we can simply set $\sigma_x^e = \sigma_y^e = 0$ in the above update equations, while inside an APML region, we need to set σ_α^e ($\alpha = x, y$), in the form given in (11.15), along the direction of wave propagation.

Plane Wave Excitation

We use a Gaussian pulse to excite a one-dimensional MRTD (1D-MRTD) time domain plane wave along a specified direction of wave propagation (ϕ_i) and derive all incident fields inside the target through a linear interpolation of the excited one-dimensional fields. The initial incident fields, induced by the Gaussian plane wave in free space, are given by

$$E^{\text{inc}(0)}(i_\eta) = E_0 \exp\left(\gamma \frac{(i_\eta - i_c)^2}{i_p^2}\right) \tag{11.33a}$$

$$H^{\text{inc}(1/2)}(i_\eta) = \frac{E_0}{Z_0} \exp\left(\gamma \frac{(i_\eta - i_c - 0.5c\ \Delta t/\Delta\eta)^2}{i_p^2}\right) \tag{11.33b}$$

where E_0 is the amplitude of the Gaussian pulse, which has an initial peak at grid i_c; i_p denotes the pulse width; γ is a constant; Z_0 is the intrinsic impedance of free space; c is the wave velocity; and $\hat{\eta}$ is the specified incident direction. Figure 11.9 shows that a normalized plane wave propagates along the direction $\phi_i = 90^\circ$.

The incident fields inside the scatterer, that is, the target region, can be obtained by linearly interpolating the 1D-MRTD incident fields ($\vec{H}^{\text{inc}}$, $\vec{E}^{\text{inc}}$). From the unit wave-vector expression

$$\hat{k}^{\text{inc}} = k_x^{\text{inc}}\hat{x} + k_y^{\text{inc}}\hat{y} = \sin\phi_i\hat{x} + \sin\phi_i\,\hat{y} \tag{11.34}$$

we obtain its projection on the scattered target, which is specified by $\vec{r}$. The distance d is therefore given as

$$d = \hat{k}^{\text{inc}} \cdot \vec{r} = (\cos\phi_i\hat{x} + \sin\phi_i\,\hat{y}) \cdot (i_c\hat{x} + j_c\hat{y}) = i_c\cos\phi_i + j_c\sin\phi_i \tag{11.35}$$

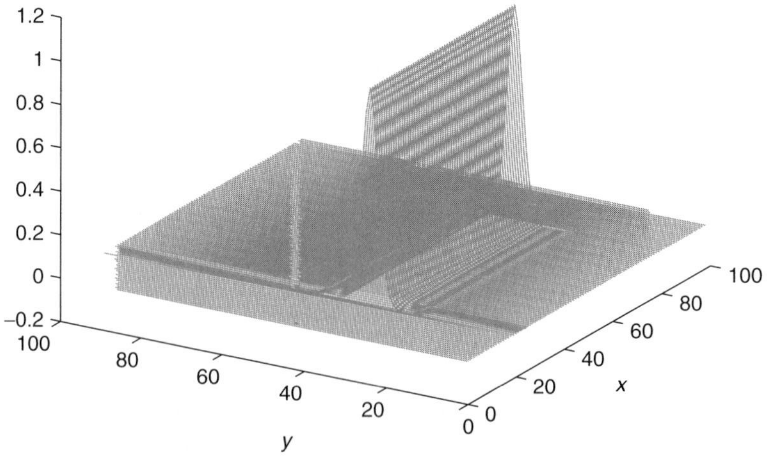

FIGURE 11.9 Gaussian incident plane wave.

where the (i_c, j_c) are the coordinates of the point of interest in the target shown in Figure 11.10. For the E-fields, the values of (i_c, j_c) are integers located at the node of a cell, and for the H-fields the values (i_c, j_c) are half-integers located at the half-node of a cell.

Let the plane wave be represented by $(E^{\text{inc}}, H^{\text{inc}})$, the electromagnetic fields at the point d be $(E^{\text{inc}}(d), H^{\text{inc}}(d))$, and let the notation int(d) denote the largest integer of the real number d. The distance d $(d > 0)$ can be discretized and a linear interpolation used to determine the electromagnetic fields at an arbitrary distance. For example, the E-field at d can be written as

$$E^{\text{inc}}(d) = [1 - d + \text{int}(d)] \cdot E^{\text{inc}}(\text{int}(d)) + [d - \text{int}(d)] \cdot E^{\text{inc}}(\text{int}(d) + 1) \tag{11.36}$$

where int(x) is the integer part of x shown in Figure 11.11.

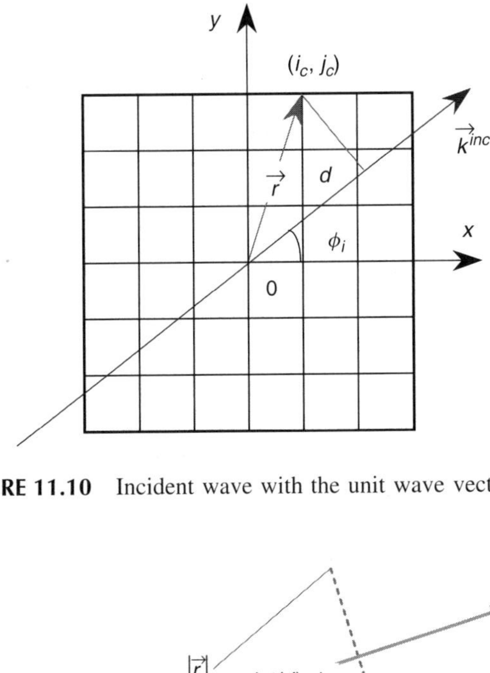

FIGURE 11.10 Incident wave with the unit wave vector $\vec{k}^{\text{inc}}$.

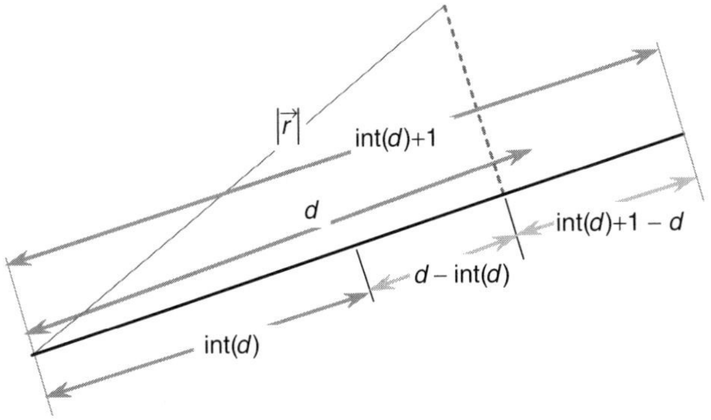

FIGURE 11.11 Explanation of linear interpolation.

Finally, we obtain the incident fields in the Cartesian system by using those defined in the polar system by employing the equations

$$E_x^{\text{inc}} = -E^{\text{inc}} \sin \phi_i \tag{11.37a}$$

$$E_y^{\text{inc}} = E^{\text{inc}} \cos \phi_i \tag{11.37b}$$

$$H_z^{\text{inc}} = H^{\text{inc}} \tag{11.37c}$$

for the TE wave, and

$$H_x^{\text{inc}} = H^{\text{inc}} \sin \phi_i \tag{11.38a}$$

$$H_y^{\text{inc}} = -H^{\text{inc}} \cos \phi_i \tag{11.38b}$$

$$E_z^{\text{inc}} = E^{\text{inc}} \tag{11.38c}$$

for the TM wave.

MRTD Near-to-Far-Zone Transform for TM$_z$ Wave

To derive the near-to-far-zone field transforms in the MRTD procedure, we use an approach similar to those employed in the FDTD and PSTD applications [13, 14]. In the two-dimensional case, the vector potentials in the frequency domain can be written as

$$\vec{A}(\vec{\rho}) = \frac{1}{4j} \int_S \vec{J}_s(\vec{\rho}\,')H_0^{(2)}(k|\vec{\rho} - \vec{\rho}\,'|)\, ds \tag{11.39a}$$

$$\vec{F}(\vec{\rho}) = \frac{1}{4j} \int_S \vec{M}_s(\vec{\rho}\,')H_0^{(2)}(k|\vec{\rho} - \vec{\rho}\,'|)\, ds \tag{11.39b}$$

where $H_0^{(2)}(k|\vec{\rho} - \vec{\rho}\,'|)$ is the *Hankel* function of the second kind and k is the wave number in free space. In the far-zone region, the vector magnetic and electric potentials are approximated as

$$\vec{A}(\vec{\rho}) = \sqrt{\frac{1}{8j\pi k\rho}} e^{-j\omega\rho/c} \int_S \vec{J}_s(\vec{\rho}\,')H_0^{(2)} e^{jk\vec{\rho}\,'\cdot\hat{\rho}}\, ds \tag{11.40a}$$

$$\vec{F}(\vec{\rho}) = \sqrt{\frac{1}{8j\pi k\rho}} e^{-j\omega\rho/c} \int_S \vec{M}_s(\vec{\rho}\,')H_0^{(2)} e^{jk\vec{\rho}\,'\cdot\hat{\rho}}\, ds \tag{11.40b}$$

The far-field distribution can be obtained from the induced currents J_z as follows:

$$E_z^{\text{far}} \approx \sqrt{\frac{1}{8j\pi k\rho}} e^{-j\omega\rho/c} \int_c J_z(x', y')e^{jk(x'\cos\phi + y'\sin\phi)}\, dl' \tag{11.41}$$

That is,

$$E_z^{\rm far} = j\omega\mu_0 A_z + jk(F_y \cos\phi_s - F_x \sin\phi_s) \tag{11.42}$$

where ϕ_s is the direction of the scattering angle, and the far-zone magnetic and electric vector potentials are given by

$$\vec{A}(\vec{\rho}) = \sqrt{\frac{1}{8j\pi k\rho}} e^{-j\omega\rho/c} \int_{S'} \vec{J}_s(\vec{\rho}\,') e^{jk\vec{\rho}\,'\cdot\hat{\rho}}\, ds' \tag{11.43a}$$

$$\vec{F}(\vec{\rho}) = \sqrt{\frac{1}{8j\pi k\rho}} e^{-j\omega\rho/c} \int_{S'} \vec{M}_s(\vec{\rho}\,') e^{jk\vec{\rho}\,'\cdot\hat{\rho}}\, ds' \tag{11.43b}$$

For computational convenience, we rewrite (11.42) as

$$E_z^{\rm far} = \sqrt{\frac{2\pi c}{j\omega}} \left\{-\eta_0 A'_z(\vec{\rho}) + \left[F'_y(\rho)\cos\phi_s - F'_x(\vec{\rho})\sin\phi_s\right]\right\} \tag{11.44}$$

with a pair of redefined potentials as follows:

$$A'_z(\vec{\rho}) = \frac{j\omega}{4\pi c\sqrt{\rho}} e^{-j\omega\rho/c} \int_{S'} J_{sz}(\vec{\rho}\,') e^{jk\vec{\rho}\,'\cdot\hat{\rho}}\, ds' \tag{11.45a}$$

$$F'_{x,y}(\vec{\rho}) = \frac{j\omega}{4\pi c\sqrt{\rho}} e^{-j\omega\rho/c} \int_{S'} M_{sx,sy}(\vec{\rho}\,') e^{jk\vec{\rho}\,'\cdot\hat{\rho}}\, ds' \tag{11.45b}$$

We can then obtain their time domain versions by taking the inverse Fourier transform:

$$A'_z(\vec{\rho}, t) = \frac{1}{4\pi c\sqrt{\rho}} \frac{\partial}{\partial t} \int_{S} J_{sz}(t-\tau)\, ds \tag{11.46a}$$

$$F'_{x,y}(\vec{\rho}, t) = \frac{1}{4\pi c\sqrt{\rho}} \frac{\partial}{\partial t} \left\{ \int_{S'} M_{sx,sy}(t-\tau)\, ds' \right\} \tag{11.46b}$$

where $\tau = \rho/c - (\vec{\rho}\,' \cdot \hat{\rho})/c$, and J_{sz} and $M_{sx,sy}$ are the equivalent electric and magnetic current densities on an artificially enclosed boundary contour S' defined in the near-zone region:

$$\vec{J}_s = \hat{n} \times \vec{H}^{\rm scat} \tag{11.47a}$$

$$M_s = -\hat{n} \times \vec{E}^{\rm scat} \tag{11.47b}$$

where $\hat{n}$ is the unit vector normally outward from S' as shown in Figure 11.12.

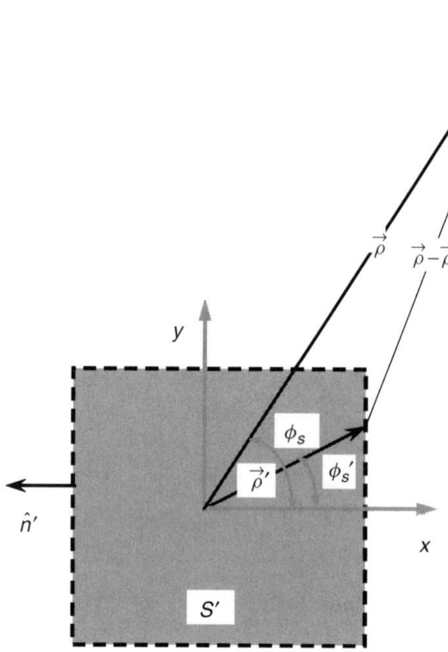

FIGURE 11.12 Geometry of the MRTD near-to-far-zone field transformation in the two-dimensional case.

Note that both $\vec{H}^{\text{scat}}$ and $\vec{E}^{\text{scat}}$ in (11.47) are defined as the scattered fields, as opposed to the expansion coefficients. To construct the scattered fields at the node (i_0, j_0), we must include the contributions from the node itself, as well as those from its neighboring nodes within a compact region. For example, we can construct the total field E_z^{scat} at (i_0, j_0, t_0) from the expansion

$$
\begin{aligned}
E_z^{\text{scat}}(i_0, j_0, t_0) &= \iiint_{\infty} E_z^{\text{scat}}(x, y, t)\delta(x - x_0)\delta(y - y_0)\delta(t - t_0)\, dx\, dy\, dt \\
&= \sum_{i,j=-\infty}^{\infty} {}_{\phi z}^{\text{scat}}E_{i,j}^{n}\,\phi_i(x_0)\phi_j(y_0) \\
&= \sum_{i,j=-\infty}^{\infty} {}_{\phi z}^{\text{scat}}E_{i,j}^{n}\,\phi(i - i_0)\phi(j - j_0) \\
&\approx \sum_{i,j=-m_x,m_y}^{m_x,m_y} {}_{\phi z}^{\text{scat}}E_{i,j}^{n}\,\phi(i - i_0)\phi(j - j_0)
\end{aligned}
\qquad (11.48)
$$

where we have chosen both the finite integers m_x and m_y to be 9 for the cubic spline Battle–Lemarié wavelet basis to ensure computational accuracy.

Finally, we extract the scattering width (SW) for the TM$_z$ scattering wave from

$$\text{RCS} = \lim_{\rho\to\infty}\left(2\pi\rho\frac{|E_z^{\text{far}}|^2}{|E^{\text{inc}}|^2}\right) \tag{11.49}$$

11.5 MRTD SCATTERING ALGORITHM FOR TE$_Z$ WAVE

MRTD Update Equations Inside the Target

As in the TM$_z$ case, we again expand all of the scattered field quantities for the TE$_z$ case in terms of the scaling function in space and pulse function in time [3]:

$$E_z^{\text{scat}}(x,y,t) = \sum_{n,i,j=-\infty}^{+\infty} {}_{\phi x}^{z}E_{i+1/2,j}^{n}\phi_{i+1/2}(x)\phi_j(y)(x)h_n(t) \tag{11.50a}$$

$$E_y^{\text{scat}}(x,y,t) = \sum_{n,i,j=-\infty}^{+\infty} {}_{\phi x}^{\text{scat}}E_{i,j+1/2}^{n}\phi_i(x)\phi_{j+1/2}(y)(x)h_n(t) \tag{11.50b}$$

$$H_z^{\text{scat}}(x,y,t) = \sum_{n,i,j=-\infty}^{+\infty} {}_{\phi z}^{\text{scat}}H_{i+1/2,j+1/2}^{n+1/2}\phi_{i+1/2}(x)\phi_{j+1/2}(y)h_{n+1/2}(t) \tag{11.50c}$$

A substitution of the above MRTD field expansions into the time domain Maxwell equations leads to the update equations for the H-field:

$$\begin{aligned}{}_{\phi z}^{\text{scat}}H_{i+1/2,j+1/2}^{n+1/2} &= {}_{\phi z}^{\text{scat}}H_{i+1/2,j+1/2}^{n-1/2} + \frac{1}{\mu_0}\sum_{\nu}a(\nu)\\ &\times\left({}_{\phi x}^{\text{scat}}E_{i,j+\nu+1}^{n}\frac{\Delta t}{\Delta x} - {}_{\phi y}^{\text{scat}}E_{i,j+\nu+1}^{n}\frac{\Delta t}{\Delta y}\right)\end{aligned} \tag{11.51}$$

For a lossless dielectric target ($\sigma = 0$), we can use (11.12) to write the time domain equations as

$$\varepsilon\frac{\partial E_x^{\text{scat}}}{\partial t} = -(\varepsilon-\varepsilon_0)\frac{\partial E_x^{\text{inc}}}{\partial t} + \frac{\partial H_z^{\text{scat}}}{\partial y} \tag{11.52a}$$

$$\varepsilon\frac{\partial E_y^{\text{scat}}}{\partial t} = -(\varepsilon-\varepsilon_0)\frac{\partial E_y^{\text{inc}}}{\partial t} + \frac{\partial H_z^{\text{scat}}}{\partial x} \tag{11.52b}$$

Similarly, we obtain two E-field update equations:

$$\begin{aligned}{}_{\phi x}^{\text{scat}}E_{i+1/2,j}^{n+1} &= {}_{\phi x}^{\text{scat}}E_{i+1/2,j}^{n} - \frac{1}{\left[1+\sum\limits_{\kappa=1}^{N}(\varepsilon_r^{\kappa}-1)\alpha_{i,i}^{\kappa}\beta_{j,j}^{\kappa}\right]}\\ &\times\sum_{i'=i-\Lambda}^{i'=i+\Lambda}\sum_{j'=j-\Lambda}^{j'=j+\Lambda}\sum_{\kappa=1}^{N}(\varepsilon_r^{\kappa}-1)\alpha_{i,i'}^{\kappa}\beta_{j,j'}^{\kappa}\left({}_{\phi x}^{\text{inc}}E_{i'+1/2,j'}^{n+1} - {}_{\phi x}^{\text{inc}}E_{i'+1/2,j'}^{n}\right)\end{aligned} \tag{11.53}$$

$$+\frac{1}{\varepsilon_0\left[1+\sum_{\kappa=1}^{N}(\varepsilon_r^\kappa-1)\alpha_{i,i}^\kappa\beta_{j,j}^\kappa\right]}\sum_\nu \alpha(\nu)\,{}^{\text{scat}}_{\phi z}H^{n+1/2}_{i+\nu+1/2,j+1/2}\frac{\Delta t}{\Delta x}$$

$$\begin{aligned}{}^{\text{scat}}_{\phi y}E^{n+1}_{i,j+1/2} &= {}^{\text{scat}}_{\phi y}E^{n}_{i,j+1/2}-\frac{1}{\left[1+\sum_{\kappa=1}^{N}(\varepsilon_r^\kappa-1)\alpha_{i,i}^\kappa\beta_{j,j}^\kappa\right]}\\ &\times\sum_{i'=i-\Lambda}^{i'=i+\Lambda}\sum_{j'=j-\Lambda}^{j'=j+\Lambda}\sum_{\kappa=1}^{N}(\varepsilon_r^\kappa-1)\alpha_{i,i'}^\kappa\beta_{j,j'}^\kappa\left({}^{\text{inc}}_{\phi y}E^{n+1}_{i',j'+1/2}-{}^{\text{inc}}_{\phi y}E^{n}_{i',j'+1/2}\right)\\ &+\frac{1}{\varepsilon_0\left[1+\sum_{\kappa=1}^{N}(\varepsilon_r^\kappa-1)\alpha_{i,i}^\kappa\beta_{j,j}^\kappa\right]}\sum_\nu \alpha(\nu)\,{}^{\text{scat}}_{\phi z}H^{n+1/2}_{i+1/2,j+\nu+1/2}\frac{\Delta t}{\Delta y}\end{aligned} \tag{11.54}$$

For a lossy dielectric target, we have $\sigma \neq 0$ in (11.12), and the corresponding update equations are given by

$$\begin{aligned}{}^{\text{scat}}_{\phi x}E^{n+1}_{i+1/2,j} &= \frac{C_-}{C_+}\,{}^{\text{scat}}_{\phi x}E^{n}_{i+1/2,j}\\ &-\frac{C_0}{C_+}\sum_{i'=i-\Lambda}^{i'=i+\Lambda}\sum_{j'=j-\Lambda}^{j'=j+\Lambda}\sum_{\kappa=1}^{N}\alpha_{i,i'}^\kappa\beta_{j,j'}^\kappa\left({}^{\text{inc}}_{\phi x}E^{n+1}_{i'+1/2,j'}+{}^{\text{inc}}_{\phi x}E^{n}_{i'+1/2,j'}\right)\\ &-\frac{1}{C_+}\sum_{i'=i-\Lambda}^{i'=i+\Lambda}\sum_{j'=j-\Lambda}^{j'=j+\Lambda}\sum_{\kappa=1}^{N}(\varepsilon_r^\kappa-1)\alpha_{i,i'}^\kappa\beta_{j,j'}^\kappa\left({}^{\text{inc}}_{\phi x}E^{n+1}_{i'+1/2,j'}-{}^{\text{inc}}_{\phi x}E^{n}_{i'+1/2,j'}\right)\\ &+\frac{1}{C_+}\frac{1}{\varepsilon_0}\sum_\nu \alpha(\nu)\,{}^{\text{scat}}_{\phi z}H^{n+1/2}_{i+\nu+1/2,j+1/2}\frac{\Delta t}{\Delta x}\end{aligned} \tag{11.55a}$$

where $\kappa = 0, 1, 2, \ldots, N$ is the number of objects in the computational volume and the coefficients $C_\pm$ are defined as

$$C_\pm = \sum_{\kappa=1}^{N}\left((\varepsilon_r^\kappa - 1) \pm \frac{\sigma(\kappa)\Delta t}{2\varepsilon_0}\right) \tag{11.55b}$$

In a similar way, we derive

$$\begin{aligned}{}^{\text{scat}}_{\phi y}E^{n+1}_{i,j+1/2} &= \frac{C_-}{C_+}\,{}^{\text{scat}}_{\phi y}E^{n}_{i,j+1/2}\\ &-\frac{C_0}{C_+}\sum_{i'=i-\Lambda}^{i'=i+\Lambda}\sum_{j'=j-\Lambda}^{j'=j+\Lambda}\sum_{\kappa=1}^{N}\alpha_{i,i'}^\kappa\beta_{j,j'}^\kappa\left({}^{\text{inc}}_{\phi y}E^{n+1}_{i',j'+1/2}+{}^{\text{inc}}_{\phi y}E^{n}_{i',j'+1/2}\right)\\ &-\frac{1}{C_+}\sum_{i'=i-\Lambda}^{i'=i+\Lambda}\sum_{j'=j-\Lambda}^{j'=j+\Lambda}\sum_{\kappa=1}^{N}\left(\varepsilon_r^\kappa-1\right)\alpha_{i,i'}^\kappa\beta_{j,j'}^\kappa\left({}^{\text{inc}}_{\phi x}E^{n+1}_{i'+1/2,j'}-{}^{\text{inc}}_{\phi x}E^{n}_{i'+1/2,j'}\right)\end{aligned}$$

$$+\frac{1}{C_+}\frac{1}{\varepsilon_0}\sum_{v} a(v)\,{}^{\text{scat}}_{\phi z}H^{n+1/2}_{i+1/2,j+v+1/2}\frac{\Delta t}{\Delta y} \tag{11.56}$$

where the same diagonal approximation relations have been used.

MRTD Update Equations Outside the Target

As before, in regions outside the target, the computational domain consists of both the free space and APML regions. Here, we use a set of unified governing equations to include these regions:

$$\mu_0 s_x s_y s_z^{-1}\frac{\partial H_z^{\text{scat}}}{\partial t} = -\left(\frac{\partial E_y^{\text{scat}}}{\partial x} - \frac{\partial E_x^{\text{scat}}}{\partial y}\right) \tag{11.57a}$$

$$\varepsilon_0 s_x^{-1} s_y s_z \frac{\partial E_x^{\text{scat}}}{\partial t} = \frac{\partial H_z^{\text{scat}}}{\partial y} \tag{11.57b}$$

$$\varepsilon_0 s_x s_y^{-1} s_z \frac{\partial E_y^{\text{scat}}}{\partial t} = -\frac{\partial H_z^{\text{scat}}}{\partial x} \tag{11.57c}$$

Similar to the TM$_z$ case, the MRTD update equations in regions outside the target can be expressed as

$$
{}^{\text{scat}}_{\phi z}B^{n+1/2}_{i+1/2,j+1/2} = \frac{1-\dfrac{\sigma_x\Delta t}{2\varepsilon_0}}{1+\dfrac{\sigma_x\Delta t}{2\varepsilon_0}}\,{}^{\text{scat}}_{\phi z}B^{n-1/2}_{i+1/2,j+1/2} - \frac{1}{1+\dfrac{\sigma_x\Delta t}{2\varepsilon_0}}
$$

$$
\times \sum_{v} a(v)\left({}^{\text{scat}}_{\phi y}E^{n+1/2}_{i+v,j+1/2}\frac{\Delta t}{\Delta y} - {}^{\text{scat}}_{\phi x}E^{n+1/2}_{i+\frac{1}{2},j+v}\frac{\Delta t}{\Delta x}\right) \tag{11.58a}
$$

$$
{}^{\text{scat}}_{\phi z}H^{n+1/2}_{i+1/2,j+1/2} = \frac{1-\dfrac{\sigma_y\Delta t}{2\varepsilon_0}}{1+\dfrac{\sigma_y\Delta t}{2\varepsilon_0}}\,{}^{\text{scat}}_{\phi z}H^{n-1/2}_{i+1/2,j+1/2} + \frac{1}{\mu_0}\frac{1}{1+\dfrac{\sigma_y\Delta t}{2\varepsilon_0}}
$$

$$
\times\left[{}^{\text{scat}}_{\phi z}B^{n+1}_{i+1/2,j+1/2} - {}^{\text{scat}}_{\phi z}B^{n}_{i+1/2,j+1/2}\right] \tag{11.58b}
$$

$$
{}^{\text{scat}}_{\phi x}D^{n+1}_{i+1/2,j} = {}^{\text{scat}}_{\phi x}D^{n}_{i+1/2,j} + \sum_{v} a(v)\,{}^{\text{scat}}_{\phi z}H^{n+1/2}_{i+1/2,j+v+1/2}\frac{\Delta t}{\Delta x} \tag{11.58c}
$$

$$
{}^{\text{scat}}_{\phi x}E^{n+1}_{i+1/2,j} = \frac{1-\dfrac{\sigma_x\Delta t}{2\varepsilon_0}}{1+\dfrac{\sigma_x\Delta t}{2\varepsilon_0}}\,{}^{\text{scat}}_{\phi x}E^{n}_{i+1/2,j} + \frac{1}{\varepsilon_0}\frac{1}{1+\dfrac{\sigma_x\Delta t}{2\varepsilon_0}}
$$

$$
\times\left[\left(1+\frac{\sigma_y\Delta t}{2\varepsilon_0}\right){}^{\text{scat}}_{\phi x}D^{n+1}_{i+1/2,j} - \left(1-\frac{\sigma_y\Delta t}{2\varepsilon_0}\right){}^{\text{scat}}_{\phi x}D^{n}_{i+1/2,j}\right] \tag{11.58d}
$$

$$ {}^{\text{scat}}_{\phi y}D^{n+1}_{i,j+1/2} = {}^{\text{scat}}_{\phi y}D^{n}_{i,j+1/2} - \sum_{v} a(v)\,{}^{\text{scat}}_{\phi z}H^{n}_{i+v+1/2,j+1/2}\frac{\Delta t}{\Delta y} \tag{11.58e} $$

$$ {}^{\text{scat}}_{\phi y}E^{n+1}_{i,j+1/2} = \frac{1-\dfrac{\sigma_x \Delta t}{2\varepsilon_0}}{1+\dfrac{\sigma_x \Delta t}{2\varepsilon_0}}\,{}^{\text{scat}}_{\phi y}D^{n}_{i,j+1/2} + \frac{1}{\varepsilon_0}\frac{1}{1+\dfrac{\sigma_x \Delta t}{2\varepsilon_0}} $$

$$ \times\left[\left(1+\frac{\sigma_y \Delta t}{2\varepsilon_0}\right){}^{\text{scat}}_{\phi y}D^{n+1}_{i,j+1/2} - \left(1-\frac{\sigma_y \Delta t}{2\varepsilon_0}\right){}^{\text{scat}}_{\phi y}D^{n}_{i,j+1/2}\right] \tag{11.58f} $$

MRTD Near-to-Far-Zone Transform for TE$_z$ Wave

Similar to the TM$_z$ case, we again first express the far-zone scattered fields in the frequency domain for the TE$_z$ case as follows:

$$ H_z^{\text{scat}} = -j\omega\varepsilon_0 F_z + jk(A_x \sin\varphi_s - A_y \cos\varphi_s) \tag{11.59} $$

In the time domain, we rewrite (11.59) as

$$ H_z^{\text{scat}} = \sqrt{\frac{2\pi c}{j\omega}}\left\{-\frac{1}{\eta_0}F_z'(\rho) + [A_x'(\vec{\rho})\sin\varphi_s - A_y'(\vec{\rho})\cos\varphi_s]\right\} \tag{11.60} $$

in the direction of the scattering angle ϕ_s, where the far-zone magnetic and electric vector potentials are given by

$$ A_x'(\vec{\rho}) = \frac{j\omega}{4\pi c\sqrt{\rho}}e^{-j\omega\rho/c}\int_S J_{sx}(\vec{\rho}\,')e^{jk\vec{\rho}\,'\cdot\hat{\rho}}\,ds \tag{11.61a} $$

$$ A_y'(\vec{\rho}) = \frac{j\omega}{4\pi c\sqrt{\rho}}e^{-j\omega\rho/c}\int_S J_{sy}(\vec{\rho}\,')e^{jk\vec{\rho}\,'\cdot\hat{\rho}}\,ds \tag{11.61b} $$

$$ F_z'(\vec{\rho}) = \frac{j\omega}{4\pi c\sqrt{\rho}}e^{-j\omega\rho/c}\int_S M_{sz}(\vec{\rho}\,')e^{jk\vec{\rho}\,'\cdot\hat{\rho}}\,ds \tag{11.61c} $$

We can obtain their time domain versions by taking the inverse Fourier transform, yielding

$$ A_x'(\vec{\rho},t) = \frac{1}{4\pi c\sqrt{\rho}}\frac{\partial}{\partial t}\left\{\int_S J_{sx}\left(t-\frac{\rho-\vec{\rho}\,'\cdot\hat{\rho}}{c}\right)ds\right\} \tag{11.62a} $$

$$ A_y'(\vec{\rho},t) = \frac{1}{4\pi c\sqrt{\rho}}\frac{\partial}{\partial t}\left\{\int_S J_{sy}\left(t-\frac{\rho-\vec{\rho}\,'\cdot\hat{\rho}}{c}\right)ds\right\} \tag{11.62b} $$

$$F'_z(\vec{\rho}, t) = \frac{1}{4\pi c\sqrt{\rho}} \frac{\partial}{\partial t} \left\{ \int_S M_{sz} \left(t - \frac{\rho - \vec{\rho}' \cdot \widehat{\rho}}{c} \right) ds \right\} \tag{11.62c}$$

Finally, we can extract the SW for the TE_z scattering wave by using

$$\text{RCS} = \lim_{\rho \to \infty} \left(2\pi\rho \frac{|H_z^{\text{scat}}|^2}{|H^{\text{inc}}|^2} \right) \tag{11.63}$$

11.6 APPLICATION RESULTS

As in the first example, we analyze a square PEC cylinder with a TM_z polarized plane wave incident along the $\phi_i = 90^\circ$ direction, where the dimension of the PEC cylinder is set to be $k_0 A_s = 1$ and A_s is half the width of the side of the cylinder. The entire computational domain is discretized into 42×42 cells, including a six-layer APML in all directions, while the PEC cylindrical target is discretized by using 20×20 cells. As shown in Figure 11.13, the magnitude of the normalized surface electric current derived from the MRTD scheme using the Battle–Lemarié wavelet basis shows good agreement with that obtained using the FDTD method [1]. Figure 11.14 shows that we obtain good agreement between these two methods for the SW, with the scattering angle ranging from 90° to 270°.

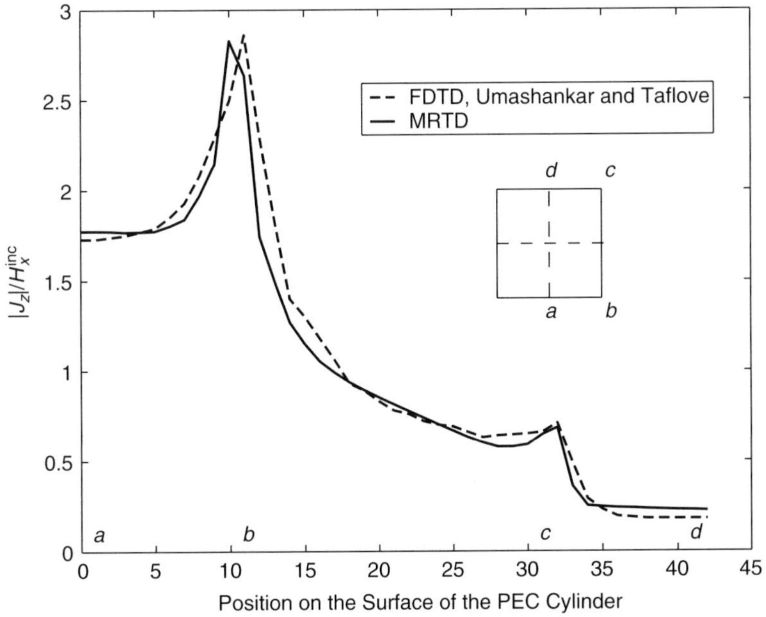

FIGURE 11.13 Normalized electric current for TM_z wave scattering for a square PEC cylinder ($\phi_i = 90^\circ$, $f = 1$ GHz).

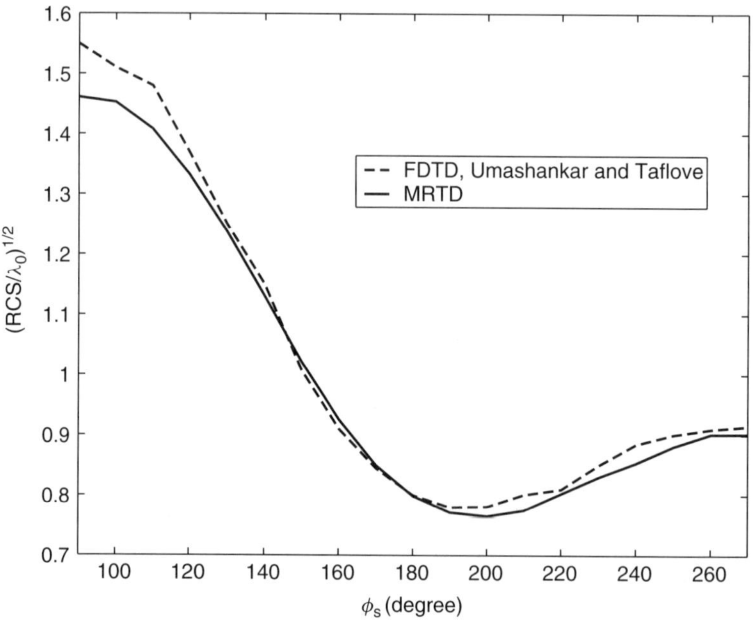

FIGURE 11.14 SW of TM_z wave scattering for a square PEC cylinder ($\phi_i = 90°$, $f = 1$ GHz).

TABLE 11.2 Discretization of Square PEC and Dielectric Cylinders Using MRTD Scheme and FDTD Method

Target / Discretization	Dielectric Cylinder (0.18×0.18 m^2, $\varepsilon_r = 4$)		PEC Cylinder (0.6×0.6 m^2)	
	MRTD	FDTD	MRTD	FDTD
$\Delta x = \Delta y$ (m)	0.0075	0.0015	0.01	0.003
Δt (10^{-12} s)	5.52	2.5	8.6	5
Target dimensions (cells)	24×24	120×120	60×60	200×200
Computational space (cells)	64×64	160×160	100×100	240×240
Total time step	4000	4000	2000	2000
APML (cells)	10	10	10	10

Next, we investigate the SW for a square dielectric cylinder with a plane wave incidence. Table 11.2 lists the dimensions of the target and the discretization used in the MRTD scheme based on the Battle–Lemarié wavelet basis and the FDTD method [14]. As seen from Figures 11.15–11.18 for the TE_z and TM_z waves, respectively, there is good agreement in the scattering width for both frequency responses at a specified angle; also, the entire scattering spans angles at a specified frequency. However, the computational space required in the MRTD scheme is only about 16% of that employed in the FDTD approach.

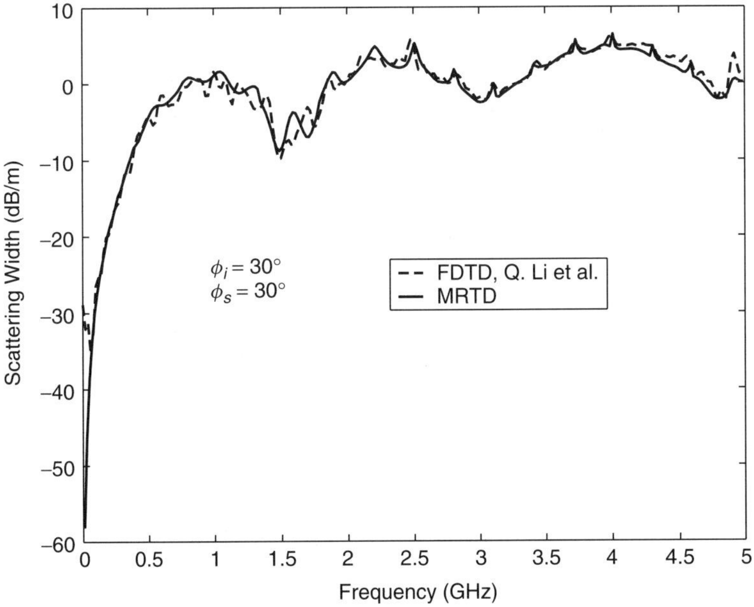

FIGURE 11.15 SW as a function of frequency ($\phi_i = 30°, \phi_s = 30°$) for a TE_z wave incident on a rectangular dielectric cylinder.

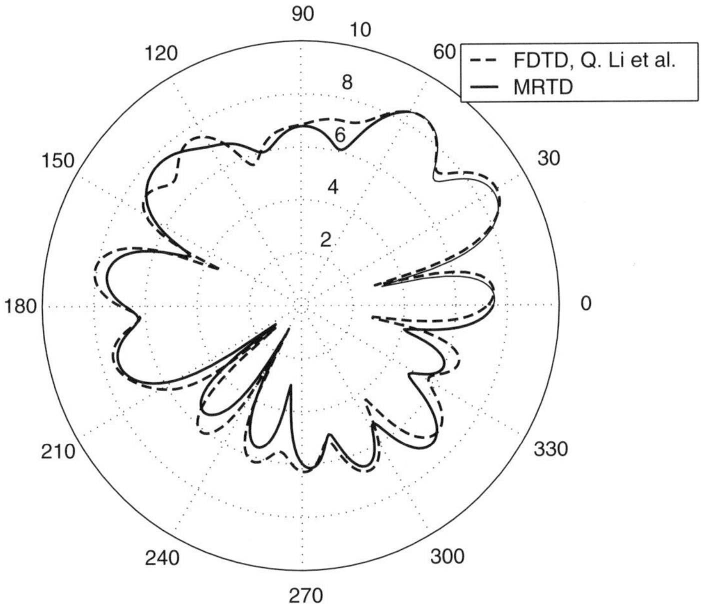

FIGURE 11.16 SW as a function of scattering angle ($\phi_i = 30°$, $f = 3.13$ GHz) for a TE_z wave incident on a rectangular dielectric cylinder.

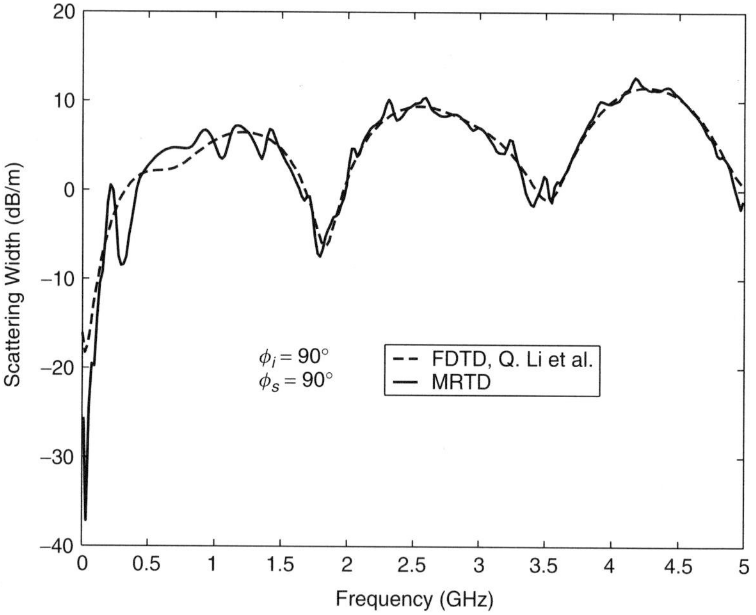

FIGURE 11.17 SW as a function of frequency ($\phi_i = 30^\circ, \phi_s = 30^\circ$) for a TM_z wave incident on a rectangular dielectric cylinder.

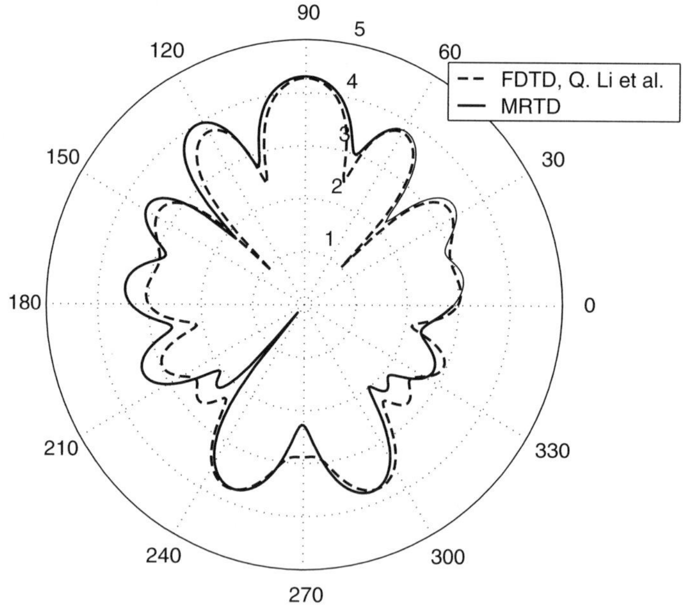

FIGURE 11.18 SW as a function of scattering angle ($\phi_i = 30^\circ$, $f = 3.13$ GHz) for a TM_z wave incident on a rectangular dielectric cylinder.

We further investigate the differences in the MRTD results for the plane wave scattering width of a square dielectric cylinder derived by using different wavelet basis functions. The dimensions of the target are $0.18 \times 0.18\ \mathrm{m}^2$ and the relative permittivity is $\varepsilon_r = 4$. The number of APMLs used is 10, the time step used is $\Delta t = 2.5 \times 10^{-12}$ s, and the total number of steps used is 4000 for the MRTD scheme as well as for the FDTD method. Figure 11.19 compares the SW results for the cylinder, computed via the MRTD scheme using different wavelet bases as well as with the FDTD method for the TM_z mode. The results of the MRTD scheme are seen to be in good agreement with those derived from the FDTD for a specified frequency and angle. However, the computational space for the MRTD scheme is $60\ \Delta x \times 60\ \Delta y$ and is only about 14.1% of that employed in the FDTD, which is $160\ \Delta x \times 160\ \Delta y$. Figures 11.20–11.24 plot the SW results for various cell sizes of different MRTD wavelet bases for comparison.

In order to compare the accuracy of the computed SW of the MRTD scheme based on different wavelet bases, we use the measure of the relative peak error defined by $\Sigma_p = [\sigma_{\mathrm{FDTD}}(n) - \sigma_{\mathrm{MRTD}}(n)]/\sigma_{\mathrm{FDTD}}(n)$, where $\sigma(n)$ is the value of the SW at the specified nth peak of the curve, as well as the relative shifting position error $\Sigma_s = (f_0 - f_{\mathrm{MRTD}})/f_0$, where f_0 and f_{MRTD} are the corresponding frequency values of the FDTD method and the MRTD scheme on the horizontal axis for the nth peak in the SW curve, respectively. Table 11.3 lists the two relative errors, in which the relative peak errors are for the third peak curve and the shifting position errors are for the specified frequency position.

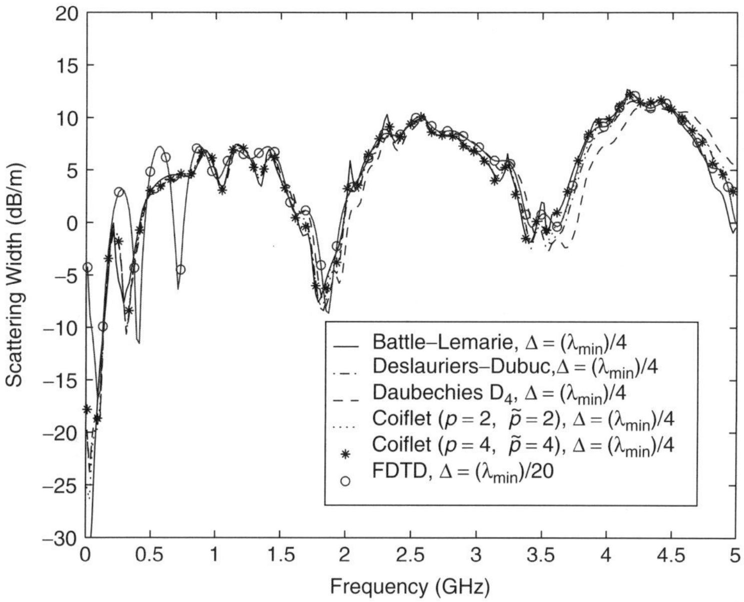

FIGURE 11.19 SW as a function of frequency ($\phi_i = 30^\circ, \phi_s = 30^\circ$) for a TM_z wave incident on a square dielectric cylinder.

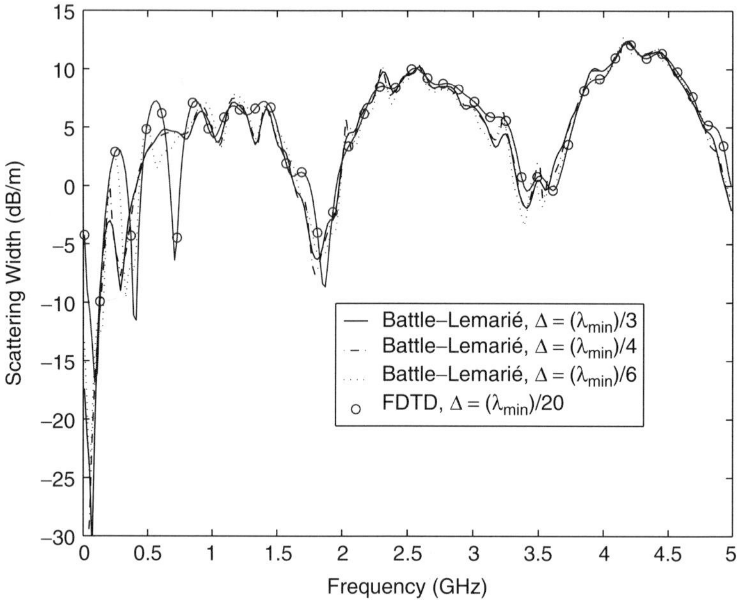

FIGURE 11.20 SW as a function of frequency ($\phi_i = 30^\circ, \phi_s = 30^\circ$) for a TM_z wave incident on a square dielectric cylinder using the cubic spline Battle–Lemarié basis.

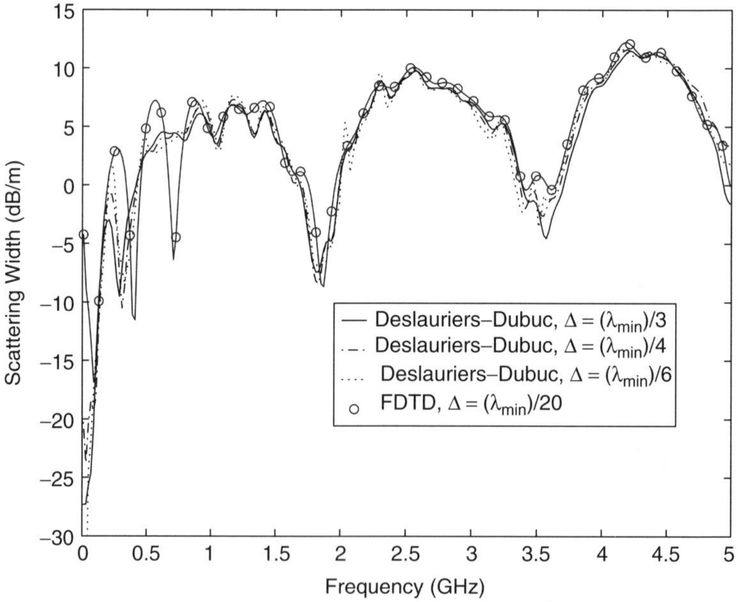

FIGURE 11.21 SW as a function of frequency ($\phi_i = 30^\circ, \phi_s = 30^\circ$) for a TM_z wave incident on a square dielectric cylinder using the Deslauriers–Dubuc basis.

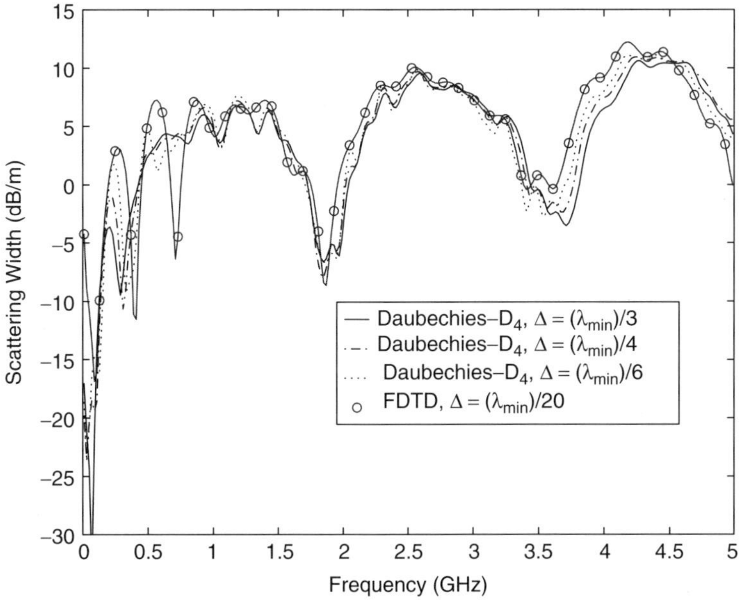

FIGURE 11.22 SW as a function of frequency ($\phi_i = 30^\circ, \phi_s = 30^\circ$) for a TM_z wave incident on a square dielectric cylinder using the Daubechies D_4 basis.

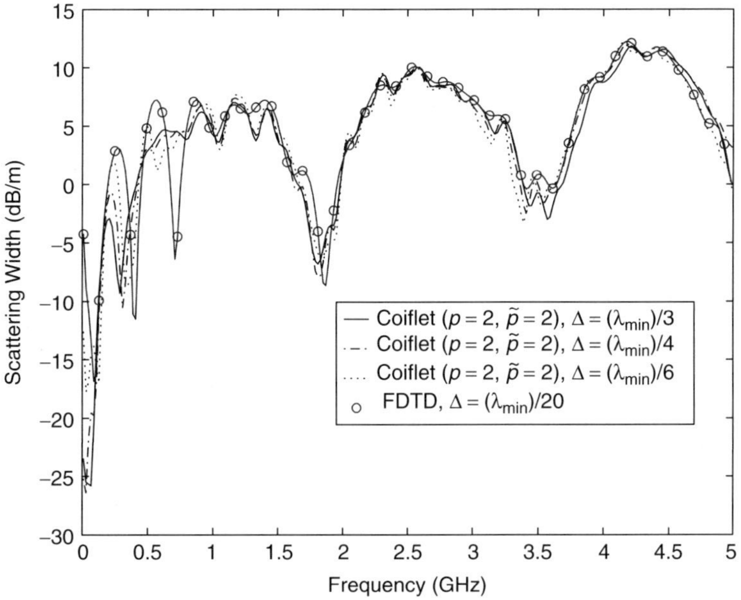

FIGURE 11.23 SW as a function of frequency ($\phi_i = 30^\circ, \phi_s = 30^\circ$) for a TM_z wave incident on a square dielectric cylinder using the biorthogonal Coiflet basis ($p = 2, \tilde{p} = 2$).

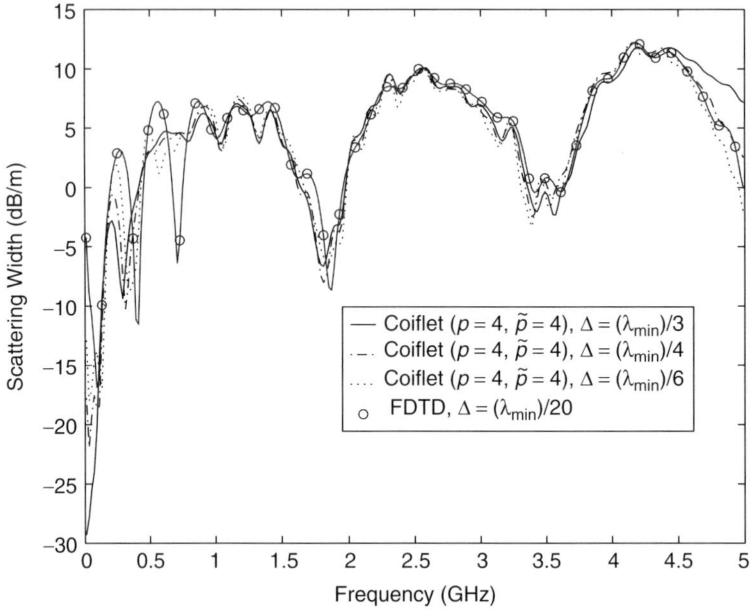

FIGURE 11.24 SW as a function of frequency ($\phi_i = 30°, \phi_s = 30°$) for a TM_z wave incident on a square dielectric cylinder using the Coiflet basis ($p = 4, \widetilde{p} = 4$).

TABLE 11.3 Relative Peak and Shifting Position Errors of SW for the Third Peak Curve

	Battle–Lemarié (%)		Daubechies D_4 (%)		Coiflet ($p = 2, \widetilde{p} = 2$) (%)		Coiflet ($p = 4, \widetilde{p} = 4$) (%)		Deslauriers–Dubuc (%)	
Δx(cm)	Σ_f	Σ_s	Σ_f	Σ_s	Σ_f	Σ_s	Σ_f	Σ_s	Σ_f	Σ_s
1.00	−1.88	0.78	13.11	1.92	3.28	1.02	3.69	−3.22	5.98	−0.26
0.75	−1.00	0.73	10.49	2.90	2.05	1.01	−0.33	−0.30	4.92	−0.74
0.50	−0.57	0.39	9.18	−1.02	0.41	0.17	−1.97	0.56	3.44	−0.04

For example, for the FDTD method the value of frequency position is about $f_0 = 4.646$ GHz. Figure 11.19 plots the SW results for various cell sizes and different MRTD wavelet bases for the sake of comparison. From Table 11.3 and Figures 11.20–11.24, we find that the scattering widths of the cubic spline Battle–Lemarié wavelet basis compare favorably with the FDTD results and are more accurate than those obtained by using other wavelet bases. From Table 11.3, we note that the results of the Daubechies D_4 basis show larger peaks and shifting position errors than those of other wavelet bases.

11.7 CONCLUSIONS

This chapter developed a multiresolution time domain scheme based on different wavelet bases for the scattering analysis of two-dimensional targets. In this research, we used the pure scattered field formulation to construct incident fields inside the target region and developed the TE_z and TM_z 2D-MRTD near-to-far-zone field transforms for computing the scattering widths of the targets. We validated the technique for both the PEC and dielectric cylinders and observed good agreement between the MRTD and FDTD results. The results lead us to conclude that the MRTD scheme is more efficient than the FDTD method for the cases investigated here for solving electromagnetic scattering problems, particularly for large objects of the type discussed here. Also, we compared different MRTD-based schemes from the points of view of computational accuracy and efficiency, scattering width relative error, and relative shifted position. We found that the MRTD results based on the Battle–Lemarié wavelets are the most accurate.

REFERENCES

[1] K. Umashankar and A. Taflove, "A novel method to analyze electromagnetic scattering of complex objects," *IEEE Trans. Electromagn. Compat.*, vol. EMC-24, pp. 397–405, Nov. 1982.

[2] R. J. Luebbers, Karl S. Kunz, M. Schneider, and F. Hunsberger, "A finite-difference time-domain near zone to far zone transformation," *IEEE Trans. Antenna Propag.*, vol. 39, no. 4, pp. 429–433, Apr. 1991.

[3] M. Krumpholz and L. P. B. Katehi, "MRTD: New time domain schemes based on multiresolution analysis," *IEEE Trans. Microwave Theory Tech.*, vol. 44, no. 4, pp. 555–571, Apr. 1996.

[4] K. S. Kunz and R. J. Luebbers, *The Finite Difference Time Domain Method for Electromagnetic*, CRC Press, Boca Raton, FL, 1993.

[5] Q. Cao, Y. Chen, and P. K. A. Wai, "MRTD electromagnetic scattering analysis," *Microwave Opt. Technol. Lett.*, vol., 28, no. 3, pp. 189–195, Feb. 2001.

[6] S. D. Gedney, "An anisotropic perfectly matched layered-absorbing medium for the truncation of FDTD lattice," *IEEE Trans. Antennas Propag.*, vol. 44, no. 12, pp. 1630–1939, Dec. 1996.

[7] Y. Chen, K. Sun, B. Beker, and R. Mittra, "Unified matrix presentation of Maxwell's and wave equations using generalized differential matrix operator," *IEEE Trans. Educ.*, vol. 41, no. 1, pp. 61–69, Feb. 1998.

[8] I. Daubechies, *Ten Lectures on Wavelet*, CBMS-NSF Series in Applied Mathematics. SIAM, Philadelphia, 1992.

[9] S. G. Mallat, *A Wavelet Tour of Signal Processing*, Academic Press, San Diego, 1998.

[10] G. Deslauriers and S. Dubuc, "Symmetric iterative interpolation processes," *Constr. Approx.*, vol. 5, pp. 49–68, 1989.

[11] W. Sweldens and R. Piessens, "Wavelet sampling techniques," in *1993 Proceedings of the Statistical Computing Section, American Statistical Association*, San Francisco, pp. 20–29, Aug. 1993.

[12] M. Fujii and W. J. R. Hoefer, "Application of biorthogonal interpolating wavelets to the Galerkin scheme of time dependent Maxwell's equations," *IEEE Microwave Wireless Components Lett.*, vol. 11, no. 1, pp. 22–24, Jan. 2001.

[13] R. J. Luebbers, D. Ryan, and J. Beggs, "A two-dimensional time domain near zone to far zone transformation," *IEEE Trans. Antennas Propag.*, vol. 40, no. 7, pp. 848–851, July 1992.

[14] Q. Li, Y. Chen, and D. Ge, "Comparison study of the PSTD and FDTD methods for scattering analysis," *Microwave Opt. Technol. Lett.*, vol. 25, no. 3, pp. 220–226, May 2000.

CHAPTER TWELVE

MRTD Scattering Analysis: 3D Cases

12.1 INTRODUCTION

As mentioned in the previous chapter, the finite difference time domain (FDTD) method has been used extensively to analyze electromagnetic scattering problems involving both dielectric and perfectly electric conducting (PEC) objects that are three-dimensional in nature [1–3]. To improve the efficiency of the FDTD techniques, the multiresolution time domain (MRTD) scheme has been employed for the analysis of a number of practical electromagnetic problems [4–6].

In this chapter, we further extend the MRTD to three dimensions and augment it with a near-to-far-zone transformation scheme for simulating the problem of scattering for PEC and dielectric targets. We adopt the pure scattered field formulation, as mentioned in Chapter 11, so that only the incident fields appear inside the scattered targets. Initially, we summarize the governing equations of the MRTD scheme, and then we derive the generalized MRTD field update equations involving both PEC and dielectric scattering targets. We further develop a 3D-MRTD near-to-far-zone field transformation for calculation of the radar scattering cross section (RCS) of targets. Finally, we compare the numerical results obtained from the MRTD to those derived from the FDTD and method of moments (MoM).

12.2 GOVERNING EQUATIONS OF MRTD

In this chapter, we adopt a pure scattered field formulation in the MRTD similar to the technique employed in the FDTD method [7]. Note that both scattered and

Multiresolution Time Domain Scheme for Electromagnetic Engineering
By Yinchao Chen, Qunsheng Cao, and Raj Mittra
ISBN 0-471-27230-2

incident fields exist inside the target region while only the scattered field exists outside the target. Thus, inside the target region the governing equations are

$$j\omega\varepsilon\vec{E}^{\text{scat}} = -\sigma\vec{E}^{\text{scat}} - \sigma\vec{E}^{\text{inc}} - j\omega(\varepsilon - \varepsilon_0)\vec{E}^{\text{inc}} + \nabla\times\vec{H}^{\text{scat}} \tag{12.1a}$$

$$j\omega\mu\vec{H}^{\text{scat}} = -\sigma^*\vec{H}^{\text{scat}} - \sigma^*\vec{H}^{\text{inc}} - j\omega(\mu - \mu_0)\vec{H}^{\text{inc}} - \nabla\times\vec{E}^{\text{scat}} \tag{12.1b}$$

where $(\vec{E}^{\text{inc}}, \vec{H}^{\text{inc}})$ and $(\vec{E}^{\text{scat}}, \vec{H}^{\text{scat}})$ are the incident and scattered fields, and ε, μ, σ, and σ^* denote permittivity, permeability, electric conductivity, and magnetic conductivity for the scattering target, respectively.

Using (12.1), for example, the x-components of the governing equations can be written

$$j\omega\varepsilon E_x^{\text{scat}} = -\sigma E_x^{\text{scat}} - \sigma E_x^{\text{inc}} - j\omega(\varepsilon - \varepsilon_0)E_x^{\text{inc}} + \left[\frac{\partial H_z^{\text{scat}}}{\partial y} - \frac{\partial H_y^{\text{scat}}}{\partial z}\right] \tag{12.2a}$$

$$j\omega\mu H_x^{\text{scat}} = -\sigma^* H_x^{\text{scat}} - \sigma^* H_x^{\text{inc}} - j\omega(\mu - \mu_0)E_x^{\text{inc}} + \left[\frac{\partial E_z^{\text{scat}}}{\partial y} - \frac{\partial E_y^{\text{scat}}}{\partial z}\right] \tag{12.2b}$$

Since there is only the scattered field outside the target region, we can apply an anisotropic perfectly matched layer (APML) [8, 9] as the truncation boundary condition for the computational boundaries. The governing equations in the regions external to the target and in the APMLs are expressed as

$$j\omega\varepsilon[\Lambda]\vec{E}^{\text{scat}} = \nabla\times\vec{H}^{\text{scat}} \tag{12.3a}$$

$$j\omega\mu[\Lambda]\vec{H}^{\text{scat}} = -\nabla\times\vec{E}^{\text{scat}} \tag{12.3b}$$

where

$$\begin{aligned}[\Lambda] &= [\Lambda_x][\Lambda_y][\Lambda_z] \\ &= \begin{bmatrix} s_x^{-1} & 0 & 0 \\ 0 & s_x & 0 \\ 0 & 0 & s_x \end{bmatrix}\begin{bmatrix} s_y & 0 & 0 \\ 0 & s_y^{-1} & 0 \\ 0 & 0 & s_y \end{bmatrix}\begin{bmatrix} s_z & 0 & 0 \\ 0 & s_z & 0 \\ 0 & 0 & s_z^-1 \end{bmatrix} \\ &= \begin{bmatrix} s_x^{-1}s_y s_z & 0 & 0 \\ 0 & s_x s_y^{-1} s_z & 0 \\ 0 & 0 & s_x s_y s_z^{-1} \end{bmatrix}\end{aligned} \tag{12.4}$$

with

$$s_\alpha = \begin{cases} \left(1 + \dfrac{\sigma_\alpha^e}{j\omega\varepsilon_0}\right), & \text{APML regions} \\ 1, & \text{non-APML region } (\sigma_\alpha^e = 0) \end{cases} \qquad (\alpha = x, y, z) \tag{12.5}$$

and σ_α^e is the spatially varying conductivity along the absorption axis. In order to improve the absorption performance in the APML regions, the conductivity σ_α^e is designated as

$$\sigma_\alpha^e = \sigma_{\alpha\,\max}^e \left|\frac{\alpha - \alpha_0}{d}\right|^2, \quad \alpha = x, y, z \tag{12.6}$$

where d is the thickness of the APML region. The constant $\sigma_{\alpha\,\max}^e$ is determined based on the acceptable level of the reflection error for normal incidence on an APML wall [8]:

$$\begin{aligned} R(0) &= \exp\left[-\frac{2}{\varepsilon_0 v}\int_0^d \sigma_\alpha^e(\alpha)d\alpha\right] = \exp\left[-\frac{2\sigma_{\alpha\,\max}^e d}{3\varepsilon_0 v}\right] \\ &= \exp\left[-\frac{2\sigma_{\alpha\,\max}^e d\sqrt{\varepsilon_{r\ \mathrm{eff}}}}{3\varepsilon_0 c}\right] \end{aligned} \tag{12.7}$$

Equation (12.7) can be solved for $\sigma_{\alpha\,\max}^e$ to yield

$$\sigma_{\alpha\,\max}^e = -\frac{3\varepsilon_0 c}{2d\sqrt{\varepsilon_{r\ \mathrm{eff}}}}\ln(R(0)), \quad \alpha = x, y, z \tag{12.8}$$

where v and c denote the speeds of electromagnetic wave propagation in the media and free space, respectively, and $\varepsilon_{r\ \mathrm{eff}}$ is the effective relative permittivity.

12.3 MRTD IMPLEMENTATIONS

MRTD Update Equations Inside the Target

In the MRTD scheme, we expand all the scattering field quantities in terms of the scaling functions in space and the pulse functions in time [4] as follows: =0

$$E_x^{\mathrm{scat}}(x, y, z, t) = \sum_{n,i,j,k=-\infty}^{+\infty} {}_{\phi x}^{\mathrm{scat}}E_{i+1/2,j,k}^n \phi_{i+1/2}(x)\phi_j(y)\phi_k(z)h_n(t) \tag{12.9a}$$

$$E_y^{\mathrm{scat}}(x, y, z, t) = \sum_{n,i,j,k=-\infty}^{+\infty} {}_{\phi y}^{\mathrm{scat}}E_{i,j+1/2,k}^n \phi_i(x)\phi_{j+1/2}(y)\phi_k(z)h_n(t) \tag{12.9b}$$

$$E_z^{\mathrm{scat}}(x, y, z, t) = \sum_{n,i,j,k=-\infty}^{+\infty} {}_{\phi z}^{\mathrm{scat}}E_{i,j,k+1/2}^n \phi_i(x)\phi_j(y)\phi_{k+1/2}(z)h_n(t) \tag{12.9c}$$

$$H_x^{\mathrm{scat}}(x, y, z, t) = \sum_{n,i,j,k=-\infty}^{+\infty} {}_{\phi x}^{\mathrm{scat}}H_{i,j+1/2,k+1/2}^{n+1/2} \phi_i(x)\phi_{j+1/2}(y)\phi_{k+1/2}(z)h_{n+1/2}(t) \tag{12.10a}$$

$$H_y^{\text{scat}}(x, y, z, t) = \sum_{n,i,j,k=-\infty}^{+\infty} {}^{\text{scat}}_{\phi y}H^{n+1/2}_{i+1/2,j,k+1/2}\phi_{i+1/2}(x)\phi_j(y)\phi_{k+1/2}(z)h_{n+1/2}(t) \tag{12.10b}$$

$$H_z^{\text{scat}}(x, y, z, t) = \sum_{n,i,j,k=-\infty}^{+\infty} {}^{\text{scat}}_{\phi z}H^{n+1/2}_{i+1/2,j+1/2,k}\phi_{i+1/2}(x)\phi_{j+1/2}(y)\phi_k(z)h_{n+1/2}(t) \tag{12.10c}$$

where $\phi_i(x)$, $\phi_j(y)$, and $\phi_k(z)$ denote the space cubic spline Battle–Lemarié scaling functions, and $h_n(t)$ is a rectangular pulse function. Substituting (12.9) and (12.10) into the time domain Maxwell equations, and using the orthogonal and integral relations (see Appendix B), we obtain three update equations for the H-fields:

$$\begin{aligned} {}^{\text{scat}}_{\phi x}H^{n+1/2}_{i,j+1/2,k+1/2} &= {}^{\text{scat}}_{\phi x}H^{n-1/2}_{i,j+1/2,k+1/2} \\ &+ \frac{1}{\mu_0}\sum_{\nu} a(\nu)\left({}^{\text{scat}}_{\phi y}E^n_{i,j+1/2,k+\nu+1}\frac{\Delta t}{\Delta z}{}^{\text{scat}}_{\phi z}E^n_{i,j+\nu+1,k+1/2}\frac{\Delta t}{\Delta y}\right) \end{aligned} \tag{12.11a}$$

$$\begin{aligned} {}^{\text{scat}}_{\phi y}H^{n+1/2}_{i+1/2,j,k+1/2} &= {}^{\text{scat}}_{\phi y}H^{n-1/2}_{i+1/2,j,k+1/2} \\ &+ \frac{1}{\mu_0}\sum_{\nu} a(\nu)\times\left({}^{\text{scat}}_{\phi z}E^n_{i+\nu+1,j,k+1/2}\frac{\Delta t}{\Delta x}{}^{\text{scat}}_{\phi x}E^n_{i+1/2,j,k+\nu+1}\frac{\Delta t}{\Delta z}\right) \end{aligned} \tag{12.11b}$$

$$\begin{aligned} {}^{\text{scat}}_{\phi z}H^{n+1/2}_{i+1/2,j+1/2,k} &= {}^{\text{scat}}_{\phi z}H^{n-1/2}_{i+1/2,j+1/2,k} \\ &+ \frac{1}{\mu_0}\sum_{\nu} a(\nu)\times\left({}^{\text{scat}}_{\phi x}E^n_{i+1/2,j+\nu+1,k}\frac{\Delta t}{\Delta y} - {}^{\text{scat}}_{\phi y}E^n_{i+\nu+1,j+1/2,k}\frac{\Delta t}{\Delta x}\right) \end{aligned} \tag{12.11c}$$

where the values of coefficient $a(\nu)$ are given in Chapter 4 [4].

Lossless Target. If the target is lossless, then due to the nonlocal property of the cubic spline Battle–Lemarié scaling functions, the MRTD expansions may cover both the target and free space. Consequently, the derivation of the update equations for the E-fields is rather lengthy. For simplicity, we first consider a lossless dielectric target ($\sigma = 0$). Starting with the time domain equations, the x-components of the governing equations, for example, are

$$D_x^{\text{scat}} = \varepsilon_0\varepsilon_r E_x^{\text{scat}} \tag{12.12a}$$

$$\frac{\partial D_x^{\text{scat}}}{\partial t} = -\varepsilon_0(\varepsilon_r - 1)\frac{\partial E_x^{\text{inc}}}{\partial t} + \left[\frac{\partial H_z^{\text{scat}}}{\partial y} - \frac{\partial H_y^{\text{scat}}}{\partial z}\right] \tag{12.12b}$$

From (12.12), we can derive the update equations for the E-field component, which read

$$
\begin{aligned}
{}^{\text{scat}}_{\phi x}D^{n+1}_{i+1/2,j,k} &= {}^{\text{scat}}_{\phi x}D^{n}_{i+1/2,j,k} \\
&\quad - \varepsilon_0 \sum_{i',j',k'=i_1^\kappa,j_1^\kappa,k_1^\kappa}^{i_2^\kappa,j_2^\kappa,k_2^\kappa} \sum_{\kappa=1}^{N} \left(\varepsilon_r^\kappa - 1\right) \alpha^\kappa_{i+1/2,i'+1/2}\alpha^\kappa_{j,j'}\alpha^\kappa_{k,k'} \\
&\quad \times \left({}^{\text{inc}}_{\phi x}E^{n+1}_{i'+1/2,j',k'} - {}^{\text{inc}}_{\phi x}E^{n}_{i'+1/2,j',k'}\right) + \sum_{\nu} a(\nu) \\
&\quad \times \left({}^{\text{scat}}_{\phi z}H^{n+1/2}_{i+1/2,j+\nu+1/2,k}\frac{\Delta t}{\Delta y}\,{}^{\text{scat}}_{\phi y}H^{n+1/2}_{i+1/2,j,k+\nu+1/2}\frac{\Delta t}{\Delta z}\right)
\end{aligned}
\tag{12.13a}
$$

$$
\begin{aligned}
{}^{\text{scat}}_{\phi x}D^{n}_{i+1/2,j,k} &= \varepsilon_0 \sum_{i',j',k'=-\infty}^{+\infty} \left[\delta_{i+1/2,i'+1/2}\delta_{j,j'}\delta_{k,k'}\right. \\
&\quad \left. + \sum_{\kappa=1}^{N}(\varepsilon_r^\kappa - 1)\alpha^\kappa_{i+1/2,i'+1/2}\alpha^\kappa_{j,j'}\alpha^\kappa_{k,k'}\right] {}^{\text{scat}}_{\phi x}E^{n}_{i'+1/2,j',k'}
\end{aligned}
\tag{12.13b}
$$

where the coefficients $\alpha^\kappa_{i+1/2,i'+1/2}$, $\alpha^\kappa_{j,j'}$, and $\alpha^\kappa_{k,k'}$ are defined as

$$
\alpha^\kappa_{i+1/2,i'+1/2} = \int_{x_2^\kappa}^{x_2^\kappa} \phi_{i'+1/2}(x)\phi_{i+1/2}(x)\frac{dx}{\Delta x} \tag{12.14a}
$$

$$
\alpha^\kappa_{j,j'} = \int_{y_1^\kappa}^{y_2^\kappa} \phi_{j'}(y)\phi_j(y)\frac{dy}{\Delta y} \tag{12.14b}
$$

$$
\alpha^\kappa_{k,k'} = \int_{z_1^\kappa}^{z_2^\kappa} \phi_{k'}(z)\phi_k(z)\frac{dz}{\Delta z} \tag{12.14c}
$$

The parameters (x_1^κ, x_2^κ), (y_1^κ, y_2^κ), and (z_1^κ, z_2^κ) are the lower and upper limits of the κth dielectric target region defined along the x-, y- and z-directions, respectively; and N is the total number of dielectric targets. In the above derivation, we have used the main diagonal approximation for the product of the coefficients, that is, replaced $\alpha^\kappa_{i+1/2,i'+1/2}\alpha^\kappa_{j,j'}\alpha^\kappa_{k,k'}$ by $\alpha^\kappa_{i+1/2,i'+1/2}\alpha^\kappa_{j,j'}\alpha^\kappa_{k,k'}\delta_{i+1/2,i'+1/2}\delta_{j,j'}\delta_{k,k'}$ in (12.13b). The approximation is justified because of the compact support of the scaling functions. The updated quantity in (12.13b) becomes

$$
{}^{\text{scat}}_{\phi x}D^{n}_{i+1/2,j,k} = \varepsilon_0 \left[1 + \sum_{\kappa=1}^{N}(\varepsilon_r^\kappa - 1)\alpha^\kappa_{i+1/2,i+1/2}\alpha^\kappa_{j,j}\alpha^\kappa_{k,k}\right] {}^{\text{scat}}_{\phi x}E^{n}_{i+1/2,j,k} \tag{12.15}
$$

Finally, the x-component of the update equation becomes

$$
\begin{aligned}
{}^{\text{scat}}_{\phi x}E^{n+1}_{i+1/2,j,k} &= {}^{\text{scat}}_{\phi x}E^{n}_{i+1/2,j,k} - \frac{1}{C_x}\sum_{i',j',k'=i_1,j_1,k_1}^{i_2,j_2,k_2}\sum_{\kappa=1}^{N}(\varepsilon_r^{\kappa}-1)\alpha^{\kappa}_{i+1/2,i'+1/2}\alpha^{\kappa}_{j,j'}\alpha^{\kappa}_{k,k'} \\
&\quad \times \left({}^{\text{inc}}_{\phi x}E^{n+1}_{i'+1/2,j',k'} - {}^{\text{inc}}_{\phi x}E^{n}_{i'+1/2,j',k'}\right) + \frac{1}{C_x}\frac{1}{\varepsilon_0}\sum_{\nu}a(\nu) \qquad (12.16\text{a}) \\
&\quad \times \left({}^{\text{scat}}_{\phi z}H^{n+1/2}_{i+1/2,j+\nu+1/2,k}\frac{\Delta t}{\Delta y} - {}^{\text{scat}}_{\phi y}H^{n+1/2}_{i+1/2,j,k+\nu+1/2}\frac{\Delta t}{\Delta z}\right)
\end{aligned}
$$

where

$$
C_x = \left[1 + \sum_{\kappa=1}^{N}(\varepsilon_r^{\kappa}-1)\alpha^{\kappa}_{i+1/2,i+1/2}\alpha^{\kappa}_{j,j}\alpha^{\kappa}_{k,k}\right] \qquad (12.16\text{b})
$$

We can derive the remaining two update equations in a similar manner, and the y-component of the update equation for the D-fields is given by

$$
\begin{aligned}
{}^{\text{scat}}_{\phi y}D^{n+1}_{i,j+1/2,k} &= {}^{\text{scat}}_{\phi y}D^{n}_{i,j+1/2,k} - \varepsilon_0\sum_{i',j',k'=i_1,j_1,k_1}^{i_2,j_2,k_2}\sum_{\kappa=1}^{N}(\varepsilon_r^{\kappa}-1)\alpha^{\kappa}_{i,i'}\alpha^{\kappa}_{j+1/2,j'+1/2}\alpha^{\kappa}_{k,k'} \\
&\quad \times \left({}^{\text{inc}}_{\phi y}E^{n+1}_{i',j'+1/2,k'} - {}^{\text{inc}}_{\phi y}E^{n}_{i',j'+1/2,k'}\right) + \sum_{\nu}a(\nu) \\
&\quad \times \left({}^{\text{scat}}_{\phi x}H^{n+1/2}_{i,j+1/2,k+\nu+1/2}\frac{\Delta t}{\Delta z} - {}^{\text{scat}}_{\phi z}H^{n+1/2}_{i+\nu+1/2,j+1/2,k}\frac{\Delta t}{\Delta x}\right) \qquad (12.17\text{a})
\end{aligned}
$$

$$
\begin{aligned}
{}^{\text{scat}}_{\phi y}D^{n}_{i,j+1/2,k} &= \varepsilon_0\sum_{i',j',k'=-\infty}^{+\infty}\left[\delta_{i,i'}\delta_{j+1/2,j'+1/2}\delta_{k,k'}\right. \\
&\quad \left. + \sum_{\kappa=1}^{N}\left(\varepsilon_r^{\kappa}-1\right)\alpha^{\kappa}_{i,i'}\alpha^{\kappa}_{j+1/2,j'+1/2}\alpha^{\kappa}_{k,k'}\right]{}^{\text{scat}}_{\phi y}E^{n}_{i',j'+1/2,k'} \qquad (12.17\text{b})
\end{aligned}
$$

Also, using the main diagonal approximation as before, that is, replacing $\alpha^{\kappa}_{i,i'}\alpha^{\kappa}_{j+1/2,j'+1/2}\alpha^{\kappa}_{k,k'}$ by $\alpha^{\kappa}_{i,i'}\alpha^{\kappa}_{j+1/2,j'+1/2}\alpha^{\kappa}_{k,k'}\delta_{i,i'}\delta_{j+1/2,j'+1/2}\delta_{k,k'}$, we obtain the y-component of the update equation:

$$
\begin{aligned}
{}^{\text{scat}}_{\phi y}E^{n+1}_{i,j+1/2,k} &= {}^{\text{scat}}_{\phi y}E^{n}_{i,j+1/2,k} - \frac{1}{C_y}\sum_{i',j',k'=i_1,j_1,k_1}^{i_2,j_2,k_2}\sum_{\kappa=1}^{N}(\varepsilon_r^{\kappa}-1)\alpha^{\kappa}_{i,i'}\alpha^{\kappa}_{j+1/2,j'+1/2}\alpha^{\kappa}_{k,k'} \\
&\quad \times \left({}^{\text{inc}}_{\phi y}E^{n+1}_{i',j'+1/2,k'} - {}^{\text{inc}}_{\phi y}E^{n}_{i',j'+1/2,k'}\right) + \frac{1}{C_y}\frac{1}{\varepsilon_0}\sum_{\nu}a(\nu) \qquad (12.18\text{a}) \\
&\quad \times \left({}^{\text{scat}}_{\phi x}H^{n+1/2}_{i,j+1/2,k+\nu+1/2}\frac{\Delta t}{\Delta z} - {}^{\text{scat}}_{\phi z}H^{n+1/2}_{i+\nu+1/2,j+1/2,k}\frac{\Delta t}{\Delta x}\right)
\end{aligned}
$$

where

$$C_y = \left[1 + \sum_{\kappa=1}^{N} (\varepsilon_r^\kappa - 1)\alpha_{i,i}^\kappa \alpha_{j,j}^\kappa \alpha_{k+1/2,k+1/2}^\kappa\right] \tag{12.18b}$$

Finally, we have

$$\begin{aligned} {}^{\text{scat}}_{\phi z}D^{n+1}_{i,j,k+1/2} &= {}^{\text{scat}}_{\phi y}D^{n}_{i,j,k+1/2} - \varepsilon_0 \sum_{i',j',k'=i_1,j_1,k_1}^{i_2,j_2,k_2} \sum_{\kappa=1}^{N} (\varepsilon_r^\kappa - 1)\alpha_{i,i'}^\kappa \alpha_{j,j'}^\kappa \alpha_{k+1/2,k'+1/2}^\kappa \\ &\quad \times \left({}^{\text{inc}}_{\phi z}E^{n+1}_{i',j',k'+1/2} - {}^{\text{inc}}_{\phi z}E^{n}_{i',j',k'+1/2}\right) \\ &\quad + \sum_{\nu} a(\nu) \left({}^{\text{scat}}_{\phi y}H^{n+1/2}_{i+\nu+1/2,j,k+1/2}\frac{\Delta t}{\Delta x} - {}^{\text{scat}}_{\phi x}H^{n+1/2}_{i,j+1/2,k+\nu+1/2}\frac{\Delta t}{\Delta z}\right) \end{aligned} \tag{12.19a}$$

$$\begin{aligned} {}^{\text{scat}}_{\phi z}D^{n}_{i,j,k+1/2} &= \varepsilon_0 \sum_{i',j',k'=-\infty}^{+\infty} \left[\delta_{i,i'}\delta_{j,j'}\delta_{k+1/2,k'+1/2}\right. \\ &\quad \left. + \sum_{\kappa=1}^{N} (\varepsilon_r^\kappa - 1)\alpha_{i,i'}^\kappa \alpha_{j,j'}^\kappa \alpha_{k+1/2,k'+1/2}^\kappa\right] {}^{\text{scat}}_{\phi z}E^{n}_{i',j',k'+1/2} \end{aligned} \tag{12.19b}$$

Replacing $\alpha_{i,i'}^\kappa \alpha_{j,j'}^\kappa \alpha_{k+1/2,k'+1/2}^\kappa$ by $\alpha_{i,i'}^\kappa \alpha_{j,j'}^\kappa \alpha_{k+1/2,k'+1/2}^\kappa \delta_{i,i'}\delta_{j,j'}\delta_{k+1/2,k'+1/2}$, we obtain the z-component of the E-field update equation:

$$\begin{aligned} {}^{\text{scat}}_{\phi z}E^{n+1}_{i,j,k+1/2} &= {}^{\text{scat}}_{\phi z}E^{n}_{i,j,k+1/2} - \frac{1}{C_z} \sum_{i',j',k'=i_1,j_1,k_1}^{i_2,j_2,k_2} \sum_{\kappa=1}^{N} (\varepsilon_r^\kappa - 1)\alpha_{i,i'}^\kappa \alpha_{j,j'}^\kappa \alpha_{k+1/2,k'+1/2}^\kappa \\ &\times \left({}^{\text{inc}}_{\phi z}E^{n+1}_{i',j',k'+1/2} - {}^{\text{inc}}_{\phi z}E^{n}_{i',j',k'+1/2}\right) + \frac{1}{C_z}\frac{1}{\varepsilon_0}\sum_{\nu} a(\nu) \\ &\times \left({}^{\text{scat}}_{\phi y}H^{n+1/2}_{i+\nu+1/2,j,k+1/2}\frac{\Delta t}{\Delta x} - {}^{\text{scat}}_{\phi x}H^{n+1/2}_{i,j+1/2,k+\nu+1/2}\frac{\Delta t}{\Delta z}\right) \end{aligned} \tag{12.20a}$$

where

$$C_z = \left[1 + \sum_{\kappa=1}^{N} (\varepsilon_r^\kappa - 1)\alpha_{i,i}^\kappa \alpha_{j,j}^\kappa \alpha_{k+1/2,k+1/2}^\kappa\right] \tag{12.20b}$$

PEC Target. Next, we consider a PEC target, which is described by $\sigma = \sigma^\kappa$ if a point is inside the κ th target, and otherwise, $\sigma = 0$. The x-component of the time domain equation is expressed as

$$\frac{\partial E_x^{\text{scat}}}{\partial t} = -\frac{\sigma}{\varepsilon_0}E_x^{\text{scat}} - \frac{\sigma}{\varepsilon_0}E_x^{\text{inc}} + \frac{1}{\varepsilon_0}\left[\frac{\partial H_z^{\text{scat}}}{\partial y} - \frac{\partial H_y^{\text{scat}}}{\partial z}\right] \tag{12.21}$$

Using the MRTD expansions and main diagonal approximation, the corresponding update equation becomes

$$
\begin{aligned}
{}^{\text{scat}}_{\phi x}E^{n+1}_{i+1/2,j,k} &= \frac{C_{x-}}{C_{x+}}\,{}^{\text{scat}}_{\phi x}E^{n}_{i+1/2,j,k} - \frac{1}{C_{x+}} \sum_{i',j',k'=i_1,j_1,k_1}^{i_2,j_2,k_2} \sum_{\kappa=1}^{N} \\
&\times \frac{\sigma^{\kappa}\Delta t}{2\varepsilon_0}\alpha^{\kappa}_{i+1/2,i'+1/2}\alpha^{\kappa}_{j,j'}\alpha^{\kappa}_{k,k'}\left({}^{\text{inc}}_{\phi x}E^{n+1}_{i'+1/2,j',k'} + {}^{\text{inc}}_{\phi x}E^{n}_{i'+1/2,j',k'}\right) \\
&+ \frac{1}{C_{x+}}\frac{1}{\varepsilon_0}\sum_{\nu} a(\nu)\left({}^{\text{scat}}_{\phi z}H^{n+1/2}_{i+1/2,j+\nu+1/2,k}\frac{\Delta t}{\Delta y} - {}^{\text{scat}}_{\phi y}H^{n+1/2}_{i+1/2,j,k+\nu+1/2}\frac{\Delta t}{\Delta z}\right)
\end{aligned}
\tag{12.22a}
$$

with

$$
\begin{aligned}
C_{x-} &= 1 - \frac{\sum_{\kappa=1}^{N}\sigma^{\kappa}\alpha^{\kappa}_{i+1/2,i+1/2}\alpha^{\kappa}_{j,j}\alpha^{\kappa}_{k,k}\Delta t}{2\varepsilon_0} \\
C_{x+} &= 1 + \frac{\sum_{\kappa=1}^{N}\sigma^{\kappa}\alpha^{\kappa}_{i+1/2,i+1/2}\alpha^{\kappa}_{j,j}\alpha^{\kappa}_{k,k}\Delta t}{2\varepsilon_0}
\end{aligned}
\tag{12.22b}
$$

In a similar manner, we can obtain the update equations in the y- and z-directions, which read

$$
\begin{aligned}
{}^{\text{scat}}_{\phi y}E^{n+1}_{i,j+1/2,k} &= \frac{C_{y-}}{C_{y+}}\,{}^{\text{scat}}_{\phi y}E^{n}_{i,j+1/2,k} - \frac{1}{C_{y+}} \sum_{i',j',k'=i_1,j_1,k_1}^{i_2,j_2,k_2} \sum_{\kappa=1}^{N} \\
&\times \frac{\sigma^{\kappa}\Delta t}{2\varepsilon_0}\alpha^{\kappa}_{i,i'}\alpha^{\kappa}_{j+1/2,j'+1/2}\alpha^{\kappa}_{k,k'}\left({}^{\text{inc}}_{\phi y}E^{n+1}_{i',j'+1/2,k'} + {}^{\text{inc}}_{\phi y}E^{n}_{i',j'+1/2,k'}\right) \\
&+ \frac{1}{C_{y+}}\frac{1}{\varepsilon_0}\sum_{\nu} a(\nu)\left({}^{\text{scat}}_{\phi x}H^{n+1/2}_{i,j+1/2,k+\nu+1/2}\frac{\Delta t}{\Delta z} - {}^{\text{scat}}_{\phi z}H^{n+1/2}_{i+\nu+1/2,j+1/2,k}\frac{\Delta t}{\Delta x}\right)
\end{aligned}
\tag{12.23a}
$$

with

$$
\begin{aligned}
C_{y-} &= 1 - \frac{\sum_{\kappa=1}^{N}\sigma^{\kappa}\alpha^{\kappa}_{i,i}\alpha^{\kappa}_{j+1/2,j+1/2}\alpha^{\kappa}_{k,k}\Delta t}{2\varepsilon_0} \\
C_{x+} &= 1 + \frac{\sum_{\kappa=1}^{N}\sigma^{\kappa}\alpha^{\kappa}_{i,i}\alpha^{\kappa}_{j+1/2,j+1/2}\alpha^{\kappa}_{k,k}\Delta t}{2\varepsilon_0}
\end{aligned}
\tag{12.23b}
$$

and

$$
{}^{\text{scat}}_{\phi z}E^{n+1}_{i,j,k+1/2} = \frac{C_{z-}}{C_{z+}}\,{}^{\text{scat}}_{\phi z}E^{n}_{i,j,k+1/2} - \frac{1}{C_{z+}} \sum_{i',j',k'=i_1,j_1,k_1}^{i_2,j_2,k_2} \sum_{\kappa=1}^{N}
$$

$$\times \frac{\sigma^{\kappa}\Delta t}{2\varepsilon_0}\alpha^{\kappa}_{i,i'}\alpha^{\kappa}_{j,j'}\alpha^{\kappa}_{k+1/2,k'+1/2}\left({}^{\text{inc}}_{\phi z}E^{n+1}_{i',j',k'+1/2} + {}^{\text{inc}}_{\phi z}E^{n}_{i',j',k'+1/2}\right) \tag{12.24a}$$

$$+ \frac{1}{C_{z+}}\frac{1}{\varepsilon_0}\sum_{\nu} a(\nu)\left({}^{\text{scat}}_{\phi y}H^{n+1/2}_{i+\nu+1/2,j,k+1/2}\frac{\Delta t}{\Delta x} - {}^{\text{scat}}_{\phi x}H^{n+1/2}_{i,j+\nu+1/2,k+1/2}\frac{\Delta t}{\Delta y}\right)$$

with

$$C_{z-} = 1 - \frac{\sum\limits_{\kappa=1}^{N}\sigma^{\kappa}\alpha^{\kappa}_{i,i}\alpha^{\kappa}_{j,j}\alpha^{\kappa}_{k+1/2,k+1/2}\Delta t}{2\varepsilon_0}$$

$$C_{z+} = 1 + \frac{\sum\limits_{\kappa=1}^{N}\sigma^{\kappa}\alpha^{\kappa}_{i,i}\alpha^{\kappa}_{j,j}\alpha^{\kappa}_{k+1/2,k+1/2}\Delta t}{2\varepsilon_0} \tag{12.24b}$$

In practice, if the targets can be approximated as PECs, the update equations reduce to a very simple form for regions inside the PEC targets:

$${}^{\text{scat}}_{\phi x}E^{n+1}_{i+1/2,j,k} = -{}^{\text{scat}}_{\phi x}E^{n}_{i+1/2,j,k} \tag{12.25}$$

$${}^{\text{scat}}_{\phi y}E^{n+1}_{i,j+1/2,k} = -{}^{\text{scat}}_{\phi y}E^{n}_{i,j+1/2,k} \tag{12.26}$$

$${}^{\text{scat}}_{\phi z}E^{n+1}_{i,j,k+1/2} = -{}^{\text{scat}}_{\phi z}E^{n}_{i,j,k+1/2} \tag{12.27}$$

Lossy Target. Finally, let us turn to a more general case for a lossy dielectric target ($\sigma \neq \infty$). We use a procedure similar to the one for the PEC problem to develop the update equations for this case. For example, the update equations for the E-fields inside the lossy dielectric medium are given by

$${}^{\text{scat}}_{\phi x}E^{n+1}_{i+1/2,,j,k} = \frac{C_{x-}}{C_{x+}}{}^{\text{scat}}_{\phi x}E^{n}_{i+1/2,,j,k} - \frac{1}{C_{x+}}\sum_{i',j',k'=-\infty}^{\infty}\sum_{\kappa=1}^{N}$$

$$\times \frac{\sigma^{\kappa}\Delta t}{2\varepsilon_0}\alpha^{\kappa}_{i+1/2,i'+1/2}\alpha^{\kappa}_{j,j'}\alpha^{\kappa}_{k,k'}\left({}^{\text{inc}}_{\phi x}E^{n+1}_{i'+1/2,j',k'} + {}^{\text{inc}}_{\phi xa}E^{n}_{i'+1/2,j',k'}\right)$$

$$- \frac{1}{C_{x+}}\sum_{i',j',k'=-\infty}^{+\infty}\sum_{\kappa=1}^{N}(\varepsilon^{\kappa}_{r} - 1)\alpha^{\kappa}_{i+1/2,i'+1/2}\alpha^{\kappa}_{j,j'}\alpha^{\kappa}_{k,k'} \tag{12.28a}$$

$$\times \left({}^{\text{inc}}_{\phi x}E^{n+1}_{i'+1/2,j',k'} - {}^{\text{inc}}_{\phi x}E^{n}_{i'+1/2,j',k'}\right)$$

$$+ \frac{1}{\varepsilon_0}\frac{1}{C_{x+}}\sum_{\nu} a(\nu)\left({}^{\text{scat}}_{\phi z}H^{n+1/2}_{i+1/2,j+\nu+1/2,k}\frac{\Delta t}{\Delta y} - {}^{\text{scat}}_{\phi y}H^{n+1/2}_{i+1/2,j,k+\nu+1/2}\frac{\Delta t}{\Delta z}\right)$$

where

$$C_{x-} = \left[1 + \sum_{\kappa=1}^{N}\left((\varepsilon_r^{\kappa} - 1) - \frac{\sigma^{\kappa}\Delta t}{2\varepsilon_0}\right)\alpha^{\kappa}_{i+1/2,i+1/2}\alpha^{\kappa}_{j,j}\alpha^{\kappa}_{k,k}\right]$$
$$C_{x+} = \left[1 + \sum_{\kappa=1}^{N}\left((\varepsilon_r^{\kappa} - 1) + \frac{\sigma^{\kappa}\Delta t}{2\varepsilon_0}\right)\alpha^{\kappa}_{i+1/2,i+1/2}\alpha^{\kappa}_{j,j}\alpha^{\kappa}_{k,k}\right] \tag{12.28b}$$

and

$$
\begin{aligned}
{}^{\text{scat}}_{\phi y}E^{n+1}_{i,j+1/2,k} &= \frac{C_{y-}}{C_{y+}}\,{}^{\text{scat}}_{\phi y}E^{n}_{i,j+1/2,k} - \frac{1}{C_{y+}}\sum_{i',j',k'=-\infty}^{\infty}\sum_{\kappa=1}^{N} \\
&\times \frac{\sigma^{\kappa}\Delta t}{2\varepsilon_0}\alpha^{\kappa}_{i,i'}\alpha^{\kappa}_{j+1/2,j'+1/2}\alpha^{\kappa}_{k,k'}\left({}^{\text{inc}}_{\phi y}E^{n+1}_{i',j'+1/2,k'} + {}^{\text{inc}}_{\phi y}E^{n}_{i',j'+1/2,k'}\right) \\
&- \frac{1}{C_{y+}}\sum_{i',j',k'=-\infty}^{+\infty}\sum_{\kappa=1}^{N}(\varepsilon_r^{\kappa} - 1)\alpha^{\kappa}_{i,i'}\alpha^{\kappa}_{j+1/2,j'+1/2}\alpha^{\kappa}_{k,k'} \\
&\times \left({}^{\text{inc}}_{\phi y}E^{n+1}_{i',j'+1/2,k'} - {}^{\text{inc}}_{\phi y}E^{n}_{i',j'+1/2,k'}\right) \\
&+ \frac{1}{\varepsilon_0}\frac{1}{C_{y+}}\sum_{\nu}a(\nu)\left({}^{\text{scat}}_{\phi x}H^{n+1/2}_{i,j+1/2,k+\nu+1/2}\frac{\Delta t}{\Delta z} - {}^{\text{scat}}_{\phi z}H^{n+1/2}_{i+\nu+1/2,j+1/2,k}\frac{\Delta t}{\Delta x}\right)
\end{aligned} \tag{12.29a}
$$

where

$$C_{y-} = \left[1 + \sum_{\kappa=1}^{N}\left((\varepsilon_r^{\kappa} - 1) - \frac{\sigma^{\kappa}\Delta t}{2\varepsilon_0}\right)\alpha^{\kappa}_{i,i}\alpha^{\kappa}_{j+1/2,j+1/2}\alpha^{\kappa}_{k,k}\right]$$
$$C_{y+} = \left[1 + \sum_{\kappa=1}^{N}\left((\varepsilon_r^{\kappa} - 1) + \frac{\sigma^{\kappa}\Delta t}{2\varepsilon_0}\right)\alpha^{\kappa}_{i,i}\alpha^{\kappa}_{j+1/2,j+1/2}\alpha^{\kappa}_{k,k}\right] \tag{12.29b}$$

Also,

$$
\begin{aligned}
{}^{\text{scat}}_{\phi z}E^{n+1}_{i,j,k+1/2} &= \frac{C_{z-}}{C_{z+}}\,{}^{\text{scat}}_{\phi z}E^{n}_{i,j,k+1/2} - \frac{1}{C_{z+}}\sum_{i',j',k'=-\infty}^{\infty}\sum_{\kappa=1}^{N} \\
&\times \frac{\sigma^{\kappa}\Delta t}{2\varepsilon_0}\alpha^{\kappa}_{i,i'}\alpha^{\kappa}_{j,j'}\alpha^{\kappa}_{k+1/2,k'+1/2}\left({}^{\text{inc}}_{\phi z}E^{n+1}_{i',j',k'+1/2} + {}^{\text{inc}}_{\phi z}E^{n}_{i',j',k'+1/2}\right) \\
&- \frac{1}{C_{z+}}\sum_{i',j',k'=-\infty}^{+\infty}\sum_{\kappa=1}^{N}(\varepsilon_r^{\kappa} - 1)\alpha^{\kappa}_{i,i'}\alpha^{\kappa}_{j,j'}\alpha^{\kappa}_{k+1/2,k'+1/2} \\
&\times \left({}^{\text{inc}}_{\phi z}E^{n+1}_{i',j',k'+1/2} - {}^{\text{inc}}_{\phi z}E^{n}_{i',j',k'+1/2}\right) \\
&+ \frac{1}{\varepsilon_0}\frac{1}{C_{z+}}\sum_{\nu}a(\nu)\left({}^{\text{scat}}_{\phi y}H^{n+1/2}_{i+\nu+1/2,j,k+1/2}\frac{\Delta t}{\Delta x} - {}^{\text{scat}}_{\phi x}H^{n+1/2}_{i,j+\nu+1/2,k+1/2}\frac{\Delta t}{\Delta y}\right)
\end{aligned} \tag{12.30a}
$$

where

$$C_{z-} = \left[1 + \sum_{\kappa=1}^{N} \left((\varepsilon_r^{\kappa} - 1) - \frac{\sigma^{\kappa}\Delta t}{2\varepsilon_0}\right) \alpha_{i,i}^{\kappa}\alpha_{j,j}^{\kappa}\alpha_{k+1/2,k+1/2}^{\kappa}\right]$$
$$C_{z+} = \left[1 + \sum_{\kappa=1}^{N} \left((\varepsilon_r^{\kappa} - 1) + \frac{\sigma^{\kappa}\Delta t}{2\varepsilon_0}\right) \alpha_{i,i}^{\kappa}\alpha_{j,j}^{\kappa}\alpha_{k+1/2,k+1/2}^{\kappa}\right] \tag{12.30b}$$

If we assume $\sigma = 0$, the update equations (12.28)–(12.30) can reduce to the lossless dielectric target case; and, if we assume $\varepsilon_r^{\kappa} = 1$, they become the same as for the case of a PEC target.

MRTD Update Equations Outside the Target

In the APML regions, there are six face-APMLs, twelve edge-APMLs, and eight corner-APMLs. As an example, we consider a corner-APML region and derive the MRTD update equations for the x-component of the E-fields:

$$j\omega\varepsilon_0 \frac{\left(1 + \dfrac{\sigma_y^e}{j\omega\varepsilon_0}\right)\left(1 + \dfrac{\sigma_z^e}{j\omega\varepsilon_0}\right)}{\left(1 + \dfrac{\sigma_x^e}{j\omega\varepsilon_0}\right)} E_x^{\text{scat}} = \left[\frac{\partial H_z^{\text{scat}}}{\partial y} - \frac{\partial H_y^{\text{scat}}}{\partial z}\right] \tag{12.31}$$

We use a two-step method in regions outside the target region as detailed in Appendix C, yielding

$$\begin{aligned} {}_{\phi x}D_{i+1/2,j,k}^{n+1} &= \left(\frac{1 - \dfrac{\sigma_y^e\,\Delta t}{2\varepsilon_0}}{1 + \dfrac{\sigma_y^e\,\Delta t}{2\varepsilon_0}}\right) {}_{\phi x}D_{i+1/2,j,k}^{n} + \left(1 + \frac{\sigma_y^e\,\Delta t}{2\varepsilon_0}\right)^{-1} \sum_{\nu} a(\nu) \\ &\times \left({}_{\phi z}H_{i+1/2,j+\nu+1/2,k}^{n+1/2} \frac{\Delta t}{\Delta y} - {}_{\phi y}H_{i+1/2,j,k+\nu+1/2}^{n+1/2} \frac{\Delta t}{\Delta z}\right) \end{aligned} \tag{12.32a}$$

$$\begin{aligned} {}_{\phi x}E_{i+1/2,j,k}^{n+1} &= \left(\frac{1 - \dfrac{\sigma_z^e\,\Delta t}{2\varepsilon_0}}{1 + \dfrac{\sigma_z^e\,\Delta t}{2\varepsilon_0}}\right) {}_{\phi x}E_{i+1/2,j,k}^{n} + \frac{1}{\varepsilon_0}\left(1 + \frac{\sigma_z^e\,\Delta t}{2\varepsilon_0}\right)^{-1} \\ &\times \left[\left(1 + \frac{\sigma_x^e\,\Delta t}{2\varepsilon_0}\right) {}_{\phi x}D_{i+1/2,j,k}^{n+1} - \left(1 - \frac{\sigma_x^e\,\Delta t}{2\varepsilon_0}\right) {}_{\phi x}D_{i+1/2,j,k}^{n}\right] \end{aligned} \tag{12.32b}$$

In free space (i.e., non-APML regions), we simply set $\sigma_x^e = \sigma_y^e = \sigma_z^e = 0$ in (12.32), while inside APML regions, we use σ_α^e ($\alpha = x, y, z$) along the directions of wave propagation.

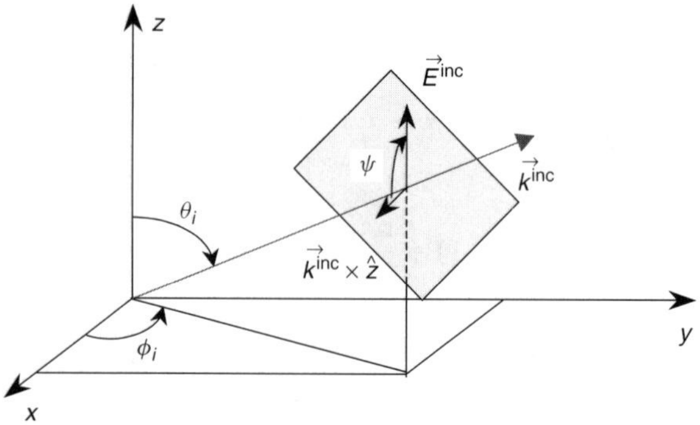

FIGURE 12.1 Definition of direction of propagation of the incident wave and polarization.

Incident Plane-Wave and Linear Interpolation

Assume that a TEM pulse is incident along a specified direction $\hat{\eta}$. Figure 12.1 shows the unit wave vector, $\vec{k}^{\text{inc}}$, which can be decomposed into three components.

Assuming that the incident wave is a plane-wave Gaussian pulse, we can express the incident electromagnetic fields as

$$E^{\text{inc}(0)}(i_\eta) = E_0 \exp\left(\gamma \frac{(i_\eta - i_c)^2}{i_p^2}\right) \tag{12.33a}$$

$$H^{\text{inc}(1/2)}(i_\eta) = \frac{E_0}{Z_0} \exp\left(\gamma \frac{(i_\eta - i_c - 0.5c\Delta t/\Delta\eta)^2}{i_p^2}\right) \tag{12.33b}$$

where E_0 is the amplitude of the Gaussian pulse with a peak at grid i_c; i_p denotes the pulse width; γ is a constant; Z_0 is the impedance of free space; and c denotes the velocity of the propagating wave.

Next, using the initial condition and the MRTD scheme, we build a one-dimensional incident wave propagation model. The incident fields inside the scattered region, that is, the target region, can be obtained by linear interpolation of the one-dimensional MRTD incident total fields ($\vec{H}^{\text{inc}}$, $\vec{E}^{\text{inc}}$). The expression

$$\vec{k}^{\text{inc}} = k_x^{\text{inc}}\hat{x} + k_y^{\text{inc}}\hat{y} + k_z^{\text{inc}}\hat{z} = \sin\theta_i \cos\phi_i \hat{x} + \sin\theta_i \sin\phi_i \hat{y} + \cos\theta_i \hat{z} \tag{12.34}$$

defines the unit wave-vector projection on the scattered target, which is specified by $\vec{r}$. The distance d is given by

$$d = \vec{k}^{\text{inc}} \cdot \vec{r} = \sin\theta_i \cos\phi_i \hat{x} + \sin\theta_i \sin\phi_i \hat{y} + \cos\theta_i \hat{z} \tag{12.35}$$

The above distance d can be discretized and a linear interpolation can be used to determine the electromagnetic fields at the distance d ($d > 0$):

$$E^{\text{inc}}(d) = [1 - d + \text{int}(d)] \cdot E^{\text{inc}}[\text{int}(d)] + [d - \text{int}(d)] \cdot E^{\text{inc}}[\text{int}(d) + 1] \tag{12.36}$$

where $\text{int}(x)$ is the integer part of x.

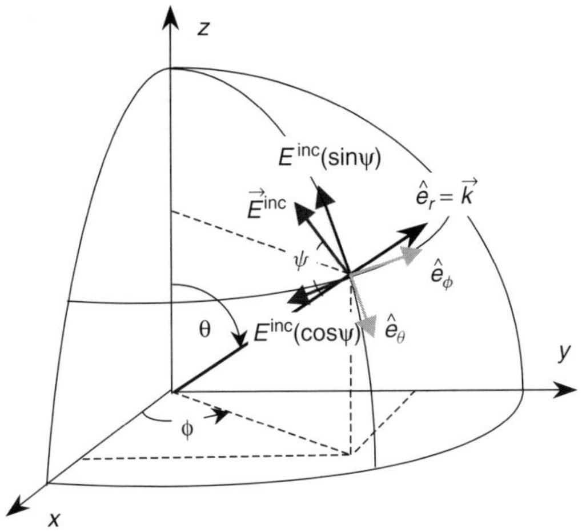

FIGURE 12.2 Orthogonal scattering coordinate system.

In order to obtain the incident field expressions in terms of the Cartesian coordinate system, we need to obtain the coordinate system transformation from spherical coordinates to Cartesian coordinates. We express the unit vectors in the spherical coordinate system $(\hat{e}_r, \hat{e}_\theta, \hat{e}_\phi)$, as shown in Figure 12.2. The polarized incident fields $(\vec{E}^{\text{inc}}, \vec{H}^{\text{inc}})$ can be represented as the superposition of the two components along $(\hat{e}_\theta, \hat{e}_\phi)$ directions, since the incident fields $\vec{E}^{\text{inc}}$ and $\vec{H}^{\text{inc}}$ have no radial components: that is, $\vec{E}^{\text{inc}} = E_\theta^{\text{inc}}\hat{e}_\theta + E_\phi^{\text{inc}}\hat{e}_\phi$, and $\vec{H}^{\text{inc}} = H_\theta^{\text{inc}}\hat{e}_\theta + H_\phi^{\text{inc}}\hat{e}_\phi$. Since $E_\theta^{\text{inc}} = -E^{\text{inc}}\sin\psi$, $E_\phi^{\text{inc}} = -E^{\text{inc}}\cos\psi$, $H_\theta^{\text{inc}} = E^{\text{inc}}\cos\psi$, and $H_\phi^{\text{inc}} = -H^{\text{inc}}\sin\psi$ (see Figure 12.2), we can express the incident fields in the Cartesian system by transforming fields in the spherical system:

$$
\begin{aligned}
E_x^{\text{inc}} &= E_\theta^{\text{inc}}\cos\theta_i\cos\phi_i - E_\phi^{\text{inc}}\sin\phi_i \\
E_y^{\text{inc}} &= E_\theta^{\text{inc}}\cos\theta_i\sin\phi_i + E_\phi^{\text{inc}}\cos\phi_i \\
E_z^{\text{inc}} &= -E_\theta^{\text{inc}}\sin\theta_i
\end{aligned}
\tag{12.37a}
$$

or

$$
\begin{aligned}
E_x^{\text{inc}} &= -E^{\text{inc}}\sin\psi\cos\theta_i\cos\phi_i + E^{\text{inc}}\cos\psi\sin\phi_i \\
E_y^{\text{inc}} &= -E^{\text{inc}}\sin\psi\cos\theta_i\sin\phi_i - E^{\text{inc}}\cos\psi\cos\phi_i \\
E_z^{\text{inc}} &= E^{\text{inc}}\sin\psi\sin\theta_i
\end{aligned}
\tag{12.37b}
$$

or

$$
\begin{aligned}
E_x^{\text{inc}} &= E^{\text{inc}}(\cos\psi\sin\phi_i - \sin\psi\cos\theta_i\cos\phi_i) \\
E_y^{\text{inc}} &= E^{\text{inc}}(-\cos\psi\cos\phi_i + \sin\psi\cos\theta_i\sin\phi_i) \\
E_z^{\text{inc}} &= E^{\text{inc}}(\sin\psi\sin\theta_i)
\end{aligned}
\tag{12.37c}
$$

In a similar manner, we can express the incident magnetic fields as

$$
\begin{aligned}
H_x^{\text{inc}} &= H^{\text{inc}}(\sin\psi \sin\phi_i + \cos\psi \cos\theta_i \cos\phi_i) \\
H_y^{\text{inc}} &= E^{\text{inc}}(-\sin\psi \cos\phi_i + \cos\psi \cos\theta_i \sin\phi_i) \\
H_z^{\text{inc}} &= E^{\text{inc}}(-\cos\psi \sin\theta_i)
\end{aligned}
\tag{12.38}
$$

where θ_i is the polar angle, ϕ_i is the azimuthal angle, and ψ is the angle between the electric field $\vec{E}^{\text{inc}}$ and the unit wave vector $\vec{k}^{\text{inc}} \times \hat{z}$, as shown in Figure 12.2.

Therefore, at location d, the incident fields $(\vec{H}^{\text{inc}}, \vec{E}^{\text{inc}})$ can be decomposed into six components as follows:

$$E_x^{\text{inc}}(d) = E^{\text{inc}}(d)(\cos\psi \sin\phi_i - \sin\psi \cos\theta_i \cos\phi_i) \tag{12.39a}$$

$$E_y^{\text{inc}}(d) = E^{\text{inc}}(d)(-\cos\psi \cos\phi_i - \sin\psi \cos\theta_i \sin\phi_i) \tag{12.39b}$$

$$E_z^{\text{inc}}(d) = E^{\text{inc}}(d)(\sin\psi \sin\theta_i) \tag{12.39c}$$

$$H_x^{\text{inc}}(d) = H^{\text{inc}}(d)(\sin\psi \sin\phi_i + \cos\psi \cos\theta_i \cos\phi_i) \tag{12.40a}$$

$$H_y^{\text{inc}}(d) = H^{\text{inc}}(d)(-\sin\psi \cos\phi_i + \cos\psi \cos\theta_i \sin\phi_i) \tag{12.40b}$$

$$H_z^{\text{inc}}(d) = H^{\text{inc}}(d)(-\cos\psi \sin\theta_i) \tag{12.40c}$$

MRTD Near-to-Far-Zone Transformation

The radiated electric field can be evaluated numerically from the field on an imaginary surface S [9]:

$$
\begin{aligned}
\vec{E}^{\text{far}}(\vec{r}, t) &= -\frac{1}{4\pi rc}\hat{r} \times \int_s \left[\hat{n} \times \frac{\partial \vec{E}^{\text{scat}}(\vec{r}', t')}{\partial t}\right] ds \\
&\quad + \frac{\mu_0}{4\pi r}\hat{r} \times \int_s \left[\hat{r} \times \hat{n} \times \frac{\partial \vec{H}^{\text{scat}}(\vec{r}', t')}{\partial t}\right] ds \\
&= \frac{1}{4\pi rc}\hat{r} \times \int_s \frac{\partial[\vec{M}(\vec{r}')]}{\partial t} ds + \frac{\mu_0}{4\pi r}\hat{r} \times \int_s \left(\hat{r} \times \frac{\partial[\vec{J}(\vec{r}')]}{\partial t}\right) ds
\end{aligned}
\tag{12.41}
$$

where the unit vector $\hat{r}$ is associated with the field point; $\hat{r}'$ is the radius vector of the source point; $\vec{R} = \hat{r} - \hat{r}'$; and the square brackets ([ ]) indicate that the variables contained within them are evaluated at the retarded times $t' = t - R/c$. The variables $\vec{J}(\vec{r}')$ and $\vec{M}'(\vec{r}')$ are the equivalent scattered electric and magnetic source currents, given by $\vec{J}(\vec{r}') = \hat{n} \times \vec{H}(\vec{r}')$ and $\vec{M}(\vec{r}') = -\hat{n} \times \vec{E}(\vec{r}')$, respectively. The outward unit vector $\hat{n}$ is normal to each face on the surface S, and $\vec{E}(\vec{r}')$ and $\vec{H}(\vec{r}')$ are the near electromagnetic scattered fields on S.

In common with the FDTD method developed in [10], we can define the 'dual' variables as follows:

$$\vec{W}(t) = \frac{\mu_0}{4\pi r}\frac{\partial}{\partial t}\int_s \vec{J}\left[t - \frac{r - \vec{r}'\cdot\hat{r}}{c}\right]ds' \tag{12.42a}$$

$$\vec{U}(t) = \frac{1}{4\pi r c}\frac{\partial}{\partial t}\int_s \vec{M}\left[t - \frac{r - \vec{r}'\cdot\hat{r}}{c}\right]ds' \tag{12.42b}$$

In the time domain, the electric fields in the far zone can be determined by using

$$E_\theta^{\text{far}} = -W_\theta - U_\phi \tag{12.43a}$$

$$E_\phi^{\text{far}} = -W_\phi + U_\theta \tag{12.43b}$$

where

$$W_\theta = W_x \cos\theta_s \cos\phi_s + W_y \cos\theta_s \sin\phi_s - W_z \sin\theta_s \tag{12.44a}$$

$$W_\phi = -W_x \sin\phi_s + W_y \cos\phi_s \tag{12.44b}$$

$$U_\theta = U_x \cos\theta_s \cos\phi_s + U_y \cos\theta_s \sin\phi_s - U_z \sin\theta_s \tag{12.45a}$$

$$U_\phi = -U_x \sin\phi_s + U_y \cos\phi_s \tag{12.45b}$$

and θ_s and ϕ_s are the scattered polarization angles [1–3].

The apportionment for the running sum of the W_z component is

$$W_z|_{\vec{r}}^{nn} = W_z|_{\vec{r}}^{nn} - (1-\alpha)\frac{\mu_0}{4\pi r}\frac{\Delta x\ \Delta z}{\Delta t}\left(H_x\Big|_{\vec{r}'}^{n+1/2} - H_x\Big|_{\vec{r}'}^{n-1/2}\right) \tag{12.46a}$$

$$W_z|_{\vec{r}}^{nn+1} = W_z|_{\vec{r}}^{nn+1} - (\alpha)\frac{\mu_0}{4\pi r}\frac{\Delta x\ \Delta z}{\Delta t}\left(H_x\Big|_{\vec{r}'}^{n+1/2} - H_x\Big|_{\vec{r}'}^{n-1/2}\right) \tag{12.46b}$$

where α is the time interpolation factor $\alpha = (n_t + (r - r'\cos\zeta)/c\ \Delta t) - \text{int}[n_t + (r - r'\cos\zeta)/c\ \Delta t]$; n_t is the current time step; and ζ is the angle between $\vec{r}$ and $\vec{r}'$. Thus, $r'\cos\zeta = x'\sin\theta_s\cos\phi_s + y'\sin\theta_s\sin\phi_s + z'\cos\phi_s$ [6, 10].

Finally, we can extract the radar cross section (RCS) for the scattering target by using

$$\text{RCS} = \lim_{r\to\infty}\left(4\pi r^2 \frac{E_\theta^2 + E_\phi^2}{(E^{\text{inc}})^2}\right) \tag{12.47}$$

12.4 APPLICATION RESULTS

Co-, Cross-, and ϕ-Polarizations

In scattering analysis, we often use the terms co- and cross-polarizations. In this section, we first clarify these concepts, which are frequently used in the following numerical analysis. As seen in (12.37), the unit vector of the $\vec{E}^{\text{inc}}$ field is given as

$$\begin{aligned}\hat{E}^{\text{inc}} &= \hat{x}(\cos\psi\sin\phi_i - \sin\psi\cos\phi_i\cos\theta_i) \\ &\quad + \hat{y}(-\cos\psi\cos\phi_i - \sin\psi\sin\phi_i\cos\theta_i) + \hat{z}(\sin\psi\sin\theta_i)\end{aligned} \tag{12.48}$$

By using the transformation between the Cartesian and spherical coordinates, the projection of the far-zone scattered E-fields upon the incident E-field direction, namely, the polarized component, can be written

$$\begin{aligned}(\vec{E}_\theta^{\text{far}} + \vec{E}_\phi^{\text{far}}) \cdot \hat{E}^{\text{inc}} \\ &= [E_\theta^{\text{far}}(\hat{x}\cos\theta_i\cos\phi_i + \hat{y}\cos\theta_i\sin\phi_i - \hat{z}\sin\theta_i) \\ &\quad + E_\phi^{\text{far}}(-\hat{x}\sin\phi_i + \hat{y}\cos\phi_i)] \cdot \hat{E}^{\text{inc}} \\ &= (E_\theta^{\text{far}}\cos\theta_i\cos\phi_i - E_\phi^{\text{far}}\sin\phi_i)(\cos\psi\sin\phi_i - \sin\psi\cos\phi_i\cos\theta_i) \\ &\quad + (E_\theta^{\text{far}}\cos\theta_i\sin\phi_i + E_\phi^{\text{far}}\cos\phi_i)(-\cos\psi\cos\phi_i - \sin\psi\sin\phi_i\cos\theta_i) \\ &\quad + (-E_\theta^{\text{far}}\sin\theta_i)(\sin\psi\sin\theta_i) \\ &= -E_\theta^{\text{far}}\sin\psi - E_\phi^{\text{far}}\cos\psi\end{aligned} \tag{12.49}$$

Thus, when $\psi = 0^\circ$, the co-polarized component is defined as E_ϕ and the cross-polarized component is E_θ; conversely, for $\psi = 90^\circ$, the co-polarized component is E_θ and the cross-polarized component is E_ϕ.

Note that the unit reference vector $\hat{k}^{\text{inc}} \times \hat{z}$ is in the same direction as $-\hat{\phi}$, since

$$\begin{aligned}\hat{k}^{\text{inc}} \times \hat{z} &= -\hat{y}\sin\theta_i\cos\phi_i + \hat{x}\sin\theta_i\sin\phi_i \\ &= \sin\theta_i(\hat{x}\sin\phi_i - \hat{y}\cos\phi_i) \quad q\end{aligned} \tag{12.50a}$$

$$\hat{\phi} = -\hat{x}\sin\phi + \hat{y}\cos\phi \tag{12.50b}$$

$$\hat{k}^{\text{inc}} \times \hat{z} = -\sin\theta_i\hat{\phi} \tag{12.50c}$$

When $\psi = 180^\circ$, the incident wave is sometimes considered ϕ-polarized.

Validation of Near Fields for a PEC Cube

In the first example, we compute the magnitude of the surface electric current along the perfectly conducting cube using the MRTD scheme with the validation

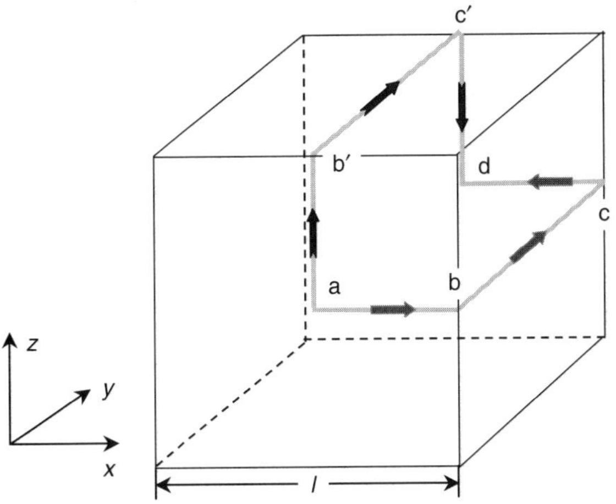

FIGURE 12.3 Geometry of a PEC cube scatterer.

of the FDTD method [1]. Each side of the PEC cube is divided into 20 intervals for both the FDTD and the MRTD. The loci are along two different straight lines traversing the cube surface, as shown in Figure 12.3. The field components of the plane-wave excitation are $E^{\text{inc}} = E_z^{\text{inc}}$ and $H^{\text{inc}} = H_x^{\text{inc}}$. The incident wave propagates in the positive y-direction; that is, $\theta_i = 90^\circ$, $\phi_i = 90^\circ$, and $\psi = 90^\circ$. At the frequency of 3 GHz, the electric size of the PEC cube is $kl = 2$ (l is the side length of the PEC cube). The total computational domain (excluding APML regions) contains $30 \times 30 \times 30$ cells for both cases, and its target dimensions are $20 \times 20 \times 20$ cells. The surface current is given by $\vec{J} = \hat{n} \times \vec{H}$, where $\vec{H}$ is the total magnetic field, similar to that used in [5].

To obtain the current density for the loop $\overline{ab'c'd}$, the tangential magnetic field H_{tan} is taken as H_x along the entire path. To obtain the z-directed current density J_z for the loop $\overline{abcd}$, the tangential magnetic field H_{tan} is taken as H_x, H_y, H_z for paths $\overline{ab}$, $\overline{bc}$, and $\overline{cd}$, respectively. Figures 12.4–12.7 show the magnitudes of the looping currents and the phases of the electric current on the surfaces of the two loops. The results from the MRTD agree with those derived from the FDTD method [1] in terms of accuracy, along the two loops $\overline{ab'c'd}$ and $\overline{abcd}$.

Scattering Analysis for a PEC Plate

We now analyze a PEC plate to illustrate the application of the MRTD method. The dimensions of the plate are 29 cm × 29 cm × 1 cm, as shown in Figure 12.8. Table 12.1 lists the dimensions of the target discretization and the computation grid size used in the MRTD scheme and the FDTD method [3]. We first consider a Gaussian pulse incident along the $+z$-direction. The incident wave is normal to the plate, that is, $\theta_i = 0^\circ$, $\phi_i = 0^\circ$, $\psi = 90^\circ$, $\theta_s = 0^\circ$, and $\phi_s = 180^\circ$, and Figure 12.9 plots the monostatic RCS versus frequency. Next, we consider

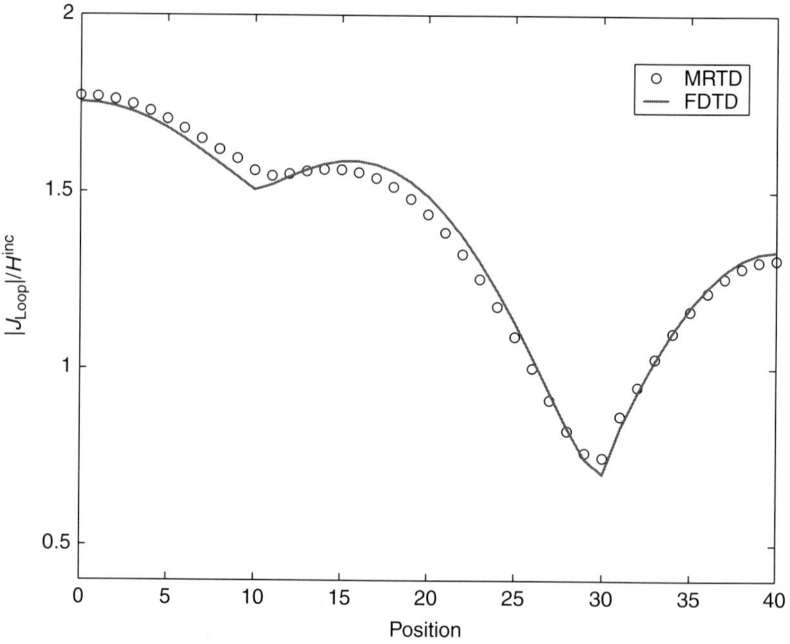

FIGURE 12.4 Magnitude of the loop $\overline{ab'c'd}$ current.

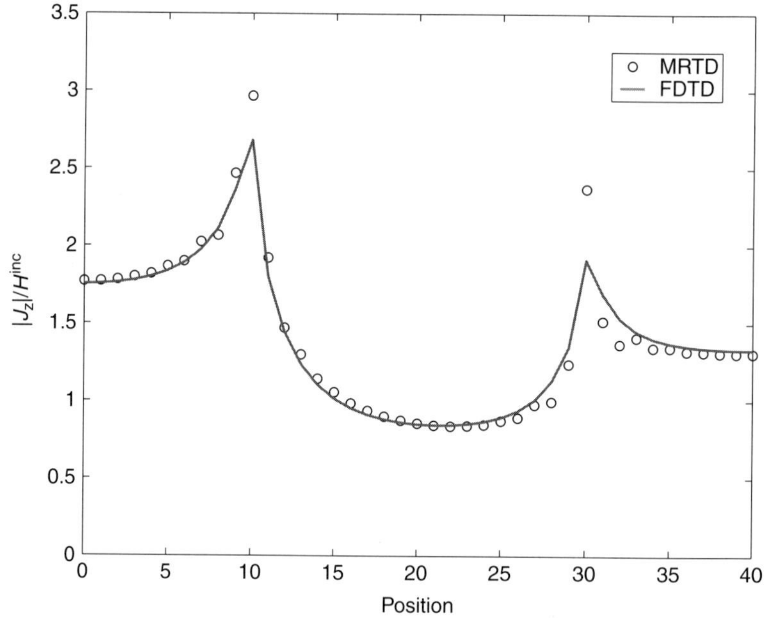

FIGURE 12.5 Magnitude of the looping $\overline{abcd}$.

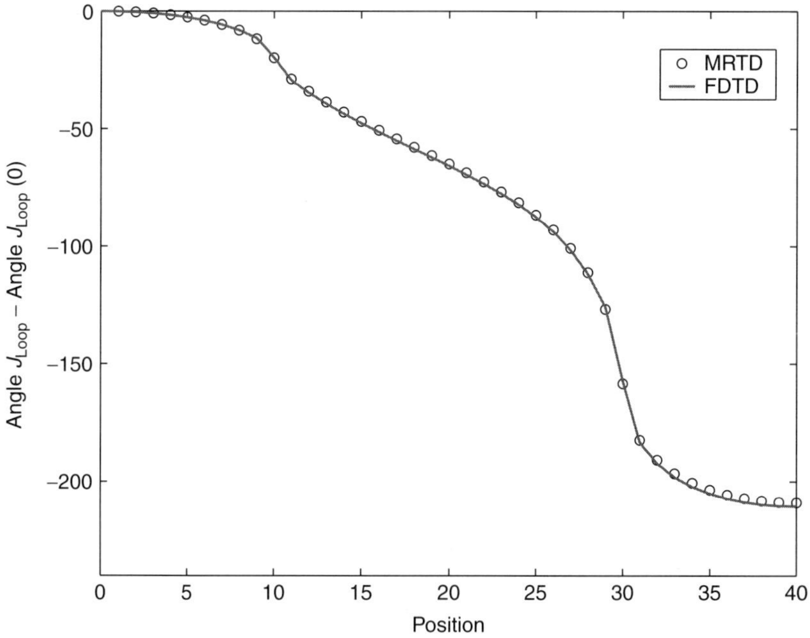

FIGURE 12.6 Phase of looping $\overline{ab'c'd}$ current.

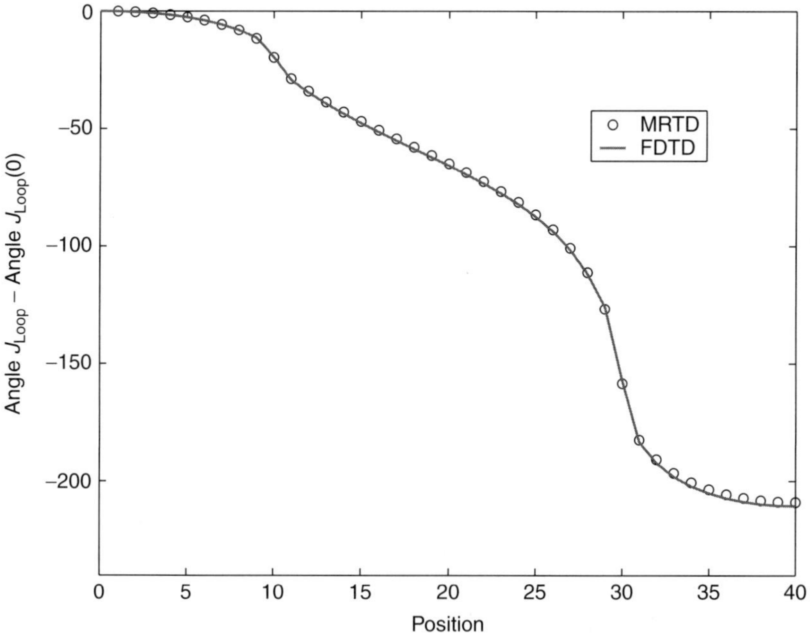

FIGURE 12.7 Phase of looping $\overline{abcd}$ current.

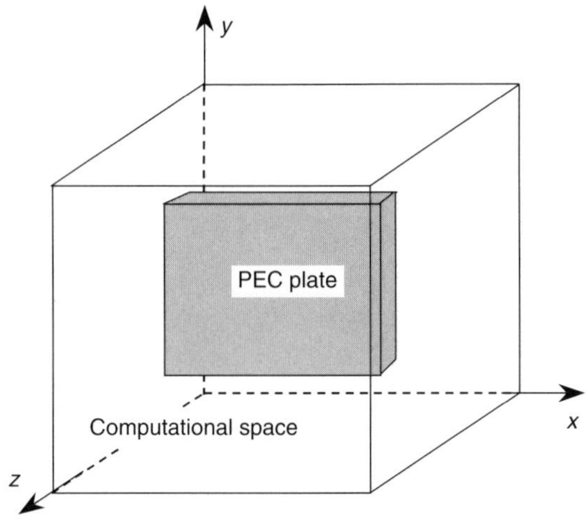

FIGURE 12.8 Geometry of a square PEC plate scatterer.

TABLE 12.1 Discretization of PEC Plate and Dielectric Flat Plate Using the MRTD Scheme and the FDTD Method

Target / Discretization	PEC Plate ($29 \times 29 \times 1$ cm^3)		Lossless and Lossy Dielectric Plates ($18 \times 18 \times 6$ cm^3)	
	MRTD	FDTD	MRTD	FDTD
$\Delta x = \Delta y$ (cm), Δz (cm)	2.64, 1.0	1.0, 1.0	0.90, 0.86	0.3, 0.3
$\Delta t = (10^{-11}$ s)	1.18	1.92	0.626	0.5
Target dimensions (cells)	$11 \times 11 \times 1$	$29 \times 29 \times 1$	$20 \times 20 \times 7$	$60 \times 60 \times 20$
Total computation space (containing APMLs) (cells)	$35 \times 35 \times 25$	$55 \times 55 \times 25$	$44 \times 44 \times 31$	$84 \times 84 \times 44$

an incident plane wave with $\theta_i = 45^\circ, \phi_i = 30^\circ$, and $\psi = 0^\circ$, and we calculate the bistatic RCS with $\theta_s = 45^\circ$ and $\phi_s = 210^\circ$. Figures 12.10 and 12.11 compare the numerical results of the MRTD co-polarized and cross-polarized RCSs in the frequency domain with those of the FDTD method and the MoM [11], respectively. They show good agreement among all the methods. Here, an eight-layer AMPL is used in the boundary truncation. The computational CPU time used for normal incidence on the plate is about 1125 seconds for the MRTD and 1208 seconds for the FDTD method, using a 500-MHz Alpha digital workstation.

The total number of time steps in both cases is 1000. Although the CPU times for the two cases are comparable, the MRTD scheme uses less memory space ($V_{\mathrm{MRTD}}/V_{\mathrm{FDTD}} \approx 40.5\%$), because the sampling rate of the MRTD ($\Delta x = \Delta y \approx \lambda_{\min}/4.74$) is lower than that in the FDTD ($\Delta x = \Delta y = \lambda_{\min}/10.0$), where $\lambda_{\min}$ is the wavelength of the highest frequency of interest. In fact, the grid density used in the FDTD method is rather "coarse."

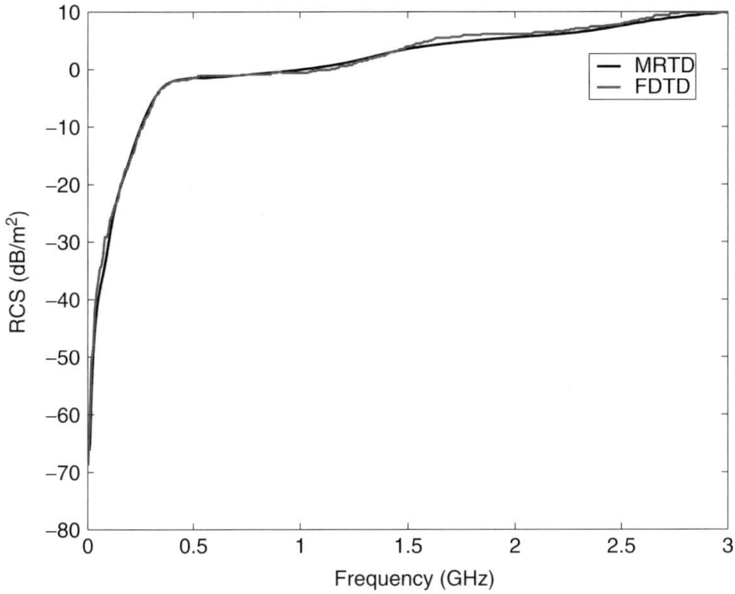

FIGURE 12.9 Backscattering RCS as a function of frequency ($\theta_i = 0^\circ$, $\phi_i = 0^\circ$, $\psi = 90^\circ$, $\theta_s = 0^\circ$, and $\phi_s = 180^\circ$) for a square PEC plate scatterer.

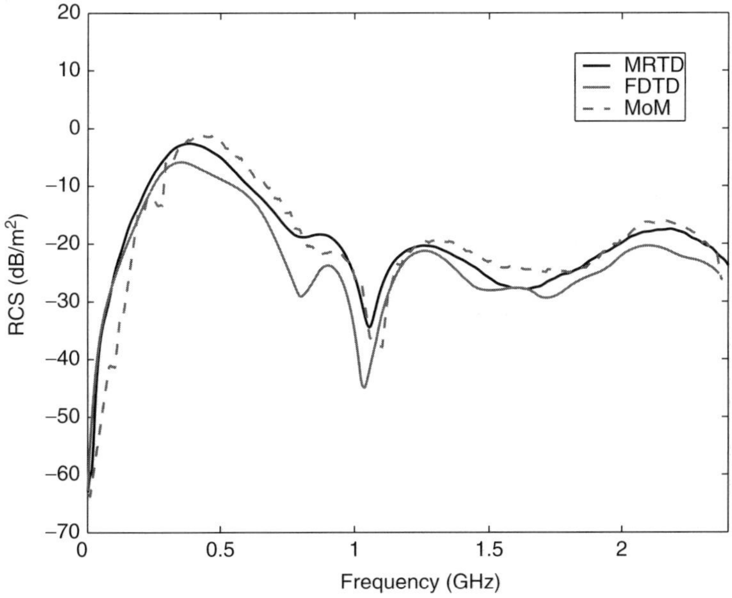

FIGURE 12.10 Co-polarized backscattering RCS as a function of frequency ($\theta_i = 45^\circ$, $\phi_i = 30^\circ$, $\psi = 0^\circ$, $\theta_s = 45^\circ$, and $\phi_s = 210^\circ$) for a square PEC plate scatterer.

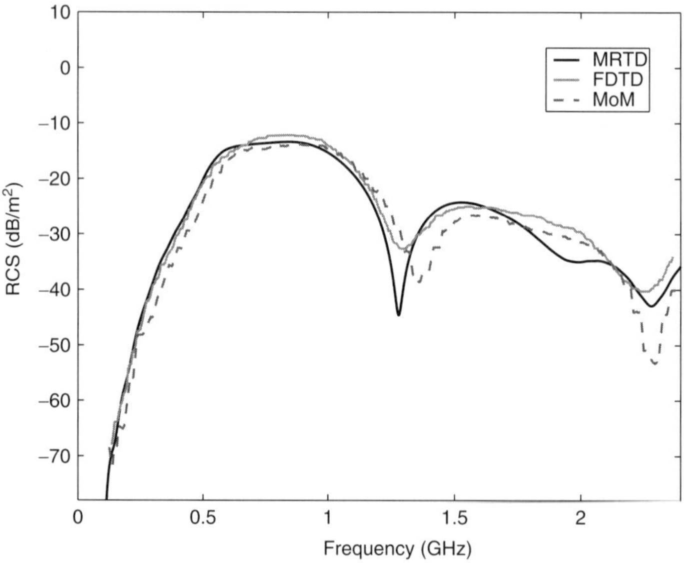

FIGURE 12.11 Cross-polarized backscattering RCS as a function of frequency ($\theta_i = 45°$, $\phi_i = 30°$, $\psi = 0°$, $\theta_s = 45°$, and $\phi_s = 210°$) for a square PEC plate scatterer.

Scattering Analysis for a Lossless Dielectric Flat Plate

Next, we apply the MRTD scheme to analyze a lossless dielectric ($\varepsilon_r = 2.0$) flat plate (18 cm × 18 cm × 6 cm). A Gaussian pulse with a frequency bandwidth of 5 GHz is incident on the target with $\theta_i = 0°$, $\phi_i = 0°$, and $\psi = 90°$. Table 12.1 also lists the dimensions of the structure discretization used in the MRTD scheme and the FDTD method, respectively [12]. Figure 12.12 shows the results for the forward scattering ($\theta_s = 0°$ and $\phi_s = 0°$) RCS as a function of frequency, while Figure 12.13 displays the RCS over the entire span of angles (θ_s). The eight-layer AMPL is used to truncate the boundary. We note that there is a good agreement between the RCS results derived by using the MRTD and the FDTD methods; however, the MRTD scheme uses much less space and computational time. For forward scattering ($\theta_s = \phi_s = 0°$), the CPU times for the MRTD and the FDTD are about 2525 seconds and 4692 seconds, respectively, when the same 500-MHz Alpha digital workstation is used. The total number of time steps in both cases is 1000. The CPU time and memory for the MRTD simulation are about 54% and 19.3% of the FDTD, respectively.

Scattering from a Lossy Dielectric Flat Plate

Finally, we apply the MRTD scheme to the analysis of a lossy dielectric plate, which is characterized by $\varepsilon_r = 2.0$ and $\sigma = 0.01$ s/m, whose dimensions are 18 cm × 18 cm × 6 cm. A Gaussian pulse, with a frequency bandwidth of 5 GHz, is incident from the direction of $\theta_i = 0°$, $\phi_i = 0°$, and $\psi = 90°$. Table 12.1 lists the dimensions of the discretization used in the MRTD scheme and FDTD modeling [12], and Figure 12.14 shows the monostatic RCS ($\theta_s = 30°$ and $\phi_s = 0°$)

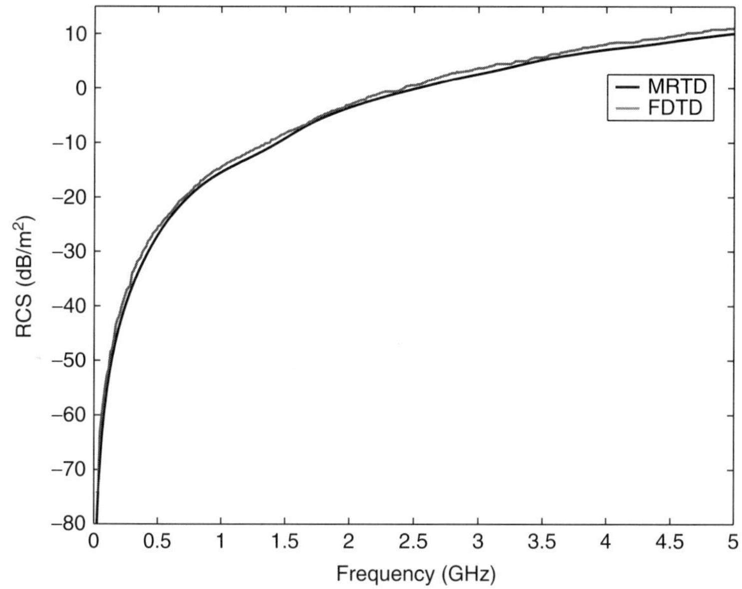

FIGURE 12.12 Forward scattering RCS for a lossless dielectric flat plate as a function of frequency ($\theta_i = 0°$, $\phi_i = 0°$, $\psi = 90°$, $\theta_s = 0°$, and $\phi_s = 0°$).

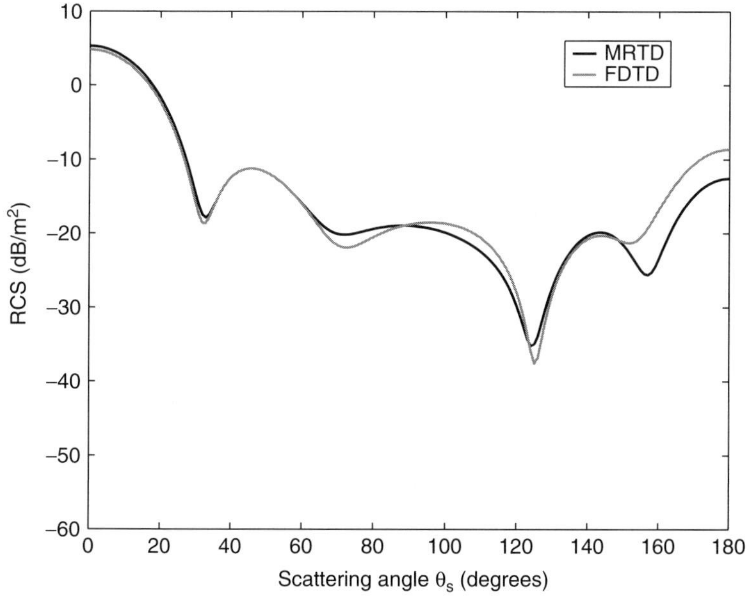

FIGURE 12.13 RCS for a lossless dielectric flat plate as a function of θ_s at $f = 3.03$ GHz and $\phi_s = 0°$.

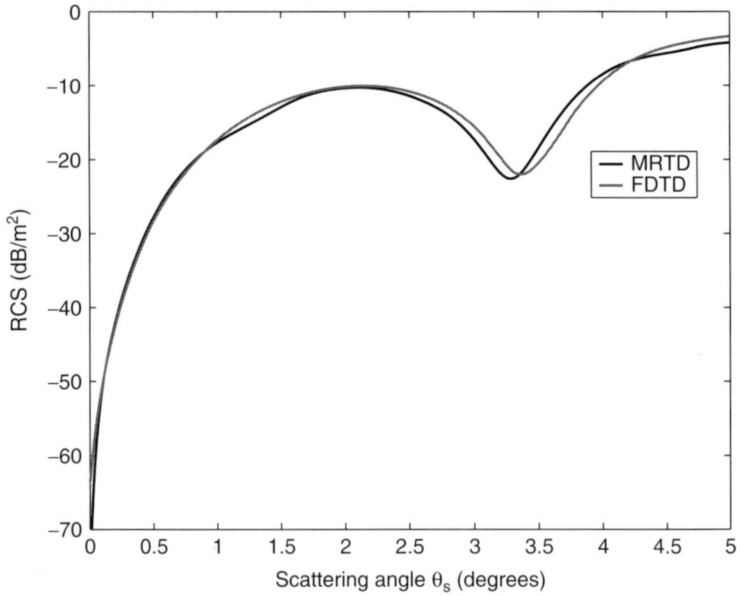

FIGURE 12.14 RCS for a lossy dielectric flat plate as a function of frequency ($\theta_i = 0°$, $\phi_i = 0°$, $\psi = 90°$, $\theta_s = 30°$, and $\phi_s = 0°$).

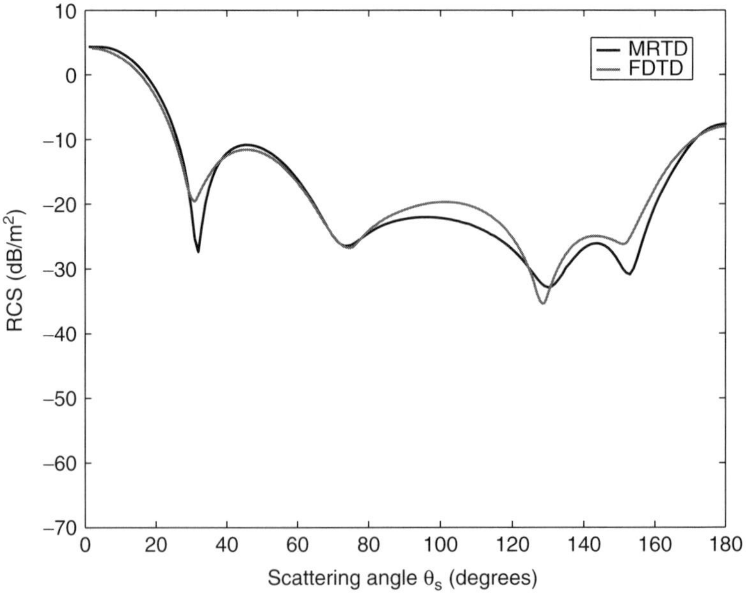

FIGURE 12.15 RCS for a lossy dielectric flat plate as a function of θ_s at $f = 3.15$ GHz and $\phi_s = 0°$.

as a function of frequency, obtained from the above two schemes. Figure 12.15 compares the RCS results over the entire span of angles (θ_s) and shows that there is good agreement between the MRTD and FDTD methods.

The MRTD CPU time for the case is approximately 2488 seconds, while it is 4734 seconds on the same computer for a total of 1000 time steps. The MRTD CPU time is approximately 53% of that used in the FDTD method and the MRTD computational space is only 19.3% of that in the FDTD.

12.5 CONCLUSIONS

We developed a MRTD scheme for analyzing the problem of scattering from three-dimensional targets. We used the pure scattered field formulation to construct the incident fields inside the target region and developed a 3D-MRTD near-to-far-zone field transformation to compute the scattering RCS for various target objects. We validated the near-fields, the far-fields, and the RCS for a number of test objects involving PEC and dielectric (lossless and lossy) cylinders and plates. The results from the MRTD method are in good agreement with those derived from the FDTD. Since the MRTD scheme can predict the electromagnetic field distribution and calculate the RCS in an efficient manner compared to the FDTD method, it possesses great potential for solving electromagnetic problems of a general nature, particularly those involving large objects.

REFERENCES

[1] K. Umashankar and A. Taflove, "A novel method to analyze electromagnetic scattering of complex objects," *IEEE Trans. Electromagn. Compat.*, vol. EMC-24, pp. 397–405, Nov. 1982.

[2] A. Taflove and K. Umashankar, "Review of FD-TD numerical modeling of electromagnetic wave scattering and radar cross section," *Proc. IEEE*, vol. 77, pp. 682–698, May 1989.

[3] R. J. Luebbers, Karl S. Kunz, M. Schneider, and F. Hunsberger, "A finite-difference time-domain near zone to far zone transformation," *IEEE Trans. Antenna Propag.*, vol. 39, no. 4, pp. 429–433, Apr. 1991.

[4] M. Krumpholz and L. P. B. Katehi, "MRTD: New time domain schemes based on multiresolution analysis," *IEEE Trans. Microwave Theory Tech.*, vol. 44, no. 4, pp. 555–571, Apr. 1996.

[5] Q. Cao, Y. Chen, and P. K. A. Wai, "MRTD electromagnetic scattering analysis," *Microwave Opt. Technol. Lett.*, vol. 28, no. 3, pp. 189–195, Feb. 2001.

[6] Q. Cao, K. Kamma, P. K. A. Wai, and Y. Chen, "Three-dimensional electromagnetic scattering analysis using MRTD scheme," *J. Electromagn. Waves Appl.*, vol. 17, no. 12, 1683–1701, 2003.

[7] K. S. Kunz and R. J. Luebbers, *The Finite Difference Time Domain Method for Electromagnetic*, CRC Press, Boca Raton, FL, 1993.

[8] S. D. Gedney, "An anisotropic perfectly matched layered-absorbing medium for the truncation of FDTD lattice," *IEEE Trans. Antennas Propag.*, vol. 44, no. 12, pp. 1630–1939, Dec. 1996.

[9] C. A. Balanis, *Advanced Engineering Electromagnetic*, Wiley, Hoboken, NJ, 1989.

[10] A. Taflove, *Computational Electrodynamics—The Finite-Difference Time-Domain Method*, Artech House, Norwood, MA, 1995.

[11] E. Newman, "A user's manual for the electromagnetic surface patch code: ESP version IV," preliminary version, Ohio State Univ. Res. Foundation, ElectroSci. Lab., Dept. Elec. Eng., Columbus, OH 43212, Aug. 1988.

[12] Q. Li and Y. Chen, "Applications of the PSTD for scattering analysis," *IEEE Trans. Antennas Propag.*, vol. 50, no. 9, pp. 1317–1319, Sept. 2002.

APPENDIX A

Generalized Differential Matrix Operators

A.1 GENERALIZED DIFFERENTIAL MATRIX OPERATORS

The operations involved in the ∇ entail complex mathematical manipulations that include the gradient, divergence, and curl operations. All these vector operators can be expressed as generalized differential matrix operators (GDMOs), and the required manipulation can be simplified by taking advantage of the properties of matrix operations [1]. For an arbitrary orthogonal coordinate system (q_1, q_2, q_3), we use the following notation to define a vector $\vec{A}$:

$$|A\rangle := \begin{bmatrix} A_1 & A_2 & A_3 \end{bmatrix}^t \tag{A.1}$$

Next, the gradient, divergence, and curl operators are defined in the GDMO format as

$$|\nabla\rangle := \begin{bmatrix} \frac{1}{h_1}\partial_{q_1} & \frac{1}{h_2}\partial_{q_2} & \frac{1}{h_3}\partial_{q_3} \end{bmatrix}^t \tag{A.2}$$

$$\langle\nabla\cdot| := \frac{1}{h_1 h_2 h_3}[\, \partial_{q_1} h_2 h_3 \quad \partial_{q_2} h_3 h_1 \quad \partial_{q_3} h_1 h_2 \,] \tag{A.3}$$

$$[\nabla\times] := \frac{1}{h_1 h_2 h_3}\begin{bmatrix} 0 & -h_1 \partial_{q_3} h_2 & h_1 \partial_{q_2} h_3 \\ h_2 \partial_{q_3} h_1 & 0 & -h_2 \partial_{q_1} h_3 \\ -h_3 \partial_{q_2} h_1 & h_3 \partial_{q_1} h_2 & 0 \end{bmatrix} \tag{A.4}$$

where the metric coefficients $h_i S$ are given by

$$h_i := \sqrt{(\partial_{q_i} x)^2 + (\partial_{q_i} y)^2 + (\partial_{q_i} z)^2} \qquad \partial_{q_i} := \frac{\partial}{\partial q_i} \quad (i = 1, 2, 3) \tag{A.5}$$

Multiresolution Time Domain Scheme for Electromagnetic Engineering
By Yinchao Chen, Qunsheng Cao, and Raj Mittra
ISBN 0-471-27230-2

In the above equations, $| \, \rangle$, $\langle \, |$, and $[\,]$ refer to the column, row, and square matrices, respectively, with t indicating the transpose of the matrix. A summary of the metric coefficients of orthogonal coordinates [2] is given in Table A.2 for the most commonly used coordinate systems [3]. Using the above definitions of the GDMOs, complex vector expressions can be handled by using simple matrix techniques. For example, Laplace and curl operations can be rewritten in a matrix product form as

$$\nabla^2 := \langle \nabla \cdot \| \nabla \rangle = \frac{1}{h_1 h_2 h_3} [\, \partial_{q_1} h_2 h_3 \quad \partial_{q_2} h_3 h_1 \quad \partial_{q_3} h_1 h_2 \,] \times \left[\frac{1}{h_1}\partial_{q_1} \quad \frac{1}{h_2}\partial_{q_2} \quad \frac{1}{h_3}\partial_{q_3} \right]^t \tag{A.6}$$

$$[\nabla \times] | A \rangle := \frac{1}{h_1 h_2 h_3} \begin{bmatrix} 0 & -h_1 \partial_{q_3} h_2 & h_1 \partial_{q_2} h_3 \\ h_2 \partial_{q_3} h_1 & 0 & -h_2 \partial_{q_1} h_3 \\ -h_3 \partial_{q_1} h_1 & h_3 \partial_{q_1} h_2 & 0 \end{bmatrix} \begin{bmatrix} A_1 \\ A_2 \\ A_3 \end{bmatrix} \tag{A.7}$$

Note that the GDMOs obey the matrix manipulation rules in that the differential operators are not permitted to commute arbitrarily.

More complete forms of the GDMOs are defined in Table A.1. Also, the GDMOs for an arbitrary orthogonal coordinate system can be obtained by substituting the appropriate coordinates and metric coefficients, which are listed in Table A.2. Finally, we mention that multiple curl operations, which are difficult to deal with using the conventional determinant form of curl equations, can readily be handled in the context of the GDMOs by cascading matrices.

As an illustration, let us consider the representation of the components of the electric field in terms of magnetic and electric potentials [4] as follows:

$$|E\rangle = -j\omega |A\rangle - \frac{j}{\omega\mu\varepsilon} |\nabla\rangle (\langle \nabla \cdot \| A \rangle) - \frac{1}{\varepsilon} [\nabla \times] | F \rangle$$

$$= -j\omega \begin{bmatrix} A_x \\ A_y \\ A_z \end{bmatrix} - \frac{j}{\omega\mu\varepsilon} \begin{bmatrix} \partial_x \\ \partial_y \\ \partial_z \end{bmatrix} \left([\, \partial_x \quad \partial_y \quad \partial_z \,] \begin{bmatrix} A_x \\ A_y \\ A_z \end{bmatrix} \right)$$

$$- \frac{1}{\varepsilon} \begin{bmatrix} 0 & -\partial_z & \partial_y \\ \partial_z & 0 & -\partial_x \\ -\partial_y & \partial_x & 0 \end{bmatrix} \begin{bmatrix} F_x \\ F_y \\ F_z \end{bmatrix}$$

$$= \begin{Bmatrix} -j\omega A_x - \dfrac{j}{\omega\mu\varepsilon} \partial_x [\partial_x A_x + \partial_y A_y + \partial_z A_z] - \dfrac{1}{\varepsilon} [\partial_y F_z - \partial_z F_y] \\ -j\omega A_y - \dfrac{j}{\omega\mu\varepsilon} \partial_y [\partial_x A_x + \partial_y A_y + \partial_z A_z] - \dfrac{1}{\varepsilon} [\partial_z F_x - \partial_x F_z] \\ -j\omega A_z - \dfrac{j}{\omega\mu\varepsilon} \partial_z [\partial_x A_x + \partial_y A_y + \partial_z A_z] - \dfrac{1}{\varepsilon} [\partial_x F_y - \partial_y F_x] \end{Bmatrix}$$

TABLE A.1 Generalized Differential Matrix Operators

Vector	GDMOs
$\vec{A}, \vec{B}$	$\lvert A\rangle = [\,A_1 \quad A_2 \quad A_3\,]^t, \lvert B\rangle = [\,B_1 \quad B_2 \quad B_3\,]^t$
$\vec{A}\cdot$	$\langle A\cdot\rvert = [\,A_1 \quad A_2 \quad A_3\,] \Rightarrow \vec{A}\cdot\vec{B} = \langle A\cdot\Vert B\rangle$
$\vec{A}\times$	$[A\times] = \begin{bmatrix} 0 & -A_3 & A_2 \\ A_3 & 0 & -A_1 \\ -A_2 & A_1 & 0 \end{bmatrix} \Rightarrow \vec{A}\times\vec{B} = [A\times]\lvert B\rangle$
∇	$\lvert\nabla\rangle = \left[\frac{1}{h_1}\partial_{q_1} \quad \frac{1}{h_2}\partial_{q_2} \quad \frac{1}{h_3}\partial_{q_3}\right]^t \Rightarrow \nabla u = \lvert\nabla\rangle u$
$\nabla\cdot$	$\langle\nabla\cdot\rvert = \frac{1}{h_1h_2h_3}[\,\partial_{q_1}h_2h_3 \quad \partial_{q_2}h_3h_1 \quad \partial_{q_3}h_1h_2\,] \Rightarrow \nabla\cdot\vec{A} = \langle\nabla\cdot\Vert A\rangle$
∇^2	$\langle\nabla\cdot\Vert\nabla\rangle = \frac{1}{h_1h_2h_3}[\,\partial_{q_1}h_2h_3 \quad \partial_{q_2}h_3h_1 \quad \partial_{q_3}h_1h_2\,]\left[\frac{1}{h_1}\partial_{q_1} \quad \frac{1}{h_2}\partial_{q_2} \quad \frac{1}{h_3}\partial_{q_3}\right]^t$
$\nabla\times$	$[\nabla\times] = \frac{1}{h_1h_2h_3}\begin{bmatrix} 0 & -h_1\partial_{q_3}h_2 & h_1\partial_{q_2}h_3 \\ h_2\partial_{q_3}h_1 & 0 & -h_2\partial_{q_1}h_3 \\ -h_3\partial_{q_2}h_1 & h_3\partial_{q_1}h_2 & 0 \end{bmatrix} \Rightarrow \nabla\times\vec{A} = [\nabla\times]\lvert A\rangle$

In the above formulations, the h_i are metric coefficients and the q_i are generalized orthogonal coordinates. Under the specified coordinate system, we can easily find the corresponding GDMO forms by substituting (q_i, h_i) in the corresponding equations.

As seen above, the GDMOs provide an efficient and convenient way to carry out the vector differential operation.

A.2 GDMO REPRESENTATION OF MAXWELL AND WAVE EQUATIONS

Let us now proceed to the use of the GDMOs to derive the Maxwell and vector wave equations for the fields in the spatial domain in any orthogonal coordinate system. Toward this end, we use the GDMOs to replace the operators in the Maxwell as well as wave equations by matrix equations. We do this by choosing the appropriate coordinates and the corresponding metric coefficients listed in Tables A.1 and A.2. We consider inhomogeneous Maxwell equations including the electric and magnetic source terms ($\vec{J}$, ρ_e; $\vec{M}$, ρ_m) in a material medium characterized by full permittivity and permeability tensors. Using the GDMO notation, we can express the unified Maxwell equations as

$$[\nabla\times]\lvert E\rangle = -j\omega\mu_0[\mu]\lvert H\rangle - \lvert M\rangle \tag{A.8a}$$

$$[\nabla\times]\lvert H\rangle = +j\omega\varepsilon_0[\varepsilon]\lvert E\rangle + \lvert J\rangle \tag{A.8b}$$

$$\varepsilon_0\langle\nabla\cdot\Vert[\varepsilon_r]\rvert E\rangle = \rho_e \tag{A.9a}$$

$$\mu_0\langle\nabla\cdot\Vert[\mu_r]\rvert H\rangle = \rho_m \tag{A.9b}$$

TABLE A.2 Orthogonal Coordinate Systems

Coordinates	Variable	Metric Coefficient	Relation
Cartesian	(x, y, z)	$(1,1,1)$	$x = x$ $y = y$ $z = z$
Cylindrical	(r, ϕ, z)	$(1, r, 1)$	$x = r\cos\phi$ $y = r\sin\phi$ $z = z$
Spherical	(R, θ, ϕ)	$(1, R, R\sin\theta)$	$x = R\sin\theta\cos\phi$ $y = R\sin\theta\sin\phi$ $z = R\cos\theta$
Parabolic cylinder	(η, ξ, z)	$\left[\sqrt{\eta^2+\xi^2},\ \sqrt{\eta^2+\xi^2},\ 1\right]$	$x = (\eta^2 - \xi^2)/2$ $y = \eta\xi$ $z = z$
Elliptic cylinder	(η, ξ, z)	$\left[c_e\sqrt{\dfrac{\xi^2-\eta^2}{1-\eta^2}},\ c_e\sqrt{\dfrac{\xi^2-\eta^2}{\xi^2-1}},\ 1\right]$	$x = c_e\eta\xi$ $y = c_e\sqrt{(1-\eta^2)(\xi^2-1)}$ $z = z$
Prolate spheroidal	(η, ξ, z)	$\left[c_p\sqrt{\dfrac{\xi^2-\eta^2}{1-\eta^2}},\ c_p\sqrt{\dfrac{\xi^2-\eta^2}{\xi^2-1}},\ c_p\sqrt{(1-\eta^2)(\xi^2-1)}\right]$	$x = c_p\cos\phi\sqrt{(1-\eta^2)(\xi^2-1)}$ $y = c_p\sin\phi\sqrt{(1-\eta^2)(\xi^2-1)}$ $z = c_p\eta\xi$
Oblate spheroidal	(ξ, η, ϕ)	$\left[c_o\sqrt{\dfrac{\xi^2-\eta^2}{\xi^2-1}},\ c_o\sqrt{\dfrac{\xi^2-\eta^2}{1-\eta^2}},\ c_p\xi\eta\right]$	$x = c_o\xi\eta\cos\phi$ $y = c_o\xi\eta\sin\phi$ $z = c_o\sqrt{(\xi^2-1)(1-\eta^2)}$

In the above formulations, c_i ($i = e, p$, and o) are constants given in [3].

or explicitly as

$$\begin{bmatrix} 0 & -\partial_z & \partial_y \\ \partial_z & 0 & -\partial_x \\ -\partial_y & \partial_x & 0 \end{bmatrix} \begin{bmatrix} E_x \\ E_y \\ E_z \end{bmatrix} = -j\omega\mu_0 \begin{bmatrix} \mu_{xx} & \mu_{xy} & \mu_{xz} \\ \mu_{yx} & \mu_{yy} & \mu_{yz} \\ \mu_{zx} & \mu_{zy} & \mu_{zz} \end{bmatrix} \begin{bmatrix} H_x \\ H_y \\ H_z \end{bmatrix} - \begin{bmatrix} M_x \\ M_y \\ M_z \end{bmatrix} \tag{A.10a}$$

$$\begin{bmatrix} 0 & -\partial_z & \partial_y \\ \partial_z & 0 & -\partial_x \\ -\partial_y & \partial_x & 0 \end{bmatrix} \begin{bmatrix} H_x \\ H_y \\ H_z \end{bmatrix} = j\omega\varepsilon_0 \begin{bmatrix} \varepsilon_{xx} & \varepsilon_{xy} & \varepsilon_{xz} \\ \varepsilon_{yx} & \varepsilon_{yy} & \varepsilon_{yz} \\ \varepsilon_{zx} & \varepsilon_{zy} & \varepsilon_{zz} \end{bmatrix} \begin{bmatrix} E_x \\ E_y \\ E_z \end{bmatrix} + \begin{bmatrix} J_x \\ J_y \\ J_z \end{bmatrix} \tag{A.10b}$$

$$\varepsilon_o[\,\partial_x \quad \partial_y \quad \partial_z\,]\begin{bmatrix} \varepsilon_{xx} & \varepsilon_{xy} & \varepsilon_{xz} \\ \varepsilon_{yx} & \varepsilon_{yy} & \varepsilon_{yz} \\ \varepsilon_{zx} & \varepsilon_{zy} & \varepsilon_{zz} \end{bmatrix}\begin{bmatrix} E_x \\ E_y \\ E_z \end{bmatrix} = \rho_e \tag{A.11a}$$

$$\mu_o[\,\partial_x \quad \partial_y \quad \partial_z\,]\begin{bmatrix} \mu_{xx} & \mu_{xy} & \mu_{xz} \\ \mu_{yx} & \mu_{yy} & \mu_{yz} \\ \mu_{zx} & \mu_{zy} & \mu_{zz} \end{bmatrix}\begin{bmatrix} H_x \\ H_y \\ H_z \end{bmatrix} = \rho_m \tag{A.11b}$$

where the time-dependent factor $e^{j\omega t}$ is implicit. In a source-free region, the vector wave equations can readily be written in matrix form as

$$[\nabla\times][\mu_r]^{-1}[\nabla\times]|E\rangle - k_0^2[\varepsilon_r]|E\rangle = |0\rangle \tag{A.12a}$$

$$[\nabla\times][\varepsilon_r]^{-1}[\nabla\times]|H\rangle - k_0^2[\mu_r]|H\rangle = |0\rangle \tag{A.12b}$$

In an expanded form they read

$$\begin{bmatrix} 0 & -\partial_z & \partial_y \\ \partial_z & 0 & -\partial_x \\ -\partial_y & \partial_x & 0 \end{bmatrix}\begin{bmatrix} \mu_{xx} & \mu_{xy} & \mu_{xz} \\ \mu_{yx} & \mu_{yy} & \mu_{yz} \\ \mu_{zx} & \mu_{zy} & \mu_{zz} \end{bmatrix}^{-1}\begin{bmatrix} 0 & -\partial_z & \partial_y \\ \partial_z & 0 & -\partial_x \\ -\partial_y & \partial_x & 0 \end{bmatrix}\begin{bmatrix} E_x \\ E_y \\ E_z \end{bmatrix}$$
$$- k_0^2\begin{bmatrix} \varepsilon_{xx} & \varepsilon_{xy} & \varepsilon_{xz} \\ \varepsilon_{yx} & \varepsilon_{yy} & \varepsilon_{yz} \\ \varepsilon_{zx} & \varepsilon_{zy} & \varepsilon_{zz} \end{bmatrix}\begin{bmatrix} E_x \\ E_y \\ E_z \end{bmatrix} = \begin{bmatrix} 0 \\ 0 \\ 0 \end{bmatrix} \tag{A.13a}$$

$$\begin{bmatrix} 0 & -\partial_z & \partial_y \\ \partial_z & 0 & -\partial_x \\ -\partial_y & \partial_x & 0 \end{bmatrix}\begin{bmatrix} \varepsilon_{xx} & \varepsilon_{xy} & \varepsilon_{xz} \\ \varepsilon_{yx} & \varepsilon_{yy} & \varepsilon_{yz} \\ \varepsilon_{zx} & \varepsilon_{zy} & \varepsilon_{zz} \end{bmatrix}^{-1}\begin{bmatrix} 0 & -\partial_z & \partial_y \\ \partial_z & 0 & -\partial_x \\ -\partial_y & \partial_x & 0 \end{bmatrix}\begin{bmatrix} H_x \\ H_y \\ H_z \end{bmatrix}$$
$$- k_0^2\begin{bmatrix} \mu_{xx} & \mu_{xy} & \mu_{xz} \\ \mu_{yx} & \mu_{yy} & \mu_{yz} \\ \mu_{zx} & \mu_{zy} & \mu_{zz} \end{bmatrix}\begin{bmatrix} H_x \\ H_y \\ H_z \end{bmatrix} = \begin{bmatrix} 0 \\ 0 \\ 0 \end{bmatrix} \tag{A.13b}$$

The advantage of using the above differential matrix equations in place of the complex vector wave equations is that we can formulate complex boundary-value problems in a matrix form relatively easily; and, furthermore, many fundamental matrix properties such as addition, subtraction, and multiplication can be used to simplify the solution of the resulting equations. Also, as mentioned above, the three components of the vector equation can be obtained concurrently. This is especially convenient and efficient for problems dealing with anisotropic media. Since the derivative relations are included in the differential matrix operators, in principle, they cannot communicate with the functions or other related variables. Furthermore, the GDMOs are not limited in their application to the spatial domain alone, and they are equally useful in the spectral domain, which is often utilized in the formulation of a class of electromagnetic boundary-value problems.

TABLE A.3 Comparison of the Required Number of Steps in Conventional and GDMO Formulations

Step	Conventional $\nabla \times (\overset{\rightrightarrows}{\mu}_r^{-1} \cdot \nabla \times \vec{E}) - k_0^2 \overset{\rightrightarrows}{\varepsilon}_r \cdot \vec{E} = \vec{0}$	GDMO $[\nabla\times][\mu_r]^{-1}[\nabla\times]\|E\rangle - k_0^2[\varepsilon]\|E\rangle = \|0\rangle$
1	$\overset{\rightrightarrows}{\mu}^{-1}$	$[\mu_r]^{-1}$
2	$\vec{F}_1 = \nabla \times \vec{E}$	$[a]\|E\rangle = \|0\rangle$
3	$\vec{F}_2 = \overset{\rightrightarrows}{\mu}_r^{-1} \cdot \vec{F}_1$	
4	$\vec{F}_3 = \nabla \times \vec{F}_2$	
5	$\hat{a}_i \cdot (\vec{F}_3 - k_0^2 \overset{\rightrightarrows}{\varepsilon}_r \cdot \vec{E}) = 0$	

Illustrating the efficiency of GDMOs, Table A.3 compares the steps for obtaining the electric field component equations using conventional vector differential operators versus the GDMOs. Obviously, the two steps required by GDMOs provide a much more efficient calculation; otherwise, five steps are needed when using the conventional method.

A.3 GDMOs FOR DIFFERENTIAL AND INTEGRAL FORMULATIONS

In electromagnetics, vector differential and integral operators are frequently encountered and are typically handled via the use of vector and tensor analysis. Here, we present an alternative technique based on application of the GDMOs that provides a simple yet efficient way to handle complex differential and integral operators.

In this section, we consider vector identities and integral theorems that find frequent application in electromagnetics to illustrate the usefulness of the GDMO approach.

Differential Formulations

Let us begin with examining the vector identity

$$\nabla \times \nabla \times \vec{A} = \nabla(\nabla \cdot \vec{A}) - \nabla^2 \vec{A} \tag{A.14}$$

Using the definitions in Table A.1, we can rewrite (A.14) in GDMO notation as

$$[\nabla\times][\nabla\times]|A\rangle = |\nabla\rangle(\langle\nabla \cdot ||A\rangle) - \langle\nabla \cdot ||\nabla\rangle|A\rangle \tag{A.15a}$$

or more explicitly,

$$
\frac{1}{h_1h_2h_3}\begin{bmatrix} 0 & -h_1\partial_{q_3}h_2 & h_1\partial_{q_2}h_3 \\ h_2\partial_{q_3}h_1 & 0 & -h_2\partial_{q_1}h_3 \\ -h_3\partial_{q_1}h_1 & h_3\partial_{q_1}h_2 & 0 \end{bmatrix}
$$

$$
\times\left\{\frac{1}{h_1h_2h_3}\begin{bmatrix} 0 & -h_1\partial_{q_3}h_2 & h_1\partial_{q_2}h_3 \\ h_2\partial_{q_3}h_1 & 0 & -h_2\partial_{q_1}h_3 \\ -h_3\partial_{q_1}h_1 & h_3\partial_{q_1}h_2 & 0 \end{bmatrix}\begin{bmatrix} A_1 \\ A_2 \\ A_3 \end{bmatrix}\right\}
$$

$$
= \begin{bmatrix} \frac{1}{h_1}\partial_{q_1} \\ \frac{1}{h_2}\partial_{q_2} \\ \frac{1}{h_3}\partial_{q_3} \end{bmatrix}\left\{\frac{1}{h_1h_2h_3}[\,\partial_{q_1}h_2h_3 \quad \partial_{q_2}h_3h_1 \quad \partial_{q_3}h_1h_2\,]\begin{bmatrix} A_1 \\ A_2 \\ A_3 \end{bmatrix}\right\}
$$

$$
- \left\{\frac{1}{h_1h_2h_3}[\,\partial_{q_1}h_2h_3 \quad \partial_{q_2}h_3h_1 \quad \partial_{q_3}h_1h_2\,]\begin{bmatrix} \frac{1}{h_1}\partial_{q_1} \\ \frac{1}{h_2}\partial_{q_2} \\ \frac{1}{h_3}\partial_{q_3} \end{bmatrix}\right\}\begin{bmatrix} A_1 \\ A_2 \\ A_3 \end{bmatrix} \tag{A.15b}
$$

Equation (A.15) illustrates the advantage of the GDMO approach over the use of conventional vector formulation. It is difficult to find any explicit form for (A.14) like (A.15b) using vector algebra manipulation, since in the latter approach the curl operator on the left-hand side (LHS) of (A.14) is expressed in the form of a determinant, and an explicit form of the $\nabla \times \nabla\times$ similar to the one in (A.15b) is difficult to obtain. In contrast, the GDMO technique can readily accommodate additional curl operations that are merely series operations in the context of GDMOs. The same is true for the right-hand side (RHS) of (A.14) as well, and the GDMO approach is found to be efficient for this case too. Furthermore, when there are three or more curl operations, the workload reductions offered by GDMOs appear even more attractive since they treat the curls as series operations. An additional list of the GDMOs for vector identities is given in Table A.4.

Integral Formulations

The GDMOs can also be applied to integral theorems, which are valid in arbitrary, orthogonal coordinate systems. We assume that $\vec{A}$ and Φ are continuous functions with first derivatives within a volume V and on the surface S bounded by a contour C. A well-known integral relationship is the divergence theorem,

TABLE A.4 GDMOs for the Vector Identities

Vector Form	GDMO Form
$\vec{a}\cdot(\vec{b}\times\vec{c})=\vec{b}\cdot(\vec{c}\times\vec{a})=\vec{c}\cdot(\vec{a}\times\vec{b})$	$\langle a\cdot\|\{[b\times]\|c\rangle\}=\langle b\cdot\|\{[c\times]\|a\rangle\}=\langle c\cdot\|\{[a\times]\|b\rangle\}$
$\vec{a}\times(\vec{b}\times\vec{c})=(\vec{a}\cdot\vec{c})\vec{b}-(\vec{a}\cdot\vec{b})\vec{c}$	$[a\times]\{[b\times]\|c\rangle\}=\{\langle a\cdot\|\|c\rangle\}\|b\rangle-\{\langle a\cdot\|\|b\rangle\}\|c\rangle$
$\nabla(ab)=a\nabla b+b\nabla a$	$\|\nabla\rangle(ab)=a\|\nabla\rangle b+b\|\nabla\rangle a$
$\nabla\cdot(a\vec{b})=a\nabla\cdot\vec{b}+\vec{b}\cdot\nabla a$	$\langle\nabla\cdot\|\|ab\rangle=a\langle\nabla\cdot\|\|b\rangle+\langle b\cdot\|\|\nabla\rangle a$
$\nabla\times(a\vec{b})=a\nabla\times\vec{b}-\vec{b}\times\nabla a$	$[\nabla\times](a\|b\rangle)=a[\nabla\times]\|b\rangle-[b\times]\|\nabla\rangle a$
$\nabla\cdot(\vec{a}\times\vec{b})=\vec{b}\cdot\nabla\times\vec{a}-\vec{a}\cdot\nabla\times\vec{b}$	$\langle\nabla\cdot\|\{[a\times]\|b\rangle\}=\langle b\cdot\|[\nabla\times]\|a\rangle-\langle a\cdot\|[\nabla\times]\|b\rangle$
$\nabla(\vec{a}\cdot\vec{b})=\vec{a}\times\nabla\times\vec{b}$ $+\vec{b}\times\nabla\times\vec{a}+(\vec{a}\cdot\nabla)\vec{b}+(\vec{b}\cdot\nabla)\vec{a}$	$\|\nabla\rangle\{\langle a\cdot\|\|b\rangle\}=[a\times][\nabla\times]\|b\rangle$ $+[b\times][\nabla\times]\|a\rangle+\{\langle a\cdot\|\|\nabla\rangle\}\|b\rangle+\{\langle b\cdot\|\|\nabla\rangle\}\|a\rangle$
$\nabla\times(\vec{a}\times\vec{b})=\vec{a}\nabla\cdot\vec{b}$ $-\vec{b}\nabla\cdot\vec{a}-(\vec{a}\cdot\nabla)\vec{b}+(\vec{b}\cdot\nabla)\vec{a}$	$[\nabla\times]\{[a\times]\|b\rangle\}=\|a\langle(\langle\nabla\cdot\|\|b\rangle)$ $-\|b\langle(\langle\nabla\cdot\|\|a\rangle)-(\langle a\cdot\|\|\nabla\rangle)\|b\rangle+(\langle b\cdot\|\|\nabla\rangle)\|a\rangle$
$\nabla\times(\nabla\times\vec{a})=\nabla(\nabla\cdot\vec{a})-\nabla^2\vec{a}$	$[\nabla\times]\{[\nabla\times]\|a\rangle\}=\|\nabla\langle(\langle\nabla\cdot\|\|a\rangle)-(\langle\nabla\cdot\|\|\nabla\rangle)\|a\rangle$
$\nabla\times(\nabla a)=\vec{0}$	$[\nabla\times](\|\nabla\rangle a)=\|0\rangle$
$\nabla\cdot(\nabla\times\vec{a})=0$	$\langle\nabla\cdot\|\{[\nabla\times]\|a\rangle\}=0$

which reads

$$\int_V \nabla\cdot\vec{A}\,dV=\oint_S \vec{A}\cdot\hat{n}\,dS \tag{A.15}$$

where $\hat{n}$ is the unit normal vector associated with the surface S. Using the GDMOs, (A.15) can be rewritten as

$$\int_V \langle\nabla\cdot||A\rangle dV=\oint_S \langle A\cdot||n\rangle\,dS \tag{A.16a}$$

or, equivalently,

$$\int_V \frac{1}{h_1h_2h_3}\left[\,\partial_{q_1}h_2h_3\quad \partial_{q_2}h_3h_1\quad \partial_{q_3}h_1h_2\,\right]\begin{bmatrix}A_1\\A_2\\A_3\end{bmatrix}dV$$

$$=\oint_S [\,A_1\quad A_2\quad A_3\,]\begin{bmatrix}n_1\\n_2\\n_3\end{bmatrix}dS \tag{A.16b}$$

Through this simple example, we have shown that GDMOs are convenient to use, and their advantage becomes more evident when applied to complex expressions, such as, vector forms of Green's identities. For the sake of completeness we provide in Table A.5 a comparison of the frequently used integral theorems in vector and GDMO forms.

TABLE A.5 GDMOs for Integral Theorems

Theorem	Vector Form	GDMO Form
Gradient	$\int_V \nabla\Phi dV = \oint_S \Phi\hat{n}\, dS$	$\int_V \vert\nabla\rangle\Phi dV = \oint_S \Phi\vert n\rangle dS$
Divergence	$\int_V \nabla\cdot\vec{A}\, dV = \oint_S \vec{A}\cdot\hat{n}\, dS$	$\int_V \langle\nabla\cdot\vert\vert A\rangle dV = \oint_S \langle A\cdot\vert n\rangle dS$
Curl	$\int_V \nabla\times\vec{A}\, dV = \oint_S \hat{n}\times\vec{A}\, dS$	$\int_V [\nabla\times]\vert A\rangle dV = \oint_S [n\times]\vert A\rangle dS$
Cross-gradient	$\int_S \hat{n}\times\nabla\Phi dS = \oint_C \Phi\hat{t}\, dl$	$\int_S [n\times]\vert\nabla\rangle\Phi dS = \oint_C \Phi\vert t\rangle dl$
Stokes's	$\int_S \hat{n}\cdot\nabla\times\vec{A}\, dS = \oint_C \vec{A}\cdot\hat{t}\, dl$	$\int_S \langle n\cdot\vert[\nabla\times]\vert A\rangle dS = \oint_C \langle A\cdot\vert\vert t\rangle dl$
First Green's	$\int_V (\nabla\Phi\cdot\nabla\Psi + \Psi\nabla^2\Phi)dV$ $= \oint_S \Psi\nabla\Phi\cdot\hat{n}\, dS$	$\int_V (\langle\nabla\vert\Phi\cdot\vert\nabla\rangle\Psi + \Psi\langle\nabla\cdot\vert\vert\nabla\rangle\Phi)dV$ $= \oint_S \Psi\langle\vert\nabla\rangle\Phi\cdot\vert\vert n\rangle dS$
Second Green's	$\int_V (\Psi\nabla^2\Phi - \Phi\nabla^2\Psi)dV$ $= \oint_S (\Psi\nabla\Phi - \Phi\nabla\Psi)\cdot\hat{n}\, dS$	$\int_V (\Psi\langle\nabla\cdot\vert\vert\nabla\rangle\Phi - \Phi\langle\nabla\cdot\vert\vert\nabla\rangle\Psi)dV$ $= \oint_S \{\langle(\Psi\vert\nabla\rangle\Phi - \Phi\vert\nabla\rangle\Psi)\cdot\vert\}\vert n\rangle dS$
First vector Green's	$\int_V \begin{bmatrix} (\nabla\times\vec{A})\cdot(\nabla\times\vec{B}) \\ -\vec{A}\cdot\nabla\times\nabla\times\vec{B} \end{bmatrix} dV$ $= \oint_S \hat{n}\cdot(\vec{A}\times\nabla\times\vec{B})dS$	$\int_V \begin{bmatrix} \langle([\nabla\times]\vert A\rangle)\cdot\vert([\nabla\times]\vert B\rangle) \\ -\langle A\cdot\vert[\nabla\times][\nabla\times]\vert B\rangle \end{bmatrix} dV$ $= \oint_S \langle n\cdot\vert([A\times][\nabla\times]\vert B\rangle)dS$
Second vector Green's	$\int_V \begin{pmatrix} \vec{B}\cdot\nabla\times\nabla\times\vec{A} \\ -\vec{A}\cdot\nabla\times\nabla\times\vec{B} \end{pmatrix} dV$ $= \oint_S \begin{pmatrix} \vec{A}\times\nabla\times\vec{B} \\ -\vec{B}\times\nabla\times\vec{A} \end{pmatrix} dS$	$\int_V \begin{pmatrix} \langle B\cdot\vert[\nabla\times][\nabla\times]\vert A\rangle \\ -\langle A\cdot\vert[\nabla\times][\nabla\times]\vert B\rangle \end{pmatrix} dV$ $= \oint_S ([A\times][\nabla\times]\vert B\rangle - [B\times][\nabla\times]\vert A\rangle)dS$

REFERENCES

[1] Y. Chen, K. Sun, B. Beker, and R. Mittra, "Unified matrix presentation of Maxwell's and wave equations using generalized differential matrix operators," *IEEE Trans. Educ.*, vol. 41, no. 1, pp. 61–69, Feb. 1998.

[2] G. Arfken, *Mathematical Methods for Physicists*, Academic Press, San Diego, 1985.

[3] C. T. Tai, *Generalized Vector and Dyadic Analysis*, chap. 2, IEEE Press, New York, 1992.

[4] C. A. Balanis, *Advanced Engineering Electromagnetics*, p. 262, Wiley, Hoboken, NJ, 1989.

APPENDIX B

MRTD Orthogonal and Integral Relations

B.1 ORTHOGONAL AND INTEGRAL RELATIONS

In the MRTD scheme, we expand the electromagnetic fields as a summation of the scaling functions $\{\phi_i(x)\}$ and wavelet functions $\{\psi_{s,i+2^{-s}(l+1/2)}(x)\}$ in space, and rectangular pulse $h_n(t)$ in time. In deriving the field update equations, we apply Galerkin's method and frequently use various orthogonal and integral relationships [1, 2]. For the convenience of readers, we summarize some major, frequently used relations in this appendix, particularly those used in conjunction with the cubic spline Battle–Lemarié scaling and wavelet functions. The method developed here can be applied directly to other wavelet families.

B.2 ORTHOGONAL RELATIONS

We now summarize some frequently used orthogonal scaling and wavelet relations as follows:

$$\int_{-\infty}^{+\infty} \phi_i(\varsigma)\phi_{i'}(\varsigma)\, d\varsigma = \delta_{i,i'}\Delta\varsigma \tag{B.1}$$

$$\int_{-\infty}^{+\infty} \psi_{s,i}(\varsigma)\psi_{s',i'}(\varsigma)\, d\varsigma = \delta_{s,s'}\delta_{i,i'}\Delta\varsigma \tag{B.2}$$

Multiresolution Time Domain Scheme for Electromagnetic Engineering
By Yinchao Chen, Qunsheng Cao, and Raj Mittra
ISBN 0-471-27230-2

$$\int_{-\infty}^{+\infty} \phi_i(\varsigma)\psi_{i+1/2}(\varsigma)\,d\varsigma = 0 \tag{B.3}$$

$$\int_{-\infty}^{+\infty} \psi_{s,i+2^{-s}(l+1/2)}(\varsigma)\psi_{s',i'+2^{-s'}(l'+1/2)}(\varsigma)\,d\varsigma = \delta_{s,s'}\delta_{l,l'}\delta_{i,i'}\Delta\varsigma \tag{B.4}$$

$$\int_{-\infty}^{+\infty} \phi_i(\varsigma)\psi_{i'+2^{-s}(l+1/2)}(\varsigma)\,d\varsigma = 0 \tag{B.5}$$

$$\int_{-\infty}^{+\infty} h_{n'}(t)h_n(t)\,dt = \delta_{n,n'}\Delta t \tag{B.6}$$

As an example, we now prove the relation given in (B.6). The pulse $h_n(t)$ is defined as

$$h(t) = \begin{cases} 1, & |t| < \frac{1}{2} \\ 1/2, & |t| = \frac{1}{2} \\ 0, & |t| > \frac{1}{2} \end{cases}$$

which is shown in Figure B.1.

By using the notation of the MRTD scheme, we can express the shifted rectangular pulse function $h_n(t)$ located at n as

$$h_n(t) = h\left(\frac{t}{\Delta t} - n\right)$$

Obviously, if $n \neq n'$, we have

$$\int_{-\infty}^{+\infty} h_{n'}(t)h_n(t)\,dt = 0$$

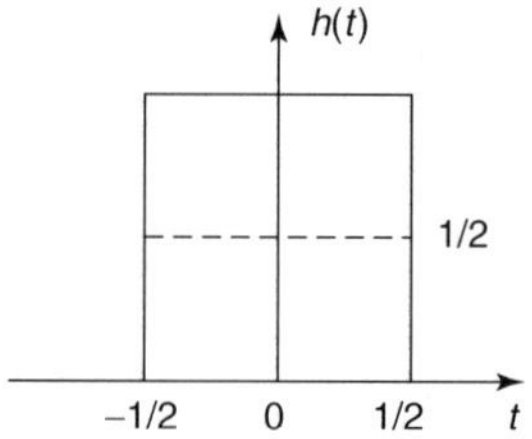

FIGURE B.1 Distribution of the rectangular pulse function.

and if $n = n'$, we have

$$\int_{-\infty}^{+\infty} h_n(t)h_n(t)\,dt = \int_{-\infty}^{+\infty} h_n^2(t)\,dt = \int_{(m-1/2)\Delta t}^{(m+1/2)\Delta t} h^2\left(\frac{t}{\Delta t} - n\right) dt$$

$$= \int_{-1/2}^{1/2} h^2(x)\,dx = \Delta t \cdot x \Big|_{-1/2}^{1/2} = \Delta t$$

B.3 INTEGRAL RELATIONS OF THE PULSE FUNCTION

In the MRTD formulation, we have to use various integral relations for rectangular pulse functions. Typically, we have the following relations:

$$\int_{-\infty}^{+\infty} h_{n'}(t)\frac{\partial}{\partial t}h_{n+1/2}(t)\,dt = \delta_{n',n} - \delta_{n',n+1} \tag{B.7}$$

$$\int_{-\infty}^{+\infty} h_{n'+1/2}(t)\frac{\partial}{\partial t}h_n(t)\,dt = \delta_{n',n+1} - \delta_{n',n} \tag{B.8}$$

$$\int_{-\infty}^{+\infty} h_n(t)\frac{\partial}{\partial t}h_{n'}(t)\,dt = \tfrac{1}{2}(\delta_{n',n+1} - \delta_{n',n-1}) \tag{B.9}$$

This time we derive the relation (B.7) for the purpose of illustration. We first expand the left-hand side (LHS) of (B.7) in accordance with the definition of the function:

$$\int_{-\infty}^{+\infty} h_{n'}(t)\frac{\partial}{\partial t}h_{n+1/2}(t)\,dt = \int_{-\infty}^{+\infty} h\left(\frac{t}{\Delta t} - n'\right)\frac{\partial}{\partial t}h\left(\frac{t}{\Delta t} - n - \frac{1}{2}\right) dt$$

and then introduce a normalized variable $x = t/\Delta t$:

$$\frac{\partial}{\partial t}h\left(\frac{t}{\Delta t} - n - \frac{1}{2}\right) = \frac{1}{\Delta t}\frac{\partial}{\partial x}h\left(x - n - \frac{1}{2}\right)$$

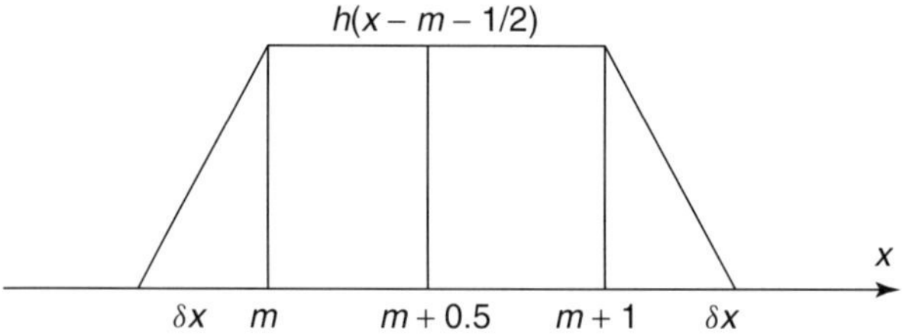

FIGURE B.2 Exaggerated shape of a rectangular pulse function.

From Figure B.2, we can write

$$\frac{\partial}{\partial x} h\left(x - m - \frac{1}{2}\right) = \begin{cases} \lim\limits_{\delta x \to 0} \dfrac{1}{\delta x}, & \text{if } x = m \\ -\lim\limits_{\delta x \to 0} \dfrac{1}{\delta x}, & \text{if } x = m + 1 \\ 0, & \text{otherwise} \end{cases}$$

$$= \begin{cases} \delta(x - m), & \text{if } x = m \\ -\delta(x - m - 1), & \text{if } x = m + 1 \\ 0, & \text{otherwise} \end{cases}$$

Finally, we have

$$\int_{-\infty}^{+\infty} h_{n'}(t) \frac{\partial}{\partial t} h_{n+1/2}(t)\, dt$$

$$= \int_{-\infty}^{+\infty} h\left(\frac{t}{\Delta t} - n'\right) \frac{1}{\Delta t}\left[\delta\left(\frac{t}{\Delta t} - n\right) - \delta\left(\frac{t}{\Delta t} - n - 1\right)\right] dt$$

$$= \int_{-\infty}^{+\infty} h(x - n')[\delta(x - n) - \delta(x - n - 1)]\, dt$$

$$= \delta_{n',n} - \delta_{n',n+1}$$

We can apply the same technique to derive the relations (B.8) and (B.9).

B.4 INTEGRAL RELATIONS FOR THE SCALING FUNCTIONS

The integral relationships pertaining to the scaling functions are very important in formulating the MRTD update equations. Typically, these relationships take the form

$$\int_{-\infty}^{+\infty} \phi_i(x) \frac{\partial}{\partial x} \phi_{i'+1/2}(x)\, dx = \sum_{\nu=-\infty}^{+\infty} a(\nu) \delta_{i+\nu,i'} \tag{B.10}$$

$$\int_{-\infty}^{+\infty} \phi_{i+1/2}(x) \frac{\partial}{\partial x} \phi_{i'}(x)\, dx = \sum_{\nu=-\infty}^{+\infty} a(\nu) \delta_{i+\nu+1,i'} \tag{B.11}$$

We now prove (B.10). Using the translation relation, we have

$$\phi_i(t) = \phi\left(\frac{x}{\Delta x} - i\right)$$

Then the LHS of (B.10) becomes

$$\begin{aligned}
I &= \int_{-\infty}^{+\infty} \phi\left(\frac{x}{\Delta x} - i\right) \frac{\partial}{\partial (x/\Delta x)} \phi\left(\frac{x}{\Delta x} - i' - \frac{1}{2}\right) d\left(\frac{x}{\Delta x}\right) \\
&= \int_{-\infty}^{+\infty} \phi(x' - i) \frac{\partial}{\partial x'} \phi\left(x' - i' - \frac{1}{2}\right) dx' \\
&= \int_{-\infty}^{+\infty} \frac{1}{2\pi} \int_{-\infty}^{+\infty} \widetilde{\phi}(\omega) e^{-j\omega(x'-i)} \frac{1}{2\pi} \frac{\partial}{\partial x'} \int_{-\infty}^{+\infty} \widetilde{\phi}^*(\omega') e^{j\omega'(x'-i'-1/2)}\, d\omega\, d\omega'\, dx' \\
&= \int_{-\infty}^{+\infty} \frac{1}{2\pi} \int_{-\infty}^{+\infty} \widetilde{\phi}(\omega) e^{-j\omega(x'-i)} \left(\frac{1}{2\pi} \int_{-\infty}^{+\infty} \widetilde{\phi}^*(\omega') e^{j\omega'(x'-i'-1/2)} (j\omega')\, d\omega' \right) d\omega\, dx' \\
&= \int_{-\infty}^{+\infty} \frac{1}{2\pi} \int_{-\infty}^{+\infty} \widetilde{\phi}(\omega) e^{j\omega i} \widetilde{\phi}^*(\omega') e^{-j\omega'(i'+1/2)} \\
&\quad \times \left(\frac{1}{2\pi} \int_{-\infty}^{+\infty} e^{-jx'(\omega-\omega')}\, dx' \right) (j\omega')\, d\omega\, d\omega'
\end{aligned}$$

where we have used the definition of the inverse Fourier transform and the orthogonal relationship

$$\frac{1}{2\pi} \int_{-\infty}^{+\infty} e^{-jx'(\omega-\omega')}\, dx' = \delta_{\omega',\omega} = \delta(\omega' - \omega)$$

Next, we can write

$$I = \frac{1}{2\pi} \int_{-\infty}^{+\infty} \tilde{\phi}(\omega) e^{j\omega i} \left(\int_{-\infty}^{+\infty} \tilde{\phi}^*(\omega') e^{-j\omega'(\omega'+1/2)} (j\omega') \delta(\omega' - \omega)\, d\omega \right) d\omega'$$

$$= \frac{1}{2\pi} \int_{-\infty}^{+\infty} \tilde{\phi}(\omega) e^{j\omega i} \tilde{\phi}^*(\omega) e^{-j\omega(i'+1/2)} (j\omega)\, d\omega$$

$$= \frac{1}{2\pi} \int_{-\infty}^{+\infty} \left|\tilde{\phi}(\omega)\right|^2 (j\omega) e^{-j\omega(i'-i+1/2)}\, d\omega$$

$$= \frac{1}{2\pi} \int_{-\infty}^{+\infty} \left|\tilde{\phi}(\omega)\right|^2 (j\omega) \left[\cos\omega \left(i' - i + \frac{1}{2} \right) - j \sin\omega \left(i' - i + \frac{1}{2} \right) \right] d\omega$$

Since $\int_{-\infty}^{+\infty} \omega \cos\omega(i' + i + \frac{1}{2})\, d\omega = 0$ for the odd function of $\omega \cos\omega(i' + i + \frac{1}{2})$, the above equation can be simplified as

$$I = -\frac{1}{2\pi} \int_{-\infty}^{+\infty} \left|\tilde{\phi}(\omega)\right|^2 (j\omega) j \sin\omega(i' - i + \tfrac{1}{2})\, d\omega$$

$$= \frac{1}{2\pi} \int_{-\infty}^{+\infty} \left|\tilde{\phi}(\omega)\right|^2 \omega \sin\omega(i' - i + \tfrac{1}{2})\, d\omega$$

$$= \frac{1}{\pi} \int_{0}^{+\infty} \left|\tilde{\phi}(\omega)\right|^2 \omega \sin\omega(i' - i + \tfrac{1}{2})\, d\omega$$

Let

$$\left|\tilde{\phi}(\omega)\right|^2 e^{-j\omega/2} (j\omega) = \sum_{\nu=-\infty}^{+\infty} a(\nu) e^{j\nu\omega} \tag{B.12}$$

Then we have

$$I = \frac{1}{2\pi} \int_{-\infty}^{+\infty} \left|\tilde{\phi}(\omega)\right|^2 (j\omega) e^{-j\omega/2} e^{-j\omega(i'-i)}\, d\omega$$

$$= \frac{1}{2\pi} \int_{-\infty}^{+\infty} \sum_{\nu} a(\nu) e^{j\nu\omega} e^{-j\omega(i'-i)}\, d\omega$$

$$= \sum_{\nu} a(\nu) \frac{1}{2\pi} \int_{-\infty}^{+\infty} e^{-j\omega(i'-i-\nu)}\, d\omega$$

$$= \sum_{\nu} a(\nu)\delta(i' - i - \nu)$$

$$= \sum_{\nu} a(\nu)\delta_{i+\nu,i'}$$

B.5 INTEGRAL RELATIONS FOR MIXED FUNCTIONS

The integral relations of mixed scaling and wavelet functions are also very important in formulating the MRTD update equations. We now summarize these relations:

$$\int_{-\infty}^{+\infty} \phi_i(x) \frac{\partial}{\partial x} \psi_{i'}(x)\, dx = \sum_{\nu=-\infty}^{+\infty} d_0(\nu)\delta_{i',i+\nu} \tag{B.13}$$

$$\int_{-\infty}^{+\infty} \psi_{i+1/2}(x) \frac{\partial}{\partial x} \phi_{i'+1/2}(x)\, dx = \sum_{\nu=-\infty}^{+\infty} c_0(\nu)\delta_{i',i+\nu} \tag{B.14}$$

$$\int_{-\infty}^{+\infty} \phi_i(x) \frac{\partial \psi_{s,i'-1/2+2^{-s}(l+1/2)}(x)}{\partial x}\, dx = \sum_{\nu=-\infty}^{\nu=+\infty} d_{s,l}(\nu)\delta_{i',i+\nu} \tag{B.15}$$

$$\int_{-\infty}^{+\infty} \phi_{i+1/2}(x) \frac{\partial \psi_{s,i'+2^{-s}(l+1/2)}(x)}{\partial x}\, dx = \sum_{\nu=-\infty}^{\nu=\infty} d_{s,l}(\nu)\delta_{i',i+\nu} \tag{B.16}$$

$$\int_{-\infty}^{+\infty} \psi_{s,i+2^{-s}(l+1/2)}(x) \frac{\partial \phi_{i'+1/2}(x)}{\partial x}\, dx = \sum_{\nu=-\infty}^{\nu=+\infty} c_{s,l}(\nu)\delta_{i',i+\nu} \tag{B.17}$$

$$\int_{-\infty}^{+\infty} \psi_{s,i-1/2+2^{-s}(l+1/2)}(x) \frac{\partial \phi_{i'}(x)}{\partial x}\, dx = \sum_{\nu=-\infty}^{\nu=+\infty} c_{s,l}(\nu)\delta_{i',i+\nu} \tag{B.18}$$

Proof of (B.13)

We prove (B.13) by deriving the relationship. Using the dilation property of the scaling and wavelet functions, we have

$$\phi_i(t) = \phi\left(\frac{x}{\Delta x} - i\right)$$

$$\psi_{i+1/2}(x) = \psi\left(\frac{x}{\Delta x} - i\right)$$

$$\psi_i(x) = \psi\left(\frac{x}{\Delta x} - i + \frac{1}{2}\right)$$

$$\psi_{s,i+2^{-s}(l+1/2)}(x) = 2^{s/2}\psi\left[2^s\left(\frac{x}{\Delta x} - i\right) - l\right]$$

Thus, we can write

$$I = \int_{-\infty}^{+\infty} \phi_i(x)\frac{\partial}{\partial x}\psi_{i'}(x)\,dx$$

$$= \int_{-\infty}^{+\infty} \phi(x-i)\frac{\partial}{\partial x}\psi(x-i'+\tfrac{1}{2})\,dx$$

$$= \int_{-\infty}^{+\infty} \frac{1}{2\pi}\int_{-\infty}^{+\infty} \widetilde{\phi}(\omega)e^{-j\omega(x-i)}\frac{\partial}{\partial x}\frac{1}{2\pi}\int_{-\infty}^{+\infty} \widetilde{\psi}^*(\omega')e^{j\omega'(x-i'+1/2)}\,d\omega\,d\omega'\,dx'$$

$$= \frac{1}{2\pi}\int_{-\infty}^{+\infty} \widetilde{\phi}(\omega)e^{j\omega i}\int_{-\infty}^{+\infty} \widetilde{\psi}^*(\omega')e^{-j\omega'(i'-1/2)}(j\omega')\delta(\omega-\omega')\,d\omega\,d\omega'$$

$$= \frac{1}{2\pi}\int_{-\infty}^{+\infty} \widetilde{\phi}(\omega)e^{j\omega i}\widetilde{\psi}^*(\omega)e^{-j\omega(i'-1/2)}(j\omega)\,d\omega$$

$$= \frac{1}{2\pi}\int_{-\infty}^{+\infty} \widetilde{\phi}(\omega)\widetilde{\psi}^*(\omega)(j\omega)e^{-j\omega(i'-i-1/2)}\,d\omega$$

$$= \frac{1}{2\pi}\int_{-\infty}^{+\infty} \widetilde{\phi}(\omega)\left|\widetilde{\psi}(\omega)\right|(j\omega)e^{-j\omega(i'-i)}\,d\omega$$

where $\widetilde{\psi}^*(\omega) = e^{-j\omega/2}\left|\widetilde{\psi}(\omega)\right|$.

Let

$$\widetilde{\phi}(\omega)\left|\widetilde{\psi}(\omega)\right|(j\omega) = \sum_{\nu=-\infty}^{+\infty} d_0(\nu)e^{j\nu\omega} \tag{B.19}$$

We then have

$$I = \frac{1}{2\pi}\int_{-\infty}^{+\infty}\sum_{\nu=-\infty}^{+\infty} d_0(\nu)e^{j\nu\omega}e^{-j\omega(i'-i)}\,d\omega$$

$$= \sum_{\nu=-\infty}^{+\infty} d_0(\nu)\delta(i' - i - \nu)$$

$$= \sum_{\nu=-\infty}^{+\infty} d_0(\nu)\delta_{i',i+\nu}$$

Proof of (B.14)

Next, we derive the relationship given in (B.14), whose LHS reads

$$I = \int_{-\infty}^{+\infty} \psi_{i+1/2}(x)\frac{\partial}{\partial x}\phi_{i'+1/2}(x)\,dx$$

$$= \int_{-\infty}^{+\infty} \psi(x - i)\frac{\partial}{\partial x}\phi(x - i' + \tfrac{1}{2})\,dx$$

$$= \int_{-\infty}^{+\infty} \frac{1}{2\pi}\int_{-\infty}^{+\infty} \widetilde{\psi}(\omega)e^{-j\omega(x-i)}\frac{1}{2\pi}\int_{-\infty}^{+\infty} \widetilde{\phi}(\omega')e^{j\omega'(x-i'+1/2)}(j\omega')\,d\omega\,d\omega'\,dx$$

$$= \frac{1}{2\pi}\int_{-\infty}^{+\infty} \widetilde{\psi}(\omega)e^{j\omega i}\int_{-\infty}^{+\infty} \widetilde{\phi}(\omega')e^{-j\omega'(i'-1/2)}(j\omega')\delta(\omega - \omega')\,d\omega\,d\omega'$$

$$= \frac{1}{2\pi}\int_{-\infty}^{+\infty} \widetilde{\psi}(\omega)\widetilde{\phi}(\omega)(j\omega)e^{-j\omega(i'-i-1/2)}\,d\omega$$

$$= \frac{1}{2\pi}\int_{-\infty}^{+\infty} \left|\widetilde{\psi}(\omega)\right|\widetilde{\phi}(\omega)(j\omega)e^{-j\omega(i'-i)}\,d\omega$$

where $\widetilde{\psi}(\omega) = e^{j\omega/2}\left|\widetilde{\psi}(\omega)\right|$.

Similarly, let

$$\left|\widetilde{\psi}(\omega)\right|\widetilde{\phi}(\omega)(j\omega) = \sum_{\nu=-\infty}^{+\infty} c_0(\nu)e^{j\nu\omega} = \sum_{\nu=-\infty}^{+\infty} d_0(\nu)e^{j\nu\omega} \qquad \text{(B.20)}$$

We then have

$$I = \frac{1}{2\pi}\int_{-\infty}^{+\infty}\sum_{\nu=-\infty}^{+\infty} c_0(\nu)e^{j\nu\omega}e^{-j\omega(i'-i)}\,d\omega$$

$$= \sum_{\nu=-\infty}^{+\infty} c_0(\nu)\delta(i' - i - \nu)$$

$$= \sum_{\nu=-\infty}^{+\infty} c_0(\nu)\delta_{i',i+\nu}$$

Proof of (B.15)

Now we prove the relation (B.15). We write its LHS as

$$I = \int_{-\infty}^{+\infty} \phi_i(x) \frac{\partial \psi_{s,i'-1/2+2^{-s}(l+1/2)}(x)}{\partial x}\, dx$$

$$= \int_{-\infty}^{+\infty} \phi\left(\frac{x}{\Delta x} - i\right) \frac{\partial}{\partial x}\left(2^{s/2}\psi\left[2^s\left(\frac{x}{\Delta x} - i' + \frac{1}{2}\right) - l\right]\right) dx$$

$$= \int_{-\infty}^{+\infty} \phi(x - i) \frac{\partial}{\partial x}(2^{s/2}\psi[2^s(x - i' + \tfrac{1}{2}) - l])\, dx$$

$$= \int_{-\infty}^{+\infty} \frac{1}{2\pi} \int_{-\infty}^{+\infty} \widetilde{\phi}(\omega)e^{-j\omega(x-i)}\, d\omega \frac{\partial}{\partial x}$$

$$\times \left[2^{s/2}\frac{1}{2\pi}\int_{-\infty}^{+\infty} \widetilde{\psi}^*(\omega')e^{j\omega'[2^s(x-i'+2^{-1})-l]}\, d\omega'\right] dx$$

$$= \int_{-\infty}^{+\infty} \frac{1}{2\pi} \int_{-\infty}^{+\infty} \widetilde{\phi}(\omega)e^{-j\omega(x-i)}\, d\omega$$

$$\times \left[2^{s/2}\frac{1}{2\pi}\int_{-\infty}^{+\infty} \widetilde{\psi}^*(\omega')e^{j\omega'[2^s(x-i'+2^{-1})-l]}(j2^s\omega')\, d\omega'\right] dx$$

$$= \frac{1}{2\pi}\int_{-\infty}^{+\infty} \widetilde{\phi}(\omega)e^{j\omega i}2^{s/2}\, d\omega \int_{-\infty}^{+\infty} \widetilde{\psi}^*(\omega')e^{-j\omega'[2^s(i'-2^{-1})+l]}$$

$$\times \left(\frac{1}{2\pi}\int_{-\infty}^{+\infty} e^{-jx(\omega-2^s\omega')}\, dx\right)(j2^s\omega')\, d\omega'$$

$$= \frac{1}{2\pi} \int_{-\infty}^{+\infty} \widetilde{\phi}(\omega) e^{j\omega i} 2^{s/2}$$

$$\times \left[\int_{-\infty}^{+\infty} \widetilde{\psi}^*(\omega') e^{-j\omega'[2^s(i'-2^{-1})+l]} \delta(\omega - 2^s\omega')(j2^s\omega')\, d\omega' \right] d\omega$$

We can simplify the integral part inside the above square brackets as

$$\int_{-\infty}^{+\infty} \widetilde{\psi}^*(\omega') e^{-j\omega'[2^s(i'-2^{-1}+l)]} \delta(\omega - 2^s\omega')(j2^s\omega')\, d\omega'$$

$$= \int_{-\infty}^{+\infty} \widetilde{\psi}^*(\omega') e^{-j\omega'[2^s(i'-2^{-1}+l)]} (j2^s\omega') 2^{-s} \delta(\omega' - 2^{-s}\omega)\, d\omega'$$

$$= \widetilde{\psi}^*(2^{-s}\omega) e^{-j2^{-s}\omega[2^s(i'-2^{-1}+l)]} (j\omega) 2^{-s}$$

where we have used the following properties of the Dirac delta function:

$$\delta(x-\xi) = \delta(\xi - x) \quad \text{and} \quad \delta(ax) = \frac{1}{|a|}\delta(x)$$

$$\delta(\omega - 2^s\omega') = \delta(2^s\omega' - \omega) = \delta\lfloor 2^{s'}(\omega' - 2^{-s'}\omega)\rfloor = 2^{-s'}\delta(\omega' - 2^{-s'}\omega)$$

and

$$\int_{-\infty}^{+\infty} f(x)\delta(x - x_0)\, dx = f(x_0)$$

The integral now becomes

$$I = \frac{1}{2\pi} \int_{-\infty}^{+\infty} \widetilde{\phi}(\omega) e^{j\omega i} 2^{s/2} \widetilde{\psi}^*(2^{-s}\omega) e^{-j2^{-s}\omega[2^s(i'-2^{-1})+l]} (j\omega) 2^{-s}\, d\omega$$

$$= \frac{1}{2\pi} \int_{-\infty}^{+\infty} \widetilde{\phi}(\omega) 2^{-s/2} \left|\widetilde{\psi}(2^{-s}\omega)\right| e^{-j2^{-s-1}\omega} e^{-j\omega(i'-2^{-1}+2^{-s}l-i)} (j\omega)\, d\omega$$

$$= \frac{1}{2\pi} \int_{-\infty}^{+\infty} \widetilde{\phi}(\omega) \left|\widetilde{\psi}_s(\omega)\right| \exp[-j\omega(2^{-s-1} + i' - 2^{-1} + 2^{-s}l - i)](j\omega)\, d\omega$$

$$= \frac{1}{2\pi} \int_{-\infty}^{+\infty} \widetilde{\phi}(\omega) \left|\widetilde{\psi}_s(\omega)\right| (j\omega) \exp[-j\omega(-2^{-1} + 2^{-s}l + 2^{-s-1})] e^{-j\omega(i'-i)}\, d\omega$$

where we have used the property of the wavelet function in the Fourier domain:

$$\widetilde{\psi}_s(\omega) = 2^{-s/2}\widetilde{\psi}(2^{-s}\omega) = e^{j2^{-S}\omega/2}\left|\widetilde{\psi}_s(\omega)\right|$$
$$\widetilde{\psi}_s^*(\omega) = e^{-j2^{-S}\omega/2}\left|\widetilde{\psi}_s(\omega)\right|$$

Let

$$\widetilde{\phi}(\omega)\left|\widetilde{\psi}_s(\omega)\right|(j\omega)\exp[-j\omega(-2^{-1}+2^{-s}l+2^{-s-1})] = \sum_{\nu=-\infty}^{+\infty} d_{s,l}(\nu)e^{j\nu\omega} \tag{B.21}$$

Finally, we have

$$\begin{aligned} I &= \frac{1}{2\pi}\int_{-\infty}^{+\infty}\sum_{\nu=-\infty}^{+\infty} d_{s,l}(\nu)e^{j\nu\omega}e^{-j\omega(i'-i)}\,d\omega \\ &= \sum_{\nu=-\infty}^{+\infty} d_{s,l}(\nu)\delta(i'-i-\nu) \\ &= \sum_{\nu=-\infty}^{+\infty} d_{s,l}(\nu)\delta_{i',i+\nu} \end{aligned}$$

Proof of (B.17)

To prove the relation (B.17), we start with the LHS of the equation:

$$\begin{aligned} I &= \int_{-\infty}^{+\infty} \psi_{s,i'+2^{-S}(l+1/2)}(x)\frac{\partial\phi_{i'+1/2}(x)}{\partial x}\,dx \\ &= \int_{-\infty}^{+\infty} 2^{s/2}\psi[2^s(x-i)-l]\frac{\partial}{\partial x}\phi(x-i'-\tfrac{1}{2})\,dx \\ &= \int_{-\infty}^{+\infty}\left[2^{s/2}\frac{1}{2\pi}\int_{-\infty}^{+\infty}\widetilde{\psi}(\omega)e^{-j\omega[2^s(x-i)-l]}\,d\omega\frac{1}{2\pi}\int_{-\infty}^{+\infty}\widetilde{\phi}(\omega')e^{j\omega'(x-i'-1/2)}\,d\omega'\right]dx \\ &= 2^{s/2}\frac{1}{2\pi}\int_{-\infty}^{+\infty}\widetilde{\psi}(\omega)e^{j\omega(2^si+l)}\,d\omega \\ &\quad\times\int_{-\infty}^{+\infty}\left(\frac{1}{2\pi}\int_{-\infty}^{+\infty}[e^{-jx(2^s\omega-\omega')}]\,dx\right)\widetilde{\phi}(\omega')e^{j\omega'(i'+2^{-1})}(j\omega')\,d\omega' \end{aligned}$$

$$= 2^{s/2}\frac{1}{2\pi}\int_{-\infty}^{+\infty}\widetilde{\psi}(\omega)e^{j\omega(2^s i+l)}\int_{-\infty}^{+\infty}\widetilde{\phi}(\omega')e^{-j\omega'(i'+2^{-1})}\delta(\omega'-2^s\omega)(j\omega')\,d\omega'\,d\omega$$

$$= 2^{s/2}\frac{1}{2\pi}\int_{-\infty}^{+\infty}\widetilde{\psi}(\omega)e^{j\omega(2^s i+l)}\widetilde{\phi}(2^s\omega)e^{-j2^s\omega(i'+2^{-1})}(j2^s\omega)\,d\omega$$

$$= 2^{s/2}\frac{1}{2\pi}\int_{-\infty}^{+\infty}\widetilde{\psi}(2^{-s}\omega_0)e^{j2^{-s}\omega_0(2^s i+l)}\widetilde{\phi}(\omega_0)e^{-j\omega_0(i'+2^{-1})}(j\omega_0)2^{-s}\,d\omega_0$$

$$= 2^{-s/2}\frac{1}{2\pi}\int_{-\infty}^{+\infty}\widetilde{\psi}(2^{-s}\omega)e^{j\omega(i+2^{-s}l)}\widetilde{\phi}(\omega)e^{-j\omega(i'+2^{-1})}(j\omega)\,d\omega$$

$$= \frac{1}{2\pi}\int_{-\infty}^{+\infty}\left|\widetilde{\psi}_s(\omega)\right| e^{j\omega 2^{-s}/2}\widetilde{\phi}(\omega)\exp[-j\omega(i'+2^{-1}-i-2^{-s}l)](j\omega)\,d\omega$$

$$= \frac{1}{2\pi}\int_{-\infty}^{+\infty}\left|\widetilde{\psi}_s(\omega)\right|\widetilde{\phi}(\omega)(j\omega)$$

$$\times\exp[-j\omega(2^{-1}-2^{-s}l-2^{-s-1})]\exp[-j\omega(i'-i)]\,d\omega$$

where we have applied the property of the wavelet function in the Fourier domain,

$$\widetilde{\psi}_s(\omega) = 2^{-s/2}\widetilde{\psi}(2^{-s}\omega) = e^{j2^{-S}\omega/2}\left|\widetilde{\psi}_s(\omega)\right|$$

By letting

$$\widetilde{\phi}(\omega)\left|\widetilde{\psi}_s(\omega)\right|(j\omega)\exp[j\omega(2^{-1}-2^{-s}l-2^{-s-1})] = \sum_{\nu=-\infty}^{+\infty} c_{s,l}(\nu)e^{j\nu\omega} \qquad \text{(B.22)}$$

we then have

$$I = \frac{1}{2\pi}\int_{-\infty}^{+\infty}\sum_{\nu=-\infty}^{+\infty} c_{s,l}(\nu)e^{j\nu\omega}e^{-j\omega(i'-i)}\,d\omega$$

$$= \sum_{\nu=-\infty}^{+\infty} c_{s,l}(\nu)\delta(i'-i-\nu)$$

$$= \sum_{\nu=-\infty}^{+\infty} c_{s,l}(\nu)\delta_{i',i+\nu}$$

B.6 INTEGRAL RELATIONS FOR WAVELET FUNCTIONS

In deriving the MRTD update equations, the integral relations pertaining to wavelet functions play the same role as their counterparts for the scaling functions. We frequently use the following two equations in MRTD applications:

$$\int_{-\infty}^{+\infty} \psi_{s,i+2^{-s}(l+1/2)}(x) \frac{\partial \psi_{s',i'-1/2+2^{-s}(l'+1/2)}(x)}{\partial x} dx = \sum_{\nu=-\infty}^{+\infty} b_{l,l'}^{s,s'}(\nu)\delta_{i',i+\nu+1} \quad \text{(B.23)}$$

$$\int_{-\infty}^{+\infty} \psi_{s,i-1/2+2^{-s}(l+1/2)}(x) \frac{\partial \psi_{s',i'+2^{-s}(l'+1/2)}(x)}{\partial x} dx = \sum_{\nu=-\infty}^{+\infty} b_{l,l'}^{s,s'}(\nu)\delta_{i',i+\nu} \quad \text{(B.24)}$$

Here we prove both of the above relations. Starting with (B.23), we express its LHS as

$$\begin{aligned}
I &= \int_{-\infty}^{+\infty} \psi_{s,i+2^{-s}(l+1/2)}(x) \frac{\partial \psi_{s',i'-1/2+2^{-s}(l'+1/2)}(x)}{\partial x} dx \\
&= \int_{-\infty}^{+\infty} 2^{s/2}\psi\left[2^s(x-i)-l\right] \frac{\partial}{\partial x} 2^{s'/2}\psi[2^{s'}(x-i'+\tfrac{1}{2})-l']\,dx \\
&= \int_{-\infty}^{+\infty} \left[2^{s/2}\frac{1}{2\pi} \int_{-\infty}^{+\infty} \tilde{\psi}(\omega) e^{-j\omega[2^s(x-i)-l]}\, d\omega 2^{s/2}\frac{1}{2\pi} \right. \\
&\quad \left. \times \int_{-\infty}^{+\infty} \tilde{\psi}^*(\omega') e^{j\omega'[2^{s'}(x-i'+1/2)-l']}(j2^{s'}\omega')\, d\omega' \right] dx \\
&= 2^{s/2}\frac{1}{2\pi} \int_{-\infty}^{+\infty} \tilde{\psi}(\omega) e^{j\omega[2^s i+l]} 2^{s'/2}\, d\omega \int_{-\infty}^{+\infty} \tilde{\psi}^*(\omega') e^{-j\omega'[2^{s'}(i'-2^{-1})+l']} \\
&\quad \times \left(\frac{1}{2\pi} \int_{-\infty}^{+\infty} [e^{-jx(2^s\omega-2^{s'}\omega')}]\,dx \right) (j2^{s'}\omega')\, d\omega' \\
&= 2^{s/2}\frac{1}{2\pi} \int_{-\infty}^{+\infty} \tilde{\psi}(\omega) e^{j\omega[2^s i+l]} 2^{s'/2} \\
&\quad \times \int_{-\infty}^{+\infty} \tilde{\psi}^*(\omega') e^{-j\omega'[2^{s'}(i'-2^{-1})+l']} 2^{-s'} \delta(\omega'-2^{s-s'}\omega)(j2^{s'}\omega')\, d\omega'\, d\omega
\end{aligned}$$

$$= 2^{s/2}\frac{1}{2\pi}\int_{-\infty}^{+\infty}\widetilde{\psi}(\omega)e^{j\omega[2^s i+l]}2^{s'/2}\widetilde{\psi}^*(2^{s-s'}\omega)e^{-j2^{s-s'}\omega[2^{s'}(i'-2^{-1})+l]}(j2^s\omega)2^{-s'}\,d\omega$$

$$= 2^{s/2}\frac{1}{2\pi}\int_{-\infty}^{+\infty}\widetilde{\psi}(2^{-s}\omega_0)e^{j2^{-s}\omega_0[2^s i+l]}2^{s'/2}\widetilde{\psi}^*$$

$$\times\,(2^{-s'}\omega_0)e^{-j2^{-s'}\omega_0[2^{s'}(i'-2^{-1})+l']}(j\omega_0)2^{-s'}2^{-s}\,d\omega_0$$

$$= 2^{-s/2}\frac{1}{2\pi}\int_{-\infty}^{+\infty}\widetilde{\psi}(2^{-s}\omega)e^{j\omega[i+2^{-s}l]}2^{-s'/2}\widetilde{\psi}^*(2^{-s'}\omega)e^{-j\omega[(i'-2^{-1})+2^{s'}l]}(j\omega)\,d\omega$$

$$= \frac{1}{2\pi}\int_{-\infty}^{+\infty}\widetilde{\psi}_s(\omega)e^{j\omega[i+2^{-s}l]}\widetilde{\psi}^*_{s'}(\omega)e^{-j\omega[(i'-2^{-1})+2^{s'}l']}(j\omega)\,d\omega$$

$$= \frac{1}{2\pi}\int_{-\infty}^{+\infty}\left|\widetilde{\psi}_s(\omega)\right|e^{j2^{-s-1}\omega}\left|\widetilde{\psi}_{s'}(\omega)\right|e^{-j2^{-s'-1}\omega}$$

$$\times\,\exp[-j\omega(i'-2^{-1}+2^{-s'}l'-i-2^{-s}l)](j\omega)\,d\omega$$

$$= \frac{1}{2\pi}\int_{-\infty}^{+\infty}\left(\left|\widetilde{\psi}_s(\omega)\right|\left|\widetilde{\psi}_{s'}(\omega)\right|e^{-j\omega(i'-i-1)}(j\omega)\right)$$

$$\times\,\exp[-j\omega(2^{-1}+2^{-s'}l'-2^{-s}l+2^{-s'-1}-2^{-s-1})]\,d\omega$$

Letting

$$\left|\widetilde{\psi}_s(\omega)\right|\left|\widetilde{\psi}_s(\omega)\right|\exp\lfloor-j\omega(2^{-1}+2^{-s'}l'-2^{-s}l-2^{-s'-1}+2^{-s-1})\rfloor(j\omega)$$
$$= \sum_{\nu=-\infty}^{+\infty} b_{l,l'}^{s,s'}(\nu)e^{j\nu\omega} \tag{B.25}$$

we then obtain

$$I = \frac{1}{2\pi}\int_{-\infty}^{+\infty}\sum_{\nu=-\infty}^{+\infty} b_{l,l'}^{s,s'}(\nu)e^{j\nu\omega}e^{-j\omega(i'-i-1)}d\omega$$

$$= \sum_{\nu=-\infty}^{+\infty} b_{l,l'}^{s,s'}(\nu)\delta_{i',i+\nu+1}$$

Next, we work on the relation (B.24), whose LHS is

$$
\begin{aligned}
I &= \int_{-\infty}^{+\infty} \psi_{s,i-1/2+2^{-s}(l+1/2)}(x) \frac{\partial \psi_{s',i'+2^{-s}(l'+1/2)}(x)}{\partial x}\, dx \\
&= \int_{-\infty}^{+\infty} 2^{s/2}\psi\left[2^{s}\left(x-i+\tfrac{1}{2}\right)-l\right] \frac{\partial}{\partial x} 2^{s'/2}\psi[2^{s'}(x-i')-l']\, dx \\
&= \int_{-\infty}^{+\infty} \left[2^{s/2}\frac{1}{2\pi}\int_{-\infty}^{+\infty} \widetilde{\psi}(\omega) e^{-j\omega[2^{s}(x-i+1/2)-l]}\, d\omega 2^{s'/2}\frac{1}{2\pi} \right. \\
&\qquad \left. \times \int_{-\infty}^{+\infty} \widetilde{\psi}^{*}(\omega') e^{j\omega'[2^{s'}(x-i')-l']}(j2^{s'}\omega')\, d\omega' \right] dx \\
&= 2^{s/2}\frac{1}{2\pi}\int_{-\infty}^{+\infty} \widetilde{\psi}(\omega) e^{j\omega[2^{s}(i-1/2)+l]}\, d\omega 2^{s'/2} \int_{-\infty}^{+\infty} \widetilde{\psi}^{*}(\omega') e^{-j\omega'[2^{s'}i'+l']} \\
&\qquad \times \left(\frac{1}{2\pi}\int_{-\infty}^{+\infty}\left[e^{-jx(2^{s}\omega-2^{s'}\omega')}\right] dx \right)(j2^{s'}\omega')\, d\omega' \\
&= 2^{s/2}\frac{1}{2\pi}\int_{-\infty}^{+\infty} \widetilde{\psi}(\omega) e^{j\omega[2^{s}(i-1/2)+l]} \\
&\qquad \times \left(2^{s'/2}\int_{-\infty}^{+\infty} \widetilde{\psi}^{*}(\omega') e^{-j\omega'[2^{s'}i'+l']} 2^{-s'}\delta(\omega'-2^{s-s'}\omega)(j2^{s'}\omega')\, d\omega' \right) d\omega \\
&= 2^{s/2}\frac{1}{2\pi}\int_{-\infty}^{+\infty} \widetilde{\psi}(\omega) e^{j\omega[2^{s}(i-1/2)+l]} (2^{s'/2}\widetilde{\psi}^{*}(2^{s-s'}\omega) e^{-j2^{s-s'}\omega[2^{s'}i'+l']}(j2^{s}\omega)2^{-s'})\, d\omega \\
&= 2^{s/2}\frac{1}{2\pi}\int_{-\infty}^{+\infty} \widetilde{\psi}(2^{-s}\omega_0) e^{j2^{-s}\omega_0[2^{s}(i-1/2)+l]} 2^{-s'/2}\widetilde{\psi}^{*} \\
&\qquad \times (2^{-s'}\omega_0) e^{-j\omega_0(i'+2^{-s'}l')}(j\omega_0)2^{-s}\, d\omega_0 \\
&= 2^{-s/2}\frac{1}{2\pi}\int_{-\infty}^{+\infty} \widetilde{\psi}(2^{-s}\omega) e^{j\omega[i-1/2+2^{-s}l]} 2^{-s'/2}\widetilde{\psi}^{*}(2^{-s'}\omega) e^{-j\omega(i'+2^{-s'}l')}(j\omega)\, d\omega \\
&= \frac{1}{2\pi}\int_{-\infty}^{+\infty} \widetilde{\psi}_{s}(\omega) e^{j\omega[i-1/2+2^{-s}l]} \widetilde{\psi}_{s'}^{*}(\omega) e^{-j\omega(i'+2^{-s'}l')}(j\omega)\, d\omega
\end{aligned}
$$

$$= \frac{1}{2\pi} \int_{-\infty}^{+\infty} |\widetilde{\psi}_s(\omega)||\widetilde{\psi}_{s'}(\omega)|$$

$$\times \exp[-j\omega(i' + 2^{-s'}l' - i + 2^{-1} - 2^{-s}l)]e^{[j\omega(2^{-s-1})]}e^{[-j\omega(2^{-s'-1})]}(j\omega)\,d\omega$$

$$= \frac{1}{2\pi} \int_{-\infty}^{+\infty} (|\widetilde{\psi}_s(\omega)||\widetilde{\psi}_{s'}(\omega)|e^{-j\omega(i'-i)}(j\omega))$$

$$\times \exp[-j\omega(2^{-1} + 2^{-s'}l' - 2^{-s}l + 2^{-s'-1} - 2^{-s-1})]\,d\omega$$

where $\delta(2^s\omega - 2^{s'}\omega') = \delta(2^{s'}\omega' - 2^s\omega) = \delta\lfloor 2^{s'}(\omega' - 2^{s-s'}\omega)\rfloor = 2^{-s'}\delta(\omega' - 2^{s-s'}\omega)$.

In a similar way, we let

$$|\widetilde{\psi}_s(\omega)||\widetilde{\psi}_s(\omega)| \exp[-j\omega(2^{-1} + 2^{-s'}l' - 2^{-s}l + 2^{-s'-1} - 2^{-s-1})](j\omega)$$

$$= \sum_{\nu=-\infty}^{+\infty} b_{l,l'}^{s,s'}(\nu)e^{j\nu\omega}$$

and obtain

$$I = \frac{1}{2\pi} \int_{-\infty}^{+\infty} \sum_{\nu=-\infty}^{+\infty} b_{l,l'}^{s,s'}(\nu)e^{j\nu\omega}e^{-j\omega(i'-i)}\,d\omega$$

$$= \sum_{\nu=-\infty}^{+\infty} b_{l,l'}^{s,s'}(\nu)\delta_{i',i+\nu}$$

B.7 INTEGRAL COEFFICIENTS

In the above derivation, the integral coefficients $a(\omega)$, $d_{s,l}(\nu)$, $c_{s,l}(\nu)$, and $b_{l,l'}^{s,s'}(\nu)$ are not yet resolved in (B.12), (B.21), (B.22), and (B.25), respectively. Now, it is time to complete this remaining task to validate the following expressions for the integral coefficients:

$$a(\omega) = \frac{1}{\pi} \int_0^{+\infty} |\widetilde{\phi}(\omega)|^2 \omega \sin[\omega(\nu + \tfrac{1}{2})]\,d\omega \tag{B.26}$$

$$d_{s,l}(\nu) = \frac{1}{\pi} \int_0^{\infty} \widetilde{\phi}(\omega|\widetilde{\psi}_s(\omega)|\omega \sin\left[\omega\left(\nu - \frac{1}{2} + \frac{l}{2^s} + \frac{1}{2^{s+1}}\right)\right] d\omega \tag{B.27}$$

$$c_{s,l}(\nu) = \frac{1}{\pi} \int_0^{\infty} \widetilde{\phi}(\omega|\widetilde{\psi}_s(\omega)|\omega \sin\left[\omega\left(\nu + \frac{1}{2} - \frac{l}{2^s} - \frac{1}{2^{s+1}}\right)\right] d\omega \tag{B.28}$$

$$b_{l,l'}^{s,s'}(\nu) = \frac{1}{\pi}\int_0^{\infty} |\widetilde{\psi}_s(\omega)||\widetilde{\psi}_{s'}(\omega)|\omega \times \sin\left[\omega\left(\nu + \frac{1}{2} + \frac{l'}{2^{s'}} - \frac{l}{2^s} + \frac{1}{2^{s'+1}} - \frac{1}{2^{s+1}}\right)\right] d\omega \qquad \text{(B.29)}$$

Proof of (B.26)

We start with the expression of $a(\omega)$ appearing in (B.12), which actually can be found as an expansion coefficient of a Fourier series. Thus, we have

$$\begin{aligned}\frac{1}{2\pi}\int_{-\infty}^{+\infty} |\widetilde{\phi}(\omega)|^2 e^{-j\omega/2}(j\omega)e^{-j\nu'\omega}\, d\omega &= \sum_{\nu'=-\infty}^{+\infty} a(\nu')\frac{1}{2\pi}\int_{-\infty}^{+\infty} e^{j\nu'\omega}e^{-j\nu\omega}\, d\omega \\ &= \sum_{\nu'=-\infty}^{+\infty} a(\nu')\frac{1}{2\pi}\int_{-\infty}^{+\infty} e^{-j(\nu-\nu')\omega}\, d\omega \\ &= \sum_{\nu'=-\infty}^{+\infty} a(\nu')\delta_{\nu'\nu'} = a(\nu)\end{aligned}$$

$$\begin{aligned}a(\nu) &= \frac{1}{2\pi}\int_{-\infty}^{+\infty} |\widetilde{\phi}(\omega)|^2 e^{-j\omega(\nu+1/2)}(j\omega)\, d\omega \\ &= \frac{1}{2\pi}\int_{-\infty}^{+\infty} |\widetilde{\phi}(\omega)|^2 (j\omega)[\cos\omega(\nu+\tfrac{1}{2}) - j\sin\omega(\nu+\tfrac{1}{2})]\, d\omega \\ &= \frac{1}{2\pi}\int_{-\infty}^{+\infty} |\widetilde{\phi}(\omega)|^2 \omega\sin\omega(\nu+\tfrac{1}{2})\, d\omega \\ &= \frac{1}{\pi}\int_{0}^{\infty} |\widetilde{\phi}(\omega)|^2 \omega\sin\omega(\nu+\tfrac{1}{2})\, d\omega\end{aligned}$$

Proof of (B.27)

Next, we go on to prove (B.27), where the coefficient $d_{s,l}(\omega)$ is defined in (B.21). Also, using the expression of the expansion coefficient of the Fourier series, we obtain

$$\begin{aligned}\frac{1}{2\pi}\int_{-\infty}^{+\infty}\sum_{\nu'=-\infty}^{\nu'=+\infty} d_{s,l}(\nu')e^{j\nu'\omega}e^{-j\nu\omega}\, d\omega &= \sum_{\nu'=-\infty}^{\nu'=+\infty} d_{s,l}(\nu')\left(\frac{1}{2\pi}\int_{-\infty}^{+\infty} e^{-j\omega(\nu-\nu')}\, d\omega\right) \\ &= \sum_{\nu'=-\infty}^{\nu'=+\infty} d_{s,l}(\nu')\delta_{\nu',\nu} = d_{s,l}(\nu)\end{aligned}$$

$$
\begin{aligned}
d_{s,l}(\nu) &= \frac{1}{2\pi}\int_{-\infty}^{+\infty} \widetilde{\phi}(\omega|\widetilde{\psi}_s(\omega)|e^{-j\omega(2^{-s}l+2^{-s-1}+2^{-1})}e^{-j\omega\nu}(j\omega)\,d\omega \\
&= \frac{1}{2\pi}\int_{-\infty}^{+\infty} \widetilde{\phi}(\omega|\widetilde{\psi}_s(\omega)|(j\omega)\left[\cos\omega\left(\nu-\frac{1}{2}+\frac{l}{2^s}+\frac{1}{2^{s+1}}\right)\right. \\
&\qquad\qquad \left. - j\sin\omega\left(\nu-\frac{1}{2}+\frac{l}{2^s}+\frac{1}{2^{s+1}}\right)\right]d\omega \\
&= \frac{1}{2\pi}\int_{-\infty}^{+\infty} \widetilde{\phi}(\omega|\widetilde{\psi}_s(\omega)|(\omega)\sin\left[\omega\left(\nu-\frac{1}{2}+\frac{l}{2^s}+\frac{1}{2^{s+1}}\right)\right]d\omega \\
&= \frac{1}{\pi}\int_{0}^{\infty} \widetilde{\phi}(\omega|\widetilde{\psi}_s(\omega)|\omega\sin\left[\omega\left(\nu-\frac{1}{2}+\frac{l}{2^s}+\frac{1}{2^{s+1}}\right)\right]d\omega
\end{aligned}
$$

Proof of (B.28)

In a similar way, we can find the following expression for $c_{s,l}(\omega)$, which has been defined in (B.22):

$$
\begin{aligned}
\frac{1}{2\pi}\int_{-\infty}^{+\infty}\sum_{\nu'=-\infty}^{\nu'=+\infty} c_{s,l}(\nu')e^{j\nu'\omega}e^{-j\nu\omega}\,d\omega &= \sum_{\nu'=-\infty}^{\nu'=+\infty} c_{s,l}(\nu')\left(\frac{1}{2\pi}\int_{-\infty}^{+\infty} e^{-j\omega(\nu-\nu')}\,d\omega\right) \\
&= \sum_{\nu'=-\infty}^{\nu'=+\infty} c_{s,l}(\nu')\delta_{\nu',\nu} \\
&= c_{s,l}(\nu)
\end{aligned}
$$

$$
\begin{aligned}
c_{s,l}(\nu) &= \frac{1}{2\pi}\int_{-\infty}^{+\infty} \widetilde{\phi}(\omega|\widetilde{\psi}_s(\omega)|e^{-j\omega(2^{-1}-2^{-s}l-2^{-s-1})}e^{-j\omega\nu}(j\omega)\,d\omega \\
&= \frac{1}{2\pi}\int_{-\infty}^{+\infty} \widetilde{\phi}(\omega|\widetilde{\psi}_s(\omega)|(j\omega)\left[\cos\omega\left(\nu+\frac{1}{2}-\frac{l}{2^s}-\frac{1}{2^{s+1}}\right)\right. \\
&\qquad\qquad \left. - j\sin\omega\left(\nu+\frac{1}{2}-\frac{l}{2^s}-\frac{1}{2^{s+1}}\right)\right]d\omega \\
&= \frac{1}{2\pi}\int_{-\infty}^{+\infty} \widetilde{\phi}(\omega|\widetilde{\psi}_s(\omega)|(\omega)\sin\left[\omega\left(\nu+\frac{1}{2}-\frac{l}{2^s}-\frac{1}{2^{s+1}}\right)\right]d\omega \\
&= \frac{1}{\pi}\int_{0}^{\infty} \widetilde{\phi}(\omega|\widetilde{\psi}_s(\omega)|\omega\sin\left[\omega\left(\nu+\frac{1}{2}-\frac{l}{2^s}-\frac{1}{2^{s+1}}\right)\right]d\omega
\end{aligned}
$$

Proof of (B.29)

Finally, we can obtain the expression for $b_{l,l'}^{s,s'}(\omega)$, defined in (B.25), by using the same procedure developed above:

$$\frac{1}{2\pi}\int_{-\infty}^{+\infty}\sum_{\nu=-\infty}^{\nu=\infty} b_{l,l'}^{s,s'}(\nu)e^{j\nu'\omega}e^{-j\nu\omega}\,d\omega = \sum_{\nu=-\infty}^{\nu=\infty} b_{l,l'}^{s,s'}(\nu)\left(\frac{1}{2\pi}\int_{-\infty}^{+\infty} e^{-j\omega(\nu-\nu')}\,d\omega\right)$$

$$= \sum_{\nu'=-\infty}^{\nu'=+\infty} b_{l,l'}^{s,s'}(\nu')\delta_{\nu',\nu}$$

$$= b_{l,l'}^{s,s'}(\nu)$$

$$b_{l,l'}^{s,s'}(\nu) = \frac{1}{2\pi}\int_{-\infty}^{+\infty}\sum_{\nu'=-\infty}^{\nu'=+\infty} b_{l,l'}^{s,s'}(\nu')$$

$$\times \exp[-j\omega(2^{-1}+2^{-s'}l'-2^{-s}l-2^{-s'-1}+2^{-s-1})]e^{-j\nu\omega}\,d\omega$$

$$= \frac{1}{2\pi}\int_{-\infty}^{+\infty}\frac{1}{2\pi}\int_{-\infty}^{+\infty}|\widetilde{\psi}_s(\omega)||\widetilde{\psi}_{s'}(\omega)|(j\omega)$$

$$\times \exp[-j\omega(2^{-1}+2^{-s'}l'-2^{-s}l-2^{-s'-1}+2^{-s-1}+\nu)]\,d\omega$$

$$= \frac{1}{2\pi}\int_{-\infty}^{+\infty}|\widetilde{\psi}_s(\omega)||\widetilde{\psi}_{s'}(\omega)|(j\omega)$$

$$\times\left[\cos\omega\left(\nu+\frac{1}{2}+\frac{l'}{2^{s'}}-\frac{l}{2^{s}}+\frac{1}{2^{s'+1}}-\frac{1}{2^{s+1}}\right)\right.$$
$$\left.-\,j\sin\omega\left(\nu+\frac{1}{2}+\frac{l'}{2^{s'}}-\frac{l}{2^{s}}+\frac{1}{2^{s'+1}}-\frac{1}{2^{s+1}}\right)\right]d\omega$$

$$= \frac{1}{2\pi}\int_{-\infty}^{+\infty}|\widetilde{\psi}_s(\omega)||\widetilde{\psi}_{s'}(\omega)|(\omega)$$

$$\times \sin\left[\omega\left(\nu+\frac{1}{2}+\frac{l'}{2^{s'}}-\frac{l}{2^{s}}+\frac{1}{2^{s'+1}}-\frac{1}{2^{s+1}}\right)\right]d\omega$$

$$= \frac{1}{\pi}\int_{0}^{\infty}|\widetilde{\psi}_s(\omega)||\widetilde{\psi}_{s'}(\omega)|\omega$$

$$\times \sin\left[\omega\left(\nu+\frac{1}{2}+\frac{l'}{2^{s'}}-\frac{l}{2^{s}}+\frac{1}{2^{s'+1}}-\frac{1}{2^{s+1}}\right)\right]d\omega$$

If $s = s'$, we have

$$b_{l,l}^{s,s}(\nu) = \frac{1}{\pi} \int_0^\infty |\widetilde{\psi}_s(\omega)|^2 \omega \sin\left[\omega\left(\nu + \frac{1}{2}\right)\right] d\omega$$

and if $s = 0$, we obtain

$$d_0(\nu) = \frac{1}{\pi} \int_0^\infty \widetilde{\phi}(\omega|\widetilde{\psi}(\omega)|\omega \sin[\omega(\nu)]\, d\omega$$

$$c_0(\nu) = \frac{1}{\pi} \int_0^\infty \widetilde{\phi}(\omega|\widetilde{\psi}(\omega)|\omega \sin[\omega(\nu)]\, d\omega$$

$$d_0 = c_0$$

$$b_0(\nu) = \frac{1}{\pi} \int_0^\infty |\widetilde{\psi}(\omega)|^2 \omega \sin \omega(\nu + \tfrac{1}{2})\, d\omega$$

REFERENCES

[1] I. Daubechies, *Ten Lectures on Wavelets*, SIAM, Philadelphia, 1992.

[2] Charles K. Chui, *An Introduction to Wavelets*, Academic Press, San Diego, 1992.

APPENDIX C

Update Equations in APML Regions

C.1 MAXWELL EQUATIONS IN APML REGIONS

In this appendix, we derive the update equations in the unsplit anisotropic perfectly matched layer (APML) regions, for all H- and E-field components. The original concept of the perfectly matched layers (PMLs) was introduced by Berenger [1]. It has successfully been extended and applied to scattering and guided-wave problems with excellent results [2–10].

By using the generalized differential matrix operators (GDMOs), we can write the Maxwell equations in the APML regions as

$$\begin{bmatrix} 0 & -\partial_z & \partial_y \\ \partial_z & 0 & -\partial_x \\ -\partial_y & \partial_x & 0 \end{bmatrix} \begin{bmatrix} H_x \\ H_y \\ H_z \end{bmatrix} = j\omega\varepsilon_0\varepsilon_r \begin{bmatrix} S_{xx} & 0 & 0 \\ 0 & S_{yy} & 0 \\ 0 & 0 & S_{zz} \end{bmatrix} \begin{bmatrix} E_x \\ E_y \\ E_z \end{bmatrix} \tag{C.1a}$$

$$\begin{bmatrix} 0 & -\partial_z & \partial_y \\ \partial_z & 0 & -\partial_x \\ -\partial_y & \partial_x & 0 \end{bmatrix} \begin{bmatrix} E_x \\ E_y \\ E_z \end{bmatrix} = -j\omega\mu_0\mu_r \begin{bmatrix} S_{xx} & 0 & 0 \\ 0 & S_{yy} & 0 \\ 0 & 0 & S_{zz} \end{bmatrix} \begin{bmatrix} H_x \\ H_y \\ H_z \end{bmatrix} \tag{C.1b}$$

where the APML parameter matrix $[S]$ is the product of three submatrices

$$[S] = [S_x][S_y][S_z] = \begin{bmatrix} s_x^{-1} & 0 & 0 \\ 0 & s_x & 0 \\ 0 & 0 & s_x \end{bmatrix} \begin{bmatrix} s_y & 0 & 0 \\ 0 & s_y^{-1} & 0 \\ 0 & 0 & s_y \end{bmatrix} \begin{bmatrix} s_z & 0 & 0 \\ 0 & s_z & 0 \\ 0 & 0 & s_z^{-1} \end{bmatrix} \tag{C.2}$$

Multiresolution Time Domain Scheme for Electromagnetic Engineering
By Yinchao Chen, Qunsheng Cao, and Raj Mittra
ISBN 0-471-27230-2

with

$$s_\alpha = 1 + \frac{\sigma_\alpha}{j\omega\varepsilon_0}, \qquad \alpha = x, y, \text{or } z \tag{C.3}$$

where σ_α is the spatially varying conductivity along the absorption axis. The relative permittivity (μ_r) is usually set to $\mu_r = 1$ in practice.

C.2 FIELD EXPANSIONS

For the S-MRTD scheme, we expand all field quantities in terms of the scaling functions in space and pulse functions in time as follows:

$$E_x(\vec{r}, t) = \sum_{i,j,k,n=-\infty}^{\infty} {}_{\phi x}E^n_{i+1/2,j,k}\phi_{i+1/2}(x)\phi_j(y)\phi_k(z)h_n(t) \tag{C.4a}$$

$$E_y(\vec{r}, t) = \sum_{i,j,k,n=-\infty}^{\infty} {}_{\phi y}E^n_{i,j+1/2,k}\phi_i(x)\phi_{j+1/2}(y)\phi_k(z)h_n(t) \tag{C.4b}$$

$$E_z(\vec{r}, t) = \sum_{i,j,k,n=-\infty}^{\infty} {}_{\phi z}E^n_{i,j,k+1/2}\phi_i(x)\phi_j(y)\phi_{k+1/2}(z)h_n(t) \tag{C.4c}$$

$$H_x(\vec{r}, t) = \sum_{i,j,k,n=-\infty}^{\infty} {}_{\phi x}H^n_{i,j+1/2,k+1/2}\phi_i(x)\phi_{j+1/2}(y)\phi_{k+1/2}(z)h_{n+1/2}(t) \tag{C.5a}$$

$$H_y(\vec{r}, t) = \sum_{i,j,k,n=-\infty}^{\infty} {}_{\phi y}H^n_{i+1/2,j,k+1/2}\phi_{i+1/2}(x)\phi_j(y)\phi_{k+1/2}(z)h_{n+1/2}(t) \tag{C.5b}$$

$$H_z(\vec{r}, t) = \sum_{i,j,k,n=-\infty}^{\infty} {}_{\phi z}H^n_{i+1/2,j+1/2,k}\phi_{i+1/2}(x)\phi_{j+1/2}(y)\phi_k(z)h_{n+1/2}(t) \tag{C.5c}$$

where $\phi(x)$ denotes the cubic spline Battle–Lemarié scaling function, and $h_n(t)$ is a rectangular pulse function. To derive the update equations in the APML regions, we usually adopt a two-step approach and introduce two sets of intermediate parameters, namely, D_x, D_y, D_z and B_x, B_y, B_z, respectively, which share the same forms as their corresponding field components; that is,

$$D_x(\vec{r}, t) = \sum_{i,j,k,n=-\infty}^{\infty} {}_{\phi x}D^n_{i+1/2,j,k}\phi_{i+1/2}(x)\phi_j(y)\phi_k(z)h_n(t) \tag{C.6a}$$

$$D_y(\vec{r}, t) = \sum_{i,j,k,n=-\infty}^{\infty} {}_{\phi y}D^n_{i,j+1/2,k}\phi_i(x)\phi_{j+1/2}(y)\phi_k(z)h_n(t) \tag{C.6b}$$

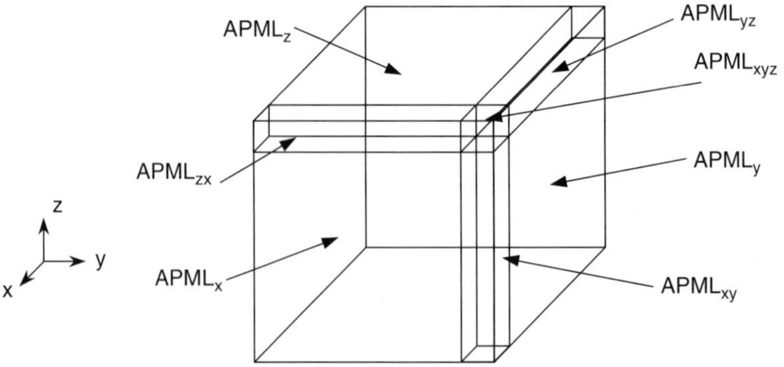

FIGURE C.1 A rectangular computational volume with different APML orientations.

$$D_z(\vec{r}, t) = \sum_{i,j,k,n=-\infty}^{\infty} {}_{\phi z}D^n_{i,j,k+1/2}\phi_i(x)\phi_j(y)\phi_{k+1/2}(z)h_n(t) \tag{C.6c}$$

$$B_x(\vec{r}, t) = \sum_{i,j,k,n=-\infty}^{\infty} {}_{\phi x}B^n_{i,j+1/2,k+1/2}\phi_i(x)\phi_{j+1/2}(y)\phi_{k+1/2}(z)h_{n+1/2}(t) \tag{C.7a}$$

$$B_y(\vec{r}, t) = \sum_{i,j,k,n=-\infty}^{\infty} {}_{\phi y}B^n_{i+1/2,j,k+1/2}\phi_{i+1/2}(x)\phi_j(y)\phi_{k+1/2}(z)h_{n+1/2}(t) \tag{C.7b}$$

$$B_z(\vec{r}, t) = \sum_{i,j,k,n=-\infty}^{\infty} {}_{\phi z}B^n_{i+1/2,j+1/2,k}\phi_{i+1/2}(x)\phi_{j+1/2}(y)\phi_k(z)h_{n+1/2}(t) \tag{C.7c}$$

A typical computational domain of rectangular shape is shown in Figure C.1. If an APML is to be applied to all six side walls, the APML region basically consists of three different boundary orientations—face-APMLs, edge-APMLs, and corner-APMLs. Hence, we have to handle all these boundary orientations.

C.3 UPDATE EQUATIONS FOR FACE-APML REGIONS

The face-APMLs are defined at the interfaces of two ends of each Cartesian axis, and thus there are six of them in total.

Face-APML$_x$ Regions

For the face-APML$_x$ regions, $[S] = [S_x]$ and $[S_y] = [S_z] = [I]$, where $[I]$ is the unitary matrix. The Maxwell equations in the frequency domain become

$$\frac{\partial H_z}{\partial y} - \frac{\partial H_y}{\partial z} = j\omega\varepsilon_0\varepsilon_r \left(1 + \frac{\sigma_x}{j\omega\varepsilon_0}\right)^{-1} E_x \tag{C.8a}$$

$$\frac{\partial H_x}{\partial z} - \frac{\partial H_z}{\partial x} = j\omega\varepsilon_0\varepsilon_r \left(1 + \frac{\sigma_x}{j\omega\varepsilon_0}\right) E_y \tag{C.8b}$$

$$\frac{\partial H_y}{\partial x} - \frac{\partial H_x}{\partial y} = j\omega\varepsilon_0\varepsilon_r \left(1 + \frac{\sigma_x}{j\omega\varepsilon_0}\right) E_z \tag{C.8c}$$

$$\frac{\partial E_z}{\partial y} - \frac{\partial E_y}{\partial z} = -j\omega\mu_0 \left(1 + \frac{\sigma_x}{j\omega\varepsilon_0}\right)^{-1} H_x \tag{C.9a}$$

$$\frac{\partial E_x}{\partial z} - \frac{\partial E_z}{\partial x} = -j\omega\mu_0 \left(1 + \frac{\sigma_x}{j\omega\varepsilon_0}\right) H_y \tag{C.9b}$$

$$\frac{\partial E_y}{\partial x} - \frac{\partial E_x}{\partial y} = -j\omega\mu_0 \left(1 + \frac{\sigma_x}{j\omega\varepsilon_0}\right) H_z \tag{C.9c}$$

We can directly convert these equations into the time domain by using the frequency–time converter, $j\omega \Leftrightarrow \partial/\partial t$. To illustrate the procedure for deriving the update equations using a two-step approach, we now work on (C.8a) and introduce a parameter D_x:

$$D_x = \varepsilon_0\varepsilon_r \left(1 + \frac{\sigma_x}{j\omega\varepsilon_0}\right)^{-1} E_x \tag{C.10}$$

This enables us to convert (C.8a) into two equations as follows:

$$\frac{\partial H_z}{\partial y} - \frac{\partial H_y}{\partial z} = \frac{\partial D_x}{\partial t} \tag{C.11}$$

$$\varepsilon_0\varepsilon_r \frac{\partial E_x}{\partial t} = \frac{\partial D_x}{\partial t} + \left(\frac{\sigma_x}{\varepsilon_0}\right) D_x \tag{C.12}$$

Similarly, for (C.9a), we can use a two-step approach by introducing a parameter B_x,

$$B_x = \mu_0 \left(1 + \frac{\sigma_x}{j\omega\varepsilon_0}\right)^{-1} H_x \tag{C.13}$$

We then split (C.9a) into the following two equations:

$$\frac{\partial E_z}{\partial y} - \frac{\partial E_y}{\partial z} = -\frac{\partial H_x}{\partial t} \tag{C.14}$$

$$\mu_0 \frac{\partial H_x}{\partial t} = \frac{\partial B_x}{\partial t} + \left(\frac{\sigma_x}{\varepsilon_0}\right) B_x \tag{C.15}$$

We now derive the MRTD update equation for (C.11) using the standard Galerkin's method. First, we substitute the field expansion equations for H_y, H_z, and D_x in (C.5b) (C.5c), and (C.6a) into (C.11); then, we test both sides of the equation by multiplying by $\phi_{i+1/2}(x)\phi_j(y)\phi_k(z)h_{n+1/2}(t)$ and integrating in both time and space. Using the orthogonal and integral relations (see Appendix B), we can easily obtain the update equations for the D_x component:

$$
\begin{aligned}
{}_{\phi x}D^{n+1}_{i+1/2,j,k} &= {}_{\phi z}D^{n}_{i+1/2,j,k} \\
&\quad + \sum_{\nu} a(\nu)\left({}_{\phi z}H^{n+1/2}_{i+1/2,j+m+1/2,k}\frac{\Delta t}{\Delta y} - {}_{\phi y}H^{n+1/2}_{i+1/2,j,k+m+1/2}\frac{\Delta t}{\Delta z}\right)
\end{aligned}
\tag{C.16}
$$

To obtain the update equation for (C.12), we substitute field expansions H_x and D_x of (C.5a) and (C.6a) into the equations, and test both sides of the resulting equation. Its left-hand side (LHS) yields

$$
\begin{aligned}
&\int_{-\infty}^{+\infty} \varepsilon_0\varepsilon_r \frac{\partial E_x}{\partial t}\phi_{i+1/2}(x)\phi_j(y)\phi_k(y)\,dx\,dy\,dz\,dt \\
&= \varepsilon_0\varepsilon_r \int_{-\infty}^{+\infty} \frac{\partial}{\partial t}\left[\sum_{i',j',k',n'=-\infty}^{\infty} {}_{\phi x}E^{n'}_{i'+1/2,j',k'}\phi_{i'+1/2}(x)\phi_{j'}(y)\phi_{k'}(z)h_{n'}(t)\right] \\
&\qquad\qquad \cdot \phi_{i+1/2}(x)\phi_j(y)\phi_k(y)h_{n+1/2}(t)\,dx\,dy\,dz\,dt \\
&= \varepsilon_0\varepsilon_r \sum_{i',j',k',n'=-\infty}^{\infty} {}_{\phi x}E^{n'}_{i'+1/2,j',k'}\,\Delta x\ \Delta y\ \Delta z\ \delta_{i,i'}\delta_{j,j'}\delta_{k,k'}\int_{-\infty}^{\infty} h_{n+1/2}(t)\frac{\partial}{\partial t}h_{n'}(t)\,dt \\
&= \varepsilon_0\varepsilon_r({}_{\phi x}E^{n+1}_{i+1/2,j,k} - {}_{\phi x}E^{n-1}_{i+1/2,j,k})\Delta x\ \Delta y\ \Delta z
\end{aligned}
\tag{C.17}
$$

while its right-hand side (RHS) leads to

$$
\begin{aligned}
&\int_{-\infty}^{+\infty}\left(\frac{\partial D_x}{\partial t} + \sigma_x D_x\right)\phi_{i+1/2}(x)\phi_j(y)\phi_k(y)\,dx\,dy\,dz\,dt \\
&= \int_{-\infty}^{+\infty}\left(\frac{\partial}{\partial t} + \sigma_x\right)\left[\sum_{i',j',k',n'=-\infty}^{\infty} {}_{\phi x}D^{n'}_{i'+1/2,j',k'}\phi_{i'+1/2}(x)\phi_{j'}(y)\phi_{k'}(z)h_{n'}(t)\right] \\
&\qquad\qquad \cdot \phi_{i+1/2}(x)\phi_j(y)\phi_k(y)h_{n+1/2}(t)\,dx\,dy\,dz\,dt \\
&= \sum_{i',j',k',n'=-\infty}^{\infty} {}_{\phi x}D^{n'}_{i'+1/2,j',k'}\,\Delta x\ \Delta y\ \Delta z\ \delta_{i,i'}\delta_{j,j'}\delta_{k,k'}(\delta_{n+1,n'} - \delta_{n-1,n'}) \\
&\quad + \sum_{i',j',k',n'=-\infty}^{\infty} {}_{\phi x}D^{n'}_{i'+1/2,j',k'}\,\Delta y\ \Delta z\ \delta_{j,j'}\delta_{k,k'}\delta_{n+1/2,n'} \\
&\quad \times \int_{-\infty}^{+\infty} \sigma_x\phi_{i+1/2}(x)\phi_{i'+1/2}(x)\,dx
\end{aligned}
\tag{C.18}
$$

$$= ({}_{\phi x} D^{n+1}_{i+1/2,j,k} - {}_{\phi x} D^{n-1}_{i+1/2,j,k}) \Delta x \ \Delta y \ \Delta z$$

$$+ \sum_{i'=-\infty}^{\infty} {}_{\phi x} D^{n+1/2}_{i'+1/2,j,k} \Delta y \ \Delta z \ (\sigma_x)_{i,i'} \Delta x$$

where the coefficient $(\sigma_x)_{i,i'}$ is given by

$$(\sigma_x)_{i,i'} = \frac{1}{\Delta x} \int_{-\infty}^{+\infty} \sigma_x(x) \phi_{i+1/2}(x) \phi_{i'+1/2}(x) \, dx$$

$$= \frac{1}{\Delta x} \int_{\text{APML}} \phi(x/\Delta x - i - \tfrac{1}{2}) \sigma_x(i' \ \Delta x) \phi(x/\Delta x - i' - \tfrac{1}{2}) \, dx$$

$$= \int_{\text{APML}} \phi(x' - i - \tfrac{1}{2}) \sigma_x(x' \ \Delta x) \phi(x' - i' - \tfrac{1}{2}) \, dx'$$

By using the diagonal approximation, namely, replacing $(\sigma_x)_{i,i'}$ with $(\sigma_x)_{i,i'}\delta_{i,i'}$, we have

$$(\sigma_x)_{i,i} = \int_{\text{APML}} \phi^2(x' - i - \tfrac{1}{2}) \sigma_x(x' \ \Delta x) \, dx' \equiv \sigma_x^i = \sigma_{\max} \left| \frac{i - i_0}{N_{\text{APML}}} \right|^m \tag{C.19}$$

where N_{APML} is the cell number of the APML region, i_0 is the starting position of the APML region in the x-direction, and m is the order of polynomial variation.

Finally, we obtain the update equation by combining (C.17)–(C.19):

$$\begin{aligned} {}_{\phi x} E^{n+1}_{i+1/2,j,k} &= {}_{\phi z} E^{n}_{i+1/2,j,k} \\ &+ \frac{1}{\varepsilon_0 \varepsilon_r} \left[\left(1 + \frac{\sigma_x^i \ \Delta t}{2\varepsilon_0}\right) {}_{\phi x} D^{n+1}_{i+1/2,j,k} - \left(1 - \frac{\sigma_x^i \ \Delta t}{2\varepsilon_0}\right) {}_{\phi x} D^{n}_{i+1/2,j,k} \right] \end{aligned} \tag{C.20}$$

Similarly, we can derive the update equations for the remaining components:

$$\begin{aligned} {}_{\phi y} E^{n+1}_{i,j+1/2,k} &= \left(1 - \frac{\sigma_x^i \ \Delta t}{2\varepsilon_0}\right) \left(1 + \frac{\sigma_x^i \ \Delta t}{2\varepsilon_0}\right)^{-1} {}_{\phi y} E^{n}_{i,j+1/2,k} + \frac{1}{\varepsilon_0 \varepsilon_r} \left(1 + \frac{\sigma_x^i \ \Delta t}{2\varepsilon_0}\right)^{-1} \\ &\times \left[\sum_{\nu} a(\nu) \left({}_{\phi x} H^{n+1/2}_{i,j+1/2,k+\nu+1/2} \frac{\Delta t}{\Delta z} - {}_{\phi z} H^{n+1/2}_{i+\nu+1/2,j+1/2,k} \frac{\Delta t}{\Delta x} \right) \right] \end{aligned} \tag{C.21}$$

$$\begin{aligned} {}_{\phi z} E^{n+1}_{i,,kj+1/2} &= \left(1 - \frac{\sigma_x^i \ \Delta t}{2\varepsilon_0}\right) \left(1 + \frac{\sigma_x^i \ \Delta t}{2\varepsilon_0}\right)^{-1} {}_{\phi z} E^{n}_{i,j,kj+1/2} + \frac{1}{\varepsilon_0 \varepsilon_r} \left(1 + \frac{\sigma_x^i \ \Delta t}{2\varepsilon_0}\right)^{-1} \\ &\times \left[\sum_{\nu} a(\nu) \left({}_{\phi y} H^{n+1/2}_{i+\nu+1/2,j,k+1/2} \frac{\Delta t}{\Delta x} - {}_{\phi z} H^{n+1/2}_{i,j+\nu+1/2,k+1/2} \frac{\Delta t}{\Delta y} \right) \right] \end{aligned} \tag{C.22}$$

$$
{}_{\phi x}B^{n+1/2}_{i,j+1/2,k+1/2} = {}_{\phi x}B^{n-1/2}_{i,j+1/2,k+1/2}
+ \sum_{\nu} a(\nu)\left({}_{\phi y}E^{n}_{i,j+1/2,k+\nu+1}\frac{\Delta t}{\Delta z} - {}_{\phi z}E^{n}_{i,j+\nu+1,k+1/2}\frac{\Delta t}{\Delta y}\right) \quad \text{(C.23)}
$$

$$
{}_{\phi x}H^{n+1/2}_{i,j+1/2,k+1/2} = {}_{\phi x}H^{n-1/2}_{i,j+1/2,k+1/2}
+ \frac{1}{\mu_0}\left[\left(1+\frac{\sigma_x^i\,\Delta t}{2\varepsilon_0}\right){}_{\phi x}B^{n+1/2}_{i,j+1/2,k+1/2} - \left(1-\frac{\sigma_x^i\,\Delta t}{2\varepsilon_0}\right){}_{\phi x}B^{n-1/2}_{i,j+1/2,k+1/2}\right] \quad \text{(C.24)}
$$

$$
{}_{\phi x}H^{n+1/2}_{i,j+1/2,k+1/2} = {}_{\phi x}H^{n-1/2}_{i,j+1/2,k+1/2}
+ \frac{1}{\mu_0}\left[\left(1+\frac{\sigma_x^i\,\Delta t}{2\varepsilon_0}\right){}_{\phi x}B^{n+1/2}_{i,j+1/2,k+1/2} - \left(1-\frac{\sigma_x^i\,\Delta t}{2\varepsilon_0}\right){}_{\phi x}B^{n-1/2}_{i,j+1/2,k+1/2}\right] \quad \text{(C.25)}
$$

$$
{}_{\phi y}H^{n+1/2}_{i+1/2,j,k+1/2} = \left(1-\frac{\sigma_x^i\,\Delta t}{2\varepsilon_0}\right)\left(1+\frac{\sigma_x^i\,\Delta t}{2\varepsilon_0}\right)^{-1} {}_{\phi y}H^{n-1/2}_{i+1/2,j,k+1/2}
+ \frac{1}{\mu_0}\left(1+\frac{\sigma_x^i\,\Delta t}{2\varepsilon_0}\right)^{-1}\sum_{\nu} a(\nu)\left({}_{\phi z}E^{n}_{i+\nu+1,j,k+1/2}\frac{\Delta t}{\Delta x} - {}_{\phi x}E^{n}_{i+1/2,j,k+\nu+1}\frac{\Delta t}{\Delta z}\right) \quad \text{(C.26)}
$$

$$
{}_{\phi z}H^{n+1/2}_{i+1/2,j+1/2,k} = \left(1-\frac{\sigma_x^i\,\Delta t}{2\varepsilon_0}\right)\left(1+\frac{\sigma_x^i\,\Delta t}{2\varepsilon_0}\right)^{-1} {}_{\phi z}H^{n-1/2}_{i+1/2,j+1/2,k}
+ \frac{1}{\mu_0}\left(1+\frac{\sigma_x^i\,\Delta t}{2\varepsilon_0}\right)^{-1}\sum_{\nu} a(\nu)\left({}_{\phi x}E^{n}_{i+1/2,j+\nu+1,k}\frac{\Delta t}{\Delta y} - {}_{\phi y}E^{n}_{i+\nu+1,j+1/2,k}\frac{\Delta t}{\Delta x}\right) \quad \text{(C.27)}
$$

Face-APML$_y$ Regions

For the face-APML$_y$ regions, $[S] = [S_y]$, $[S_x] = [S_z] = [I]$, and the Maxwell equations can be written as

$$
\frac{\partial H_z}{\partial y} - \frac{\partial H_y}{\partial z} = j\omega\varepsilon_0\varepsilon_r\left(1+\frac{\sigma_y}{j\omega\varepsilon_0}\right)E_x \quad \text{(C.28a)}
$$

$$
\frac{\partial H_x}{\partial z} - \frac{\partial H_z}{\partial x} = j\omega\varepsilon_0\varepsilon_r\left(1+\frac{\sigma_y}{j\omega\varepsilon_0}\right)^{-1}E_y \quad \text{(C.28b)}
$$

$$
\frac{\partial H_y}{\partial x} - \frac{\partial H_x}{\partial y} = j\omega\varepsilon_0\varepsilon_r\left(1+\frac{\sigma_y}{j\omega\varepsilon_0}\right)E_z \quad \text{(C.28c)}
$$

$$
\frac{\partial E_z}{\partial y} - \frac{\partial E_y}{\partial z} = -j\omega\mu_0\left(1+\frac{\sigma_y}{j\omega\varepsilon_0}\right)H_x \quad \text{(C.29a)}
$$

$$
\frac{\partial E_x}{\partial z} - \frac{\partial E_z}{\partial x} = -j\omega\mu_0\left(1+\frac{\sigma_y}{j\omega\varepsilon_0}\right)^{-1}H_y \quad \text{(C.29b)}
$$

$$\frac{\partial E_y}{\partial x} - \frac{\partial E_x}{\partial y} = -j\omega\mu_0 \left(1 + \frac{\sigma_y}{j\omega\varepsilon_0}\right) H_z \tag{C.29c}$$

Once again, we introduce the parameter

$$D_y = \varepsilon_0 \varepsilon_r \left(1 + \frac{\sigma_y}{j\omega\varepsilon_0}\right)^{-1} E_y \tag{C.30}$$

and convert (C.28b) into the following two equations:

$$\frac{\partial H_x}{\partial z} - \frac{\partial H_z}{\partial x} = \frac{\partial D_y}{\partial t} \tag{C.31}$$

$$\varepsilon_0 \varepsilon_r \frac{\partial E_y}{\partial t} = \frac{\partial D_y}{\partial t} + \left(\frac{\sigma_x}{\varepsilon_0}\right) D_y \tag{C.32}$$

For (C.29b), we let

$$B_y = \left(1 + \frac{\sigma_y}{j\omega\varepsilon_0}\right)^{-1} H_y \tag{C.33}$$

Correspondingly, (C.31b) becomes

$$\frac{\partial E_x}{\partial z} - \frac{\partial E_z}{\partial x} = -\frac{\partial H_y}{\partial t} \tag{C.34}$$

$$\mu_0 \frac{\partial H_y}{\partial t} = \frac{\partial B_y}{\partial t} + \left(\frac{\sigma_x}{\varepsilon_0}\right) B_y \tag{C.35}$$

Using Galerkin's method with applications of the orthogonal and integral relations (see Appendix B), we can obtain the update equations:

$$\begin{aligned} {}_{\phi x}E^{n+1}_{i+1/2,j,k} = {} & \left(1 - \frac{\sigma_y^j \,\Delta t}{2\varepsilon_0}\right)\left(1 + \frac{\sigma_y^j \,\Delta t}{2\varepsilon_0}\right)^{-1} {}_{\phi x}E^{n}_{i+1/2,j,k} + \frac{1}{\varepsilon_0}\left(1 + \frac{\sigma_y^j \,\Delta t}{2\varepsilon_0}\right)^{-1} \\ & \times \left[\sum_{\nu} a(\nu)\left({}_{\phi z}H^{n+1/2}_{i+1/2,j+\nu+1/2,k}\frac{\Delta t}{\Delta y} - {}_{\phi y}H^{n+1/2}_{i+1/2,j,k+\nu+1/2}\frac{\Delta t}{\Delta z}\right)\right] \end{aligned} \tag{C.36}$$

$$\begin{aligned} {}_{\phi y}D^{n+1}_{i,j+1/2,k} = {} & {}_{\phi y}D^{n}_{i,j+1/2,k} \\ & + \sum_{\nu} a(\nu)\left({}_{\phi z}H^{n+1/2}_{i+m+1/2,j+1/2,k}\frac{\Delta t}{\Delta x} - {}_{\phi x}H^{n+1/2}_{i,j+1/2,k+m+1/2}\frac{\Delta t}{\Delta z}\right) \end{aligned} \tag{C.37}$$

$$\begin{aligned} {}_{\phi y}E^{n+1}_{i,j+1/2,k} = {} & {}_{\phi y}E^{n}_{i,j+1/2,k} \\ & + \frac{1}{\varepsilon_0}\left[\left(1 + \frac{\sigma_y^j \,\Delta t}{2\varepsilon_0}\right) {}_{\phi y}D^{n+1}_{i,j+1/2,k} - \left(1 - \frac{\sigma_y^j \,\Delta t}{2\varepsilon_0}\right) {}_{\phi y}D^{n}_{i,j+1/2,k}\right] \end{aligned} \tag{C.38}$$

$$
\begin{aligned}
{}_{\phi z}E^{n+1}_{i,j,k+1/2} &= \left(1-\frac{\sigma_y^j\ \Delta t}{2\varepsilon_0}\right)\left(1+\frac{\sigma_y^j\ \Delta t}{2\varepsilon_0}\right)^{-1}{}_{\phi z}E^{n}_{i,j,k+1/2} \\
&+\frac{1}{\varepsilon_0}\left(1+\frac{\sigma_y^j\ \Delta t}{2\varepsilon_0}\right)^{-1}\left[\sum_{\nu}a(\nu)\left({}_{\phi y}H^{n+1/2}_{i+\nu+1/2,j+1/2,k}\frac{\Delta t}{\Delta y}-{}_{\phi z}H^{n+1/2}_{i+1/2,j+\nu,k}\frac{\Delta t}{\Delta y}\right)\right]
\end{aligned}
\tag{C.39}
$$

where σ_y^j is defined as follows:

$$
(\sigma_y)_{j,j'} = \int_{\text{APML}} \phi^2(y'-j)\sigma_y(y'\ \Delta y)\,dy' \equiv \sigma_y^j = \sigma_{\max}\left|\frac{j-j_0}{N_{\text{APML}}}\right|^m \tag{C.40}
$$

$$
\begin{aligned}
{}_{\phi x}H^{n+1/2}_{i,j+1/2,k+1/2} &= \left(1-\frac{\sigma_y^j\ \Delta t}{2\varepsilon_0}\right)\left(1+\frac{\sigma_y^j\ \Delta t}{2\varepsilon_0}\right)^{-1}{}_{\phi x}H^{n-1/2}_{i,j+1/2,k+1/2} \\
&+\frac{1}{\mu_0}\left(1+\frac{\sigma_y^j\ \Delta t}{2\varepsilon_0}\right)^{-1}\sum_{\nu}a(\nu)\left({}_{\phi y}E^{n}_{i,j+1/2,k+\nu+1}\frac{\Delta t}{\Delta z}-{}_{\phi z}E^{n}_{i,j+\nu+1,k+1/2}\frac{\Delta t}{\Delta y}\right)
\end{aligned}
\tag{C.41}
$$

$$
\begin{aligned}
{}_{\phi y}B^{n+1/2}_{i+1/2,j,k+1/2} &= {}_{\phi y}B^{n-1/2}_{i+1/2,j,k+1/2} \\
&+\sum_{\nu}a(\nu)\left({}_{\phi z}E^{n}_{i+\nu+1,j,k+1/2}\frac{\Delta t}{\Delta x}-{}_{\phi x}E^{n}_{i+1/2,j,k+\nu+1}\frac{\Delta t}{\Delta z}\right)
\end{aligned}
\tag{C.42}
$$

$$
\begin{aligned}
{}_{\phi y}H^{n+1/2}_{i+1/2,j,k+1/2} &= {}_{\phi y}H^{n-1/2}_{i+1/2,j,k+1/2} \\
&+\frac{1}{\mu_0}\left[\left(1+\frac{\sigma_y^j\ \Delta t}{2\varepsilon_0}\right){}_{\phi y}B^{n+1/2}_{i+1/2,j,k+1/2}-\left(1-\frac{\sigma_y^j\ \Delta t}{2\varepsilon_0}\right){}_{\phi y}B^{n-1/2}_{i+1/2,j,k+1/2}\right]
\end{aligned}
\tag{C.43}
$$

$$
\begin{aligned}
{}_{\phi z}H^{n+1/2}_{i+1/2,j+1/2,k} &= \left(1-\frac{\sigma_y^i\ \Delta t}{2\varepsilon_0}\right)\left(1+\frac{\sigma_y^i\ \Delta t}{2\varepsilon_0}\right)^{-1}{}_{\phi z}H^{n-1/2}_{i+1/2,j+1/2,k} \\
&+\frac{1}{\mu_0}\left(1+\frac{\sigma_y^i\ \Delta t}{2\varepsilon_0}\right)^{-1}\sum_{\nu}a(\nu)\left({}_{\phi x}E^{n}_{i+1/2,j+\nu+1,k}\frac{\Delta t}{\Delta y}-{}_{\phi y}E^{n}_{i+\nu+1,j+1/2,k}\frac{\Delta t}{\Delta x}\right)
\end{aligned}
\tag{C.44}
$$

Face-APML$_z$ Regions

For the face-APML$_z$ regions, $[S]=[S_z]$, $[S_x]=[S_y]=[I]$, and the Maxwell equations can be written as

$$
\frac{\partial H_z}{\partial y}-\frac{\partial H_y}{\partial z} = j\omega\varepsilon_0\varepsilon_r\left(1+\frac{\sigma_z}{j\omega\varepsilon_0}\right)E_x \tag{C.45a}
$$

$$\frac{\partial H_x}{\partial z} - \frac{\partial H_z}{\partial x} = j\omega\varepsilon_0\varepsilon_r\left(1 + \frac{\sigma_z}{j\omega\varepsilon_0}\right)E_y \tag{C.45b}$$

$$\frac{\partial H_y}{\partial x} - \frac{\partial H_x}{\partial y} = j\omega\varepsilon_0\varepsilon_r\left(1 + \frac{\sigma_z}{j\omega\varepsilon_0}\right)^{-1}E_z \tag{C.45c}$$

$$\frac{\partial E_z}{\partial y} - \frac{\partial E_y}{\partial z} = -j\omega\mu_0\left(1 + \frac{\sigma_z}{j\omega\varepsilon_0}\right)H_x \tag{C.46a}$$

$$\frac{\partial E_x}{\partial z} - \frac{\partial E_z}{\partial x} = -j\omega\mu_0\left(1 + \frac{\sigma_z}{j\omega\varepsilon_0}\right)H_y \tag{C.46b}$$

$$\frac{\partial E_y}{\partial x} - \frac{\partial E_x}{\partial y} = -j\omega\mu_0\left(1 + \frac{\sigma_z}{j\omega\varepsilon_0}\right)^{-1}H_z \tag{C.46c}$$

By introducing an intermediate parameter

$$D_z = \varepsilon_0\varepsilon_r\left(1 + \frac{\sigma_z}{j\omega\varepsilon_0}\right)^{-1}E_z \tag{C.47}$$

we can again convert (C.45c) into two equations:

$$\frac{\partial H_y}{\partial x} - \frac{\partial H_x}{\partial y} = \frac{\partial D_z}{\partial t} \tag{C.48}$$

$$\varepsilon_0\varepsilon_r\frac{\partial E_z}{\partial t} = \frac{\partial D_z}{\partial t} + \left(\frac{\sigma_x}{\varepsilon_0}\right)D_z \tag{C.49}$$

Similarly, we let

$$B_z = \mu_0\left(1 + \frac{\sigma_z}{j\omega\varepsilon_0}\right)^{-1}H_z \tag{C.50}$$

so that (C.46c) becomes

$$\frac{\partial E_y}{\partial x} - \frac{\partial E_x}{\partial y} = -\frac{\partial B_z}{\partial t} \tag{C.51}$$

$$\mu_0\frac{\partial H_z}{\partial t} = \frac{\partial B_z}{\partial t} + \left(\frac{\sigma_x}{\varepsilon_0}\right)B_z \tag{C.52}$$

Using Galerkin's method, we obtain the update equations

$$_{\phi x}E^{n+1}_{i+1/2,j,k} = \left(1 - \frac{\sigma_z^k\,\Delta t}{2\varepsilon_0}\right)\left(1 + \frac{\sigma_z^k\,\Delta t}{2\varepsilon_0}\right)^{-1} {}_{\phi x}E^{n}_{i+1/2,j,k} + \frac{1}{\varepsilon_0}\left(1 + \frac{\sigma_z^k\,\Delta t}{2\varepsilon_0}\right)^{-1}$$

$$\times \left[\sum_{\nu} a(\nu) \left({}_{\phi z}H^{n+1/2}_{i+1/2,j'+\nu+1/2,k} \frac{\Delta t}{\Delta y} - {}_{\phi y}H^{n+1/2}_{i+1/2,j',k+\nu+1/2} \frac{\Delta t}{\Delta z} \right) \right] \tag{C.53}$$

$${}_{\phi y}E^{n+1}_{i,j+1/2,k} = \left(1 - \frac{\sigma_z^k\ \Delta t}{2\varepsilon_0}\right)\left(1 + \frac{\sigma_z^k\ \Delta t}{2\varepsilon_0}\right)^{-1} {}_{\phi y}E^{n}_{i,j+1/2,k} + \frac{1}{\varepsilon_0}\left(1 + \frac{\sigma_z^k\ \Delta t}{2\varepsilon_0}\right)^{-1}$$

$$\times \left[\sum_{\nu} a(\nu) \left({}_{\phi x}H^{n+1/2}_{i,j+1/2,k+\nu+1/2} \frac{\Delta t}{\Delta z} - {}_{\phi z}H^{n+1/2}_{i+\nu+1/2,j+1/2,k} \frac{\Delta t}{\Delta x} \right) \right] \tag{C.54}$$

$${}_{\phi z}D^{n+1}_{i,j,k+1/2} = {}_{\phi z}D^{n}_{i,j,k+1/2}$$

$$+ \sum_{\nu} a(\nu) \left({}_{\phi y}H^{n+1/2}_{i+\nu+1/2,j,k+1/2} \frac{\Delta t}{\Delta x} - {}_{\phi x}H^{n+1/2}_{i,j+\nu+1/2,k+1/2} \frac{\Delta t}{\Delta y} \right) \tag{C.55}$$

$${}_{\phi z}E^{n+1}_{i,j,k+1/2} = {}_{\phi z}E^{n}_{i,j,k+1/2}$$

$$+ \frac{1}{\varepsilon_0 \varepsilon_r} \left[\left(1 + \frac{\sigma_z^k\ \Delta t}{2\varepsilon_0}\right) {}_{\phi z}D^{n+1}_{i,j,k+1/2} - \left(1 - \frac{\sigma_z^k\ \Delta t}{2\varepsilon_0}\right) {}_{\phi z}D^{n}_{i,j,k+1/2} \right] \tag{C.56}$$

$${}_{\phi x}H^{n+1/2}_{i,j+1/2,k+1/2} = \left(1 - \frac{\sigma_z^k\ \Delta t}{2\varepsilon_0}\right)\left(1 + \frac{\sigma_z^k\ \Delta t}{2\varepsilon_0}\right)^{-1} {}_{\phi x}H^{n-1/2}_{i,j+1/2,k+1/2}$$

$$+ \frac{1}{\mu_0}\left(1 + \frac{\sigma_z^k\ \Delta t}{2\varepsilon_0}\right)^{-1} \sum_{\nu} a(\nu) \left({}_{\phi y}E^{n}_{i,j+1/2,k+\nu+1} \frac{\Delta t}{\Delta z} - {}_{\phi z}E^{n}_{i,j+\nu+1,k+1/2} \frac{\Delta t}{\Delta y} \right) \tag{C.57}$$

$${}_{\phi y}H^{n+1/2}_{i+1/2,j,k+1/2} = \left(1 - \frac{\sigma_z^k\ \Delta t}{2\varepsilon_0}\right)\left(1 + \frac{\sigma_z^k\ \Delta t}{2\varepsilon_0}\right)^{-1} {}_{\phi y}H^{n-1/2}_{i+1/2,j,k+1/2}$$

$$+ \frac{1}{\mu_0}\left(1 + \frac{\sigma_z^k\ \Delta t}{2\varepsilon_0}\right)^{-1} \sum_{\nu} a(\nu) \left({}_{\phi z}E^{n}_{i+\nu+1,j,k+1/2} \frac{\Delta t}{\Delta x} - {}_{\phi x}E^{n}_{i+1/2,j,k+\nu+1} \frac{\Delta t}{\Delta z} \right) \tag{C.58}$$

$${}_{\phi z}B^{n+1/2}_{i+1/2,j+1/2,k} = {}_{\phi z}B^{n-1/2}_{i+1/2,j+1/2,k}$$

$$+ \sum_{\nu} a(\nu) \left({}_{\phi x}E^{n}_{i+1/2,j+\nu+1,k} \frac{\Delta t}{\Delta y} - {}_{\phi y}E^{n}_{i+m+1,j+1/2,k} \frac{\Delta t}{\Delta x} \right) \tag{C.59}$$

$${}_{\phi z}H^{n+1/2}_{i+1/2,j+1/2,k} = {}_{\phi z}H^{n-1/2}_{i+1/2,j+1/2,k}$$

$$+ \frac{1}{\mu_0} \left[\left(1 + \frac{\sigma_z^k\ \Delta t}{2\varepsilon_0}\right) {}_{\phi z}B^{n+1/2}_{i+1/2,j+1/2,k} - \left(1 - \frac{\sigma_z^k\ \Delta t}{2\varepsilon_0}\right) {}_{\phi z}B^{n-1/2}_{i+1/2,j+1/2,k} \right] \tag{C.60}$$

C.4 UPDATE EQUATIONS FOR EDGE-APML REGIONS

As shown in Figure C.1, for an open rectangular computational volume, there are twelve different edge-APML regions in total. These edge-APMLs are located at the intersections between any two Cartesian axes. Here, we need to consider the case where an electromagnetic wave propagates with a direction vector that has both the x- and y-components in an APML region, say, edge-APML$_{xy}$. We discuss the cases of the edge-APML$_{xz}$ and the edge-APML$_{yz}$, respectively.

Edge-APML$_{xy}$ Regions

In the edge-APML$_{xy}$ regions, $[S] = [S_x][S_y], [S_z] = [I]$, and the Maxwell equations take the form

$$\frac{\partial H_z}{\partial y} - \frac{\partial H_y}{\partial z} = j\omega\varepsilon_0\varepsilon_r\left(1 + \frac{\sigma_x}{j\omega\varepsilon_0}\right)^{-1}\left(1 + \frac{\sigma_y}{j\omega\varepsilon_0}\right)E_x \tag{C.61a}$$

$$\frac{\partial H_x}{\partial z} - \frac{\partial H_z}{\partial x} = j\omega\varepsilon_0\varepsilon_r\left(1 + \frac{\sigma_x}{j\omega\varepsilon_0}\right)\left(1 + \frac{\sigma_y}{j\omega\varepsilon_0}\right)^{-1}E_y \tag{C.61b}$$

$$\frac{\partial H_y}{\partial x} - \frac{\partial H_x}{\partial y} = j\omega\varepsilon_0\varepsilon_r\left(1 + \frac{\sigma_x}{j\omega\varepsilon_0}\right)\left(1 + \frac{\sigma_y}{j\omega\varepsilon_0}\right)E_z \tag{C.61c}$$

$$\frac{\partial E_z}{\partial y} - \frac{\partial E_y}{\partial z} = -j\omega\mu_0\left(1 + \frac{\sigma_x}{j\omega\varepsilon_0}\right)^{-1}\left(1 + \frac{\sigma_y}{j\omega\varepsilon_0}\right)H_x \tag{C.62a}$$

$$\frac{\partial E_x}{\partial z} - \frac{\partial E_z}{\partial x} = -j\omega\mu_0\left(1 + \frac{\sigma_x}{j\omega\varepsilon_0}\right)\left(1 + \frac{\sigma_y}{j\omega\varepsilon_0}\right)^{-1}H_y \tag{C.62b}$$

$$\frac{\partial E_y}{\partial x} - \frac{\partial E_x}{\partial y} = -j\omega\mu_0\left(1 + \frac{\sigma_x}{j\omega\varepsilon_0}\right)\left(1 + \frac{\sigma_y}{j\omega\varepsilon_0}\right)H_z \tag{C.62c}$$

For (C.61a), we define

$$D_x = \varepsilon_0\varepsilon_r\left(1 + \frac{\sigma_x}{j\omega\varepsilon_0}\right)^{-1}\left(1 + \frac{\sigma_y}{j\omega\varepsilon_0}\right)E_x \tag{C.63}$$

Thus, (C.61a) becomes

$$\frac{\partial H_z}{\partial y} - \frac{\partial H_y}{\partial z} = \frac{\partial E_x}{\partial t} \tag{C.64a}$$

$$\varepsilon_0\varepsilon_r\frac{\partial E_x}{\partial t} + \varepsilon_0\varepsilon_r\left(\frac{\sigma_y}{\varepsilon_0}\right)E_x = \frac{\partial D_x}{\partial t} + \left(\frac{\sigma_x}{\varepsilon_0}\right)D_x \tag{C.64b}$$

Using Galerkin's method, we can obtain

$$
{}_{\phi x}D^{n+1}_{i+1/2,j,k} = {}_{\phi x}D^{n}_{i+1/2,j,k} + \sum_{\nu} a(\nu)\left({}_{\phi z}H^{n+1/2}_{i+1/2,j+\nu+1/2,k}\frac{\Delta t}{\Delta y} - {}_{\phi y}H^{n+1/2}_{i+1/2,j,k+\nu+1/2}\frac{\Delta t}{\Delta z}\right) \tag{C.65}
$$

$$
\begin{aligned}
{}_{\phi x}E^{n+1}_{i+1/2,j,k} &= \left(1-\frac{\sigma_y^j\,\Delta t}{2\varepsilon_0}\right)\left(1+\frac{\sigma_y^i\,\Delta t}{2\varepsilon_0}\right)^{-1} {}_{\phi x}E^{n}_{i+1/2,j,k} \\
&+ \frac{1}{\varepsilon_0\varepsilon_r}\left(1+\frac{\sigma_x^i\,\Delta t}{2\varepsilon_0}\right)^{-1}\left[\left(1+\frac{\sigma_x^i\,\Delta t}{2\varepsilon_0}\right){}_{\phi x}D^{n+1}_{i+1/2,j,k} - \left(1-\frac{\sigma_x^i\,\Delta t}{2\varepsilon_0}\right){}_{\phi x}D^{n}_{i+1/2,j,k}\right]
\end{aligned} \tag{C.66}
$$

Similarly, we derive the following update equations

$$
{}_{\phi y}D^{n+1}_{i,j+1/2,k} = {}_{\phi y}D^{n}_{i,j+1/2,k} + \sum_{\nu} a(\nu)\left({}_{\phi x}H^{n+1/2}_{i,j+1/2,k+\nu+1/2}\frac{\Delta t}{\Delta y} - {}_{\phi z}H^{n+1/2}_{i+\nu+1/2,j+1/2,k}\frac{\Delta t}{\Delta z}\right) \tag{C.67}
$$

$$
\begin{aligned}
{}_{\phi y}E^{n+1}_{i,j+1/2,k} &= \left(1-\frac{\sigma_x^i\,\Delta t}{2\varepsilon_0}\right)\left(1+\frac{\sigma_x^i\,\Delta t}{2\varepsilon_0}\right)^{-1} {}_{\phi x}E^{n}_{i,j+1/2,k} + \frac{1}{\varepsilon_0\varepsilon_r}\left(1+\frac{\sigma_x^i\,\Delta t}{2\varepsilon_0}\right)^{-1} \\
&\times \left[\left(1+\frac{\sigma_y^j\,\Delta t}{2\varepsilon_0}\right){}_{\phi y}D^{n+1}_{i,j+1/2,k} - \left(1-\frac{\sigma_y^i\,\Delta t}{2\varepsilon_0}\right){}_{\phi y}D^{n}_{i,j+1/2,k}\right]
\end{aligned} \tag{C.68}
$$

$$
\begin{aligned}
{}_{\phi z}D^{n+1}_{i,j,k+1/2} &= \left(1-\frac{\sigma_y^j\,\Delta t}{2\varepsilon_0}\right)\left(1+\frac{\sigma_y^j\,\Delta t}{2\varepsilon_0}\right)^{-1} {}_{\phi z}D^{n}_{i,j,k+1/2} + \left(1+\frac{\sigma_y^j\,\Delta t}{2\varepsilon_0}\right)^{-1} \\
&\times \sum_{\nu} a(\nu)\left({}_{\phi y}H^{n+1/2}_{i+\nu+1/2,j,k+1/2}\frac{\Delta t}{\Delta x} - {}_{\phi x}H^{n+1/2}_{i,j+\nu+1/2,k+1/2}\frac{\Delta t}{\Delta y}\right)
\end{aligned} \tag{C.69}
$$

$$
\begin{aligned}
{}_{\phi z}E^{n+1}_{i,j,k+1/2} &= \left(1-\frac{\sigma_x^i\,\Delta t}{2\varepsilon_0}\right)\left(1+\frac{\sigma_x^i\,\Delta t}{2\varepsilon_0}\right)^{-1} {}_{\phi z}E^{n}_{i,j,k+1/2} \\
&+ \frac{1}{\varepsilon_0\varepsilon_r}\left(1+\frac{\sigma_x^i\,\Delta t}{2\varepsilon_0}\right)^{-1}\sum_{j'=0}^{Ny}([\varepsilon_r^*]^{-1})_{j,j'}\left[{}_{\phi z}D^{n+1}_{i,j',k+1/2} - {}_{\phi z}D^{n}_{i,j',k+1/2}\right]
\end{aligned} \tag{C.70}
$$

$$
\begin{aligned}
&{}_{\phi x}H^{n+1/2}_{i,j+1/2,k+1/2} \\
&= \left(1-\frac{\sigma_y^j\,\Delta t}{2\varepsilon_0}\right)\left(1+\frac{\sigma_y^j\,\Delta t}{2\varepsilon_0}\right)^{-1} {}_{\phi x}H^{n-1/2}_{i,j+1/2,k+1/2} + \frac{1}{\mu_0}\left(1+\frac{\sigma_y^j\,\Delta t}{2\varepsilon_0}\right)^{-1} \\
&\times \left[\left(1+\frac{\sigma_x^i\,\Delta t}{2\varepsilon_0}\right){}_{\phi x}B^{n+1/2}_{i,j+1/2,k+1/2} - \left(1-\frac{\sigma_x^i\,\Delta t}{2\varepsilon_0}\right){}_{\phi x}B^{n-1/2}_{i,j+1/2,k+1/2}\right]
\end{aligned} \tag{C.71}
$$

$$
{}_{\phi y}B^{n+1/2}_{i+1/2,j,k+1/2} = {}_{\phi y}B^{n-1/2}_{i+1/2,j,k+1/2} + \sum_{\nu} a(\nu)\left({}_{\phi z}E^{n}_{i+\nu+1,j,k+1/2}\frac{\Delta t}{\Delta x} - {}_{\phi x}E^{n}_{i+1/2,j,k+\nu+1}\frac{\Delta t}{\Delta z}\right) \tag{C.72}
$$

$$
\begin{aligned}
{}_{\phi y}H^{n+1/2}_{i+1/2,j,k+1/2} &= \left(1-\frac{\sigma_x^i\,\Delta t}{2\varepsilon_0}\right)\left(1+\frac{\sigma_x^i\,\Delta t}{2\varepsilon_0}\right)^{-1} {}_{\phi y}H^{n-1/2}_{i+1/2,j,k+1/2} + \frac{1}{\mu_0}\left(1+\frac{\sigma_x^i\,\Delta t}{2\varepsilon_0}\right)^{-1} \\
&\times\left[\left(1+\frac{\sigma_y^j\,\Delta t}{2\varepsilon_0}\right){}_{\phi y}B^{n+1/2}_{i+1/2,j,k+1/2} - \left(1-\frac{\sigma_y^j\,\Delta t}{2\varepsilon_0}\right){}_{\phi y}B^{n-1/2}_{i+1/2,j,k+1/2}\right]
\end{aligned} \tag{C.73}
$$

$$
\begin{aligned}
{}_{\phi z}B^{n+1/2}_{i+1/2,j+1/2,k} &= \left(1-\frac{\sigma_y^i\,\Delta t}{2\varepsilon_0}\right)\left(1+\frac{\sigma_y^i\,\Delta t}{2\varepsilon_0}\right)^{-1} {}_{\phi z}B^{n-1/2}_{i+1/2,j+1/2,k} \\
&+ \left(1+\frac{\sigma_y^i\,\Delta t}{2\varepsilon_0}\right)^{-1} \sum_{\nu} a(\nu)\left({}_{\phi x}E^{n}_{i+1/2,j+\nu+1,k}\frac{\Delta t}{\Delta y} - {}_{\phi y}E^{n}_{i+\nu+1,j+1/2,k}\frac{\Delta t}{\Delta x}\right)
\end{aligned} \tag{C.74}
$$

$$
\begin{aligned}
{}_{\phi z}H^{n+1/2}_{i+1/2,j+1/2,k} &= \left(1-\frac{\sigma_x^i\,\Delta t}{2\varepsilon_0}\right)\left(1+\frac{\sigma_x^i\,\Delta t}{2\varepsilon_0}\right)^{-1} {}_{\phi z}H^{n-1/2}_{i+1/2,j+1/2,k} \\
&+ \frac{1}{\mu_0}\left(1+\frac{\sigma_x^i\,\Delta t}{2\varepsilon_0}\right)^{-1}\left({}_{\phi z}B^{n+1/2}_{i+1/2,j+1/2,k} - {}_{\phi z}B^{n-1/2}_{i+1/2,j+1/2,k}\right)
\end{aligned} \tag{C.75}
$$

Edge-APML$_{xz}$ Regions

In the edge-APML$_{xz}$ regions, $[S] = [S_x][S_z]$, $[S_y] = [I]$, and the Maxwell equations read

$$
\frac{\partial H_z}{\partial y} - \frac{\partial H_y}{\partial z} = j\omega\varepsilon_0\varepsilon_r\left(1+\frac{\sigma_x}{j\omega\varepsilon_0}\right)^{-1}\left(1+\frac{\sigma_z}{j\omega\varepsilon_0}\right)E_x \tag{C.76a}
$$

$$
\frac{\partial H_x}{\partial z} - \frac{\partial H_z}{\partial x} = j\omega\varepsilon_0\varepsilon_r\left(1+\frac{\sigma_x}{j\omega\varepsilon_0}\right)\left(1+\frac{\sigma_z}{j\omega\varepsilon_0}\right)E_y \tag{C.76b}
$$

$$
\frac{\partial H_y}{\partial x} - \frac{\partial H_x}{\partial y} = j\omega\varepsilon_0\varepsilon_r\left(1+\frac{\sigma_x}{j\omega\varepsilon_0}\right)\left(1+\frac{\sigma_z}{j\omega\varepsilon_0}\right)^{-1}E_z \tag{C.76c}
$$

$$
\frac{\partial E_z}{\partial y} - \frac{\partial E_y}{\partial z} = -j\omega\mu_0\left(1+\frac{\sigma_x}{j\omega\varepsilon_0}\right)^{-1}\left(1+\frac{\sigma_z}{j\omega\varepsilon_0}\right)H_x \tag{C.77a}
$$

$$
\frac{\partial E_x}{\partial z} - \frac{\partial E_z}{\partial x} = -j\omega\mu_0\left(1+\frac{\sigma_x}{j\omega\varepsilon_0}\right)\left(1+\frac{\sigma_z}{j\omega\varepsilon_0}\right)H_y \tag{C.77b}
$$

$$
\frac{\partial E_y}{\partial x} - \frac{\partial E_x}{\partial y} = -j\omega\mu_0\left(1+\frac{\sigma_x}{j\omega\varepsilon_0}\right)\left(1+\frac{\sigma_z}{j\omega\varepsilon_0}\right)^{-1}H_z \tag{C.77c}
$$

Following the typical Galerkin's method, we derive the following update equations:

$$
\begin{aligned}
{}_{\phi x}D^{n+1}_{i+1/2,j,k} &= {}_{\phi x}D^{n}_{i+1/2,j,k} \\
&+ \sum_{\nu} a(\nu)\left({}_{\phi z}H^{n+1/2}_{i+1/2,j+\nu+1/2,k}\frac{\Delta t}{\Delta y} - {}_{\phi y}H^{n+1/2}_{i+1/2,j,k+\nu+1/2}\frac{\Delta t}{\Delta z}\right)
\end{aligned}
\tag{C.78}
$$

$$
\begin{aligned}
{}_{\phi x}E^{n+1}_{i+1/2,j,k} &= \left(1-\frac{\sigma_z^k\,\Delta t}{2\varepsilon_0}\right)\left(1+\frac{\sigma_z^k\,\Delta t}{2\varepsilon_0}\right)^{-1} {}_{\phi x}E^{n}_{i+1/2,j,k} + \frac{1}{\varepsilon_0\varepsilon_r}\left(1+\frac{\sigma_z^k\,\Delta t}{2\varepsilon_0}\right)^{-1} \\
&\times\left[\left(1+\frac{\sigma_x^i\,\Delta t}{2\varepsilon_0}\right){}_{\phi x}D^{n+1}_{i+1/2,j,k} - \left(1-\frac{\sigma_x^i\,\Delta t}{2\varepsilon_0}\right){}_{\phi x}D^{n}_{i+1/2,j,k}\right]
\end{aligned}
\tag{C.79}
$$

$$
\begin{aligned}
{}_{\phi y}D^{n+1}_{i,j+1/2,k} &= \left(1-\frac{\sigma_z^k\,\Delta t}{2\varepsilon_0}\right)\left(1+\frac{\sigma_z^k\,\Delta t}{2\varepsilon_0}\right)^{-1} {}_{\phi y}D^{n}_{i,j+1/2,k} + \left(1+\frac{\sigma_z^k\,\Delta t}{2\varepsilon_0}\right)^{-1} \\
&\times \sum_{\nu} a(\nu)\left({}_{\phi x}H^{n+1/2}_{i,j+1/2,k+\nu+1/2}\frac{\Delta t}{\Delta z} - {}_{\phi z}H^{n+1/2}_{i+\nu+1/2,j+1/2,k}\frac{\Delta t}{\Delta x}\right)
\end{aligned}
\tag{C.80}
$$

$$
\begin{aligned}
{}_{\phi y}E^{n+1}_{i,j+1/2,k} &= \left(1-\frac{\sigma_x^i\,\Delta t}{2\varepsilon_0}\right)\left(1+\frac{\sigma_x^i\,\Delta t}{2\varepsilon_0}\right)^{-1} {}_{\phi y}E^{n}_{i,j+1/2,k} \\
&+ \frac{1}{\varepsilon_0\varepsilon_r}\left(1+\frac{\sigma_x^i\,\Delta t}{2\varepsilon_0}\right)^{-1}\left[{}_{\phi y}D^{n+1}_{i,j+1/2,k} - {}_{\phi z}D^{n}_{i,j+1/2,k}\right]
\end{aligned}
\tag{C.81}
$$

$$
\begin{aligned}
{}_{\phi z}D^{n+1}_{i,j,k+1/2} &= {}_{\phi z}D^{n}_{i,j,k+1/2} \\
&+ \sum_{\nu} a(\nu)\left({}_{\phi y}H^{n+1/2}_{i+\nu+1/2,j,k+1/2}\frac{\Delta t}{\Delta x} - {}_{\phi x}H^{n+1/2}_{i,j+\nu+1/2,k+1/2}\frac{\Delta t}{\Delta y}\right)
\end{aligned}
\tag{C.82}
$$

$$
\begin{aligned}
{}_{\phi z}E^{n+1}_{i,j,k+1/2} &= \left(1-\frac{\sigma_x^i\,\Delta t}{2\varepsilon_0}\right)\left(1+\frac{\sigma_x^i\,\Delta t}{2\varepsilon_0}\right)^{-1} {}_{\phi z}E^{n}_{i,j,k+1/2} + \frac{1}{\varepsilon_0\varepsilon_r}\left(1+\frac{\sigma_x^i\,\Delta t}{2\varepsilon_0}\right)^{-1} \\
&\times\left[\left(1+\frac{\sigma_z^k\,\Delta t}{2\varepsilon_0}\right){}_{\phi z}D^{n+1}_{i,j,k+1/2} - \left(1-\frac{\sigma_z^k\,\Delta t}{2\varepsilon_0}\right){}_{\phi z}D^{n}_{i,j,k+1/2}\right]
\end{aligned}
\tag{C.83}
$$

$$
\begin{aligned}
{}_{\phi x}B^{n+1/2}_{i,j+1/2,k+1/2} &= {}_{\phi x}B^{n-1/2}_{i,j+1/2,k+1/2} \\
&+ \sum_{\nu} a(\nu)\left({}_{\phi y}E^{n}_{i,j+1/2,k+\nu+1}\frac{\Delta t}{\Delta z} - {}_{\phi z}E^{n}_{i,j+\nu+1,k+1/2}\frac{\Delta t}{\Delta y}\right)
\end{aligned}
\tag{C.84}
$$

$$
\begin{aligned}
{}_{\phi x}H^{n+1/2}_{i,j+1/2,k+1/2} &= \left(1-\frac{\sigma_z^i\,\Delta t}{2\varepsilon_0}\right)\left(1+\frac{\sigma_z^i\,\Delta t}{2\varepsilon_0}\right)^{-1} {}_{\phi x}H^{n-1/2}_{i,j+1/2,k+1/2} \\
&+ \frac{1}{\mu_0}\left(1+\frac{\sigma_z^i\,\Delta t}{2\varepsilon_0}\right)^{-1}\left[\left(1+\frac{\sigma_x^i\,\Delta t}{2\varepsilon_0}\right){}_{\phi x}B^{n+1/2}_{i,j+1/2,k+1/2}\right. \\
&\left. - \left(1-\frac{\sigma_x^i\,\Delta t}{2\varepsilon_0}\right){}_{\phi x}B^{n-1/2}_{i,j+1/2,k+1/2}\right]
\end{aligned}
\tag{C.85}
$$

$$
{}_{\phi y}B^{n+1/2}_{i+1/2,j,k+1/2} = \left(1 - \frac{\sigma^i_z\,\Delta t}{2\varepsilon_0}\right)\left(1 + \frac{\sigma^i_z\,\Delta t}{2\varepsilon_0}\right)^{-1} {}_{\phi y}B^{n-1/2}_{i+1/2,j,k+1/2}
+ \left(1 + \frac{\sigma^i_z\,\Delta t}{2\varepsilon_0}\right)^{-1} \sum_{\nu} a(\nu)\left({}_{\phi z}E^n_{i+\nu+1,j,k+1/2}\frac{\Delta t}{\Delta x} - {}_{\phi x}E^n_{i+1/2,j,k+\nu+1}\frac{\Delta t}{\Delta z}\right) \tag{C.86}
$$

$$
{}_{\phi y}H^{n+1/2}_{i+1/2,j,k+1/2} = \left(1 + \frac{\sigma^i_x\,\Delta t}{2\varepsilon_0}\right)^{-1}\left(1 + \frac{\sigma^i_x\,\Delta t}{2\varepsilon_0}\right)^{-1} {}_{\phi y}H^{n-1/2}_{i+1/2,j,k+1/2}
+ \frac{1}{\mu_0}\left(1 + \frac{\sigma^i_x\,\Delta t}{2\varepsilon_0}\right)^{-1} \left({}_{\phi y}B^{n+1/2}_{i+1/2,j,k+1/2} - {}_{\phi y}B^{n-1/2}_{i+1/2,j,k+1/2}\right) \tag{C.87}
$$

$$
{}_{\phi z}B^{n+1/2}_{i+1/2,j+1/2,k} = {}_{\phi z}B^{n-1/2}_{i+1/2,j+1/2,k}
+ \sum_{\nu} a(\nu)\left({}_{\phi x}E^n_{i+1/2,j+\nu+1,k}\frac{\Delta t}{\Delta y} - {}_{\phi y}E^n_{i+\nu+1,j+1/2,k}\frac{\Delta t}{\Delta x}\right) \tag{C.88}
$$

$$
{}_{\phi z}H^{n+1/2}_{i+1/2,j+1/2,k} = \left(1 - \frac{\sigma^i_x\,\Delta t}{2\varepsilon_0}\right)\left(1 + \frac{\sigma^i_x\,\Delta t}{2\varepsilon_0}\right)^{-1} {}_{\phi z}H^{n-1/2}_{i+1/2,j+1/2,k}
+ \frac{1}{\mu_0}\left(1 + \frac{\sigma^i_x\,\Delta t}{2\varepsilon_0}\right)^{-1}\left[\left(1 + \frac{\sigma^k_z\,\Delta t}{2\varepsilon_0}\right){}_{\phi z}B^{n+1/2}_{i+1/2,j+1/2,k} - \left(1 - \frac{\sigma^k_z\,\Delta t}{2\varepsilon_0}\right){}_{\phi z}B^{n-1/2}_{i+1/2,j+1/2,k}\right] \tag{C.89}
$$

Edge-APML$_{yz}$ Regions

In the edge-APML$_{yz}$ regions, $[S] = [S_y][S_z]$, $[S_x] = [I]$, and the Maxwell equations now become

$$
\frac{\partial H_z}{\partial y} - \frac{\partial H_y}{\partial z} = j\omega\varepsilon_0\varepsilon_r\left(1 + \frac{\sigma_y}{j\omega\varepsilon_0}\right)\left(1 + \frac{\sigma_z}{j\omega\varepsilon_0}\right)E_x \tag{C.90a}
$$

$$
\frac{\partial H_x}{\partial z} - \frac{\partial H_z}{\partial x} = j\omega\varepsilon_0\varepsilon_r\left(1 + \frac{\sigma_y}{j\omega\varepsilon_0}\right)^{-1}\left(1 + \frac{\sigma_z}{j\omega\varepsilon_0}\right)E_y \tag{C.90b}
$$

$$
\frac{\partial H_y}{\partial x} - \frac{\partial H_x}{\partial y} = j\omega\varepsilon_0\varepsilon_r\left(1 + \frac{\sigma_y}{j\omega\varepsilon_0}\right)\left(1 + \frac{\sigma_z}{j\omega\varepsilon_0}\right)^{-1}E_z \tag{C.90c}
$$

$$
\frac{\partial E_z}{\partial y} - \frac{\partial E_y}{\partial z} = -j\omega\mu_0\left(1 + \frac{\sigma_y}{j\omega\varepsilon_0}\right)\left(1 + \frac{\sigma_z}{j\omega\varepsilon_0}\right)H_x \tag{C.91a}
$$

$$
\frac{\partial E_x}{\partial z} - \frac{\partial E_z}{\partial x} = -j\omega\mu_0\left(1 + \frac{\sigma_y}{j\omega\varepsilon_0}\right)^{-1}\left(1 + \frac{\sigma_z}{j\omega\varepsilon_0}\right)H_y \tag{C.91b}
$$

$$
\frac{\partial E_y}{\partial x} - \frac{\partial E_x}{\partial y} = -j\omega\mu_0\left(1 + \frac{\sigma_y}{j\omega\varepsilon_0}\right)\left(1 + \frac{\sigma_z}{j\omega\varepsilon_0}\right)^{-1}H_z \tag{C.91c}
$$

The derived update equations read

$$
{}_{\phi x}D^{n+1}_{i+1/2,j,k} = \left(1 - \frac{\sigma_z^k\ \Delta t}{2\varepsilon_0}\right)\left(1 + \frac{\sigma_z^k\ \Delta t}{2\varepsilon_0}\right)^{-1} {}_{\phi x}D^{n}_{i,j+1/2,k} + \left(1 + \frac{\sigma_z^k\ \Delta t}{2\varepsilon_0}\right)^{-1}
\times \sum_{\nu} a(\nu)\left({}_{\phi z}H^{n+1/2}_{i+1/2,j+\nu+1/2,k}\frac{\Delta t}{\Delta y} - {}_{\phi y}H^{n+1/2}_{i+1/2,j,k+\nu+1/2}\frac{\Delta t}{\Delta z}\right) \tag{C.92}
$$

$$
{}_{\phi x}E^{n+1}_{i+1/2,j,k} = \left(1 - \frac{\sigma_y^j\ \Delta t}{2\varepsilon_0}\right)\left(1 + \frac{\sigma_y^j\ \Delta t}{2\varepsilon_0}\right)^{-1} {}_{\phi x}E^{n}_{i+1/2,j,k}
+ \frac{1}{\varepsilon_0\varepsilon_r}\left(1 + \frac{\sigma_y^j\ \Delta t}{2\varepsilon_0}\right)^{-1}\left[{}_{\phi x}D^{n+1}_{i+1/2,j,k} - {}_{\phi x}D^{n}_{i+1/2,j,k}\right] \tag{C.93}
$$

$$
{}_{\phi y}D^{n+1}_{i,j+1/2,k} = {}_{\phi y}D^{n}_{i,j+1/2,k}
+ \sum_{\nu} a(\nu)\left({}_{\phi x}H^{n+1/2}_{i,j+1/2,k+\nu+1/2}\frac{\Delta t}{\Delta z} - {}_{\phi z}H^{n+1/2}_{i+\nu+1/2,j+1/2,k}\frac{\Delta t}{\Delta x}\right) \tag{C.94}
$$

$$
{}_{\phi y}E^{n+1}_{i,j+1/2,k} = \left(1 - \frac{\sigma_y^j\ \Delta t}{2\varepsilon_0}\right)\left(1 + \frac{\sigma_y^j\ \Delta t}{2\varepsilon_0}\right)^{-1} {}_{\phi y}E^{n}_{i,j+1/2,k} + \frac{1}{\varepsilon_0\varepsilon_r}\left(1 + \frac{\sigma_y^j\ \Delta t}{2\varepsilon_0}\right)^{-1}
\times\left[\left(1 + \frac{\sigma_z^k\ \Delta t}{2\varepsilon_0}\right){}_{\phi y}D^{n+1}_{i,j+1/2,k} - \left(1 - \frac{\sigma_z^k\ \Delta t}{2\varepsilon_0}\right){}_{\phi z}D^{n}_{i,j+1/2,k}\right] \tag{C.95}
$$

$$
{}_{\phi z}D^{n+1}_{i,j,k+1/2} = {}_{\phi z}D^{n}_{i,j,k+1/2}
+ \sum_{\nu} a(\nu)\left({}_{\phi y}H^{n+1/2}_{i+\nu+1/2,j,k+1/2}\frac{\Delta t}{\Delta x} - {}_{\phi x}H^{n+1/2}_{i,j+\nu+1/2,k+1/2}\frac{\Delta t}{\Delta y}\right) \tag{C.96}
$$

$$
{}_{\phi z}E^{n+1}_{i,j,k+1/2} = \left(1 - \frac{\sigma_y^j\ \Delta t}{2\varepsilon_0}\right)\left(1 + \frac{\sigma_y^j\ \Delta t}{2\varepsilon_0}\right)^{-1} {}_{\phi z}E^{n}_{i,j,k+1/2}
+ \frac{1}{\varepsilon_0\varepsilon_r}\left(1 + \frac{\sigma_y^j\ \Delta t}{2\varepsilon_0}\right)^{-1}\left[\left(1 + \frac{\sigma_z^k\ \Delta t}{2\varepsilon_0}\right){}_{\phi z}D^{n+1}_{i,j,k+1/2} - \left(1 - \frac{\sigma_z^k\ \Delta t}{2\varepsilon_0}\right){}_{\phi z}D^{n}_{i,j,k+1/2}\right] \tag{C.97}
$$

$$
{}_{\phi x}B^{n+1/2}_{i,j+1/2,k+1/2} = \left(1 - \frac{\sigma_z^j\ \Delta t}{2\varepsilon_0}\right)\left(1 + \frac{\sigma_z^j\ \Delta t}{2\varepsilon_0}\right)^{-1} {}_{\phi x}B^{n-1/2}_{i,j+1/2,k+1/2}
+ \left(1 + \frac{\sigma_z^j\ \Delta t}{2\varepsilon_0}\right)^{-1}\sum_{\nu} a(\nu)\left({}_{\phi y}E^{n}_{i,j+1/2,k+\nu+1}\frac{\Delta t}{\Delta z} - {}_{\phi z}E^{n}_{i,j+\nu+1,k+1/2}\frac{\Delta t}{\Delta y}\right) \tag{C.98}
$$

$$
{}_{\phi x}H^{n+1/2}_{i,j+1/2,k+1/2} = \left(1 - \frac{\sigma_y^j\,\Delta t}{2\varepsilon_0}\right)\left(1 + \frac{\sigma_y^j\,\Delta t}{2\varepsilon_0}\right)^{-1} {}_{\phi x}H^{n-1/2}_{i,j+1/2,k+1/2}
+ \frac{1}{\mu_0}\left(1 + \frac{\sigma_y^j\,\Delta t}{2\varepsilon_0}\right)^{-1}\left({}_{\phi x}B^{n+1/2}_{i,j+1/2,k+1/2} - {}_{\phi x}B^{n-1/2}_{i,j+1/2,k+1/2}\right) \quad \text{(C.99)}
$$

$$
{}_{\phi y}B^{n+1/2}_{i+1/2,j,k+1/2} = {}_{\phi y}B^{n-1/2}_{i+1/2,j,k+1/2}
+ \sum_{\nu} a(\nu)\left({}_{\phi z}E^{n}_{i+\nu+1,j,k+1/2}\frac{\Delta t}{\Delta x} - {}_{\phi x}E^{n}_{i+1/2,j,k+\nu+1}\frac{\Delta t}{\Delta z}\right) \quad \text{(C.100)}
$$

$$
{}_{\phi y}H^{n+1/2}_{i+1/2,j,k+1/2} = \left(1 - \frac{\sigma_z^j\,\Delta t}{2\varepsilon_0}\right)\left(1 + \frac{\sigma_z^j\,\Delta t}{2\varepsilon_0}\right)^{-1} {}_{\phi y}H^{n-1/2}_{i+1/2,j,k+1/2}
+ \frac{1}{\mu_0}\left(1 + \frac{\sigma_z^j\,\Delta t}{2\varepsilon_0}\right)^{-1}\left[\left(1 + \frac{\sigma_y^j\,\Delta t}{2\varepsilon_0}\right){}_{\phi y}B^{n+1/2}_{i+1/2,j,k+1/2}
- \left(1 - \frac{\sigma_y^j\,\Delta t}{2\varepsilon_0}\right){}_{\phi y}B^{n-1/2}_{i+1/2,j,k+1/2}\right] \quad \text{(C.101)}
$$

$$
{}_{\phi z}H^{n+1/2}_{i+1/2,j+1/2,k} = \left(1 - \frac{\sigma_y^j\,\Delta t}{2\varepsilon_0}\right)\left(1 + \frac{\sigma_y^j\,\Delta t}{2\varepsilon_0}\right)^{-1} {}_{\phi z}H^{n-1/2}_{i+1/2,j+1/2,k}
+ \frac{1}{\mu_0}\left(1 + \frac{\sigma_y^j\,\Delta t}{2\varepsilon_0}\right)^{-1}\left[\left(1 + \frac{\sigma_z^k\,\Delta t}{2\varepsilon_0}\right){}_{\phi z}B^{n+1/2}_{i+1/2,j+1/2,k}
- \left(1 - \frac{\sigma_z^k\,\Delta t}{2\varepsilon_0}\right){}_{\phi z}B^{n-1/2}_{i+1/2,j+1/2,k}\right] \quad \text{(C.102)}
$$

C.5 UPDATE EQUATIONS FOR CORNER-APML REGIONS

It is obvious that there are eight different corner-APML regions in total for a rectangular computational volume as shown in Figure C.1. In the corner regions, we have to simultaneously consider electromagnetic waves that decay in all three (x, y, and z) directions; thus, we specify them as corner-APML$_{\text{xyz}}$ regions.

The Maxwell equations now satisfy the following relations:

$$
\frac{\partial H_z}{\partial y} - \frac{\partial H_y}{\partial z} = j\omega\varepsilon_0\varepsilon_r\left(1 + \frac{\sigma_x}{j\omega\varepsilon_0}\right)^{-1}\left(1 + \frac{\sigma_y}{j\omega\varepsilon_0}\right)\left(1 + \frac{\sigma_z}{j\omega\varepsilon_0}\right)E_x \quad \text{(C.103a)}
$$

$$
\frac{\partial H_x}{\partial z} - \frac{\partial H_z}{\partial x} = j\omega\varepsilon_0\varepsilon_r\left(1 + \frac{\sigma_x}{j\omega\varepsilon_0}\right)\left(1 + \frac{\sigma_y}{j\omega\varepsilon_0}\right)^{-1}\left(1 + \frac{\sigma_z}{j\omega\varepsilon_0}\right)E_y \quad \text{(C.103b)}
$$

$$\frac{\partial H_y}{\partial x} - \frac{\partial H_x}{\partial y} = j\omega\varepsilon_0\varepsilon_r \left(1 + \frac{\sigma_x}{j\omega\varepsilon_0}\right)\left(1 + \frac{\sigma_y}{j\omega\varepsilon_0}\right)\left(1 + \frac{\sigma_z}{j\omega\varepsilon_0}\right)^{-1} E_z \quad \text{(C.103c)}$$

$$\frac{\partial E_z}{\partial y} - \frac{\partial E_y}{\partial z} = -j\omega\mu_0 \left(1 + \frac{\sigma_x}{j\omega\varepsilon_0}\right)^{-1}\left(1 + \frac{\sigma_y}{j\omega\varepsilon_0}\right)\left(1 + \frac{\sigma_z}{j\omega\varepsilon_0}\right) H_x \quad \text{(C.104a)}$$

$$\frac{\partial E_x}{\partial z} - \frac{\partial E_z}{\partial x} = -j\omega\mu_0 \left(1 + \frac{\sigma_x}{j\omega\varepsilon_0}\right)\left(1 + \frac{\sigma_y}{j\omega\varepsilon_0}\right)^{-1}\left(1 + \frac{\sigma_z}{j\omega\varepsilon_0}\right) H_y \quad \text{(C.104b)}$$

$$\frac{\partial E_y}{\partial x} - \frac{\partial E_x}{\partial y} = -j\omega\mu_0 \left(1 + \frac{\sigma_x}{j\omega\varepsilon_0}\right)\left(1 + \frac{\sigma_y}{j\omega\varepsilon_0}\right)\left(1 + \frac{\sigma_z}{j\omega\varepsilon_0}\right)^{-1} H_z \quad \text{(C.104c)}$$

By introducing an intermediate parameter

$$D_x = \left(1 + \frac{\sigma_x}{j\omega\varepsilon_0}\right)^{-1}\left(1 + \frac{\sigma_y}{j\omega\varepsilon_0}\right) E_x \quad \text{(C.105)}$$

(C.104a) can be split into two equations,

$$\frac{\partial H_z}{\partial y} - \frac{\partial H_y}{\partial z} = \varepsilon_0\varepsilon_r \frac{\partial D_x}{\partial t} + \varepsilon_r\sigma_z D_x \quad \text{(C.106a)}$$

$$\frac{\partial E_x}{\partial t} + \left(\frac{\sigma_y}{\varepsilon_0}\right) E_x = \frac{\partial D_x}{\partial t} + \left(\frac{\sigma_x}{\varepsilon_0}\right) D_x \quad \text{(C.106b)}$$

which yield the update equations of the x-components:

$$\begin{aligned} {}_{\phi x}D^{n+1}_{i+1/2,j,k} &= \left(1 - \frac{\sigma_z^k\,\Delta t}{2\varepsilon_0}\right)^{-1}\left(1 + \frac{\sigma_z^k\,\Delta t}{2\varepsilon_0}\right)^{-1} {}_{\phi x}D^{n}_{i+1/2,j,k} \\ &+ \left(1 + \frac{\sigma_z^k\,\Delta t}{2\varepsilon_0}\right)^{-1} \sum_\nu a(\nu)\left({}_{\phi z}H^{n+1/2}_{i+1/2,j+\nu+1/2,k}\frac{\Delta t}{\Delta y} - {}_{\phi y}H^{n+1/2}_{i+1/2,j,k+\nu+1/2}\frac{\Delta t}{\Delta z}\right) \end{aligned} \quad \text{(C.107)}$$

$$\begin{aligned} {}_{\phi x}E^{n+1}_{i+1/2,j,k} &= \left(1 - \frac{\sigma_y^j\,\Delta t}{2\varepsilon_0}\right)\left(1 + \frac{\sigma_y^j\,\Delta t}{2\varepsilon_0}\right)^{-1} {}_{\phi y}E^{n}_{i,j+1/2,k} + \frac{1}{\varepsilon_0\varepsilon_r}\left(1 + \frac{\sigma_y^j\,\Delta t}{2\varepsilon_0}\right)^{-1} \\ &\times \left[\left(1 + \frac{\sigma_x^i\,\Delta t}{2\varepsilon_0}\right) {}_{\phi x}D^{n+1}_{i+1/2,j,k} - \left(1 - \frac{\sigma_x^i\,\Delta t}{2\varepsilon_0}\right) {}_{\phi y}D^{n}_{i+1/2,j,k}\right] \end{aligned} \quad \text{(C.108)}$$

Similarly, we derive the remaining update equations:

$$\begin{aligned} {}_{\phi y}D^{n+1}_{i,j+1/2,k} &= \left(1 - \frac{\sigma_z^k\,\Delta t}{2\varepsilon_0}\right)\left(1 + \frac{\sigma_z^k\,\Delta t}{2\varepsilon_0}\right)^{-1} {}_{\phi y}D^{n}_{i,j+1/2,k} + \left(1 + \frac{\sigma_z^k\,\Delta t}{2\varepsilon_0}\right)^{-1} \\ &\times \sum_\nu a(\nu)\left({}_{\phi x}H^{n+1/2}_{i,j+1/2,k+\nu+1/2}\frac{\Delta t}{\Delta z} - {}_{\phi z}H^{n+1/2}_{i+\nu+1/2,j+1/2,k}\frac{\Delta t}{\Delta x}\right) \end{aligned} \quad \text{(C.109)}$$

$$
{}_{\phi y}E^{n+1}_{i,j+1/2,k} = \left(1 - \frac{\sigma^i_x\ \Delta t}{2\varepsilon_0}\right)\left(1 + \frac{\sigma^i_x\ \Delta t}{2\varepsilon_0}\right)^{-1} {}_{\phi y}E^{n}_{i,j+1/2,k}
$$

$$
+ \frac{1}{\varepsilon_0}\left(1+\frac{\sigma^i_x\ \Delta t}{2\varepsilon_0}\right)^{-1}\left[\left(1+\frac{\sigma^j_y\ \Delta t}{2\varepsilon_0}\right){}_{\phi y}D^{n+1}_{i,j+1/2,k} - \left(1-\frac{\sigma^j_y\ \Delta t}{2\varepsilon_0}\right){}_{\phi y}D^{n}_{i,j+1/2,k}\right] \quad \text{(C.110)}
$$

$$
{}_{\phi z}D^{n+1}_{i,j,k+1/2} = \left(1 - \frac{\sigma^j_y\ \Delta t}{2\varepsilon_0}\right)\left(1 + \frac{\sigma^j_y\ \Delta t}{2\varepsilon_0}\right)^{-1} {}_{\phi x}D^{n}_{i,j,k+1/2}
$$

$$
+ \left(1+\frac{\sigma^j_y\ \Delta t}{2\varepsilon_0}\right)^{-1}\sum_\nu a(\nu)\left({}_{\phi y}H^{n+1/2}_{i+\nu+1/2,j,k+1/2}\frac{\Delta t}{\Delta x} - {}_{\phi x}H^{n+1/2}_{i,j+\nu+1/2,k+1/2}\frac{\Delta t}{\Delta y}\right) \quad \text{(C.111)}
$$

$$
{}_{\phi z}E^{n+1}_{i,j,k+1/2} = \left(1 - \frac{\sigma^i_x\ \Delta t}{2\varepsilon_0}\right)\left(1 + \frac{\sigma^i_x\ \Delta t}{2\varepsilon_0}\right)^{-1} {}_{\phi z}E^{n}_{i,j,k+1/2}
$$

$$
+ \frac{1}{\varepsilon_0\varepsilon_r}\left(1 + \frac{\sigma^i_x\ \Delta t}{2\varepsilon_0}\right)\left[\left(1 + \frac{\sigma^k_z\ \Delta t}{2\varepsilon_0}\right){}_{\phi z}D^{n+1}_{i,j,k+1/2} - \left(1-\frac{\sigma^k_z\ \Delta t}{2\varepsilon_0}\right){}_{\phi y}D^{n}_{i,j,k+1/2}\right] \quad \text{(C.112)}
$$

$$
{}_{\phi x}B^{n+1/2}_{i,j+1/2,k+1/2} = \left(1 - \frac{\sigma_z\ \Delta t}{2\varepsilon_0}\right)\left(1 + \frac{\sigma_z\ \Delta t}{2\varepsilon_0}\right)^{-1} {}_{\phi x}B^{n-1/2}_{i,j+1/2,k+1/2}
$$

$$
+ \left(1 + \frac{\sigma_z\ \Delta t}{2\varepsilon_0}\right)^{-1}\sum_\nu a(\nu)\left({}_{\phi y}E^{n}_{i,j+1/2,k+\nu+1}\frac{\Delta t}{\Delta z} - {}_{\phi z}E^{n}_{i,j+\nu+1,k+1/2}\frac{\Delta t}{\Delta y}\right) \quad \text{(C.113)}
$$

$$
{}_{\phi x}H^{n+1/2}_{i,j+1/2,k+1/2} = \left(1 - \frac{\sigma_y\ \Delta t}{2\varepsilon_0}\right)\left(1 + \frac{\sigma_y\ \Delta t}{2\varepsilon_0}\right)^{-1} {}_{\phi x}H^{n-1/2}_{i,j+1/2,k+1/2}
$$

$$
+ \frac{1}{\mu_0}\left(1 + \frac{\sigma_y\ \Delta t}{2\varepsilon_0}\right)^{-1}\left[\left(1 + \frac{\sigma^i_x\ \Delta t}{2\varepsilon_0}\right){}_{\phi x}B^{n+1/2}_{i,j+1/2,k+1/2} - \left(1 - \frac{\sigma^i_x\ \Delta t}{2\varepsilon_0}\right){}_{\phi x}B^{n-1/2}_{i,j+1/2,k+1/2}\right] \quad \text{(C.114)}
$$

$$
{}_{\phi y}B^{n+1/2}_{i+1/2,j,k+1/2} = \left(1 - \frac{\sigma_z\ \Delta t}{2\varepsilon_0}\right)\left(1 + \frac{\sigma_z\ \Delta t}{2\varepsilon_0}\right)^{-1} {}_{\phi y}B^{n-1/2}_{i+1/2,j,k+1/2}
$$

$$
+ \left(1 + \frac{\sigma_z\ \Delta t}{2\varepsilon_0}\right)^{-1}\sum_\nu a(\nu)\left({}_{\phi z}E^{n}_{i+\nu+1,j,k+1/2}\frac{\Delta t}{\Delta x} - {}_{\phi x}E^{n}_{i+1/2,j,k+\nu+1}\frac{\Delta t}{\Delta z}\right) \quad \text{(C.115)}
$$

$$
\begin{aligned}
{}_{\phi y}H^{n+1/2}_{i+1/2,j,k+1/2} &= \left(1-\frac{\sigma_x\,\Delta t}{2\varepsilon_0}\right)\left(1+\frac{\sigma_x\,\Delta t}{2\varepsilon_0}\right)^{-1} {}_{\phi y}H^{n-1/2}_{i+1/2,j,k+1/2} \\
&+ \frac{1}{\mu_0}\left(1+\frac{\sigma_x\,\Delta t}{2\varepsilon_0}\right)^{-1}\left[\left(1+\frac{\sigma_y^j\,\Delta t}{2\varepsilon_0}\right){}_{\phi y}B^{n+1/2}_{i+1/2,j,k+1/2}\right. \\
&\left. - \left(1-\frac{\sigma_y^j\,\Delta t}{2\varepsilon_0}\right){}_{\phi y}B^{n-1/2}_{i+1/2,j,k+1/2}\right]
\end{aligned}
\tag{C.116}
$$

$$
\begin{aligned}
{}_{\phi z}B^{n+1/2}_{i+1/2,j+1/2,k} &= \left(1-\frac{\sigma_y\,\Delta t}{2\varepsilon_0}\right)\left(1+\frac{\sigma_y\,\Delta t}{2\varepsilon_0}\right)^{-1} {}_{\phi z}B^{n-1/2}_{i+1/2,j+1/2,k} \\
&+ \left(1+\frac{\sigma_y\,\Delta t}{2\varepsilon_0}\right)^{-1}\sum_{\nu} a(\nu)\left({}_{\phi x}E^{n}_{i+1/2,j+\nu+1,k}\frac{\Delta t}{\Delta y} - {}_{\phi y}E^{n}_{i+\nu+1,j+1/2,k}\frac{\Delta t}{\Delta x}\right)
\end{aligned}
\tag{C.117}
$$

$$
\begin{aligned}
{}_{\phi z}H^{n+1/2}_{i+1/2,j+1/2,k} &= \left(1-\frac{\sigma_x\,\Delta t}{2\varepsilon_0}\right)\left(1+\frac{\sigma_x\,\Delta t}{2\varepsilon_0}\right)^{-1} {}_{\phi z}H^{n-1/2}_{i+1/2,j+1/2,k} \\
&+ \frac{1}{\mu_0}\left(1+\frac{\sigma_x\,\Delta t}{2\varepsilon_0}\right)^{-1}\left[\left(1+\frac{\sigma_z^k\,\Delta t}{2\varepsilon_0}\right){}_{\phi z}B^{n+1/2}_{i+1/2,j+1/2,k}\right. \\
&\left. - \left(1-\frac{\sigma_z^k\,\Delta t}{2\varepsilon_0}\right){}_{\phi z}B^{n-1/2}_{i+1/2,j+1/2,k}\right]
\end{aligned}
\tag{C.118}
$$

REFERENCES

[1] J. P. Berenger, "A perfectly matched layer for the absorption of electromagnetic waves," *J. Comput. Phys.*, vol. 114, pp. 185–200, May 1994.

[2] D. S. Katz, E. T. Thiele, and A. Taflove, "Validation and extension to three dimensions of the Berenger absorbing boundary condition for FDTD meshes," *IEEE Microwave Guided Wave Lett.*, vol. 4, no. 8, pp. 268–270, Aug. 1994.

[3] W. C. Chew and W. H. Weedon, "A 3-D perfectly matched medium from modified Maxwell's equations with stretched coordinates," *Microwave Opt. Technol. Lett.*, vol. 7, no. 13, pp. 599–604, Sept. 1994.

[4] R. Mittra and U. Pekel, "A new look at the perfectly matched layer (PML) concept for the reflectionless absorption of electromagnetic waves," *IEEE Microwave Guided Wave Lett.*, vol. 5, no. 3, pp. 84–87, Mar. 1995.

[5] J. Fang and Z. Wu, "Generalized perfectly matched layer—an extension of Berenger's perfectly matched layer boundary condition," *IEEE Microwave Guided Wave Lett.*, vol. 5, no. 12, pp. 451–453, Dec. 1995.

[6] J. C. Veihl, *Effective Mesh Truncation Techniques for the Solution of Maxwell's Equations Using the Finite Difference Time Domain Method*, Ph.D. Dissertation, Department of Electrical Engineering, University of Illinois at Urbana-Champaign, 1996.

[7] Z. S. Sacks, D. M. Kingsland, R. Lee, and J.-F. Lee, "A perfectly matched anisotropic absorber for use as an absorbing boundary condition," *IEEE Trans. Antennas Propag.*, vol. AP-43, no. 12, pp. 1460–1463, Dec. 1995.

[8] L. Zhao and A. C. Cangellaris, "A general approach for the development of unsplit-field time-domain implementations of perfectly matched layers for FDTD grid truncation," *IEEE Microwave Guided Wave Lett.*, vol. 6, no. 5, pp. 209–211, May 1996.

[9] L. Zhao and A. C. Cangellaris, "GT-PML: Generalized theory of perfectly matched layers and its applications to the reflectionless truncation of finite-difference time-domain grids," *IEEE Trans. Microwave Theory Tech.*, vol. MTT-44, no. 12, pp. 2555–2563, Dec. 1996.

[10] S. D. Gedney, "An anisotropic perfectly matched layer-absorbing medium for the truncation of FDTD lattices," *IEEE Trans. Antennas Propag.*, vol. AP-44, no. 12, pp. 1630–1639, Dec. 1996.

APPENDIX D

Expressions and Properties of the Cubic Battle–Lemarié Functions

D.1 EXPRESSION FOR THE B-SPLINE FUNCTION IN THE FREQUENCY DOMAIN

In this appendix, we develop the expressions and properties of the spline wavelet, especially for the cubic spline Battle–Lemarié scaling and mother wavelet functions.

In the multiresolution analysis (MRA), a sequence $\{\phi_{s,l}(t) = 2^{s/2}\phi(2^s t - l)\}$, $s, l \in Z$ forms an orthogonal basis, and we can obtain the two-scale scaling equation [1, 2]:

$$\phi(x) = \sum_k h_k \phi_{1k}(x) = \sqrt{2} \sum_k h_k \phi(2x - k) \tag{D.1}$$

By taking the Fourier transform of both sides of (D.1), we further have

$$\begin{aligned}
\tilde{\phi}(\omega) &= \int_{-\infty}^{+\infty} \phi(x)\, e^{-i\omega x}\, dx = \sum_k \sqrt{2} h_k \int_{-\infty}^{+\infty} \phi(2x - k)\, e^{-i\omega x}\, dx \\
&= \left(\sum_k \frac{h_k}{\sqrt{2}} e^{-ik\omega/2} \right) \int_{-\infty}^{+\infty} \phi(2x - k)\, e^{-i(2x-k)\omega/2} d(2x) \qquad \text{(D.2)} \\
&= \frac{1}{\sqrt{2}} \sum_k h_k\, e^{-ik\omega/2} \tilde{\phi}\left(\frac{\omega}{2}\right)
\end{aligned}$$

Multiresolution Time Domain Scheme for Electromagnetic Engineering
By Yinchao Chen, Qunsheng Cao, and Raj Mittra
ISBN 0-471-27230-2

We can rewrite equation (D.2) as

$$\widetilde{\phi}(\omega) = m_0(\omega/2)\widetilde{\phi}(\omega/2) \tag{D.3}$$

with

$$m_0(\omega) = \frac{1}{\sqrt{2}} \sum_k h_k\, e^{-ik\omega} \tag{D.4}$$

where the function m_0 is a 2π-periodic function and $m_0 \in L^2$ ({0, 2π]).

We now integrate (D.1) by using the normalization property of the scaling function $\int_{\infty}^{+\infty} \phi(x)\, dx = 1$:

$$\int_{-\infty}^{+\infty} \phi(x)\, dx = 1 = \sqrt{2} \sum_k h_k \int_{-\infty}^{+\infty} \phi(2x-k)\, dx = \frac{1}{\sqrt{2}} \sum_k h_k \int_{-\infty}^{+\infty} \phi(2x-k) d(2x)$$

which leads to $\sum_k h_k = \sqrt{2}$, or

$$m_0(0) = \frac{1}{\sqrt{2}} \sum_k h_k = 1 \tag{D.5}$$

Applying (D.3) recursively for $\omega/2$, $\omega/4, \ldots$, we get

$$\widetilde{\phi}(\omega) = m_0(\omega/2) m_0(\omega/4) \widetilde{\phi}(\omega/4) \cdots$$

Eventually, when this procedure is repeated infinitely, we obtain

$$\widetilde{\phi}(\omega) = \frac{1}{\sqrt{2\pi}} \prod_{j=1}^{+\infty} m_0(2^{-j}\omega) \tag{D.6}$$

Similarly, we take the Fourier transform of the two-scale wavelet equation, as given in Chapter 2. From

$$\psi(x) = \sqrt{2} \sum_k g_k \phi(2x-k) \tag{D.7}$$

we then obtain

$$\widetilde{\psi}(\omega) = \frac{1}{\sqrt{2}} \sum_k g_k\, e^{-ik\omega/2} \widetilde{\phi}\left(\frac{\omega}{2}\right) \tag{D.8}$$

Rewriting (D.8) yields

$$\widetilde{\psi}(\omega) = m_1(\omega/2)\widetilde{\phi}(\omega/2) \tag{D.9}$$

with

$$m_1(\omega) = \frac{1}{\sqrt{2}} \sum_k g_k\, e^{-ik\omega} \tag{D.10}$$

where m_1 is a 2π-periodic function.

D.2 ORTHOGONALITY CONDITION IN THE FREQUENCY DOMAIN

First, we assume that the scaling function $\phi(x)$ is orthogonal to its translated version $\{\phi(x-k)\}$; namely,

$$\begin{aligned}\int_{-\infty}^{+\infty} \phi(x)\overline{\phi(x-k)}\,dx &= \frac{1}{2\pi}\int_{-\infty}^{+\infty} \widetilde{\phi}(\omega)\overline{\widetilde{\phi}(\omega)}\,e^{i\omega k}\,d\omega \\ &= \frac{1}{2\pi}\int_{-\infty}^{+\infty} \left|\widetilde{\phi}(\omega)\right|^2 e^{i\omega k}\,d\omega \\ &= \frac{1}{2\pi}\int_{0}^{2\pi} e^{i\omega k}\sum_{l=-\infty}^{+\infty}\left|\widetilde{\phi}(\omega+2\pi l)\right|^2 d\omega = \delta_{k,0}\end{aligned} \tag{D.11}$$

where the last expression is identical to the Fourier series coefficients for a 2π-periodic function with a period of 2π. This implies

$$\sum_{l=-\infty}^{+\infty}\left|\widetilde{\phi}(\omega+2\pi l)\right|^2 = \sum_k a_k\, e^{-ik\omega} = 1$$

Thus, we can express the orthogonality of $\{\phi(x-k)\}$ in the frequency domain as

$$\sum_{l=-\infty}^{+\infty}\left|\widetilde{\phi}(\omega+2\pi l)\right|^2 = 1 \tag{D.12}$$

Next, if the scaling function $\phi(x)$ is not orthogonal to its translated version $\{\phi(x-k)\}$ in the frequency domain, we need to divide $\widetilde{\phi}(\omega)$ by the square root of $\sum_{l=-\infty}^{+\infty}\left|\widetilde{\phi}(\omega+2\pi l)\right|^2$ to construct a new normalized function [1], which reads

$$\widetilde{\phi}_{\mathrm{orth}}(\omega) = \frac{\widetilde{\phi}(\omega)}{\sqrt{\sum_{l=-\infty}^{+\infty}\left|\widetilde{\phi}(\omega+2\pi l)\right|^2}} \tag{D.13}$$

and

$$\sum_{k=-\infty}^{+\infty}\left|\widetilde{\phi}_{\mathrm{orth}}(\omega+2\pi k)\right|^2 = \sum_{k=-\infty}^{+\infty}\left|\frac{\widetilde{\phi}(\omega+2\pi k)}{\sqrt{\sum_{l=-\infty}^{+\infty}\left|\widetilde{\phi}(\omega+2\pi l)\right|^2}}\right|^2 = 1 \tag{D.14}$$

where $\widetilde{\phi}_{\mathrm{orth}}(\omega)$ satisfies the orthonormality condition, and the basis functions $\{\phi_{\mathrm{orth}}(x-k)\}$ are orthonormal to each other.

D.3 EXPRESSION OF B-SPLINE FUNCTIONS IN THE FREQUENCY AND SPACE DOMAINS

As an illustration, we find the expression of the cubic spline Battle–Lemarié scaling function in the frequency domain. The key step here is to obtain the B-spline function in the spectral domain.

First, according to (D.13), we write the Fourier transform of the mth-order spline scaling function as

$$\widetilde{\phi}_m(\omega) = \frac{\widetilde{\beta}_m(\omega)}{\sqrt{\sum_k \left|\widetilde{\beta}(\omega + 2\pi k)\right|^2}} \tag{D.15}$$

Next, we write the spatial distribution of the zero-order B-spline function:

$$\beta_0(t) = \begin{cases} 1, & 1 \le t < 0 \\ 0, & \text{otherwise} \end{cases} \tag{D.16}$$

and then find its corresponding Fourier transform:

$$\widetilde{\beta}_0(\omega) = \int_{-\infty}^{+\infty} \beta_0(t)\, e^{-i\omega t}\, dt = \int_0^1 e^{-i\omega t}\, dt = \frac{1 - e^{-i\omega}}{i\omega} \tag{D.17}$$

For an arbitrarily positive integer ($m \ge 1$), the mth-order B-spline function is defined as

$$\beta_m(t) = \int_0^1 \beta_{m-1}(t - x)\, dx \tag{D.18}$$

whose Fourier transform is given by

$$\widetilde{\beta}_m(\omega) = (\widetilde{\beta}_0(\omega))^{m+1} = \left(\frac{1 - e^{-i\omega}}{i\omega}\right)^{m+1} = e^{-j(m+1)\omega/2}\left(\frac{\sin\omega/2}{\omega/2}\right)^{m+1} \tag{D.19}$$

For $m = 3$, we obtain

$$\widetilde{\beta}^{(3)}(\omega) = \left(\frac{1 - e^{-i\omega}}{i\omega}\right)^4 = e^{-2j\omega}\left(\frac{\sin\omega/2}{\omega/2}\right)^4 = \left(\frac{\sin\omega/2}{\omega/2}\right)^4 \tag{D.20}$$

$$\sum_k |\widetilde{\beta}_3(\omega + 2\pi k)|^2 = \sum_k \left|\frac{\sin(\omega/2 + k\pi)}{\omega/2 + k\pi}\right|^8 = \sin(\omega/2)^8 \sum_k \left|\frac{1}{\omega/2 + k\pi}\right|^8 \tag{D.21}$$

By taking the derivative of the both sides of the identity

$$\sum_k \frac{1}{(z+k\pi)^2} = \frac{1}{\sin^2 z} \tag{D.22}$$

we obtain

$$\sum_k \frac{1}{(z+k\pi)^4} = \frac{1}{\sin^4 z} - \frac{2}{3}\frac{1}{\sin^2 z} \tag{D.23}$$

Now we differentiate both sides of the above equation once more and derive

$$\begin{aligned}\sum_k \frac{1}{(z+k\pi)^8}\bigg|_{z=\omega/2} &= \left(\frac{1}{\sin^8 z} - \frac{4}{3}\frac{1}{\sin^6 z} + \frac{2}{5}\frac{1}{\sin^4 z} - \frac{4}{315}\frac{1}{\sin^2 z}\right)\bigg|_{z=\omega/2} \\ &= \frac{1}{\sin^8 \omega/2} - \frac{4}{3}\frac{1}{\sin^6 \omega/2} + \frac{2}{5}\frac{1}{\sin^4 \omega/2} - \frac{4}{315}\frac{1}{\sin^2 \omega/2}\end{aligned}$$

which leads to

$$\sum_k |\widetilde{\beta}_3(\omega+2\pi k)|^2 = 1 - \frac{4}{3}\sin^2\omega/2 + \frac{2}{5}\sin^4\omega/2 - \frac{2}{315}\sin^6\omega/2 \tag{D.24}$$

Substitution of (D.21) and (D.24) into (D.15) yields

$$\begin{aligned}\widetilde{\phi}_3(\omega) &= \frac{\widetilde{\beta}_3(\omega)}{\left(\sum_k |\widetilde{\beta}_3(\omega+2\pi k)|^2\right)^{1/2}} \\ &= \left(\frac{\sin\omega/2}{\omega/2}\right)^4 \frac{1}{\sqrt{1-\frac{4}{3}\sin^2\omega/2 + \frac{2}{5}\sin^4\omega/2 - \frac{2}{315}\sin^6\omega/2}}\end{aligned} \tag{D.25}$$

Furthermore, for the mth-order central B-spline function $\beta_m(t)$, we can extend (D.24) as

$$\begin{aligned}\sum_k |\widetilde{\beta}_m(\omega+2\pi k)|^2 &= [2\sin(\omega/2)]^{2(m+1)} \sum_k \left|\frac{1}{\omega+2k\pi}\right|^{2(m+1)} \\ &= [2\sin(\omega/2)]^{2(m+1)} \sum_k \frac{1}{(\omega+2k\pi)^{2(m+1)}}\end{aligned} \tag{D.26}$$

Similarly, by taking m time derivatives of the equation $\sum_k [1/(\omega + 2k\pi)^2]$, we are led to

$$\sum_k \frac{1}{(\omega + 2k\pi)^m} = \frac{(-1)^{m-2}}{(n-1)!} \frac{d^{m-2}}{d\omega^{m-2}} \sum_k \frac{1}{(\omega + 2k\pi)^2} \tag{D.27}$$

Finally, we obtain the Fourier transform of the mth-order spline scaling function:

$$\widetilde{\phi}(\omega) \equiv \widetilde{\phi}_m(\omega) = \frac{\widetilde{\beta}_m(\omega)}{\sqrt{[2\sin(\omega/2)]^{2(m+1)} \sum_k \dfrac{1}{(\omega + 2k\pi)^{2(m+1)}}}} \equiv M(\omega)\widetilde{\beta}_m(\omega) \tag{D.28}$$

where $M(\omega)$ is an even, 2π-periodic function.

In order to find the scaling function in the space domain, we apply the inverse Fourier transform to the function $\widetilde{\phi}(\omega)$. Expansion of $M(\omega)$ as a series

$$M(\omega) = \sum_n c_n \, e^{-in\omega}$$

leads to the scaling functions in the space domain:

$$\phi(x) = \sum_n c_n \beta_m(x - n) \tag{D.29}$$

D.4 CUBIC SPLINE BATTLE–LEMARIÉ WAVELET FUNCTION IN THE FREQUENCY DOMAIN

Recall that the dilation form of the scaling function in the frequency domain is

$$\widetilde{\phi}(\omega) = m_0(\omega/2)\widetilde{\phi}(\omega/2)$$

with

$$m_0(\omega) = \frac{1}{\sqrt{2}} \sum_k h_k \, e^{-ik\omega}$$

If $m_0(\omega)$ is known, we can obtain the kth Fourier coefficient of $m_0(\omega)$, $h_k/\sqrt{2}$. From the two-scale equation of $\phi_m(x)$, we use the new symbol P_m to replace m_0:

$$\widetilde{\phi}_m(\omega) = P_m(\omega/2)\widetilde{\phi}_m(\omega/2)$$

Note that here we have used a new symbol P_m to replace m_0 to be consistent with Daubechies [1] and Chui [2]. Thus, we obtain

$$P_m(\omega/2) = \frac{\widetilde{\phi}_m(\omega)}{\widetilde{\phi}_m(\omega/2)} \tag{D.30}$$

The Fourier transform of the mth-order Battle–Lemarié mother wavelet [2] is given by

$$\begin{aligned}\widetilde{\psi}_m(\omega) &= G(\omega/2)\widetilde{\phi}_m(\omega/2) \\ &= -e^{-i\omega/2}\ \overline{P(\omega/2+\pi)}\cdot\widetilde{\phi}_m(\omega/2) \\ &= -e^{-i\omega/2}\cdot\frac{\overline{\widetilde{\phi}_m(\omega+2\pi)}\widetilde{\phi}_m(\omega/2)}{\widetilde{\phi}_m(\omega/2+\pi)}\end{aligned} \tag{D.31}$$

where $\overline{P_m(\omega)}$ denotes the complex conjugate of $P_m(\omega)$. Let $\omega' = \omega + 2\pi$, we then have

$$\overline{P_m(e^{-i\omega'/2})} = \frac{\overline{\widetilde{\phi}_m(\omega')}}{\overline{\widetilde{\phi}_m(\omega'/2)}} = \frac{\overline{\widetilde{\phi}_m(\omega+2\pi)}}{\overline{\widetilde{\phi}_m(\omega/2+\pi)}} \tag{D.32}$$

For $m = 3$, we have the following equations derived from (D.25):

$$\widetilde{\phi}_3(\omega/2) = \left(\frac{\sin\omega/4}{\omega/4}\right)^4 \frac{1}{\left(\begin{array}{c}1-\frac{4}{3}\sin^2(\omega/4)+\frac{2}{5}\sin^4(\omega/4)\\ -\frac{4}{315}\sin^6(\omega/4)\end{array}\right)^{1/2}} \tag{D.33}$$

$$\widetilde{\phi}_3(\omega+2\pi) = \left(\frac{\sin\omega/2}{\omega/2+\pi}\right)^4 \frac{1}{\left(\begin{array}{c}1-\frac{4}{3}\sin^2(\omega/2)+\frac{2}{5}\sin^4(\omega/2)\\ -\frac{4}{315}\sin^6(\omega/2)\end{array}\right)^{1/2}} \tag{D.34}$$

$$\widetilde{\phi}_3(\omega/2+\pi) = \left(\frac{\cos\omega/4}{\omega/4+\pi/2}\right)^4 \frac{1}{\left(\begin{array}{c}1-\frac{4}{3}\cos^2(\omega/4)+\frac{2}{5}\cos^4(\omega/4)\\ -\frac{4}{315}\cos^6(\omega/4)\end{array}\right)^{1/2}} \tag{D.35}$$

Substitution of (D.33), (D.34), and (D.35) into (D.31) leads to the cubic spline Battle–Lamarié mother wavelet function expression in the frequency domain:

$$\begin{aligned}\widetilde{\psi}(\omega) = \frac{\overline{\widetilde{\phi}_3(\omega+2\pi)}}{\widetilde{\phi}_3(\omega/2+\pi)} &= \frac{-e^{-i\omega/2}\sin^4(\omega/2)\tan^4(\omega/2)(2/\omega)^4}{\left(1-\frac{4}{3}\sin^2(\omega/2)+\frac{2}{5}\sin^4(\omega/2)-\frac{4}{315}\sin^6(\omega/2)\right)^{1/2}} \\ &\cdot\left(\frac{1-\frac{4}{3}\cos^2(\omega/4)+\frac{2}{5}\cos^4(\omega/4)-\frac{4}{315}\cos^6(\omega/4)}{1-\frac{4}{3}\sin^2(\omega/4)+\frac{2}{5}\sin^4(\omega/4)-\frac{4}{315}\sin^6(\omega/4)}\right)^{1/2}\end{aligned} \tag{D.36}$$

REFERENCES

[1] I. Daubechies, *Ten Lectures on Wavelets*, SIAM, Philadelphia, 1992.
[2] Charles K. Chui, *An Introduction to Wavelets*, Academic Press, San Diego, 1992.

INDEX

Multiresolution Time Domain Scheme for Electromagnetic Engineering
By Yinchao Chen, Qunsheng Cao, and Raj Mittra
ISBN 0-471-27230-2